T0271155

The Choanoflagellates
Evolution, Biology and Ecology

Choanoflagellates have three distinctive claims to fame: they are the closest, living, unicellular relatives of animals; they are a major component of aquatic microbial foodwebs; and one group is remarkable for its siliceous basket-like coverings.

This landmark book offers a unique synthesis of over 40 years of choanoflagellate research. Key areas are covered, from the phylogenetic evidence supporting the sister-group relationship between choanoflagellates and Metazoa, to choanoflagellate distribution and diversity in marine and freshwater environments. The structure and assembly of choanoflagellate loricae is also presented together with a full discussion of a novel example of 'regulatory evolution', suggesting that the switch from nudiform to tectiform cell division and lorica production was achieved by a sudden reorganisation of existing structures and mechanisms.

Providing an authoritative summary of what is currently known about choanoflagellates, this title will serve as a foundation upon which future research and discussion can take place.

BARRY S. C. LEADBEATER is a retired Reader in Protistology at the University of Birmingham. His academic research interests include: ultrastructure, physiology and ecology of algae and protozoa, whereas his biotechnological research interests include: algae and water quality, physiology of algal/protistan biofilms and biological aspects of water treatment processes. He has authored and co-authored over 80 papers, edited five books and, presently, he is a monitoring editor of the journal *Protist*.

The Choanoflagellates

Evolution, Biology and Ecology

Barry S. C. Leadbeater
University of Birmingham, UK

CAMBRIDGE
UNIVERSITY PRESS

Shaftesbury Road, Cambridge CB2 8EA, United Kingdom

One Liberty Plaza, 20th Floor, New York, NY 10006, USA

477 Williamstown Road, Port Melbourne, VIC 3207, Australia

314–321, 3rd Floor, Plot 3, Splendor Forum, Jasola District Centre, New Delhi – 110025, India

103 Penang Road, #05–06/07, Visioncrest Commercial, Singapore 238467

Cambridge University Press is part of Cambridge University Press & Assessment,
a department of the University of Cambridge.

We share the University's mission to contribute to society through the pursuit of
education, learning and research at the highest international levels of excellence.

www.cambridge.org
Information on this title: www.cambridge.org/9780521884440

First published 2015

A catalogue record for this publication is available from the British Library

Library of Congress Cataloging-in-Publication data
Leadbeater, Barry S. C.
The choanoflagellates : evolution, biology, and ecology / Barry S. C. Leadbeater, University of Birmingham, UK.
    pages cm.
Includes bibliographical references and index.
1. Choanoflagellates. 2. Protozoa. I. Title.
QL368.C5L43 2014
579.4–dc23   2014022201

ISBN   978-0-521-88444-0   Hardback

Contents

Preface   *page* ix
Note on Terminology, Taxonomy and Nomenclature   xii

1 Historical perspectives   1
  1.1 Introduction   1
  1.2 First published record of a collar-bearing flagellate   1
  1.3 Morphology and reproduction of the 'collared flagellate'   4
  1.4 Choanoflagellates as ancestors of the sponges and lower Metazoa   10
  1.5 Haeckel's Gastraea theory and origin of the lower Metazoa (non-Bilateria)   13
  1.6 Alternative theories of metazoan evolution involving choanoflagellate ancestors   14
  1.7 The 'collar cell' as a distinctive cell type   14
  1.8 Choanoflagellate classification   15

2 The collared flagellate: functional morphology and ultrastructure   18
  2.1 Functional morphology: introduction   18
  2.2 The hydrodynamic environment of choanoflagellates   18
  2.3 Modelling water flow in response to flagellar activity   19
  2.4 Particle capture and ingestion   26
  2.5 Ultrastructure: introduction   28
  2.6 The flagellar apparatus   28
  2.7 Cytoskeleton   32
  2.8 Cytoplasmic organelles   37
  2.9 Nucleus and nuclear division (mitosis)   40

3 Craspedida: choanoflagellates with exclusively organic coverings   44
  3.1 Introduction   44
  3.2 Cell cycle of the Craspedida   44
  3.3 Species with non-restrictive cell coverings   44
  3.4 Species with restrictive cell coverings   47
  3.5 Colonial choanoflagellates   54
  3.6 Encystment   61
  3.7 Discussion   63

4 Loricate choanoflagellates: Acanthoecida   66
  4.1 Introduction   66
  4.2 Terminology and preservation of choanoflagellate loricae   67
  4.3 Lorica construction   70
  4.4 Evidence for a universal left-handed rotation in lorica construction   82
  4.5 Organic components of loricae   83
  4.6 Variations on themes   85

4.7 Discussion   98
4.8 'Rules' of lorica construction and assembly   99

5 Loricate choanoflagellates: requirement for silicon and its deposition in costal strips   100
5.1 Introduction   100
5.2 Biogeochemical cycling of silicon in seawater   101
5.3 Terminology   102
5.4 *Stephanoeca diplocostata*: the experimental species   102
5.5 Ultrastructure of costal strip production in *Stephanoeca diplocostata*   104
5.6 Effects of microtubule poisons on the cytoskeleton and costal strip development   108
5.7 Growth and silicon turnover in *Stephanoeca diplocostata*   112
5.8 Discussion   126

6 Loricate choanoflagellates: Acanthoecidae – nudiform species   130
6.1 Introduction   130
6.2 Terminology: spirals and helices   132
6.3 Nudiform choanoflagellates (Acanthoecidae)   132
6.4 *Savillea*   133
6.5 *Helgoeca*   138
6.6 The *Polyoeca–Acanthoeca* continuum   140
6.7 Logistical and mechanical aspects of lorica assembly   147
6.8 Role of the cytoskeleton during lorica assembly   147
6.9 Evolutionary significance of nudiform choanoflagellates   153

7 Loricate choanoflagellates: Stephanoecidae – tectiform species   155
7.1 Introduction   155
7.2 *Stephanoeca diplocostata*   156
7.3 *Didymoeca costata*   166
7.4 *Diaphanoeca grandis*   175
7.5 Features that characterise the tectiform condition   179
7.6 Universality of features characteristic of tectiform division   181
7.7 Are transverse costae an exclusively tectiform feature?   183

8 Loricate choanoflagellates: evolutionary relationship between the nudiform and tectiform conditions 185
8.1 Introduction: an evolutionary paradox   185
8.2 Background   185
8.3 The nudiform condition   185
8.4 The tectiform condition   187
8.5 Codifying the salient features of the nudiform and tectiform conditions   191
8.6 One jump evolution or gradual change?   192
8.7 Evolutionary implications of the relationship between the nudiform and tectiform conditions   199

9 Choanoflagellate ecology   202
9.1 Introduction   202
9.2 Systematic and functional diversity of heterotrophic flagellates   203
9.3 Ecophysiology of choanoflagellates   204
9.4 Ecology of marine and brackish water choanoflagellates   213
9.5 Variability in lorica size of *Bicosta* and *Calliacantha* species   233
9.6 Functional role of choanoflagellate loricae   236
9.7 Ecology of freshwater choanoflagellates   237
9.8 Conclusions   240

10 Choanoflagellate phylogeny: evolution of metazoan multicellularity   241

10.1 Introduction   241

10.2 Phylogeny of Opisthokonta: morphological and molecular perspectives   243

10.3 Opisthokont classification   249

10.4 Holozoa   251

10.5 Nucletmycea   261

10.6 Opisthokont evolution: a consensus between morphology and molecular phylogeny?   262

10.7 Estimated dating of opisthokont diversification   263

10.8 Evolution of metazoan multicellularity   264

10.9 Cell–cell and cell–matrix adhesion   265

10.10 Cell–cell signalling transduction pathways   269

10.11 Discussion   276

Glossary   278

References   280

Figure and table credits   305

Choanoflagellate species index   307

Other species index   309

General index   311

Preface

Choanoflagellates have enjoyed an unusual degree of celebrity since Henry James-Clark (in 1867) first showed that they bore a remarkable morphological resemblance to the choanocytes of a sponge. This similarity gave rise to the hypothesis that sponges, and, by implication, other animals, evolved from a choanoflagellate-like ancestor. For a while it appeared that colonial choanoflagellates might be a missing link between unicellular protozoa and animals. However, it is only in the last 20 years that molecular phylogenetic studies have provided substantive support for choanoflagellates being the sister group to all Metazoa, which makes them the closest, known, living, unicellular relatives of animals.

Ecologically, choanoflagellates have also enjoyed an important, if less high-profile, reputation. They are universally distributed in aquatic habitats and are a major contributor to aquatic microbial foodwebs (the microbial loop). What appears at first sight to be a rather uniform and unadventurous protistan group turns out to be a diverse and fascinating collection of organisms. In particular, variations in lorica morphology have led to the extensive distribution and diversification of loricate choanoflagellates in the oceans of the world. Detailed ultrastructural studies, particularly on lorica construction and assembly, have also revealed many novel features of cell biological interest.

It can now be said with confidence that choanoflagellates have come of age. Recent molecular studies have rekindled their fashionable status and the time has come to consolidate an extensive body of literature into book format. This is a particularly good time at which to undertake the task since, for the moment at least, the various disciplines can still be treated as a coherent whole. In the future it is likely that the subject will fragment as individual disciplines become ever more detailed and specialised. A book on choanoflagellates at this juncture will not only provide a summary of knowledge to date, but will also serve as a foundation upon which future research and discussion can take place.

The three disciplines on which this book is based have different historical backgrounds. Cell biology has grown out of traditional light microscopy dating back to the nineteenth century. The advent of electron microscopy permitted more detailed investigations of lorica morphology and the intricacies of lorica production and assembly. Ecological studies are also rooted in the descriptive work of the nineteenth and early twentieth centuries, but have been greatly augmented by quantitative and, more recently, molecular investigations. Molecular and evolutionary studies, although having their roots in the late twentieth century, are essentially a feature of the twenty-first century. Choanoflagellates, as a member of the supergroup Opisthokonta, have fared particularly well in phylogenetic analyses, being regularly recovered as the monophyletic sister group to Metazoa. Purely molecular studies, particularly those based on entire genomes, are still in their infancy.

This book has two basic aims. The first is to provide an authoritative summary of what is currently known about choanoflagellates (Chapters 1, 2, 3, 9 and 10). The second is to provide a detailed account of loricate choanoflagellates with particular reference to their requirement for silicon and the production and assembly of the basket-like lorica (Chapters 4–8). This includes a full discussion for the first time of a novel example of 'regulatory evolution', whereby it is suggested that the switch from nudiform to tectiform cell division and lorica production was achieved by a sudden reorganisation of existing structures and mechanisms rather than the emergence of new structures and mechanisms.

Serendipity has played a large part both in my research of choanoflagellates and in writing this book. My first encounter with choanoflagellates was accidental. I was a phycologist working on calcified nanoplankton when, by fortune, on a field excursion to Norway in 1970, I netted loricate choanoflagellates instead of coccolithophorids. In spite of treating choanoflagellates as a hobby I subsequently managed to isolate several species in culture and from that time on I was hooked. The writing of this book has also been dependent on good fortune. First, in receiving financial assistance from the Leverhulme Trust for an Emeritus Fellowship, and second for the generosity of the School of Biosciences, University of Birmingham in providing office and laboratory facilities long after I had retired from active teaching.

Since my interest in choanoflagellates began over 40 years ago, there are many colleagues and past students worldwide whom I wish to thank for assistance, advice, collaboration and hospitality. I thank John Dodge for having first introduced me to flagellates during my postgraduate research. The opportunity to carry out postdoctoral work at Leeds University gave me the opportunity to work with Irene Manton FRS, who had a profound influence on the course of my research career. I am grateful to Helge Thomsen, a fellow traveller who has remained a colleague and friend since we first met in Denmark in 1971. I thank Professor Sergey Karpov for research collaboration and generous hospitality on numerous exchange visits between Birmingham and St Petersburg. His excellent work in Birmingham led to three joint publications and some of his TEM images are used in this book (see Figure and Table Credits).

I am sincerely grateful to many scientific colleagues who contributed expertise in disciplines outside my own area of competence. In this category come Malcolm Davies, who collaborated with work on silicon metabolism and whose untimely death came before completion of the work and publication of the results. Jackie Parry provided me with generous assistance on choanoflagellate ecophysiology and the kinetics of continuous culture. John Blake brought a mathematical mind to bear on the hydrodynamics of filter feeding in choanoflagellates. Sandra Baldauf and Martin Carr introduced me to the world of molecular phylogeny and biosystematics. Oliver Smart, Dov Stekel and QiBin Yu kindly developed the computer graphic model used for constructing costal patterns in choanoflagellate loricae.

Choanoflagellate research has taken me to many parts of the world. I would particularly like to thank Yves LeGal and the staff at the Concarneau Marine Biology Station for so generously making available their facilities for countless field excursions to the Brittany coast over 30 years. I am grateful to Rick Wetherbee and Jeremy Pickett-Heaps FRS for hosting my 1996 sabbatical stay in the Department of Botany, University of Melbourne which proved to be one of the most productive and enjoyable periods of research during my working life. I took part on two sea-going collections in Icelandic waters arranged by the late Thorunn Thordardottir of the Marine Research Institute, Reykjavik. Ray Leakey, while working at the British Antarctic Survey, kindly arranged for me to partake in a sea-going cruise to the Southern Ocean and South Georgia.

I am thankful to Peter Holland FRS for convening the first choanoflagellate workshop in Reading in 2002; Nicole King and colleagues for hosting two subsequent workshops (2009 and 2011) at UC Berkeley; and Hartmut Arndt, Frank Nitsche and colleagues for hosting the fourth workshop in Cologne (2013). These meetings have provided a convivial forum for the exchange of results, discussion and techniques and have served to nucleate and develop a global 'choanoflagellate community'.

I gratefully acknowledge fellow scientists, doctoral students and generations of undergraduates who have contributed to my work on flagellates in general and choanoflagellates in particular. They include (in alphabetical order): Christi Adams, Lisa Armstrong, Jenny Bailey, Nik Berovic, Kurt Buck, Rong Cheng, Stephen Coupe, Sheila Crosbie, Stephen Day, Angela de Rosa, Ilana Fleetwood, Jan Frösler, Richard Geider, Anne Hartley, Ruhana Hassan, Magali Henouil, Sarah Innes, Seamus Jackson, Harriet Jones, Michelle Kelly, Christopher Kent, Vicki Lawson, Sally Leys, Malwin Malinowski, Sharon McCready, Linda Medlin, Pamela Merrick, Elaine Mitchell, Carol Morton, Rich Norris, Ken Oates, Helen Peterkin, Michala Pettitt, Julie Pickett-Heaps, Steve Price, Daniel Richter, David Roberts, Roy Sladden, Nancy Slater, David Small, Jahn Throndsen, Lesley Tomkins, Phillipa Towlson, Peter Whittle and Mehreen Yousuf.

I am indebted to countless members of the British Society for Protist Biology, the International Society of Protistologists, the British Phycological Society and the Systematics Association for a working lifetime of meetings, discussion, collaborations and hospitality. In particular I would like to thank: Keith Vickerman FRS, who on many occasions supported applications to the Royal Society and elsewhere for assistance with funding; John Green and the late Mary Parke FRS of the Marine Biological Association, Plymouth for their expertise and collaboration on flagellates in general; Bob Andersen, Hartmut Arndt, Tom Cavalier-Smith FRS, Richard Crawford, David Hibberd, Harvey Marchant, Michael Melkonian, Øjvind Moestrup, Jackie Parry, David (Paddy) Patterson, Tony Walsby FRS and the late Tyge Christensen, Dennis Greenwood, Michael Peel and Frank Round for so much help over many years as colleagues and friends.

I gratefully acknowledge the generosity of the following colleagues for providing me with illustrations or unpublished information and for giving me permission to include them in this book: Eve Boucaud, Thomas Bürglin, Martin Carr, William Gurske, David Hibberd, John van den Hoff, Sergey Karpov, Áron Kiss, David Lazarus, Ray Leakey, Sally Leys, Sunshine Menezes, Leonel Mendoza, Cynthia McKenzie, Frank Nitsche, N. V. Nemtseva, Franco Novarino, Marcus Paul,

Michala Pettitt, A. O. Plotnikov, E. A. Selivanova, Emma Steenkamp, Daniel Stoupin and Helge Thomsen. I am grateful to Chris Sheppard, Head of Special Collections at the University of Leeds, for permitting me to refer to the Irene Manton archive located in the Brotherton Library.

I owe an enormous debt of gratitude to Mrs Joyce Kent for the hours of work she devoted to organising the illustrative work for this book. Jennifer Neumann kindly assisted with the translation of German texts. I am thankful to Martin Carr, Nicole King, Nicholas Leadbeater, Frank Nitsche and Jackie Parry for reading and commenting on individual chapters. I am especially grateful to Emeritus Professor Michael Sleigh for scrupulously reading a complete draft of the script and for making many helpful comments and suggestions with respect to its scientific and editorial content. The text has been greatly improved by his wise counsel. I am thankful to colleagues at Cambridge University Press, particularly Dominic Lewis, Jessica Murphy and Ilaria Tassistro, for their patience in waiting so long for the final text.

Finally, I am deeply indebted to my wife, family, colleagues and friends for their patience and support during my research career and while writing this book. Deadlines have come and gone so often that, short of showing them a copy of the book itself, they will think this project was a figment of my imagination.

Note on Terminology, Taxonomy and Nomenclature

TERMINOLOGY

Choanoflagellate studies, starting with James-Clark's publication in 1866, are about to celebrate their sesquicentenary. In the intervening period the subject has acquired a substantial vocabulary of terminology. Since the relevant terminology has accumulated in a piecemeal fashion there are many examples of different terms with the same meaning and of general terms being used in specific contexts.

The policy that has been adopted in this book is to retain the use of general terms, such as theca (restrictive organic covering) and lorica (siliceous basket-like covering), that have been traditionally used in a 'choanoflagellate' context. Definitions of these and many other terms can be found in the appropriate chapters and/or in the glossary towards the end of this book. Creating new names or adopting infrequently used terms has been resisted on the grounds that they could confuse the reader and might distract from understanding the more important aspects of the text.

TAXONOMY AND NOMENCLATURE

The author is acutely aware that, at the time of writing, protist taxonomy and nomenclature are undergoing a major upheaval for a variety of reasons. The situation has been compounded by the phylogenetic revolution in taxonomy. Traditional taxonomy based on morphology has been the norm for choanoflagellates until recently. For craspedid choanoflagellates this has generally been unsatisfactory because many morphological features are ill-defined or variable at the light microscopical level and most species have polymorphic life cycles that are still unknown. For acanthoecids (loricate taxa) the situation is better because the pattern of costae comprising the lorica provides a set of well-defined morphological characters.

The introduction of molecular phylogeny to taxonomy will, no doubt, be of the greatest benefit in the long term. However, combining morphological and molecular phylogenies is proving much more difficult than was at first appreciated. For a start, it is most desirable that comprehensive data sets from both morphological and molecular sources should be available which, unfortunately, is often not the case. Incongruencies between morphological and molecular data are particularly problematical. If species can only be identified by molecular sequence data this not only presents problems for specialists without molecular facilities but may also invalidate the taxonomy of closely related species. As discussed in Chapter 10, phylogenetic systematics also challenges the Linnaean rank-based system of nomenclature that has underpinned much protistan taxonomy and nomenclature to date. The result is a confusion of hierarchical ranks and names. These are uncertain times and it is not obvious when or if stability will return.

Like it or not taxonomy and nomenclature underpin much biological research. Species names recorded in this text have been obtained from publications. Only occasionally have original names been changed in this text, for instance species of *Diplotheca* have been altered to *Didymoeca* for reasons explained in Chapter 7 and two species of *Savillea* have been combined because they co-occur within a single clonal culture (see Chapter 6).

The loricate choanoflagellate *Stephanoeca cupula* (Leadbeater 1972) Thomsen 1988 presents an anomalous situation that requires special mention. The original name given to this species was *Pleurasiga cupula* (Leadbeater, 1972c). The type illustrations (Figs 16 and 17 in Leadbeater, 1972c) show a cup-shaped lorica with anterior chamber comprising 8-10 longitudinal and two transverse costae, one forming the anterior ring and the second in an intermediate position. Subsequently, Thomsen (1988) transferred this species to the genus *Stephanoeca* and expanded the description by including more specimens. However, some of these specimens (Figs 16–19 in Thomsen, 1988) differed from the original material in that they possessed an additional transverse costa in the anterior chamber and the anterior intermediate transverse costa is located outside the longitudinal costae, a feature not displayed by the original specimens. It is now judged that the two forms of *S. cupula* are probably not the same species. However, until the second form can be re-named the two forms are designated in this text as *Stephanoeca cupula sensu* Leadbeater (1972c) for the original form and *Stephanoeca cupula sensu* Thomsen (1988) for the second form.

Current authorities are appended to lists of choanoflagellates and other species recorded in their respective indexes. For most names applied to higher ranks the publication by Adl *et al.* (2012) has been used. Their classification, proposed on behalf of the International Society of Protistologists, makes use of the oldest valid name that describes each group irrespective of its status. Thus the scheme incorporates an eclectic assortment of taxonomic nomenclature and spellings.

This book is about choanoflagellate evolution, cell biology and ecology and therefore taxonomy and nomenclature are not of primary importance. However, as it now seems likely that there will be major changes in nomenclature and phylogeny within the coming years, it is hoped that the underlying themes of this book will outlive whatever taxonomic and systematic changes might occur in the future.

1 • Historical perspectives

Myths, legends and reality.

1.1 INTRODUCTION

Despite a well-established tradition for protistological studies in nineteenth-century Europe, it was an American,
Henry James-Clark, who published the first correct and
unequivocal description of a 'collar-bearing' flagellate.
At the same time he also observed the morphological similarity between free-living collared flagellates and the
choanocytes of a sponge. The definitive date was 1866,
when James-Clark (1866a, b) published a summary of his
findings, to be followed a year later by a more expansive and
illustrated account under the title *On the Spongiae Ciliatae as
Infusoria Flagellata; or, observations on the structure, animality
and relationship of* Leucosolenia botryoides *Bowerbank*
(James-Clark, 1867b). While this title appears somewhat
archaic by today's standards, nevertheless it encapsulates
the significance of his findings. Carter (1857, 1859) had
previously concluded that the ampullaceous (aquiferous)
sacs of sponges were 'ciliated chambers', hence the
term *Spongiae Ciliatae*. James-Clark (1866a, 1867b, 1871b)
was now able to show that the flagella-bearing collared
monads lining the body cavity of the calcareous sponge
Leucosolenia botryoides bore a striking resemblance to the
free-living, collar-bearing flagellates he had just described,
and for this reason he considered sponges to be colonial
members of the *Infusoria Flagellata*. The historical importance of James-Clark's (1867b) observations with respect
to the study of choanoflagellates in particular, and to the
debate concerning the possible evolution of sponges and
animals in general, cannot be overestimated.

1.2 FIRST PUBLISHED RECORD OF A COLLAR-BEARING FLAGELLATE

Confusion surrounds the first published record of a collar-
bearing flagellate. There are several reasons for this,
including: the superficial similarity of stalked colonial choanoflagellates to other unrelated protists; the limitations of
early light microscopy; incomplete original taxonomic
descriptions and a general lack of coordination in the early
literature. Central to this confusion was the allocation of
'genuine' choanoflagellates to existing genera whose holo-
types, in hindsight, could not be choanoflagellates. The
two most important non-choanoflagellate genera involved
were *Anthophysa* Bory de Saint-Vincent 1822 and *Epistylis*
Ehrenberg 1830; the former is now a well-established genus
of colourless colonial chrysophytes and the latter a genus
of stalked peritrichous ciliates.

Anthophysa Bory (1822) was erected for an 'apparently'
stalked species of *Volvox* first described as *V. vegetans* by
Otto Frederik Müller (1786) (Fig. 1.1). At the same time
as introducing this new genus, Bory de Saint-Vincent
(1822) changed the specific name from *vegetans* to *mulleri*
in honour of Müller. This name change was subsequently
reversed by Stein (1878). Bory de Saint-Vincent (1822)
included a second *Anthophysa* species, *A. dichotoma*, in his
1822 work and subsequently added a third, *A. solitaria*,
with the briefest of detail and no illustrations (Bory de
Saint-Vincent 1824). Ehrenberg (1830) created the genus
Epistylis for a stalked colonial ciliate to which he subsequently added a new species, *E. botrytis* (holotype shown in
Figs 1.2–1.3) (Ehrenberg, 1831, 1838). While the cellular
details of *E. botrytis* differ from those of a choanoflagellate,
in particular they include an anterior ring of 'cilia' and lack
a single flagellum, nevertheless the overall form of the
colony is not dissimilar to that of a stalked choanoflagellate
such as a species of *Codosiga* (Fig. 3.1).

Trying to unravel the multiple confusions between specimens attributable to *Anthophysa vegetans*, *Epistylis botrytis*
and 'genuine' stalked colonial choanoflagellates is fraught
with difficulty. The limitations of nineteenth-century
microscopy led authors such as Stein (1849) to represent
anterior rings of cilia as peripheral spines which
could equally well have been the shrunken collars of

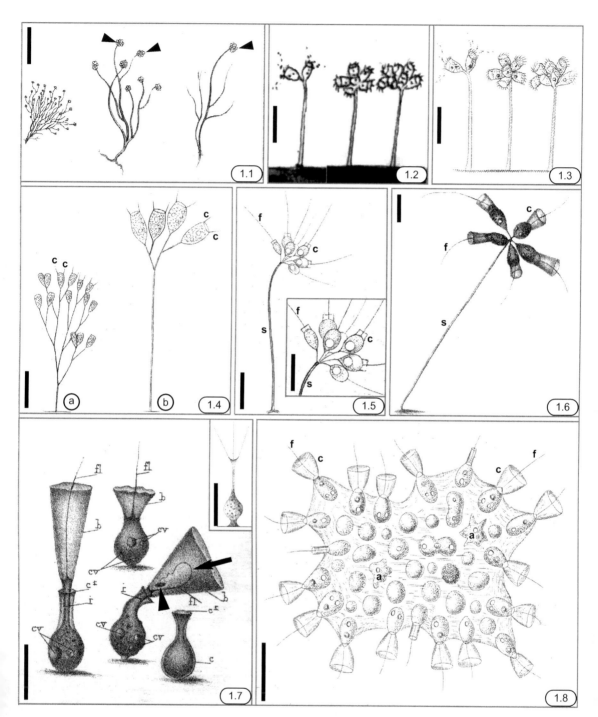

Plate 1 (Figures 1.1–1.8)

Figs 1.1–1.8 Illustrations of choanoflagellates, some with original labelling. Flagellum (f), collar (c), stalk (s). **Fig. 1.1** *Anthophysa vegetans*. Reproduced from Müller (1786). Thick-stalked protist with terminal colonies of cells (arrowheads).

choanoflagellates (Fig. 1.4a, b). What was needed at this time was recognition that 'genuine' choanoflagellate cells possessed a single anterior flagellum surrounded by a 'hyaline' collar. Fresenius (1858) must take the credit for this achievement. He was the first to illustrate unequivocally a choanoflagellate cell with a collar and single anterior flagellum (Fig. 1.5 and inset). However, he failed to recognise the collar as a novel and distinctive structure. Instead, he referred to every cell as having "an identical fine trimmed appendix out of which a long locomotor thread protrudes" (Fresenius, 1858, p. 25). He named his specimen *Anthophysa solitaria* on the basis of Bory de Saint-Vincent's (1824) sketchy specific description and he also acknowledged a resemblance to Ehrenberg's (1838) illustration of *Epistylis botrytis*. It was against this muddled background that James-Clark's (1866a *et seq.*) observations proved to be so enlightening.

1.2.1 Dates of James-Clark's publications

Confusion also surrounds the quoted dates of James-Clark's publications with respect to collar-bearing flagellates and sponges. This is for two reasons: first, his major publications were pre-empted in 1866 by a two-page printed summary of a lecture he gave to the Boston Natural History Society, in which he included descriptions of two choanoflagellate genera (James-Clark, 1866a). Second, during the following six years he wrote three substantive papers, each of which was first published in an American journal followed one year later by a verbatim copy, except for corrections, in a British journal. Thus the titles of each of these three papers have two publication dates according to whether the American or British version is being quoted (James-Clark, 1866b, 1867a, b, 1868, 1871b, 1872). Table 1.1 lists James-Clark's seven relevant papers, giving the dates of imprint. For the two choanoflagellate genera described by James-Clark, namely *Codosiga* and *Salpingoeca*, 1866 is the valid date of publication since the descriptions meet the criteria of the International Code of Zoological Nomenclature (ICZN) and this is the date recorded in Nomenclator Zoologicus (online version 0.86; 2005). The single species attributed to *Codosiga*, namely *C. pulcherrimus*, and the three *Salpingoeca* species, namely *S. gracilis*, *S. marinus* and *S. amphoridium*, were first described in James-Clark, 1867b. Two of these names, *pulcherrimus* and *marinus*, were subsequently corrected to *pulcherrima* and *marina* in the British version of this paper (James-Clark, 1868). However, according to the ICZN this does not alter the original valid date of publication as being 1867.

1.2.2 William Saville Kent; Otto Bütschli; Friedrich Ritter von Stein

James-Clark's pioneering work proved to be a catalyst for multiple independent but overlapping investigations on choanoflagellates, particularly by William Saville Kent (1871b, 1878c, 1880–2), Otto Bütschli (1878) and Friedrich Ritter von Stein (1878). Independently, apparently without collusion, these three authors came to the same conclusion that *Codosiga pulcherrima* James-Clark

---

Figs 1.2–1.3 *Epistylis* (*Codosiga*) *botrytis*. Bars = *c*.5 μm. Reproduced from the Ehrenberg Collection with permission from the Museum für Naturkunde, Berlin.

Fig. 1.2 Copy of original drawing (513) by Ehrenberg (1831).

Fig. 1.3 Illustration from *Die Infusionsthierchen* (Ehrenberg, 1838).

Fig. 1.4a–b. Stein's (1849) illustrations of *Epistylis* (*Codosiga*) *botrytis*. Individual collars appear as rigid spines (c). Bars = 5 and 10 μm, respectively. Reproduced from Stein (1849).

Fig. 1.5 *Anthophysa solitaria* (= *Codosiga botrytis*). Earliest convincing illustration of a stalked colony of choanoflagellate cells. Bars = 10 μm. Reproduced from Fresenius (1858).

Fig. 1.6 *Codosiga pulcherrima*. Bar = 10 μm. Reproduced from James-Clark (1867b).

Fig. 1.7 *Salpingoeca amphoridium*. The central cell possesses a recurved flagellum (arrow) and a bacterium near the base of the collar (arrowhead). Bar = 10 μm. Reproduced from James-Clark (1867b). **Fig. 1.7** inset: *Salpingoeca amphoridium* – drawing from notebook (1857). Reproduced from Carter (1871). Bar = 10 μm.

Fig. 1.8 *Proterospongia haeckeli*. Collar-bearing cells at the surface and amoeboid cells (a) embedded in 'zoocytium'. Bar = 10 μm. Reproduced from Kent (1880–82).

Table 1.1 *Details of James-Clark's seven important publications relating to choanoflagellates and sponges (date of imprint in brackets).*

1866a Paper presented at the Boston Society of Natural History on the animality of Sponges and their relationship with the Infusoria Flagellata. *Proceedings of the Boston Society of Natural History* 11: 16–17 (December 1866)

1866b Conclusive proofs on the animality of the ciliate sponges, and their affinities with the *Infusoria Flagellata. American Journal of Science and Arts,* Series 2, **42**: 320–5 (November 1866).

1867a Conclusive proofs on the animality of the ciliate sponges, and their affinities with the *Infusoria Flagellata. The Annals and Magazine of Natural History,* Series 3, **19**, 13–19 (January 1867).

1867b On the *Spongiae Ciliatae* as *Infusoria Flagellata*: or, observations on the structure, animality and relationship of *Leucosolenia botryoides* Bowerbank. *Memoirs of the Boston Society of Natural History* 1: 305–40 (September 1867).

1868 On the *Spongiae Ciliatae* as *Infusoria Flagellata*: or, observations on the structure, animality and relationship of *Leucosolenia botryoides* Bowerbank. *The Annals and Magazine of Natural History,* Series 4, **1**: 133–42, 188–215, 250–64 (February, March and April 1868).

1871b The American *Spongilla*, a craspedote, flagellate, infusorian. *American Journal of Science and Arts* 12: 426–36 (December 1871).

1872 The American *Spongilla*, a craspedote, flagellate, infusorian. *Monthly Microscopical Journal* 7: 104–114 (March 1872).

(Fig. 1.6) was synonymous with Ehrenberg's (1831) *Epistylis botrytis* (Fig. 1.3) (Bütschli, 1878; Kent, 1878c; Stein, 1878). As a result the combined name *Codosiga botrytis* (Ehrenberg) Bütschli came into being. The three authors introduced different collective names for colla-red flagellates, namely: Family Cylicomastiges Bütschli (1878); Order Craspedomonadina Stein (1878); and Family Choanoflagellata (Kent, 1880–2). Subsequently, the term Cylicomastiges disappeared without trace, but Craspedomonadina (craspedomonads) and Choanoflagellata (choanoflagellates, Kragenmonaden) have continued to be used interchangeably, although now choanoflagellate is the most commonly used colloquial term and is used throughout this text.

1.3 MORPHOLOGY AND REPRODUCTION OF THE 'COLLARED FLAGELLATE'

James-Clark's (1867b) illustrations of choanoflagellates are so clear and accurate that this publication alone served to establish the basic morphological features of the group (Figs 1.6–1.7, 3.11). Subsequently it has become apparent that the choanoflagellate cell plan is not only unmistakable, but also remarkably consistent, with only minor variations, such as absence of a flagellum in *Choanoeca perplexa* (Fig. 2.51). The essential features of a choanoflagellate include a radially symmetrical, spherical to ovoid cell body with a single central anterior

flagellum surrounded by a funnel-shaped collar that appears hyaline when viewed with light microscopy, but which comprises a ring of 20–50 microvilli that are held out rigidly in life (see Section 2.7.2). The flagellum undulates with a base-to-tip planar wave which creates a current of water from which prey particles are trapped on the outer surface of the collar. They are subsequently ingested at the base of the outer collar surface by pseudopodia (see Section 2.4).

Despite this relative simplicity, the early literature relating to choanoflagellate morphology, prey capture, cell coverings, cell division, sex and recombination is full of conflicting information and opinions due to the limitations of light microscopy and/or various interpretative misconceptions. Some of the more contentious issues which over time have made their way into standard textbooks are discussed below.

1.3.1 Collar morphology and the mechanism of prey capture and feeding

In spite of his excellent observations on choanoflagellate cell morphology, James-Clark (1867b, p. 315) erroneously considered that the location of prey ingestion, the mouth as he called it, was at the base of the flagellum within the confines of the collar, rather than outside the collar. He envisaged that particles of food were thrown by the flagellum "toward the mouth by vigorous spasmodic incurvations or jerks" (Fig. 1.7, arrow). This error was subsequently compounded by Kent (1878a, c, 1880–2), who envisaged the collar as being a funnel-shaped extension of the cell on the surface of which particles were trapped and "slowly, almost imperceptibly, carried along with the circulating current of the collar's substance up the outside and down the inside until, on reaching the base of its inner surface, they were engulfed within the cell" (Kent, 1880–2, p. 327) (Fig. 1.9).

While James-Clark (1867b) and Kent (1878c, 1880–2) incorrectly interpreted the details of prey capture and feeding they were, nevertheless, correct in viewing the collar as an entire funnel-shaped structure. In contrast, Entz (1883), subsequently supported by Francé (1893, 1897), Ehrlich (1908) and Schouteden (1908), regarded the collar as being the upper vertically expanded portion of a spirally coiled membrane that arises on the surface of the cell at the level of one of the two 'Schlingvacuoles' (gullet vacuoles), which had been described by earlier workers as contractile vacuoles (Fig. 1.10) (Lapage, 1925). Francé (1897) likened the collar membrane to a spirally wound piece of paper (Fig. 1.11). These authors considered that food particles followed a spiral path down the collar and were ultimately ingested by the 'gullet vacuole', which was capable of undergoing swallowing movements.

Bütschli (1878), Fisch (1885), Burck (1909), Griessmann (1913), Lapage (1925), de Saedeleer (1929) and Ellis (1929, 1935) correctly concluded that the collar was a funnel-shaped structure which served to entrap particles on its outer surface and that ingestion occurred at the base of the outer collar surface. However, opinion was divided about whether ingestion involved linguiform pseudopodia that originated from the base of the collar (Fig. 1.12) (Griessmann, 1913) or whether it occurred on the side of the cell between the plasma membrane and the surrounding organic covering (Bütschli, 1878). In fact, both observations are probably correct since in some strains of *Codosiga botrytis* prey particles are ingested at the side of the cell, although the surrounding sheath is not involved in the process as Bütschli (1878) suggested. However, in the majority of species particles are ingested by linguiform pseudopodia that rise up along the lower part of the collar (see Section 2.7.3).

Most early workers refer to the collar as being hyaline and membranous. Griessmann (1913) was the first to show unequivocally that, after fixation with osmium

tetroxide and staining with dahlia and methyl violet, the collar comprised a series of threads (Fig. 1.14). This was a relatively late record since Bidder (1895) had previously demonstrated that the choanocyte collars of the sponge *Sycon compressum*, when fixed with osmium tetroxide, embedded in paraffin wax and stained with haematoxylin, comprised a palisade of 20–30 fine threads (Fig. 1.13).

Frenzel (1891) described a new freshwater species from Argentina, *Diplosiga socialis*, which he claimed possessed two concentric collars, one outside the other. Francé (1897) subsequently added another genus, *Diplosigopsis*, for cells also with 'double collars'. However, de Saedeleer (1929) argued that the apparent existence of two collars was an observational misinterpretation. He suggested that the appearance of a second collar was either due to the remnants of pseudopodia at the base of the collar or, alternatively, the funnel-shaped anterior of the surrounding theca. Subsequent electron microscopy (EM) has confirmed de Saedeleer's (1929) opinion.

1.3.2 Terminology relating to extracellular coverings

Table 1.2 lists some of the terms that have been used to describe choanoflagellate coverings. The variety of terminology has come about for a number of reasons. First, the terms themselves, irrespective of language, are mostly non-specific and have not been used in a consistent manner. Second, standard light microscopy was often unable to resolve covering structures clearly. Electron microscopy has permitted a more thorough understanding of cell coverings, although the terminology remains subjective.

In an attempt to clarify and simplify the terminology relating to choanoflagellate coverings, three categories are currently recognised throughout this work (Table 1.2): (1) a thin, flexible extracellular organic matrix (glycocalyx) or sheath (craspedid species with non-restrictive cell division); (2) a continuous inflexible constraining organic envelope or theca (craspedid species with restricted (emergent) cell division); (3) a siliceous basket-like cage comprising a two-layered arrangement of costae made up of rod-shaped costal strips known as a lorica. Until recently, these three categories of coverings formed the basis of the three choanoflagellate families, Codonosigidae Kent, Salpingoecidae Kent and Acanthoecidae Norris, respectively (see Section 1.8.1). However, based on recent molecular

Plate 2 (Figures 1.9–1.12)

Figs 1.9–1.12 Early illustrations of the choanoflagellate collar and mechanisms of prey capture and ingestion.
Fig. 1.9 *Monosiga gracilis*. Copy of frontispiece from Kent (1880–2) showing his interpretation of the movement of carmine particles in the medium and on the collar. Arrows denote particle movement from rear of cell to their entrapment on the outer surface of the collar. Particles are then transported to the top of the collar and subsequently down the inner surface to the base, where they are ingested. Flagellum (f), nucleus (n), food vacuole (fv), contractile vacuole (cv). Bar = 2.5 μm.
Fig. 1.10 *Codosiga botrytis*. Drawing of a cell showing the flagellum (f) and spiral form of the collar (arrows c) leading to the 'Schlingvacuole' (gullet vacuole) (arrowhead). Food vacuole (fv). Bar = 2 μm. Reproduced from Burck (1909).
Fig. 1.11 Spiral form of the collar illustrated as spiral roll of paper. Reproduced from Burck (1909).
Fig. 1.12 *Salpingoeca infusionum*. Sequence of drawings showing ingestion of bacterium (b) in linguiform pseudopodium on outer surface of collar (c). Bar = 10 μm. Reproduced from Griessmann (1913).

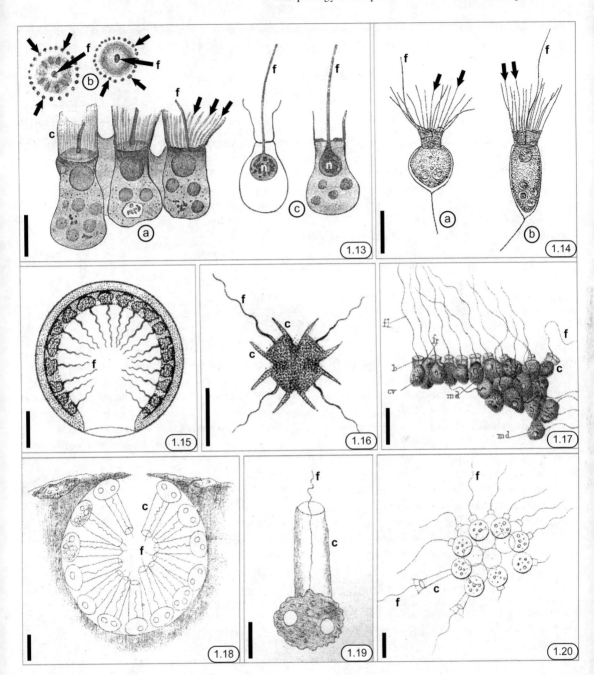

Plate 3 (Figures 1.13–1.20)

Figs 1.13–1.20 Collar-bearing sponge cells (choanocytes) and *Salpingoeca*. Flagellum (f), collar (c). **Fig. 1.13** *Sycon compressum*. Sectioned choanocytes stained with haematoxylin. a. Three cells showing collars composed of microvilli (arrows). b. Transverse section of two choanocytes showing central flagellum (arrow f) and collar comprising 20–36 microvilli (arrows). c. Two cells showing connection between flagellar base and nucleus. Bar = 5 μm. Reproduced from Bidder (1895).

Table 1.2 *Terminology that has been used to describe the coverings of choanoflagellates. Calyx (Latinised Greek) = cup-like structure; Coque (French) = shell; Gehäuse (German) = fixed envelope, shell; Hülse (German) = case, sheath, envelope; Schleimhülse (German) = mucous shell.*

| Author | Craspedida | | Acanthoecida |
	Glycocalyx (envelope)	Theca	Lorica
James-Clark (1866a)	–	Calyx	–
Bütschli (1878)	Schleimhülse	–	–
Stein (1878)	–	Hülse	–
Kent (1880–2)	–	Sheath, lorica	–
Lemmermann (1910)	–	Gehäuse	–
Saedeleer (1929)	Gel périphérique, loge	Coque	–
Ellis (1929)	Jelly-plasm	Pseudo-lorica	Loge-coque, true lorica
Boucaud-Camou (1966)	–	Coque	Coque
Norris (1965)	–	Lorica	Lorica
Bourrelly (1968)	–	Logette	–

phylogeny two major clades are now identified within the Class Choanoflagellatea; one, Craspedida, contains species with organic coverings (equivalent to the previous Codonosigidae and Salpingoecidae) and the other, Acanthoecida, contains attached and pelagic species with siliceous basket-like coverings (equivalent to Acanthoecidae) (see Section 10.4.1 and Fig. 10.3) (Nitsche *et al.*, 2011).

Preisig *et al.* (1994) surveyed the terminology used for protistan cell coverings and were critical of the use of non-specific terms, such as theca and lorica. They recommended that new terms should be created for specific structures and suggested that the term 'basket' might be used for the siliceous costal coverings of choanoflagellates. However, there is now a substantial quantity of choanoflagellate literature in which the term 'lorica' has been used to describe the silica basket. To change such an extensively used term now would not only lead to considerable confusion but would also run the risk of not becoming established in the literature.

1.3.3 Cell division

The morphology of cell division is closely allied to the categorisation of extracellular coverings (Table 1.3). In the absence of a restrictive covering, nuclear and cytoplasmic

Fig. 1.14a–b. *Salpingoeca pyxidium* and *S. infusionum*, respectively. Cells stained with dahlia and methyl violet showing single flagellum and collar comprising thread-like microvilli (arrows). Bar = 10 μm. Reproduced from Griessmann (1913).

Fig. 1.15 *Spongilla* sp. Section of ampullaceous sac showing inner lining of flagellated cells. Bar = 5 μm. Reproduced from Carter (1857).

Fig. 1.16 *Spongilla alba.* Four monociliated spiniferous (collar-bearing) cells from a spherical sac. Bar = 5 μm. Reproduced from Carter (1859).

Fig. 1.17 *Leucosolenia botryoides.* Fragment of 'monadigerous' layer showing collar-bearing cells. Bar = 10 μm. Reproduced from James-Clark (1876b).

Figs 1.18 and 1.19 *Spongilla arachnoidea.* Reproduced from James-Clark (1872).

Fig. 1.18 Section of flagellated chamber showing individual choanocytes each with a single flagellum and well-developed collar. Bar = 10 μm.

Fig. 1.19 Single choanocyte. Bar = 5 μm.

Fig. 1.20 *Grantia compressa.* Group of collar-bearing sponge cells. Bar = 5 μm. Reproduced from Carter (1871).

Table 1.3 *Terminology used in text to describe the morphology of cell division in choanoflagellates.*

Craspedida		Acanthoecida	
Non-restrictive coverings	Thecate (restrictive covering)	Nudiform	Tectiform
Longitudinal division	Emergent division	Diagonal division	Inverted division

division occur in the longitudinal (vertical) plane of the parent cell; this arrangement is termed longitudinal division. The two daughter cells that lie side-by-side until separation occurs share the cell covering equally. This category of division is typical of cells surrounded by a thin, flexible glycocalyx or sheath and is exemplified by species of *Monosiga*, *Codosiga* and *Desmarella* (Section 3.3). It is not uncommon for daughter cells resulting from longitudinal division to remain attached to each other after cytokinesis, thereby forming colonies, examples being species of *Codosiga* and the *Desmarella* and *Proterospongia* stages of craspedids (see Section 3.5). The unrestricted form of cell division was the reason why early workers used terms such as 'mucous sheath', 'gel périphérique' and 'Schleimhülse' to describe the accompanying expandable cell covering (see Table 1.2).

However, craspedid cells with restrictive coverings cannot undergo the standard process of longitudinal division because of space constraints; instead the cell becomes amoeboid and partially emerges from the parent theca. Nuclear division, which may still be in the longitudinal plane, occurs within the emergent portion of cytoplasm. One daughter nucleus remains within the cytoplasm in the parent theca, while the other passes to a developing naked 'juvenile' cell which eventually swims away, settles onto a surface and secretes a new theca. This form of division is called 'emergent' in this text (see Section 3.4). Earlier workers have referred to this type of division as 'budding' (bourgeonnement) and the motile cell has been called a 'hernia' (de Saedeleer, 1929; Ellis, 1929). The term 'juvenile' is used throughout this book to refer to a daughter cell resulting from division of a thecate or loricate cell that does not remain with the parent covering.

With respect to acanthoecids (loricate species), two forms of cell division are observed. In nudiform species, cell division is diagonal, which is a modified form of longitudinal division. The flagellar poles of both daughter cells face in an anterior direction (see Chapter 6). In tectiform species, nuclear division is more-or-less longitudinal but as cytokinesis proceeds the daughter cell that will leave the parent lorica is inverted and pushed backwards out of the lorica. The flagellar poles of the two daughter cells face each other as separation occurs. This form of division is called 'inverted' in this text (see Chapter 7).

1.3.4 Sex and recombination

Sexual reproduction can be defined as the fusion of two haploid gametic nuclei or gametes to form a single zygotic nucleus or diploid cell (zygote). Meiosis is an essential precursor to sexual reproduction. During meiosis homologous chromosomes undergo replication followed by recombination. Subsequently, two rounds of nuclear division, usually accompanied by cell division, produce four haploid nuclei or gametes. Since sexual reproduction is widespread in nature, including many unicellular organisms, it is reasonable to ask whether there is any evidence of sex in choanoflagellates.

Stein (1878) published a drawing of a stalked *Codosiga botrytis* cell with a smaller collar-bearing cell, flagellum outermost, projecting horizontally from its side (Plate VIII, Fig. 10 in Stein, 1878). He described this as "probably conjugation" between two cells. Fisch (1885) observed similar pairs of *C. botrytis* cells but considered that they were undergoing cell division as part of colony formation (Figs 74 and 75 in Fisch, 1885). There are several reports in the literature of multiphasic life cycles with unicells of varying size which might suggest variations in ploidy (Leadbeater, 1983a; Dayel *et al.*, 2011). Thomsen attributed sudden changes in lorica size (Thomsen and Larsen, 1992) and unexplained changes in costal strip morphology (Thomsen *et al.*, 1997) in *Bicosta spinifera* to the existence of complex polymorphic life cycles which might involve sexual reproduction (see Section 9.5).

Molecular studies have produced firmer evidence of sex within choanoflagellates. *Monosiga brevicollis* possesses long terminal repeat retrotransposons within its genome, which suggests that it has a sexual stage in its life cycle since asexual organisms cannot tolerate retrotransposons due to the rapid accumulation of deleterious mutations caused by their transposition (Carr *et al.*, 2008b). Furthermore, *M. brevicollis* possesses 18 of the 19 genes that comprise the 'meiotic detection toolkit' (Carr *et al.*, 2010). Eight of these genes function only in meiosis, whereas the others function in both mitosis and meiosis. This suggests that *M. brevicollis* is capable of switching between asexual and sexual reproduction.

Levin and King (2013), working with *Salpingoeca rosetta*, have reported the most convincing evidence of sex in a choanoflagellate to date. *S. rosetta* has a sexual cycle with transitions between haploid and diploid states. A haploid clonal culture was obtained that exhibited genetic stability over several months. When sub-samples of this culture were grown in unenriched (low-nutrient) seawater, after six days 89% of cells were diploid (as determined by propidium iodide staining). This change in ploidy was accompanied by instances of pairing and fusion between small rounded uniflagellated cells (tentatively called male gametes) and larger ovoid uniflagellated cells (tentatively called female gametes). Successful fusions were always initiated by contact between the basal end of the male gamete (opposite pole to the flagellum) and the base of the collar of the female gamete. This phenomenon is not too dissimilar to that illustrated by Stein (1878) for *Codosiga botrytis*.

A history of sex and recombination in *S. rosetta* is also suggested by the fact that single nucleotide polymorphisms (SNPs) were broken up into discrete haplotype blocks instead of spanning the length of each chromosome as would be expected in an asexual organism. The sharp boundaries at the edges of the segmented haplotype blocks probably mark the regions where there has been genetic exchange between homologous chromosomes (Levin and King, 2013).

Demonstrating sexual reproduction in unicellular protists is notoriously difficult. For instance, haptophycean flagellates, including coccolithophorids, are known to possess life cycle stages that vary in ploidy, but there is still a lack of hard evidence that they produce gametes and undergo meiosis.

1.4 CHOANOFLAGELLATES AS ANCESTORS OF THE SPONGES AND LOWER METAZOA

James-Clark's (1867b) publication on the possible relationship between choanoflagellates and sponges coincided with major advances in scientific knowledge, particularly in relation to animal evolution and systematics. These topics dominated biological thought in the mid nineteenth century and several powerful protagonists, including Ernst Haeckel, dominated the field. It is not surprising, therefore, that James-Clark's (1867b) observations quickly became embroiled in partisan controversy. While there is general agreement that the Metazoa are monophyletic and must have originated from a single-celled protozoan ancestor (Srivastava *et al.*, 2010), nevertheless there has been widespread debate about how multicellularity was first achieved and what sort of protozoan ancestor might have been involved (see Section 10.8).

Over the years there have been many theories relating to the possible origin of the Metazoa; the more important are reviewed by Salvini-Plawen (1978), Wilmer (1996), Nielsen (2001) and Mikhailov *et al.* (2009). Choanoflagellates have featured in many of these theories, in a morphological context starting with James-Clark's (1867b) publication and more recently in a molecular phylogenetic context (see Chapter 10). The brief discussion below is limited to theories that involve choanoflagellates or conflict directly with a choanoflagellate/sponge ancestry. This account starts with a comparison of choanoflagellate and choanocyte cell structure; reviews the case for sponges being colonial protozoa; highlights the conflicts arising from Haeckel's Gastraea theory; and summarises hypotheses that attempt to resolve outstanding conflicting views.

1.4.1 Similarity between choanoflagellates and sponge choanocytes

James-Clark's (1867b) discovery that a group of flagellated infusoria closely resembled the flagellated cells lining the chambers of sponges has stood the test of time. The morphological features shared by these cells include a single central flagellum encircled by a 'hyaline' collar. There are also similarities in their functional properties in as much as they create water currents and are able to ingest particles by means of pseudopodia, although the mechanism of the latter may not be identical in both cell types (Leys and Eerkes-Medrano, 2006). The similarity between these two cell types has been further reinforced by the common usage of the prefix *choano* (Latin = funnel); collar-bearing flagellates became choanoflagellates following the publication of Kent's (1880–2) *A Manual of the Infusoria* and collar-bearing sponge cells became choanocytes (Sollas, 1888).

The morphological similarity between the two cell types as seen with light microscopy is also apparent in cells viewed with EM (Fjerdingstad, 1961a, b; Eerkes-Medrano and Leys, 2006; Mah *et al.*, 2014). The enigmatic vane observed on the flagella of some choanoflagellates (Figs 2.18, 2.19) (Hibberd, 1975; Leadbeater, 2006) is apparently matched by equivalent structures on the flagella of some sponge choanocytes (Feige, 1969; Brill, 1973; Mehl and Reiswig, 1991). This coincidental occurrence supported the contention that there was a direct evolutionary relationship between the two cell types (Leadbeater, 2006; Mah *et al.*, 2014) (see also Sections 2.6.1 and 10.1).

While there is a striking similarity in the external morphology of the two cell types, the details of internal ultrastructure are less uniform. The general disposition of organelles is similar and, in common with choanoflagellates, choanocytes have a second 'dormant' basal body at an angle to the flagellar basal body. The flagellar transition region differs between the two cell types; choanocytes so far studied do not possess the single central filament typical of choanoflagellates (Pozdnyakov and Karpov, 2013). Information on choanocyte microtubular and fibrous root systems associated with the flagellar bases, which are generally regarded as being conservative features and therefore likely to be of phylogenetic significance, is limited to a few species and in some instances even these records are incomplete (see Gonobobleva and Maldonado, 2009). Nevertheless, no choanocyte observed so far has the typical ring arrangement of microtubules observed in choanoflagellates (Figs 2.29–2.31). Sponge choanocytes, like choanoflagellates, possess relatively inconspicuous fibrous roots associated with the basal bodies (Karpov and Leadbeater, 1998; Maldonado, 2004). This contrasts with the relatively large, and often non-striated, fibrous roots associated with the basal bodies of flagellated sponge larval cells (Woollacott and Pinto, 1995).

1.4.2 Sponges as colonial Protozoa

It was a fortuitous coincidence that James-Clark (1867b) was not only the first to recognise the distinctiveness of collar-bearing flagellates but also observed collar-bearing cells comprising the 'monadigerous layer' of the calcareous sponge *Leucosolenia botryoides* (Fig. 1.17). In his own words (James-Clark 1867b, p. 323) "If I were now to describe merely the congregated monads of this compound animal (*Leucosolenia*) without giving it a name, any one who had already become acquainted with the structure of *Codosiga* would set down the first as a colonial massive form of the latter." James-Clark (1871b) subsequently reported collared cells in a freshwater siliceous sponge, *Spongilla arachnoidea* (Figs 1.18–1.19).

Dujardin (1841, p. 306) had previously recognised flagellated and amoeboid cells in fragments of the freshwater sponge *Spongilla lacustris*. From this he concluded that sponges comprised "clusters of infusoria intermediate between amoebae and monads (flagellates)". Likewise, Carter (1857) observed the 'monociliated' cells that lined the 'ampullaceous' sacs of *Spongilla* and considered them to be most closely related to naked Rhizopoda (amoeboid cells) (Fig. 1.15). Carter (1859) subsequently illustrated 'monociliated' cells with anteriorly projecting spines either side of the lower part of the flagellum (Fig. 1.16). James-Clark (1867b) predicted that these spines were retracted collars and Carter (1871) eventually vindicated this prediction by demonstrating unequivocally that all monociliated cells lining ampullaceous chambers possessed collars (Fig. 1.20). Carter (1871) was also able to publish an illustration that he had recorded in 1855 of an epibiotic craspedid choanoflagellate on *Cladophora* in Bombay (Fig. 1.7 inset).

Kent (1870, 1871a, 1878a, b, 1880–2) became the most vociferous protagonist of the suggestion that "sponges are compound colony-building flagellate-bearing monads, exhibiting neither in their embryological nor in their adult condition phenomena that do not find their parallel among the simple unicellular Protozoa" (Kent, 1878b, p. 140). During the decade 1870–1880 he fully persuaded himself that sponges should be classified together with free-living collared flagellates in a single order, the Discostomata (disc-shaped food inception area), within the class Flagellata (Kent 1878a, 1880–2). He based this decision not only on the obvious similarities between the two types of collar-bearing cells, but also on aspects of reproduction and development. He envisaged sponges as comprising three 'simple' elements: (1) collar-bearing cells; (2) simple *Amoeba*-like elements or 'cytoblasts'; (3) a general investing gelatinous material called the 'cytoblastema' (Kent, 1880–2). He utterly refuted the suggestion that cells within the

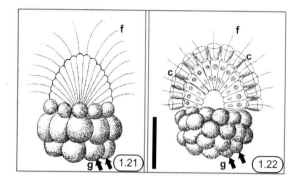

Plate 4 (Figures 1.21 and 1.22)

Figs 1.21–1.22 Conflicting morphological interpretations of sponge amphiblastula larvae. Reproduced from Kent (1880–2).
Fig. 1.21 Amphiblastula larva of the sponge *Grantia ciliata* based on Barrois (1876) showing two distinct hemispheres: the anterior hemisphere comprising 'mono-ciliated' cells (f) and the posterior hemisphere granular cells (g).
Fig. 1.22 Kent's (1880–82) drawing of an 'amphiblastuloid swarm-gemmule' of *Grantia compressa* showing collared cells on the anterior hemisphere. Flagellum (f), collar (c). Bar = 10 μm.

sponge body were organised in layers as 'tissues', which set him and his hypothesis on a direct collision course with Haeckel's view of sponges as being diploblastic (two embryonic germ layers) members of the Metazoa (Haeckel, 1872, *et seq.*).

With respect to reproduction, Kent (1878a, b, 1880–2) referred to ciliated sponge larvae as 'compound ciliated gemmules' or 'swarm gemmules' and envisaged them as being "spherical or ovate aggregations of typical collar-bearing monads connected laterally and by their bases with one another, and with their anterior flagellate and collar-bearing surface directed outwards" (Kent, 1878a, p. 11) (Fig. 1.22). In other words, Kent viewed the 'swarm-gemmules' as being colonies of collar-bearing monads, similar to the detached heads of *Codosiga* species or *Astrosiga* colonies. This again placed him at odds with Haeckel's interpretation of the sponge larva as being a gastrula, comprising two distinctive layers of cells (see Section 1.5).

Kent's (1880–2) discovery of a colonial choanoflagellate, *Proterospongia (Protospongia) haeckeli*, in a freshwater pond in Kew Gardens, was probably the most significant, and yet enigmatic, piece of supporting evidence for his contention that the sponges are closely allied to the choanoflagellates. *Proterospongia haeckeli* comprised 'extensive colony stocks' containing a mixture of collar-bearing and amoeboid cells, the latter lacking both flagella and collars (Fig. 1.8). Cells were embedded, apart from their flagella and collars, in a common mucilaginous matrix or 'zoocytium' that was a transparent film-like expansion. Cells apparently divided after becoming amoeboid and collars were regenerated on the daughter cells. Some cells appeared to be capable of 'sporulation' (Fig. 1.8). Kent (1880–2, p. 365) immediately seized on the evolutionary significance of this species as being "the nearest concatenating form (missing link) between the respective groups of the ordinary Choano-Flagellata and the Spongida". He even suggested that *Proterospongia* "could furnish a stock-form from which by a process of evolution all sponges were primarily derived".

Kent's (1880–2) exposition in support of the protozoan nature of sponges quickly attracted criticism on several counts. On the positive side most authors agreed about the morphological similarity between choanoflagellates and sponge choanocytes. However, there were at least five major points of disagreement. First, apart from James-Clark (1867b, 1871b), Carter (1871) and Kent (1880–2), the majority of authors considered that sponge larvae were differentiated into tissues (germ layers) along metazoan lines. Sollas (1884, p. 612) stated that the "individuality of component cells of the sponge is subordinated to that of the total organism to an extent and in a manner which meet with no parallel amongst the Protozoa". Second, the ciliated larvae of sponges, which Kent (1880–2) considered to be 'swarm-gemmules' or colonies of protozoan cells, are, in fact, the products of embryogenesis in which there is a reorganisation of tissues by ingression or delamination to form a multilayered larva (Leys and Eerkes-Medrano, 2005). Third, Kent's (1880–2) observation of collar-bearing cells on sponge ciliated larvae has never been confirmed and must now be taken as an observational misinterpretation (Figs 1.21-1.22) (Schulze, 1885). Fourth, Kent (1880–2) did not observe sexual reproduction in sponges that involves the production of spermatozoa and ova – another well-established metazoan character. Fifth, *Proterospongia haeckeli*, with a mixture of collar-bearing and amoeboid cells, remains an enigmatic colonial species. Not only has no authentic individual of this species containing two categories of cells been seen since (Ertl,

1968, 1981), but of the now numerous 'colonial' genera, including *Proterospongia*, *Sphaeroeca* and *Astrosiga*, that have been described, all consist entirely of collar-bearing cells (see Section 3.5). While this catalogue of shortcomings represents a substantial rebuttal of Kent's (1880–2) hypothesis, nevertheless the view persists, and is still reported in textbooks, that the sponges are derived from choanoflagellates (see also Chapter 10).

1.5 HAECKEL'S GASTRAEA THEORY AND ORIGIN OF THE LOWER METAZOA (NON-BILATERIA)

At the same time as the hypothesis that sponges were 'colonial protozoa' was being developed, an alternative and rival hypothesis was being promulgated whereby sponges were considered to be closely allied to the Cnidaria (Coelenterata) and that the metazoan diploblastic (two-tissue layered) condition was achieved by an embryonic process called 'gastrulation' (see Leys and Eerkes-Medrano, 2005). Leuckart (1854) was the first to group the sponges with the polyps and corals in the Phylum Coelenterata. In this publication, Leuckart (1854, p. 472) asserted that "if we imagine a polype-colony with imperfectly separated individuals, without tentacles, stomach sac and internal septa, we have in fact the image of a sponge with its large water canals opening outwardly". However, from the outset there were significant shortcomings to this alliance for the very reasons that Leuckart (1854) listed. Sponges are not generally considered to possess a gut in either the larval or adult stages. They are unusual within the Metazoa in that their tissues surround a series of canals and chambers through which water is filtered to feed, and they lack tentacles and cnidocysts (stinging cells) that are characteristic of the Cnidaria.

Haeckel (1866) initially classified the sponges within his newly created Kingdom Protista (Protistenreich) as distinct from the animal kingdom (Thierreich; subsequently Metazoa (Haeckel 1874a)). However, following a publication by his student Miklucho-Maclay (1868) on a calcareous sponge *Guancha blanca*, Haeckel (1870, p. 212) underwent a conversion experience to argue that the sponges were "most closely related to the corals of all organisms". He considered that "the most essential peculiarity of the sponge organization is their nutritive canal system, which is both homologous and analogous to the so-called

coelenteric vascular system, or gastrovascular apparatus of the Coelenterata". Haeckel (1870, p. 212) considered that "in the sponges, as well as in the corals and the coelenterates in general, all the different parts of the body originate by differentiation from two primitive simple formative membranes or germ lamellae, the endoderm and the ectoderm". In fact, his subsequent *magnum opus* on the calcareous sponges (*Die Kalkschwämme*) provided the first opportunity to publish what subsequently became known as the 'Gastraea theory' (Haeckel, 1872).

The basis of the Gastraea theory is heavily dependent on recapitulation theory. Thus Haeckel (1874b, 1875) not only considered that the stages in development of an embryo from a fertilised egg were universal within the Metazoa, but he also considered that the individual stages had once existed, or may still exist, as independent free-living organisms (Wilmer, 1996). The theory postulates that a fertilised egg (cytula) divides to produce a sphere of cells (morula) which then becomes hollow with a single layer of flagellated cells on the surface (blastula). The blastula subsequently undergoes invagination (gastrulation) to produce a two-layered cup, the outer cell layer comprising the ectoderm and the inner layer the endoderm, which surrounds a digestive cavity (archenteron) opening to the exterior by a blastoporal mouth for feeding.

Haeckel's Gastraea theory with respect to sponge evolution, apart from conflicting with Kent's (1880–2) colonial choanoflagellate hypothesis, also, itself, became embroiled in controversy. Haeckel based his initial description of invagination gastrulation in calcarean sponges on what is now referred to as a process of metamorphosis that occurs on settlement. During metamorphosis, calcarean amphiblastulae settle on the ciliated cells of the anterior pole, which may subsequently invaginate and contribute to the choanocyte chambers of the adult sponge. This process of invagination has been likened to gastrulation, but the homology of this process with eumetazoan gastrulation has not been widely accepted (Nielsen, 2001; Leys, 2004). Two important reasons for this are: (1) that the sponge larva is already a differentiated diploblastic or multilayered organism with distinct cell types and (2) the homology of choanocyte chambers with the eumetazoan gut is not generally accepted. Studies of sponge embryology and gastrulation still attract considerable attention and there is a substantial recent literature on these topics (Stern, 2004; Leys and Ereskovsky, 2006; Ereskovsky, 2010).

1.6 ALTERNATIVE THEORIES OF METAZOAN EVOLUTION INVOLVING CHOANOFLAGELLATE ANCESTORS

Many authors have speculated on how sponges may have evolved starting with a colonial choanoflagellate and have 'picked and mixed' between the various conflicting theories mentioned above (for references, see Salvini-Plawen, 1978; Wilmer, 1996). Hadži (1963), for instance, was sufficiently bold as to hypothesise with illustrations how a *Proterospongia* colony might undergo stepwise transformation to produce sedentary sponges of varying complexity (Fig. 1.23). Mikhailov *et al.* (2009) suggested that metazoan multicellularity resulted from the permanent combination of different stages in a multiphasic protistan (choanoflagellate) life cycle. This is an upgrading of the so-called Synzoospore hypothesis of Zakhvatkin, whereby cell differentiation leads to multicellularity rather than the more conventional view that multicellularity led to subsequent cell and tissue differentiation (see Section 10.8). Richter and King (2013), making use of comparative morphological and genomic data of extant choanoflagellates and other holozoans, have attempted to reconstruct the cell biology, life history and gene content of a hypothetical last common ancestor of the animals (urmetazoan). They envisage the urmetazoan as possessing an epithelium that contained collar cells capable of feeding on bacteria. Reproduction would have involved spermatozoa and eggs, and subsequent development of the zygote would have involved cell division, cell differentiation and probably invagination. While many of the genes involved in metazoan development, body patterning, immunity and cell-type specification must have evolved either during or after the first appearance of animals, nevertheless several gene families related to metazoan cell adhesion, signalling and gene regulation are present in extant Holozoa and therefore presumably predated the origin of animals (Richter and King, 2013; see also Section 10.8.2).

Nielsen (2001, 2008) has probably attempted the most ambitious synthesis yet by combining current morphological and molecular data with aspects of the colonial choanoflagellate and Gastraea hypotheses in a reconstruction of early metazoan evolution. Nielsen (2008) makes several assumptions; in particular he considers that sponges represent the 'blastaea' level of organisation since characteristic features of the 'gastraea' level, such as formation of an endoderm, blastopore (mouth)

and archenteron (primitive gut), are absent. He also considers that the collared units of choanoflagellates and choanocytes are homologous. Combining these assumptions, Nielsen (2008) considered that a 'choanoblastaea', a hollow sphere with a surface layer of collared cells, was the first evolutionary step leading to multicellularity and the metazoan grade of organisation. This subsequently developed to form an 'advanced choanoblastaea' that contained, in addition to the surface layer of collared cells, an inner core of non-feeding cells. This stage settled onto a substrate and the choanocytes became internalised to form the flagellated chambers typical of the sponge body (Nielsen, 2008).

Reconstruction of hypothetical 'missing links' on the basis of extant morphological and molecular data is a perfectly respectable profession. However, in the absence of a fossil record, the origin of multicellular animal life from single-celled ancestors remains one of the most enigmatic of all unresolved phylogenetic problems (Wilmer, 1996).

1.7 THE 'COLLAR CELL' AS A DISTINCTIVE CELL TYPE

'Collar cells' that possess a single central flagellum surrounded by a collar of actin-supported microvilli are a recurring feature within the Metazoa (Nørrevang and Wingstrand, 1970; Lyons, 1973; Ehlers and Ehlers, 1977; Rieger, 1976; Cantell *et al.*, 1982). Similarities in ultrastructure, which extend beyond external morphology and include a second dormant basal body and a striated fibrous root system, have led some authors to suggest that the collar cell or 'cyrtocyte' (Kummel and Brandenburg, 1961) is a fundamental cell type in metazoans, probably derived phylogenetically from a flagellate ancestor (Nørrevang and Wingstrand, 1970). Cantell *et al.* (1982) distinguish four types of collar cells: (1) choanoflagellates and sponge choanocytes whose function is to create water currents and capture particles for feeding; (2) collar cells with a sensory function – these may serve for mechano- or chemoreception and occur extensively in invertebrates, including: Hydrozoa, Scyphozoa, Anthozoa, Nemertini, Gastrotricha, Polychaeta, Oligochaeta, Priapulida and in vertebrates, for instance, as neuromasts in the lateral line systems of fish and amphibians (Ehlers and Ehlers, 1977); (3) flame cells, such as solenocytes, are excretory cells found in many invertebrate groups, for instance: Platyhelminthes, Gastrotricha, Polychaeta, Priapulida and

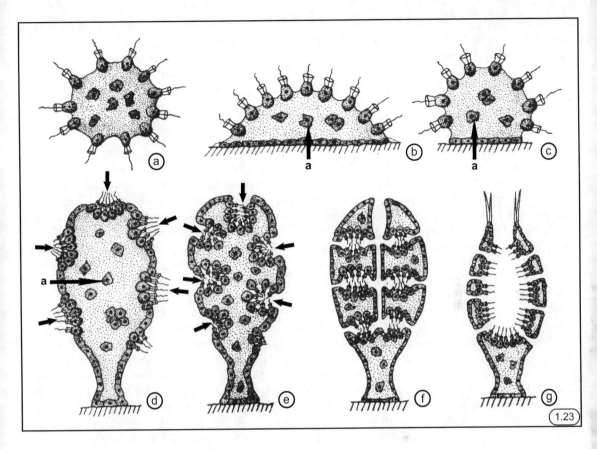

Plate 5 (Figure 1.23)

Fig. 1.23 Hypothetical representation of sponge evolution. a. Free-swimming colonial choanoflagellate stage resembling *Proterospongia*; b. and c. Sedentary stage with addition of amoeboid cells (arrow a); d–e. Erect phase with internalisation of choanocytes within chambers (arrows); f. Primitive phase of the heterocoelic state; g. Homoocoelic phase. Reproduced from Hadži (1963).

Acrania – secretion and ultrafiltration is their major function; (4) collar cells with possible adhesive functions have been described from turbellarians and bryozoans.

While cells bearing a single flagellum surrounded by a collar of microvilli are a recurring feature within the Metazoa, variations in ultrastructural detail, their possible origin from any of the three embryonic germ layers and their variety of function suggest that they have multiple origins and that their similarities are due to convergence (Wilmer, 1996). It is not, of course, impossible that choanoflagellates and choanocytes may also be of independent origin and that their similarities are also a result of convergence (Mah *et al.*, 2014).

1.8 CHOANOFLAGELLATE CLASSIFICATION

1.8.1 Traditional classification of choanoflagellates

Apart from a brief interval when choanoflagellates were mistakenly classified as algae (Section 1.8.2), the general consensus has been that choanoflagellates are heterotrophic (non-photosynthetic) flagellates and should therefore be included within the Protozoa (see also Section 10.3 and Table 10.2). Kent (1880–2) was the first to undertake a detailed classification of the group. He recognised three families: (1) Codonosigidae – cells 'naked' secreting neither a theca nor a gelatinous system, e.g. *Codosiga*; (2) Salpingoecidae – cells secreting a horny theca, e.g. *Salpingoeca*; (3) Phalansteriidae – cells secreting a gelatinous zoocytium, forming extensive social colonies, e.g. *Phalansterium* and *Proterospongia*. Subsequently, Kent's (1880–2) system has formed the basis of all classifications based on morphological data, although two major changes have occurred. The first involved the removal of Phalansteriidae and the second the addition of Acanthoecidae (see Table 1.4).

The genus *Phalansterium*, which is the basis of the Family Phalansteriidae, was created by Cienkowsky (1870) for two quite separate colonial species. *Phalansterium intestinum* comprised a surface layer of biflagellate cells and was subsequently removed from this genus by Stein (1878). The other species, *P. consociatum*, consisted of a disc-shaped zoocytium with a surface layer of uniflagellate cells each with a beak-like conical 'collar'. A second species, *P. digitatum*, was added by Stein (1878), which comprised a sub-dichotomously branched zoocytium with terminal uniflagellate cells. From the outset the systematic affinity of these species was uncertain. Stein (1878) placed *Phalansterium* close to other taxa forming colonies with a granular matrix, including *Rhipidodendron*, *Spongomonas* and *Cladomonas*. However, James-Clark (1871a) thought *P. consociatum* cells with a beak-like collar were homologous with the collared cells of *Salpingoeca*. On this basis he suggested that *Phalansterium* resembled a 'true' sponge without the spicules. Notwithstanding the differences between *Phalansterium* and choanoflagellates, Kent (1880–2) created the Family Phalansteriidae for this genus and also included *Proterospongia haeckeli* within this family since this species also consisted of collared cells embedded in a zoocytium (see Fig. 1.8). This uncertain relationship persisted until Hibberd (1983), in an ultrastructural study, showed that in all important criteria *Phalansterium* differed from choanoflagellates. The single flagellum, which lacked a second dormant flagellar base, was attached by a diverging cone of microtubules to the nucleus. The mitochondria contained unbranched tubular cristae and the beak-like collar comprised a continuous fold of cytoplasm. Subsequently, based on an 18S rDNA analysis, Cavalier-Smith *et al.* (2004) concluded that *Phalansterium* belonged to the supergroup Amoebozoa (see Section 10.1).

Norris (1965) added the Family Acanthoecidae to accommodate choanoflagellate species with siliceous basket-like loricae. Although such species had been observed previously by Kent (1880–2) and Ellis (1929), the systematic significance of the lorica was not fully appreciated (see Chapters 4–8).

1.8.2 Choanoflagellates classified as algae

For a brief period choanoflagellates were mistakenly assigned to the algae by Bourrelly (1957, 1968), Chadefaud (1960) and Christensen (1962). This resulted from an association of several errors of interpretation: (1) Lackey (1940) had described a choanoflagellate (*Stylochromonas minuta*) apparently with chloroplasts; (2) the thin lateral vane on the flagella of some choanoflagellates (see Section 2.6.1) was thought to resemble (superficially) the tubular hairs (mastigonemes) on chrysophyte flagella; (3) the single flagellum and collar of microvilli in choanoflagellates resembled similar structures in *Cyrtophora pedicellata* and *Pseudopedinella*, both representatives of the Chrysophyceae. However, each of these criteria turned out to be a misinterpretation and with the demonstration that the mitochondria of choanoflagellates contained flattened cristae, an 'apparent' opisthokont character, this group was quickly extricated from algal classifications, although the term Craspedophyceae persists in some texts (Leadbeater and Manton, 1974) (see also Section 10.2.1).

1.8.3 Classification used throughout this text

The most recent molecular phylogenetic analysis (Nitsche *et al.*, 2011) recovers two major clades within the Class Choanoflagellatea Cavalier-Smith (Fig. 10.3) (see Section 10.4.1). The first clade, Order Craspedida Cavalier-Smith 1997, includes all choanoflagellates with entirely organic coverings (equivalent to Kent's (1880–2) Codonosigidae and Salpingoecidae). The second clade, Order Acanthoecida Cavalier-Smith 1997, includes all species with siliceous loricae. In common parlance throughout this book, members of these orders are called craspedids and acanthoecids (loricate choanoflagellates) respectively.

Table 1.4 *Selection of choanoflagellate classificatory systems to illustrate historical use of nomenclature.*

Reference	Supergroup	First rank	Glycocalyx longitudinal division	Thecate species Emergent division	Loricate species	Phalansterium
Stein (1878)	**Class** Flagellata		**Family** Craspedomonadina			**Family** Spongomonadina
Kent (1880–2)	**Class** Flagellata	**Order** Choanoflagellata	**Family** Codonosigidae	**Family** Salpingoecidae		**Family** Phalansteriidae
Senn (1900)	**Order** Flagellata	**Familie** Craspedomonadaceae	**Unterfamilie** Monosigeae	**Unterfamilie** Diplosigeae		**Familie** Phalansteriaceae
Hollande (1952)		**Ordre** Choanoflagellés Craspédomandines	**Famille** Gymnocraspedidae **Sous Famille** Monosiginae Monosiginae Codonosiginae	**Famille** Salpingoecidae		**Famille** Phalansteriidae
Norris (1965)	**Order** Craspedomonadales	**Class** Pedinellaceae	**Class** Codonosigaceae	**Class** Salpingoecaceae	**Class** Acanthoecaceae	
Bourrelly (1968)	**Sous Classe** Craspédomonadophycidées	**Ordre** Monosigales	**Famille** Monosigacées	**Famille** Salpingoacées		**Famille** Phalanstériacées
Cavalier-Smith and Chao (2003)	**Subphylum** Choanozoa	**Class** Choanoflagellatea	**Order** Craspedida		**Order** Acanthoecida	
Nitsche et al. (2011)		**Class** Choanoflagellatea	**Order** Craspedida **Family** Salpingoecidae	**Order** Craspedida **Family** Salpingoecidae	**Order** Acanthoecida **Family** Acanthoecidae Stephanoecidae	
Adl et al. (2012)*	**Supergroup** Opisthokonta	**First rank** Choanomonada	**Second rank** Craspedida		**Second rank** Acanthoecida	

* See also Table 10.2.

2 • The collared flagellate: functional morphology and ultrastructure

Stability in a changing world.

2.1 FUNCTIONAL MORPHOLOGY: INTRODUCTION

Choanoflagellates are easily recognised and rarely confused with other protistan cells. This is because of their distinctive morphology, comprising an ovoid to spherical cell body with a single anterior flagellum surrounded by a collar of microvilli (Figs 2.9–2.15). Use of the descriptive adjective 'anterior' is entirely a matter of convention. The flagellar pole traditionally defines the anterior end of a choanoflagellate cell and this tradition is particularly influenced by consideration of the sedentary cell. However, confusingly, on a swimming cell the flagellum is often described as being 'posterior' because the cell is propelled from behind (see Sections 2.3.2 and 10.2.1). More correctly, a swimming cell should be described as moving with its posterior pole foremost and with its flagellum acting as a 'pulsellum'. It is the relative orientation of the cell and not the location or functioning of the flagellum that varies.

There are remarkably few variations to the choanoflagellate cell plan, the only modifications being the secondary loss of the flagellum, as in *Choanoeca perplexa* (Figs 3.27, 3.30) and the *micropora* form of *Savillea parva* (Figs 6.2 inset, 6.4), and the extreme reduction in the length of the collar as observed on some motile cells (Fig. 2.37). Functional considerations are essential to understanding this relatively straightforward cell plan. Choanoflagellates are classed as 'suspension feeders' (Fenchel, 1987); undulation of the single flagellum generates water currents from which suspended food particles are intercepted on the outer surface of the collar and subsequently ingested by means of pseudopodia at the base of the collar. For the feeding process to occur effectively the cell needs to be stationary, or its movement restrained, which is achieved either by direct attachment to an extraneous surface or attachment within a siliceous lorica. When flagellar activity

occurs on an unattached cell, for instance on a craspedid daughter cell following division, it serves for locomotion and this is important for species dispersal (see Section 3.2). The combined roles of the flagellum and the collar are intimately linked and the behaviour of these structures has been subjected to investigation by microscopists, applied mathematicians and physicists. The study of flagellar hydrodynamics with respect to choanoflagellates must begin with an appreciation of fluid flow around microscopical structures.

2.2 THE HYDRODYNAMIC ENVIRONMENT OF CHOANOFLAGELLATES

When an object moves through a fluid, the force that it experiences can be subdivided into two components: (1) an inertial force arising from the need to accelerate the fluid displaced by movement; and (2) a viscous force associated with the sliding of layers of fluid over each other. The ratio of inertial to viscous forces is determined by a dimensionless quantity called the Reynolds number (Re), which is given by the following equation:

$$Re = \rho v l / \eta$$

Where ρ and η are, respectively, the density and viscosity of the fluid through which a structure of length l is moving at velocity v.

At high Reynolds numbers, as exemplified by an adult fish swimming in water (Re > 10^4), the values of l and v are large so inertial forces predominate. In this instance, viscous forces are insignificant and can be ignored. However, for microorganisms, including choanoflagellates and bacteria, where the values of l and v are small and the Reynolds number is of the order of 10^{-2}–10^{-6}, viscous forces predominate and inertial forces will have no appreciable effect on fluid flow. In this situation movement of the cell occurs only in direct response to propulsive forces; if a cell's propulsive mechanism ceases activity the cell will rapidly come to a halt; conversely, if the cell's propulsive mechanism starts the cell will almost immediately reach its terminal velocity. In these conditions the flow of fluid is non-turbulent and small particles entrained in fluid flow have negligible momentum (Sleigh, 1991).

2.3 MODELLING WATER FLOW IN RESPONSE TO FLAGELLAR ACTIVITY

2.3.1 Background

The propulsion of microscopical organisms through liquids by means of flagellar motions has been of considerable interest to fluid dynamicists, and the development of this branch of mathematics can be traced back to the work of Taylor (1952a). Using Stokes equations for the steady flow of fluid past a sphere when viscous forces predominate, Taylor (1952a) based his initial calculations on the movement of sinusoidal transverse waves along a flexible but inextensible sheet immersed in a viscous fluid. While these gave order-of-magnitude results and the direction of propulsion of the sheet was opposite to the direction in which the waves were propagated, nevertheless the expressions he derived were only valid for combinations of displacement amplitudes and wave numbers that were unrealistically small. Taylor (1952b) subsequently extended his method of analysis to waves of lateral displacement moving down an infinite filament of circular cross-section, assuming all oscillations to be in the same plane. However, again the model lacked validity for the values of displacement amplitude and wave numbers observed on living cells. To overcome these shortcomings Hancock (1953) proposed an alternative approach. Again utilising Stokes equation for the slow, steady motion of fluid past a sphere, Hancock described the velocity field in terms of two distinct singularities situated at the centre of the sphere. Both singularities induce radial and transverse component velocities that will contribute to the resultant motion of the body. One of these singularities was a 'doublet', subsequently called a 'dipole', and the other was peculiar to viscous movements, which he called a 'stokeslet'. Hancock (1953) applied the concept of singularities to flagellar propulsion by placing a distribution of dipoles and stokeslets along the centre line of an oscillating filament. This approach subsequently became known as the 'slender-body' theory and allowed the derivation of an approximate relationship between the velocity of a body at each point along its length and the force per unit length experienced by that body at each point.

The complexity of the calculations required by the 'slender-body' theory led Gray and Hancock (1955) to adopt a simpler model based on resistance coefficients taken from Hancock's (1953) asymptotic results. With this model they examined the swimming of sea urchin spermatozoa and obtained good agreement between calculated and observed swimming speeds. After this success the 'resistive-force' model, as it became known, was adopted by many workers in the field. In the meantime attempts were made to refine the slender-body model (Lighthill, 1976; Johnson and Wu, 1979). Subsequently, Johnson and Brokaw (1979) compared the results they obtained using the resistive-force and slender-body models, respectively, for the distribution of forces, bending moments and shear moments, calculated along the length of a flagellar model generating typical planar bending waves. They found that, while the two models approximated each other for cells like spermatozoa with small bodies, for flagellates with larger bodies only the slender-body model was capable of accounting accurately for all the hydrodynamic interactions.

Motion produced by flagellar activity affects choanoflagellates in two contexts. First, it is responsible for the 'swimming' or locomotion of cells. This is particularly important for the dispersal of species and is encountered in craspedid and nudiform species when a cell divides to produce a free-swimming daughter cell (Chapters 3 and 6, respectively). It is also applicable to colonial craspedid species such as *Desmarella*, *Proterospongia*, *Sphaeroeca* and *Astrosiga* (see Section 3.5) and some acanthoecid species such as *Parvicorbicula socialis* that are capable of motility (see Section 4.6.3 and Fig. 9.8). Second, flagellar activity is responsible for the propulsion of water around stationary or restrained cells. Motion of water in this context is responsible for bringing potential prey particles to the cell. For choanoflagellates there are many studies describing and analysing flagellar activity and the creation of feeding currents, but relatively few regarding cell propulsion.

2.3.2 Hydrodynamic analysis of cell propulsion by flagellar activity

Higdon (1979a) carried out a hydrodynamic analysis of flagellar propulsion for a choanoflagellate-like cell comprising a spherical cell body propelled by base-to-tip, planar sinusoidal waves down a long, slender flagellum.

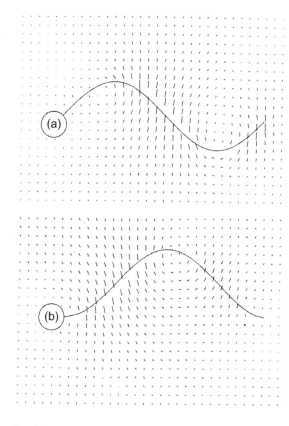

Fig. 2.1 Water propulsion around a swimming cell. (a) Streaklines around a cell at one instant during the cycle of a wave. Lines are parallel to velocity at each point and their length is proportional to the magnitude of the velocity. A line of length equal to the grid spacing represents a velocity equal to the linear wave speed. The wave is travelling to the right. (b) Streaklines around the cell one-quarter of a cycle later. Reproduced from Higdon (1979a).

Slender-body theory for Stokes flow was used whereby the flagellum was replaced with distributions of dipoles and stokeslets along its centre line. The aim of this study was to examine the influence of a range of parameters, including amplitude (a), wave number (k), flagellar length (L), flagellar radius (a) and cell body radius (A), on average swimming speed and power consumption with the object of discovering optimal swimming motions and their applicability to living cells. While the mathematical analysis relating to this study is beyond the scope of the present account, nevertheless certain of the conclusions are of relevance to choanoflagellates. The optimal sine wave for minimising power consumption is a single wave with maximum slope (ak) $\approx$ 1. The optimal ratio of flagellar length/cell radius (L/A) = 25, with an optimum L/A range of 20–40. Motile choanoflagellates, such as daughter cells resulting from division, usually have long flagella – for example, *Choanoeca perplexa* (Fig. 3.27), *Helgoeca nana* (Fig. 6.18) and *Acanthoeca spectabilis* (Fig. 6.43). Cells comprising motile colonies, in particular *Proterospongia* species, also have relatively long flagella (ratio of flagellar length/cell radius (L/A) = 12) (see Section 3.5.1) (Figs 3.43–3.45).

Higdon (1979a) illustrated two other features relevant to choanoflagellates, one being the distribution of streaklines about a motile cell at an instant during the cycle of flagellar wave (Fig. 2.1a). Streaklines in this instance have been drawn parallel to the velocity at each point and their length is proportional to the magnitude of velocity. The streaklines clearly demonstrate the dominance of the normal forces (perpendicular to the direction of the wave). They are present on the long, straight segments of the flagellum and are at a large angle to the wave direction. On the segments nearly parallel to the wave direction (near the crests of the waves), where the tangential force dominates, the fluid is moved along the tangent to the flagellum, but at a much smaller velocity. Transverse forces of the moving wave are in opposition, but longitudinal forces are not, so the overall result is that the longitudinal forces (thrust) dominate over tangential forces (drag) and the cell is propelled forwards. In the second illustration (Fig. 2.1b), one-quarter of a cycle later, there is a similar dominance of normal forces as in Fig. 2.1a, but the average flow in the wave direction is much larger, and the swimming speed faster (Higdon, 1979a). Sleigh's (1991) more simplified illustration shows how a series of highly localised circulations, extending less than a wavelength from the flagellum, link to generate axial flow in the direction of the wave along a smooth flagellum (Fig. 2.2). The second feature of interest arising from Higdon's (1979a) work is the trajectory of the organism during a wave cycle (Fig. 2.3). The positions of the cell body and flagellum are shown at intervals of one-seventh of a cycle. A feature of the motion is that the constant amplitude wave appears to have increasing amplitude owing to the yawing motion of the cell during the cycle. Unfortunately, there are few published records of choanoflagellate swimming velocites; this is because most motile cells are ephemeral dispersal stages in the life cycles of otherwise sedentary species. Fenchel (1982a) estimated a swimming velocity of about 30 μm s^{-1} for *Monosiga*,

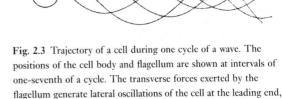

Fig. 2.2 Pattern of water flow generated by a swimming flagellate. The planar flagellum propagates sinusoidal waves from base to tip. This results in a series of local circulations that link to generate an axial flow in the direction of the wave along the smooth flagellum. The cell is propelled forwards through the medium in the direction of the larger arrow while the smaller arrows indicate the regions of maximal water propulsion by the moving wave. Reproduced from Sleigh (1991).

Fig. 2.3 Trajectory of a cell during one cycle of a wave. The positions of the cell body and flagellum are shown at intervals of one-seventh of a cycle. The transverse forces exerted by the flagellum generate lateral oscillations of the cell at the leading end, hence the seven slightly different positions of the cell body. Reproduced from Higdon (1979a).

although he cautioned against its reliability because cells rarely swim in microscopical preparations.

2.3.3 Hydrodynamic analysis of water propulsion around stationary cells

The role of the flagellum in creating flow fields of water for feeding purposes was recognised early in the history of choanoflagellate studies. Kent (1880–2) illustrated in the frontispiece of his *Manual of the Infusoria* the forward flow of water around a cell of *Monosiga gracilis* as depicted by the motion of carmine particles (Fig. 1.9). In his own words he described the flagellum as "whirling round with inconceivable rapidity" and thereby creating "a strong centrifugal current in the water setting in from behind towards the direction of its own apex, and bringing with it all such tiny organic particles as do not possess sufficient weight or power to stem the tide" (Kent, 1880–2, p. 327).

Lapage (1925) made a detailed microscopical study of the generation of feeding currents and the ingestion of food particles in *Codosiga botrytis*. Using suspensions of carmine particles, he was able to observe the trajectories of entrained particles at different positions around the collar and flagellum (Fig. 2.4). He distinguished two categories of currents: (1) primary currents (A, B and C in Fig. 2.4) that move with 'great rapidity' and are directed to the upper third of the collar; (2) secondary feeding currents (D–G in Fig. 2.4) that are slower and direct particles towards the lower part of the collar where they become entrapped and are ingested by pseudopodia. However, while Lapage's (1925) illustration of the general flow

pattern is correct, the relative velocities of the primary and secondary currents have been misinterpreted (see Higdon, 1979b and Pettitt *et al.*, 2002 below). Currents A, B and C are likely to be slower since they are on the boundary of the slow-moving toroidal circulations. On the other hand the currents D–G will be faster since they are closer to the longitudinal axis of the flagellar undulations. Lapage (1925) noted that particles adhering to the surface of the collar do not always pass directly to its base but may move up and down, sometimes becoming dislodged and swept away. This is in accordance with the observations of Boenigk and Arndt (2000) and Pettitt (2001) (see Sections 2.4 and 9.3.2).

Lighthill (1976) was the first to expand flagellar hydrodynamics to include the generation of flow fields for feeding purposes. In sedentary cells, such as stalked choanoflagellates, the entire function of flagellar thrust is to generate water movements for feeding purposes, this thrust being balanced by an equal and opposite reaction from the stalk which attaches the cell to a substrate. Instead of the localised small-scale flow fields produced by moving cells, in which velocity decays exponentially with distance (Fig. 2.2), flagellar undulation on a stationary cell produces large-scale flow fields in which velocity decays inversely with distance. The direction of fluid movement is opposite to the direction of thrust and, for a smooth flagellum, is in the same direction as the propagation of the undulation. Lighthill (1976) predicted that whereas undulations of moderate amplitude are required to generate a given swimming speed, the minimum amplitude needed to generate a given flow near the anterior end of the cell body must be much larger – a prediction in agreement with Sleigh's (1964) later observations on *Codosiga botrytis*.

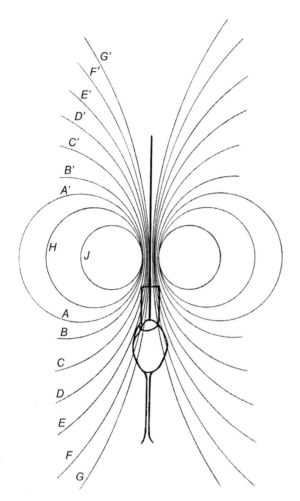

Fig. 2.4 Lapage's (1925) interpretation of the pattern of water flow created by flagellar motion on a sedentary cell of *Codosiga botrytis*. Direction of flow is from the rear of the cell to the anterior end. Individual streaklines are indicated with a capital letter (rear) and a capital letter prime (anterior). Reproduced from Lapage (1925).

Higdon (1979b), using a similar cell model to that which he had used for swimming cells (Higdon, 1979a) and applying slender-body theory, turned his attention to the generation of feeding currents by attached choanoflagellate-like cells (Fig. 2.5). He subsequently compared the results of his analysis with Sleigh's (1964) microscopical observations on *Codosiga botrytis*. As in his earlier work on swimming cells, Higdon (1979a) calculated the mean flow rate and power consumption for a wide range of parameters, including flagellar amplitude (*a*) and wave

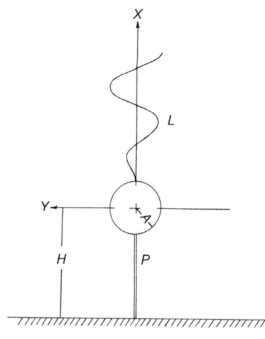

Fig. 2.5 Diagram of a sedentary organism showing relative positions of the cell body, flagellum and substratum. A = radius of the cell body; L = length of the flagellum; Y = horizontal line drawn through the centre of the cell; H = height of the centre of the cell above substratum; P = pedicel; hatched line at base indicates substratum. Reproduced from Higdon (1979b).

number (*k*). Optimal motions were determined with the criterion of minimising the power required to achieve a given flow rate. The comparisons of flagellar motion between sedentary and swimming cells are of note. The optimal wave slope (*ak*) has a maximum of 2–2.5 on sedentary cells compared to the equivalent value of ≈ 1 for swimming cells. This means that the slope of the wave on stationary cells is much steeper than that necessary for optimal swimming – a result in agreement with Lighthill's (1976) earlier prediction. The optimal ratio of flagellar length to cell radius is $L/A = 10$, which is less than the equivalent value of 20–40 for swimming cells. For optimal efficiency the length of the pedicel P (and hence height above substrate) should be greater than or equal to the length of the flagellum. When compared with Sleigh's (1964) recorded measurements for *Codosiga botrytis* there is good agreement. For *C. botrytis* the maximum slope of the wave (*ak*) is 2; the L/A ratio is 5–6; and the ratio of pedicel length/flagellar length (H/L) is 0.7–4.0.

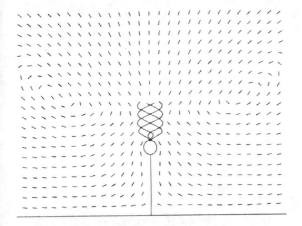

Fig. 2.6 Flow diagram showing direction of fluid motion around a sedentary cell. The sinusoidal flagellar wave propagates from base to tip. Flow is away from the substratum and along the vertical axis of the cell. Each line segment shows the direction of the time-averaged velocity at that point. With the exception of points very close to the flagellum, the lines show the trajectories particles will follow. Reproduced from Higdon (1979b).

While in general there is a difference in the ratio of flagellar length/cell radius (L/A) between motile choanoflagellate cells and sedentary feeding cells, the individual values are about one-third to a half those predicted by Higdon (1979a, b) (see Table 4.4). The cells of craspedid species generally have an L/A ratio of 10 (see Section 3.5.1), while among acanthoecids *Stephanoeca* species in general have relatively short flagella (L/A ratio 1.5–3) which are only used for generating feeding currents within the lorica and are incapable of generating locomotory movements (Figs 4.45–4.51; Table 4.4). On the other hand, some acanthoecid species, such as *Cosmoeca ventricosa, C. norvegica* and *Pleurasiga minima*, have relatively long flagella ($L/A = 7$–8) and loricae with few lightly silicified costae, which probably means that they are capable of locomotory movements (Figs 4.60–4.61). This certainly applies to the colonial species *Parvicorbicula socialis*, which has a moderately long flagellum ($L/A = 4$) but which has a lorica with few lightly silicified costae (Fig. 4.65). Colonies comprising many hundreds of cells held together by the outer anterior rims of their loricae are capable of moving as a pulsating entity within the water column (Fig. 9.8) (McKenzie *et al.*, 1997).

Higdon (1979b) also illustrated the overall flow field around a stationary cell (reproduced here as Fig. 2.6). Each

line segment shows the direction of the time-averaged velocity at that point. However, in contrast to Figs 2.1 a and b, the lines are all of equivalent length and do not reflect the magnitude of the velocity at each point, only the orientation of flow lines. The magnitude of the velocity is inversely proportional to the distance from the flagellum. Fig. 2.6 shows that the fluid is drawn into the centre along the substratum and below the cell, forced up past the cell surface and continues to move up and outward above the organism. At a height above a plane approximately level with the end of the flagellum and at a similar distance to each side of the organism is the centre of a large vortex. The fluid at the centre of each vortex remains in place for several cycles. The cell is in the centre of a stream that is constantly refreshed by fluid drawn from further away. As Higdon (1979b, p. 326) notes, "this is essential for the feeding process to be successful".

Higdon's (1979a, b) calculations, while making a valuable contribution to choanoflagellate biology, are nevertheless idealised in that they do not make allowance for the collar that surrounds the proximal part of the flagellum, nor do they take into account the possible existence of viscous eddies generated by point forces near solid boundaries (Liron and Blake, 1981). The latter are of particular relevance to stalked choanoflagellates attached to solid surfaces and specimens that are viewed in a limited volume between a coverslip and slide. Pettitt *et al.* (2002) addressed both of these limitations by undertaking a combined practical and theoretical study involving two sedentary species, *Salpingoeca amphoridium* (ATCC 50153) and *Codosiga gracilis*, the latter with short (12 μm) and long (21 μm) pedicels. The pattern of water movement generated by the two species was traced by the introduction of 0.5 μm latex spheres into the flow field. Particle movement was recorded at 0.02 sec intervals on video and the pathlines of individual microspheres were visualised by superimposing sequential video images on a single drawing (Fig 2.7 a–f). The distance between consecutive positions of the same sphere is a measure of the velocity at which the particle was travelling at a particular instant. The far-field flow generated by *Salpingoeca amphoridium* is approximately a toroidal viscous eddy (Fig. 2.7a), whereas in *Codosiga gracilis* it is more elliptical, with the long axis parallel to the substratum (Fig. 2.7b, c). The effect of a shorter pedicel on *Codosiga* is to flatten the flow on the side adjacent to the substratum (Fig. 2.7c). With respect to near-field fluid flow, close to the flagellum there is strong directional streaming and an increase in velocity

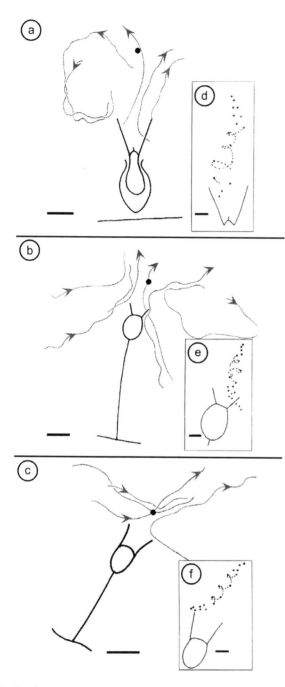

that is accompanied by oscillatory movements (Fig. 2.7d–f). Experimental and theoretical results correlate well for far-field flows. However, experimental results do not accord with the generation of smooth constant stream-lines that theoretical models predict. This discrepancy may result from the frequent changes in the direction and magnitude of the force exerted by the flagellum on the surrounding fluid and also the presence of a collar may serve to destabilise streamlines (see also Fig. 5 in Mah et al., 2014). While it has generally been assumed that the flagellar waveform of attached cells is planar and sinusoidal, Pettitt *et al.* (2002) provide some evidence for a three-dimensional motion. The later model used by Orme *et al.* (2003) makes allowance for this possibility.

Orme *et al.* (2001) expanded the numerical modelling work of Pettitt *et al.* (2002) with particular respect to the interaction of multiple toroidal eddies created by sedentary cells of *Salpingoeca amphoridium* in close proximity to one another. They suggest that the interaction of these eddies leads to 'chaotic advection', which enhances the domain of feeding for neighbouring cells (see also Roper *et al.*, 2013 and Section 3.5.1). In a further study, Orme *et al.* (2003) turned their attention to modelling the motion of individual particles around a choanoflagellate cell based on a helical beat pattern. In this work they compared some of the qualitative similarities between the numerical calculations of particle paths and experimental results in Pettitt *et al.* (2002). Three-dimensional models of individual particle trajectories have been computed and the resulting illustrations accord well with the empirical results of Pettitt *et al.* (2002).

2.3.4 Influence of the collar on water flow

The choanoflagellate collar comprises a funnel-shaped ring of actin-supported microvilli (Figs 2.9–2.12). The anterior end of the cell adjacent to the centrally located flagellum is usually flattened to give a circular platform around which the regularly spaced collar microvilli are located (Fig. 2.20). In life, the collar microvilli, which are more-or-less equivalent in length, are extended rigidly at an angle to the longitudinal axis of the cell. In fixed material the microvilli lose their rigidity and are always much shorter and thicker than in the living condition. In material sectioned for electron microscopy (EM), the diameter of each microvillus is approximately 0.1 μm (Figs 2.39–2.41). The number of microvilli comprising the collar varies according to species – for example, *Choanoeca perplexa* has approximately

Fig. 2.7a–c. Fluid flow induced by flagellar activity around: a. *Salpingoeca amphoridium*; b. *Codosiga gracilis* long pedicel; c. *C. gracilis* short pedicel. The large black dot above each cell denotes the distal extremity of the flagellum (not shown). Bar = 5 μm. **Fig. 2.7d–f.** show details of the movement of individual particles when in close proximity to the collar. Bar = 2 μm. Pathlines of individual particles have been obtained from video recordings (50 frames sec[−1]) of latex microspheres introduced into the culture medium. Reproduced from Pettitt *et al.* (2002).

18 microvilli (Fig. 2.39) whereas *Diaphanoeca grandis* has about 50 (Fig. 2.14). The height and angle of the collar also vary according to species (Figs 2.9–2.12). Actively feeding cells usually have longer microvilli than those on stationary phase cells and the collar is usually more wide-angled (Leadbeater, 1983a). The collar microvilli of *Monosiga brevicollis* cells treated with dimethyl sulphoxide before osmium tetroxide fixation and critical point drying are circled in the mid-region by a narrow ring of 'glycocalyx' (Mah *et al.*, 2014).

The precise micro-pattern of fluid flow around the collar and between the microvilli in choanoflagellates has not been observed directly. It has been generally assumed that the collar serves as a filter for removing suspended particles (Fenchel, 1986a; Pettitt *et al.*, 2002). This implies that on most living cells the spaces between neighbouring microvilli are either partially or fully open and are not occluded as they appear to be in some sponges (Leys *et al.*, 2011; Mah *et al.*, 2014). The resistance of the collar microvilli to the flow of water creates a pressure drop, the magnitude of which is proportional to the 'head-on' (unrestricted) velocity of the fluid, its viscosity and the relative spacing and radius of the microvilli (Fenchel, 1986a). In choanoflagellates, flagellar motion drives water away from the cell, which reduces the pressure within the collar. This creates a pressure drop across the collar, thereby drawing water between the microvilli. The cell can exert control over the flow of water around and through the collar by varying the parameters of flagellar motion (frequency, wavelength and amplitude) and the characteristics of the collar (microvillar length and collar angle) (Pettitt *et al.*, 2002; Mah *et al.*, 2014).

There are several published theoretical models for the prediction of pressure drop across filters at low Reynolds numbers (Jorgensen, 1983). Fenchel (1986a) used the model of Tamada and Fujikawa (1957), which was developed for fluid flow between a single layer of parallel cylinders, as a basis for calculating the pressure drop across the collars of two choanoflagellates, *Diaphanoeca grandis* and *Monosiga* sp. For porosities (free distance between collar microvilli) of 0.1–0.3 μm and 0.25–0.35 μm, respectively, and head-on fluid velocities of 10 μm s^{-1} and 25 μm s^{-1}, respectively, Fenchel (1986a) calculated pressure drops equivalent to 1.2×10^{-5} atm for *Diaphanoeca grandis* and 1.4×10^{-5} atm for *Monosiga*. These results were of the same order of magnitude (about 1.0×10^{-5} atm) as those he obtained for water filtration across the paroral membrane of the ciliates *Cyclidium* and

Vorticella (Fenchel, 1986a). A similar value (1.0×10^{-5} atm) was obtained for water filtration by choanocytes of the sponge *Haliclona permollis* (Reiswig, 1975) and other metazoan filter feeders, such as the ascidians *Ciona intestinalis* and *Ascidiella aspersa* (Jorgensen 1983). This consistency among a range of filter feeders supports the view that pressure drop is a physical constraint for this type of feeding and explains why species specialised to utilise the smallest particles generate low water velocities and consequently have relatively low values of clearance (Fenchel, 1986a).

Pettitt *et al.* (2002) undertook a more refined hydrodynamic analysis of filter feeding in choanoflagellates by applying slender-body theory and a 'control volume' approach to estimate pressure drop across the collar. This approach makes allowance for the fact that the collar microvilli are not parallel to each other throughout their length, as was assumed in the Fenchel (1986a) model, but diverge as the diameter of the collar enlarges distally. Thus, at constant fluid velocity and viscosity across the collar, pressure drop along the collar co-varies with increasing pore size between adjacent microvilli. Based on this approach, Pettitt *et al.* (2002) obtained average pressure drops of 1.3×10^{-7} atm for *Salpingoeca amphoridium*, 5.0×10^{-7} atm for *Codosiga gracilis* and 1.6×10^{-6} atm for *Stephanoeca diplocostata*. These values, although lower than those obtained by Fenchel (1986a), nevertheless confirm that, when computed in the same way, pressure drop is relatively constant in different choanoflagellate species.

2.3.5 Influence of the lorica on water flow

Relatively little work has been carried out on the effects of a basket-like lorica on water flow around the enclosed cell. Variations in the pattern of costae, location of an organic lining and positioning of the cell suggest that the lorica serves several functions, the relative importance of which varies from one species to another (see Chapter 4 and Section 9.6). Functions of the lorica that have been suggested include restriction of cell motility resulting from flagellar activity, thereby enhancing fluid propulsion feeding, direction of feeding currents to the collar, protection against predation and reduction in sinking rate for planktonic species (Andersen, 1988/9). Pettitt *et al.* (2002), using the microsphere tracer method (see above), observed the flow of water induced by flagellar motion into the lorica of *Stephanoeca diplocostata*. Water is drawn into the lorica

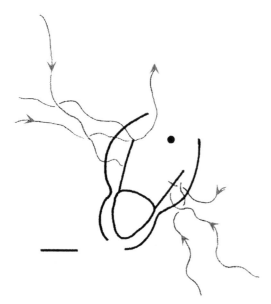

Fig. 2.8 Fluid flow induced by flagellar activity around *Stephanoeca diplocostata*. Water is drawn in horizontally through the sides of the lorica and expelled through the anterior ring. The large black dot above the cell denotes the distal extremity of the flagellum (not shown). Bar = 5 μm. Pathlines have been obtained from video recordings (50 frames sec⁻¹) of the movement of latex microspheres introduced into the culture medium. Reproduced from Pettitt *et al.* (2002).

more-or-less horizontally and exits through the anterior ring (Fig. 2.8). Recirculation of water and the associated viscous eddies so prominent in non-loricate species were not observed in *Stephanoeca*. However, the strong directional streaming of fluid along the axis of the flagellum was present, although there was no perceptible increase in fluid velocity in the region of the flagellum. A preliminary application of numerical modelling to fluid flow around a loricate cell has been published by Smith (2009). The results show that for a lorica similar in construction to that of the *micropora* form of *Savillea parva*, the near-flow field around the flagellum is relatively unchanged by the presence of the lorica. However, the remainder of the flow field is strongly suppressed in magnitude. Within the lorica chamber this component is suppressed by nearly 50% and any external flow almost completely abolished. By contrast, the hydrodynamic interaction of the flagellum and lorica is relatively weak, with force density components of the flagellum almost

unchanged by the presence of the lorica. This result is not unexpected since the lorica used in the model has so many closely positioned costae. In reality, the inner surface of the lorica of *Savillea parva* is lined by a fine mesh of microfibrils (Fig. 6.6) and so fluid motion through the walls of the lorica must be almost non-existent. The effect of the lorica on fluid flow and the possible functional role of the lorica are important topics for consideration in future research (see Section 9.6).

2.4 PARTICLE CAPTURE AND INGESTION

Early work on particle capture and ingestion in choanoflagellates became a highly contentious issue at the end of the nineteenth century (see Section 1.3.1). In summary, it was the limitations of optical microscopy at the time, as well as antagonisms between 'schools of thought', that led some workers to come to what we now know to be erroneous conclusions. However, there were notable exceptions, in particular Bütschli (1878) and later Griessmann (1913) correctly observed that food particles trapped on the outer surface of the collar are subsequently ingested by means of pseudopodia that originate from the base of the collar. In *Codosiga botrytis*, pocket-like pseudopodia are produced below the collar at the side of the cell (Lapage, 1925; Stoupin *et al.*, 2012). Engulfed particles are subsequently transported to the rear of the cell, where digestion takes place.

Pettitt (2001) made detailed video recordings of particle capture and ingestion by a relatively large freshwater species, *Salpingoeca amphoridium* (mean dimensions: cell radius = 2.31 μm; cell volume = 51.6 μm³; flagellar length = 20.65 μm; collar length = 8.05 μm) (see also Section 9.3.2). Suspensions of 0.5 μm diameter latex microspheres were used as surrogate prey. To minimise the effects of boundary interference, experiments were carried out in wells 0.5 mm deep on glass slides. Suspensions of microspheres introduced into aliquots of cell culture become entrained in flagellar-generated streamlines, and if they make contact with the collar they become 'available for capture'. Several possible fates await particles intercepted by the collar. They may become detached, usually from the distal end of the collar, and lost to the feeding process. In one experiment 98% of intercepted particles were subsequently lost in this manner. Captured particles may be translocated bi-directionally along the long axis on the collar surface; some may be transported repeatedly up and down, while others may be phagocytosed immediately. Pettitt (2001) describes the proximal quarter of the collar as being the 'phagocytosis zone', and it is from this region that particle ingestion takes place. Pseudopodia arise from the outer surface of the base of the microvilli in direct response to the presence of prey particles (Figs 2.10, 2.50). Pseudopodia are targeted precisely at prey particles, which indicates a sophisticated signalling mechanism operating between the prey–microvillus interface and the cytoskeleton involved in pseudopodial production. Phagocytosis may occur simultaneously at several locations around the base of the collar. Part of the membrane surface of the engulfing pseudopodium transforms into the bounding membrane of the enclosing food vacuole. The latter is withdrawn into the neck of the cell where it may undergo one of three possible fates. Seventy-four per cent of vacuoles containing ingested microspheres pass down the neck and move to the posterior end; 26% delay within the neck although most will eventually move to the base of the cell. However, 9% of delayed particles are subsequently ejected from the cell (Pettitt, 2001).

Pettitt (2001) also investigated whether *Salpingoeca amphoridium* was capable of selecting microspheres on the basis of either size or physicochemical surface properties. With respect to size, aliquots of known concentrations of 0.5 μm, 0.75 μm or 1.0 μm diameter microspheres were added to samples of *S. amphoridium*. Whereas the proportion of microspheres intercepted was similar for all size categories, fewer of the 0.75 μm and 1.0 μm microspheres were retained on the collar than the 0.5 μm particles. The subsequent effect on ingestion was that the intake of 0.5 μm diameter particles was significantly higher than that of either of the other two particle categories. An analysis of the position on the collar where microsphere interception occurs revealed that the majority of particles were captured on the distal half of the collar. In this region only the 1.0 μm diameter particles were larger than the respective spaces between the microvilli and thus only for these particles would the collar microvilli act as an obligate sieve. For the smaller categories of microspheres, their capture is likely to involve supplementary methods of interception including physicochemical interactions between the particles and the collar surface (Pettitt, 2001). A similar conclusion comes from the study on *Choanoeca perplexa*. Sedentary cells of this craspedid species lack a flagellum

but have a wide-angled collar with large spaces between the microvilli. Observation of *Choanoeca* cells during exponential phase growth reveals the effectiveness of the collar at entrapping prey without the assistance of flagellar-generated viscous eddies (Leadbeater, 1977). This is a classic example of 'diffusion feeding' (see also Section 9.3.2).

With respect to particle selection based on the physico-chemical surface properties, experiments were carried out in which *Salpingoeca amphoridium* was presented with three different categories of modified 0.5 μm microspheres (Pettitt, 2001). These included amine- and carboxylate-treated microspheres, which had relatively hydrophilic surfaces, and sulphate-modified spheres that were hydrophobic. All had a negative charge, the strength of which increased in the order: amine < carboxylate < sulphate. Unmodified microspheres have a large negative charge and are intermediate in their hydrophobic character. The results of this experiment reveal that the proportion of microspheres intercepted, retained and subsequently ingested by *S. amphoridium* was higher for the sulphate- and amine-modified microspheres than it was for unmodified and carboxylate modified microspheres. From this experiment it appears that hydrophilic particles with a weak negative charge or hydrophobic particles are favoured for capture and ingestion.

Little is known about the cell mechanisms involved in the adherence of particles to the collar microvilli and their subsequent bi-directional translocation. Longitudinal movement of particles is episodic, with phases of motion being separated by pauses of variable length. Neither microtubule inhibitors, such as vinblastine (25 μM) and nocodazole (100 μM), nor the actin polymerisation inhibitor cytochalsasin D (1 μg ml^{-1}) had an observable effect on microsphere movement. However, application of the myosin motor inhibitor butanedione monoxide (BDM) resulted in a reduction in particle velocity of about 50% at 10 mM concentration. BDM inhibition was reversible following removal of the drug by flushing cells with fresh medium. A preliminary conclusion to be drawn from these results is that the movement of particles is intimately associated with an acto-myosin component of the cytoskeleton. This is not surprising since individual microvilli are supported by a longitudinal bundle of actin microfilaments and the pseudopodium is known to be a microfilament-rich cytoplasmic protrusion of one or more microvilli (Leadbeater, 1983b). Microtubules, on the other hand, play a relatively minor part in collar support, except that the radial cytoskeletal microtubules are associated with the bases of the collar microfilament bundles (Figs 2.44–2.47).

2.5 ULTRASTRUCTURE: INTRODUCTION

The morphological uniformity of choanoflagellate cells is matched by the relative constancy of their internal ultrastructure. While the general disposition of organelles is standard for all species so far observed, some variations have been recorded, particularly with respect to the organisation of cytoskeletal microtubules encircling the base of the flagellum (Hibberd, 1975; Zhukov and Karpov, 1985). Loricate species (acanthoecids) are, of course, distinguished from those with thecae (craspedids) by the intracellular formation of silicified costal strips that are subsequently exocytosed onto the outer surface of the cell. The uptake of reactive silicate from the external medium and the subsequent intracellular polymerisation of biogenic silica within membrane-bounded silica deposition vesicles (SDVs) supported by microtubules is discussed in Chapter 5 (Section 5.5). In the following review of choanoflagellate cell organelles, examples and images have been selected from a range of species to emphasise the widespread constancy of ultrastructural features.

2.6 THE FLAGELLAR APPARATUS

2.6.1 External structure

A single, centrally located anterior flagellum encircled proximally by a funnel-shaped collar comprising a ring of microvilli is the hallmark of a choanoflagellate cell (Figs 2.9–2.12). Flagellar length varies considerably between species and for hydrodynamic reasons swimming cells usually have longer flagella than sedentary cells (see Section 2.3.3). Long flagella are particularly well demonstrated by the constituent cells of motile *Proterospongia* colonies (Figs 3.38, 3.43–3.45), the motile daughter cells of thecate and nudiform species (Figs 3.27, 6.18, 6.43) and some freely suspended acanthoecids (Fig. 2.17). Short flagella, whose function is restricted to creating water currents for feeding purposes, are well illustrated by acanthoecids in which the cell is completely enclosed by the lorica (Figs 4.45–4.51). Some normally sedentary species, such as *Salpingoeca amphoridium*, have long flagella and in this

example it is not uncommon to observe a thecate cell adhering to a substantial clump of bacteria being propelled through the medium. A similar situation almost certainly applies to acanthoecids with long flagella and relatively small, lightly silicified loricae (Figs 2.17, 4.60, 4.61, 4.65). Flagellar length can change in response to external conditions; for instance, stationary-phase cells usually have shorter flagella than those undergoing rapid growth. Withdrawal of the flagellum is a normal precursor to nuclear division (Figs 2.62, 2.71b–d) except for non-flagellate species such as *Choanoeca perplexa* and the *micropora* form of *Savillea parva*, which develop a flagellum immediately prior to division (Figs 3.24, 6.11).

The flagellar surface of most choanoflagellates is smooth when observed in whole mounts for EM (Figs 2.13, 2.17). However, in sectioned material the surface is usually covered by a fine fibrillar coating (Figs 2.22, 2.25). The variability in the distribution and substructure of this covering leaves some doubt about its authenticity. A systematic study has not been carried out to determine what effects different fixation protocols have on the preservation of such delicate material. An enigmatic covering, known as the flagellar 'vane' (Hibberd, 1975), has been seen on some freshwater species by a number of authors (Petersen, 1929; Vlk, 1938; Petersen and Hansen, 1954; Hibberd, 1975; Leadbeater, 2006) and on the marine species *Monosiga brevicollis* (Mah et al., 2014). Petersen's (1929) original record was based on cells of *Codosiga botrytis* and *Salpingoeca* sp. that had been treated with Loeffler's stain (see Leadbeater, 2006). Petersen (1929) referred to the lateral appendages on the flagellum of *Codosiga botrytis* as being 'fine *Flimmer* hairs' and considered the thicker part of the flagellum of *Salpingoeca* sp. as comprising many "upright, projecting, extraordinarily fine cilia being about 2 μm in length" (Fig. 2.16a, b). Vlk (1938), also working with *Codosiga botrytis* and using Loeffler's stain and nigrosine as a negative stain, was able to confirm the existence of a '*Flimmersaum*' (hairy fringe) (Fig. 2.16c) but there was always an element of inconsistency in his results (Vlk, 1938). The subsequent use of EM to some extent clarified but did not entirely resolve the situation. Some 25 years after his earlier observations, Petersen together with Hansen was able to show by means of EM that the lateral appendages were a reality on the flagella of *Codosiga botrytis* cells treated with Loeffler's stain (Petersen and Hansen, 1954). They described the vane as consisting of fine horizontal threads that

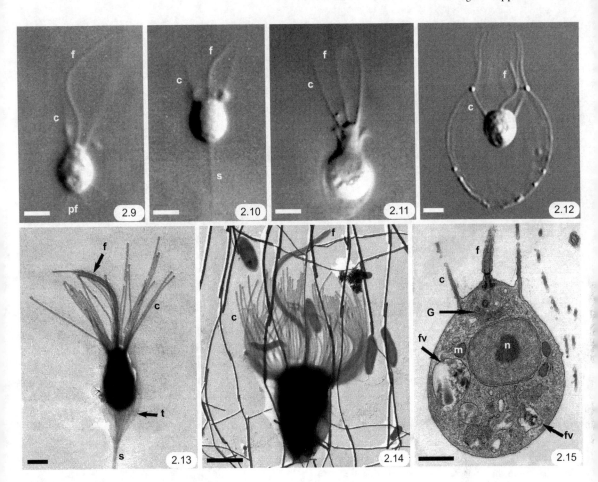

Plate 1 (Figures 2.9–2.15)

Figs 2.9–2.15 Choanoflagellate cell-body plan.

Figs 2.9–2.12 Interference contrast images of living cells of *Monosiga ovata*, *Codosiga gracilis*, *Salpingoeca amphoridium* and *Diaphanoeca grandis*, respectively, showing the spherical to ovoid cell with single anterior flagellum (f) encircled proximally by the collar (c). Posterior filopodia (pf), stalk (s). Figs 2.9–2.11: Bar = 2.5 µm; Fig. 2.12: Bar = 1 µm. Figs 2.9 and 2.10 reproduced from Karpov and Leadbeater (1998) and (1997), respectively.

Figs 2.13–2.14 TEM of whole mounts of fixed cells of *Salpingoeca* sp. and *Diaphanoeca grandis*, respectively, showing collar (c) comprising narrow microvilli and single central flagellum (f). Bar = 1 µm.

Fig. 2.15 *Monosiga ovata*. Vertical section of cell showing general disposition of organelles. Flagellum (f), collar (c), Golgi apparatus (G), nucleus (n), mitochondrial reticulum (m), food vacuoles (fv). Bar = 1 µm.

appeared to form a network. Hibberd (1975) published definitive EM images of the vane on *Codosiga botrytis* shadowcast with chromium. The vane consisted of a bilateral wing-like process approximately 2 µm wide on either side of the flagellum and contained two sets of interwoven or overlapping fine fibrils, each set of which was orientated at an angle of about 65–70° to the long axis of the flagellum. The vane was present on

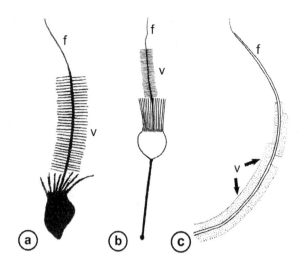

Fig. 2.16 Early illustrations of a lateral vane (v) on choanoflagellate flagella (f). a. *Codosiga botrytis* and b. *Salpingoeca* sp. Reproduced from Petersen (1929). c. *Codosiga botrytis*. Reproduced from Vlk (1938).

the proximal portion of the flagellum for approximately two-thirds its length.

Bearing in mind the uncertainty that had built up over half a century with respect to the flagellar vane, Leadbeater (2006) made a definitive attempt to resolve the 'mystery' surrounding its existence and structure. Disappointingly, after three months of work dedicated to this topic, the mystery was not entirely dispelled. A lateral vane was occasionally observed on the flagella of *Salpingoeca amphoridium* when viewed as whole mounts after fixation with osmium tetroxide and subsequent staining with 1% potassium phosphotungstate (pH 7.0). In contrast to the unpredictability of its preservation, the vane, when present, did reveal a relatively standard morphology comprising a lateral flange of mean width 1.33 μm ± 0.17 SD ($n = 26$), the overall width of the two wings and the flagellar axis being 2.96 μm ± 0.5 SD ($n = 15$). On some specimens the vane extended almost the entire length of the flagellum, but usually it was shorter and was often fragmented into segments by abrupt horizontal splits (Figs 2.18, 2.19). The region of the vane adjacent to the flagellar axis consisted of prominent fibrils usually curving away from the axis and separated by spaces. The remainder of the vane appeared either granular or fibrillar. Often a denser staining line running parallel to the edge of the vane is apparent (Fig. 2.19 arrows). Recently, following fixation with aqueous

0.8% osmium teroxide solution containing 10% ruthenium red, a vane has been observed for the first time on a marine choanoflagellate, *Monosiga brevicollis* (Mah *et al.*, 2014). Whether or not a vane will prove to be a universal feature of choanoflagellate flagella, as seems a possibility, must await future studies.

There has been much speculation about the structural and functional significance of the flagellar vane in choanoflagellates (Petersen, 1929; Vlk, 1938; Hibberd, 1975). As discussed in Section 1.8.2, the vane was erroneously mistaken by some authors for a covering of flagellar hairs, although the evidence for this conclusion was never convincing (Hibberd, 1975; Leadbeater, 2006). With respect to its possible hydrodynamic significance, no conclusive information is currently available. There is no question of the vane effecting a reversal of water flow as occurs around flagella bearing tubular hairs (Sleigh, 1991). Mah *et al.* (2014) view the collar/flagellum system as a functionally integrated unit rather than two separate components. The hydrodynamic consequences of such a relationship have yet to be addressed but are likely to be significant, depending on the mechanical properties of the vane. It is of note that the flagella of many sponge choanocytes also possess a bilateral vane-like structure, although this is usually more substantial than the structures that have been described in choanoflagellates (Mehl and Reiswig, 1991; Mah *et al.*, 2014) (see Sections 1.4.1 and 10.1).

2.6.2 Flagellar axoneme, basal body and root

The internal ultrastructure of the flagellum can be divided into three sub-compartments, namely: (1) tip and shaft (9 + 2 axoneme); (2) transitional region; (3) basal body. The arrangement of microtubules and interconnecting arms and links comprising the 9 + 2 flagellar axoneme of choanoflagellates conform to the well-established universal pattern (Fig. 2.21). It is not uncommon for the outer doublets to terminate at some distance prior to the tip so that only the two central microtubules extend the full length of the flagellum (Figs 2.13, 3.8, 3.43). The standard definition of the transitional region as the interval between the termination of the two central microtubules and the origin of the C-microtubules (triplet microtubules) of the basal body (Pitelka, 1974) requires modification for choanoflagellates. This is because the two central microtubules terminate at a considerable distance above

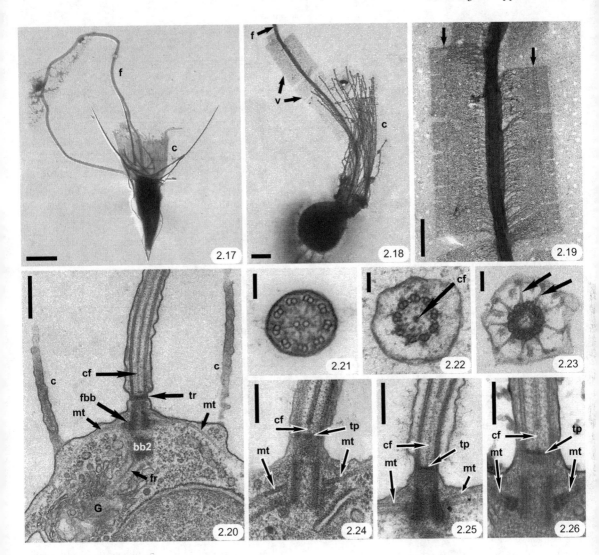

Plate 2 (Figures 2.17–2.26)

Figs 2.17–2.26 Flagellar apparatus of choanoflagellates. **Fig. 2.17** *Calliacantha natans.* Loricate cell with long (35 μm) flagellum (f) and short collar (c). Bar = 2.5 μm.

Fig. 2.18 *Salpingoeca amphoridium.* Cell with relatively long collar (c) and flagellum (f) with lateral segmented vane (v). Bar = 2.5 μm.

Fig. 2.19 *S. amphoridium.* Higher magnification of flagellar vane showing fibrillar substructure with longitudinal markings (arrows). Bar = 1 μm. Reproduced from Leadbeater (2006).

Fig. 2.20 *Monosiga ovata.* Vertical section of anterior end of cell showing flagellar apparatus and collar (c). Towards the base of the 9 + 2 axoneme of the shaft, the two central microtubules terminate and are replaced by a central filament (cf). The flagellar membrane is constricted at the level of the transitional region (tr) below which is the flagellar basal body (fbb) encircled by cytoskeletal microtubules (mt). A second basal body (bb2) is located beneath the basal body of the flagellum. A fine fibrillar root (arrow fr) links the second basal body to a cisterna of the Golgi apparatus (G). Bar = 0.5 μm. Reproduced from Karpov and Leadbeater (1998).

Figs 2.21–2.23 Three transverse sections of the flagellar axoneme proceeding proximally. Bars = 0.1 μm. **Fig. 2.21** *Stephanoeca diplocostata.* 9 + 2 axoneme of the shaft.

the dense-staining transitional plate that usually marks the distal extremity of the transitional region. In all choanoflagellates studied to date, a single central filament extends between the termination of the two central microtubules and the transitional plate (Figs 2.20, 2.22, 2.24–2.26, 2.33) (Hibberd, 1975; Karpov, 1982; Zhukov and Karpov, 1985; Karpov and Leadbeater, 1998). The distinctive central filament varies in length: 150 nm in *Codosiga botrytis* (Fig. 2.33), 60 nm in *C. gracilis* (Fig. 2.24), 400 nm in *Monosiga ovata* (Fig. 2.20), 450 nm in *Salpingoeca* sp. (Fig. 2.25) and 100 nm in *Stephanoeca diplocostata* (Fig. 2.26). In sectioned material, the flagellar membrane is tightly constricted around the outer doublets at the level of the transitional plate, forming a 'membrane collar' (Fig. 2.25). Below the transitional plate a relatively densely staining block of material fills the lumen of the axoneme (Figs 2.24–2.26). The change from the outer doublet microtubules of the transitional region to triplet microtubules of the basal body occurs at the level where the flagellar membrane deflects outwards and becomes confluent with the plasma membrane of the main cell body. At the level where the C-microtubules are added to the outer doublets a ring of nine regularly spaced transitional fibres, each of which is associated with the outer surface of a triplet, radiate outwards towards the plasma membrane (Fig. 2.23). These fibres probably serve to anchor and stabilise the base of the flagellum. The transitional region of choanoflagellate flagella accords in most respects with Pitelka's (1974) 'Type II' classification of basal bodies, the salient features of which are: (1) the termination of the C-microtubules at the level of confluence between the plasma membrane and flagellar membrane; and (2) the continuation of the outer doublets for at least 200 nm above the basal body before the appearance of the transitional plate.

A universal feature of the flagellar apparatus of choanoflagellates is the presence of a second 'dormant' (non-flagellar) basal body lying approximately at right angles to that of the basal body of the flagellum (see also Section 10.2.1). The length of the dormant basal body is similar to the flagellar basal body and comprises a full complement of triplets embedded in dense-staining amorphous material (Figs 2.20, 2.27, 2.28, 2.33). In *Monosiga ovata* the two basal bodies are held together by a linking bridge and the second basal body is further secured by means of a striated fibrous root to a membrane-bounded sac associated with cisternae of the Golgi apparatus (Figs 2.20, 2.28, 2.36) (Karpov and Leadbeater, 1998). Hibberd (1975) observed at least one further basal body in *Codosiga botrytis*. However, this is exceptional and all other species observed have been restricted to two.

2.7 CYTOSKELETON

2.7.1 Microtubular cytoskeleton

The cytoskeleton of choanoflagellates can be divided into two distinct systems, one based on microtubules (tubulin) and the other on microfilaments (actin). The microtubular cytoskeleton, excluding the flagellum and the second basal body, comprises a radial array of peripheral microtubules that converge to form a ring around the flagellar basal body. The most common arrangement, as exemplified by *Salpingoeca* sp., *Monosiga ovata* and *Stephanoeca diplocostata*, involves approximately 50 microtubules that converge to form a regular two-layered ring around the flagellar basal body (Figs 2.29–2.31). In species with larger numbers of microtubules the number of layers may increase as the basal body is approached (Fig. 2.33) (Hibberd, 1975; Zhukov and Karpov, 1985). Bands of dense-staining amorphous material comprise 2–5 concentric rings around the basal body and this material functions to hold the closely packed microtubules in place. Detergent treatment of cells to remove the background cytoplasm reveals continuity in at least one of the rings of dense-staining material in *Monosiga ovata* (Fig. 2.31). The innermost ring merges with the dense material on the surface of the basal body. Extracted material also reveals the regular manner in which the

---

Fig. 2.22 *Salpingoeca* sp. proximal end of 9 + 2 axoneme with central pair of microtubules replaced by central filament (arrow cf).

Fig. 2.23 *Monosiga ovata*. Top of flagellar basal body showing triplet microtubules each associated with a transitional fibre (arrows).

Figs 2.24–2.26 Three longitudinal sections of flagella including the proximal end of the shaft, the transitional region and basal body of *Codosiga gracilis*, *Salpingoeca* sp. and *Stephanoeca diplocostata*, respectively. Narrow central filament (arrow cf), transitional plate (arrow tp), cytoskeletal microtubules (mt). Bars = 0.25 μm.

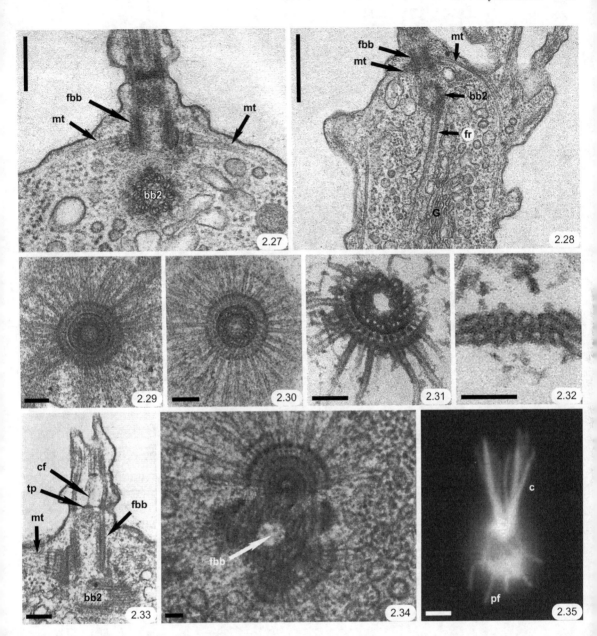

Plate 3 (Figures 2.27–2.35)

Figs 2.27–2.35 Microtubular cytoskeleton. **Figs 2.27–2.28** *Monosiga ovata*. Vertical sections through flagellar apparatus. Flagellar basal body (fbb), second basal body (bb2), fibrillar root (Fig. 2.28 arrow fr), cytoskeletal microtubules (mt). Bars = 0.25 µm and 0.5 µm, respectively. Fig. 2.28 reproduced from Karpov and Leadbeater (1998).

Figs 2.29–2.31 Transverse sections through flagellar basal body showing ring of cytoskeletal microtubules held in position by concentric rings of densely staining material.

Fig. 2.29 *Salpingoeca* sp.

Fig. 2.30 *M. ovata*. Reproduced from Karpov and Leadbeater (1998).

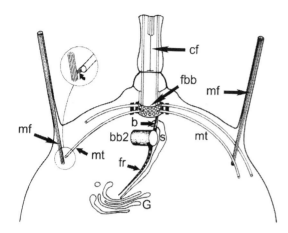

Fig. 2.36 Diagram of anterior portion of cell (based on *Monosiga ovata*) showing arrangement of basal bodies and associated cytoskeleton. Central filament (cf), flagellar basal body (fbb), fibrillar bridge linking flagellar basal body to the second basal body (b), second basal body (bb2), fibrillar root (fr), membrane-bounded cisterna (s), Golgi apparatus (G), cytoskeletal microtubules (mt), microvillar microfilaments (mf). Arrow (inset) fibrillar connection between cytoskeletal microtubule and microvillar microfilaments. Reproduced from Karpov and Leadbeater (1998).

microtubules are stacked alternately into two layers as they approach the inner circumference of the ring (Figs 2.31, 2.32). As the microtubules radiate outwards they pass immediately below the plasma membrane and extend for about one-third to one-half the length of the cell (Figs 2.20, 2.27, 2.28, 2.42). A diagrammatic reconstruction of the cytoskeleton based on *Monosiga ovata* in relation to the basal bodies is presented in Fig. 2.36 (Karpov and Leadbeater, 1998).

The relationship between the microtubular cytoskeleton and the flagellar basal body in *Codosiga botrytis*, although bearing many of the features seen in *Monosiga* and *Stephanoeca*, is considerably more complex (Fig. 2.34). The number of microtubules is greater (>60) and they converge on four or more foci. In common with species that have a single ring, concentric bands of dense-staining material hold the microtubules in position and as they approach the basal body they become stacked into five or more layers, as seen in vertical section (Fig. 2.33). Multiple foci also occur in *Sphaeroeca volvox* and *Kentrosiga thienemanni*, although the number is probably less than for *Codosiga botrytis* (Zhukov and Karpov, 1985).

Microtubules have an important role in supporting silica deposition vesicles during costal strip production in loricate species (see Sections 5.5, 5.6) and in facilitating nuclear division (Section 2.9).

2.7.2 Microfilamentous cytoskeleton

The microfilamentous cytoskeleton is actin-based and extensively distributed within the cytoplasm. In cells fixed using standard protocols and sectioned for EM, immediate visualisation of microfilaments is restricted to the microvilli comprising the collar and filopodia occurring elsewhere on the cell, particularly at the rear, where they assist in settlement (Figs 2.37–2.45). However, in detergent-extracted cells treated with a preparation of the subfragment 1 (S1) of muscle myosin, arrays of decorated microfilaments with the characteristic 'arrowhead' pattern appear extensively within the peripheral cytoplasm when viewed in section with TEM (Figs 2.48, 2.49) (Leadbeater, 1983b; Karpov and Leadbeater, 1998). Staining with rhodamine–phalloidin also reveals the distribution of actin within the cell (Fig. 2.35).

Microvilli and filopodia are narrow, hair-like, actin-supported cytoplasmic projections that extend from the cell surface (Sebé-Pedrós *et al.*, 2013; see also Section 10.9.7).

Fig. 2.31 *M. ovata* treated with detergent. Bar = 0.1 μm.

Fig. 2.32 *M. ovata*. Vertical section of two layers of microtubules close to flagellar basal body. Bar = 0.1 μm.

Fig. 2.33 *Codosiga botrytis*. Vertical section of flagellar basal body (arrow fbb) and transitional region. Second basal body (bb2), vertical stack of cytoskeletal microtubules (mt), transitional plate (arrow tp), central filament (arrow cf). Bar = 0.1 μm.

Fig. 2.34 *Codosiga botrytis*. Cytoskeletal microtubules converging on multiple foci in region of flagellar basal body (arrow fbb). Bar = 0.1 μm.

Fig. 2.35 *Monosiga ovata*. Rhodamine–phalloidin stained cell showing distribution of fluorescently labelled actin within collar microvilli (c) and posterior filopodia (pf). Bar = 2.5 μm. Figs 2.33 and 2.34 reproduced from Hibberd (1975).

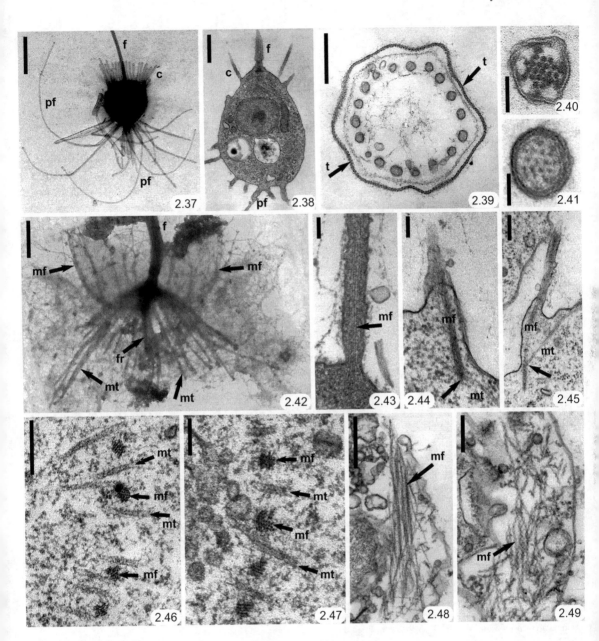

Plate 4 (Figures 2.37–2.49)

Figs 2.37–2.49 Microfilamentous cytoskeleton. Fig. 2.37 *Codosiga gracilis*. Whole mount of motile cell showing proximal portion of flagellum (f), short collar (c) and long posterior filopodia (pf). Bar = 2.5 μm.

Fig. 2.38 *Monosiga ovata*. Vertical section showing flagellum (f), collar (c) and posterior filopodia (pf). Bar = 1 μm.

Fig. 2.39 *Choanoeca perplexa*. Transverse section (TS) through ring of 18 collar microvilli and neck of theca (arrows t). Bar = 0.5 μm. Reproduced from Leadbeater (1977).

Figs 2.40–2.41 *Monosiga ovata* and *Stephanoeca diplocostata*, respectively. TS through collar microvilli showing hexagonal packing of discrete bundles of actin microfilaments. Bar = 0.05 μm.

While the ultrastructure of these two categories of cellular projections is similar, the term microvilli (singular
microvillus) is used to describe a stable and regular
palisade of projections approximately equivalent in
length, whereas the term filopodia (singular filopodium)
refers to projections that are relatively few in number,
irregularly located and various in length. In life, choanoflagellate collar microvilli project rigidly from the anterior end of the cell but following treatment with fixatives
or detergent they immediately shrink in length and lose
their rigidity. Thus in EM preparations microvilli are
always in a flaccid state (Figs 2.13, 2.14, 2.17, 2.18). Each
microvillus is supported internally by a longitudinal
bundle of cross-linked, parallel actin microfilaments that
extend the full length of the microvillus (Fig. 2.43). In
favourable transverse sections, bundles of microfilaments
can be visualised as dense-staining rods with hexagonal
packing (Figs 2.40, 2.41). At the base of each microvillus
the core of microfilaments penetrates for some distance
into the main body of the cell, although shrinkage of the
microvilli during preservation may accentuate the depth
to which the bundles penetrate (Figs 2.44, 2.45). In
longitudinal section, a close connection between the core
of microvillar microfilaments and one or more cytoskeletal microtubules is apparent (Figs 2.44, 2.45). Frequently,
cytoplasmic microtubules appear to be displaced from
their normal position immediately beneath the plasma
membrane due to the downward movement of one or
more bundles of microvillar microfilaments (Fig. 2.44).
These observations demonstrate the close connection
between the two cytoskeletal components in this region.
In tangential sections of the anterior end of the cell,
the regular spatial relationship between microfilament
bundles seen in transverse section and microtubules seen

in glancing tangential section is apparent (Figs 2.46,
2.47). In some instances individual bundles of microfilaments alternate with microtubules on a one-to-one basis.
A regular relationship can also be seen in well-preserved
detergent-extracted cytoskeletons where the microfilaments of the microvilli are deflected upwards and
the cytoskeletal microtubules are deflected downwards
(Fig. 2.42).

The close relationship between microfilaments of
the collar microvilli and the cytoskeletal microtubules
suggests that the latter probably provide a framework
for maintaining the regular positioning of the microvilli
around the anterior end of the cell. Although sometimes
there is a one-to-one ratio of microtubules to microfilamentous bundles, this is not always the case. For
instance, in *Stephanoeca diplocostata* approximately
50 cytoskeletal microtubules interdigitate with the bases
of 40 collar microvilli. Nevertheless, a close relationship
between microtubules and microvillar microfilaments is
probably important in ensuring the equitable distribution
of microvilli during cell division and in coordinating
the clockwise rotation of the ring of microvilli during
lorica assembly (see Chapters 6 and 7).

In life the collar microvilli can alter their length and
angle of inclination rapidly in response to external stimuli.
Filopodia issuing from the rear of the cell can glide
over surfaces such as those of a glass slide or coverslip,
They can also perform coordinated 'walking' movements
that are probably important in a cell 'seeking out' a suitable
surface for ultimate settlement. Collar microvilli play a
key role in lorica assembly; they provide the forward movement necessary to extend the longitudinal costae and
clockwise rotation of the collar provides the lateral movement required to extend the helical and transverse costae

Fig. 2.42 Isolated flagellar apparatus from detergent-treated cell. Flagellum (f), collar microvillar microfilaments (mf), cytoskeletal
microtubules (mt), fibrous root (fr). Bar = 0.5 μm. Reproduced from Karpov and Leadbeater (1998).

Fig. 2.43 *Proterospongia choanojuncta*. Longitudinal section (LS) of collar microvillus showing inner core of actin microfilaments
(arrow mf). Bar = 0.1 μm. Reproduced from Leadbeater (1983b).

Figs 2.44–2.45 *Monosiga ovata*. LS of base of microvillar microfilaments (mf) showing points of contact (arrows) with cytoskeletal
microtubules (mt). Bars = 0.1 μm and 0.25 μm, respectively.

Figs 2.46–2.47 *M. ovata*. Tangential sections of anterior end of cell showing the interdigitation between microvillar
microfilaments (arrows mf), sectioned in TS, and cytoskeletal microtubules (arrows mt), sectioned in LS. Bars = 0.25 μm and 0.1 μm,
respectively.

Figs 2.48–2.49 *M. ovata* and *Proterospongia choanojuncta*, respectively. Myosin S1 fragment-labelled microfilaments (arrow mf) within
peripheral cytoplasm showing arrowhead decoration. Bars = 0.5 μm and 0.25 μm, respectively. Fig. 2.48 reproduced from Karpov and
Leadbeater (1998).

(Figs 6.53–6.56, 7.19–7.21, 7.31) (see Chapters 6 and 7 for discussion of lorica assembly)

2.7.3 Ingestion of prey particles

The ingestion of prey particles by linguiform pseudopodia that move in close association with the collar microvilli is exclusively a microfilament-controlled activity. As discussed in Section 2.4, pseudopodia are produced in response to particles adhering to the proximal quarter of the collar. Prey ingestion is a precisely targeted process and takes several seconds. Pseudopodia arise as protuberances of cytoplasm from the cell body adjacent to the base of the microvilli (Figs 2.50, 2.51). They approach a prey particle from below and immediately upon contact they engulf the particle and pull it towards the collar base. The core of microfilaments supporting the microvilli is unaffected by the increase in size, but the cytosol of the pseudopodium is also rich in microfilaments (Fig. 2.52) (Leadbeater, 1983b). Once the particle has been engulfed, the membrane-bounded vacuole severs all connection with the external medium. The vacuole containing the particle is withdrawn into the cell and passes to the rear, where digestion takes place. Digestion of bacteria follows a standard pattern; initially the surface capsule disintegrates into fragments (Fig. 2.53), the internal region including the nucleoid swells and becomes translucent and finally the bacterial cytoplasm disintegrates (Leadbeater, 1977). In actively feeding cells, there are usually 4–6 food vacuoles with prey particles at varying stages of digestion (Fig. 2.15).

2.8 CYTOPLASMIC ORGANELLES

2.8.1 Mitochondrial reticulum

Currently, no detailed analysis of a choanoflagellate mitochondrial reticulum is available. The most common appearance of mitochondria in sectioned material comprises discrete spherical to elliptical profiles (Figs 2.15, 2.54), although elongated profiles are occasionally observed, particularly in cells undergoing division (Fig. 2.56). In most species the mitochondrial reticulum is located around the lower half of the nucleus and occasionally in a transverse section of a cell the mitochondrion can be seen to almost encircle the nucleus (Fig. 2.54). Frequently, the distal surface of the mitochondrial reticulum is overlain by cisternae of the endoplasmic reticulum

(ER) (Figs 2.55, 2.56). In some instances this can be almost continuous, giving the impression that the mitochondrion is located in a compartment bounded proximally by the nuclear envelope and distally by the ER (Fig. 2.57). The mitochondrial surface is bounded by two membranes, the inner of which is folded to give distinctive flattened cristae (Fig. 2.56) which may be organised into loosely arranged stacks (Fig. 2.55). Prior to the advent of molecular phylogenetic studies (Wainright *et al.*, 1993, 1994), the occurrence of flattened (non-discoidal) mitochondrial cristae was one of the most important ultrastructural indicators that choanoflagellates were allied to the Metazoa and chytrid fungi (see Section 10.2.1). In dividing cells of *Stephanoeca diplocostata* and *Didymoeca costata*, mitochondrial profiles with constrictions are common (Fig. 2.56), but no information is currently available on mitochondrial division. In dividing cells of *Monosiga ovata* the mitochondrial reticulum becomes extremely elongated and spans the constriction between two daughter cells (Fig. 2.63). Recently, Wylezich *et al.* (2012) have observed 'tubular' cristae in two craspedid species, *Codosiga minima* and *C. balthica*, collected from hypoxic water in the Baltic Sea.

2.8.2 Golgi apparatus

The Golgi apparatus is typically located immediately below or adjacent to the second basal body. The number of cisternae and their dimensions differ between species; in thecate cells with narrow necks, such as *Choanoeca perplexa* and *Proterospongia choanojuncta*, the stack may consist of only 4–6 small cisternae (Fig. 3.35). Towards the end of interphase the Golgi apparatus increases in size and activity (Fig. 2.57). In many species the Golgi apparatus is involved in producing dense-staining fibrillar material which accumulates within vesicles at the sides of the cisternal stack (Figs 2.57, 2.59). While in vertical section individual cisternae appear as discrete flattened discs, in tangential section they look more like a reticulum with meandering peripheral tubules and lacunae in the central region (Fig. 2.59).

The secretion of fibrillar material by the Golgi apparatus prior to cell division occurs in both thecate and loricate species (Figs 3.36, 7.33). A good indicator of imminent cell division is the substantial accumulation of regular-sized vesicles containing fibrillar material at the anterior end of the cell (Figs 2.60, 2.61, 3.36). When the cell divides, the vesicles are passed on to the daughter cell (juvenile) that will have the responsibility

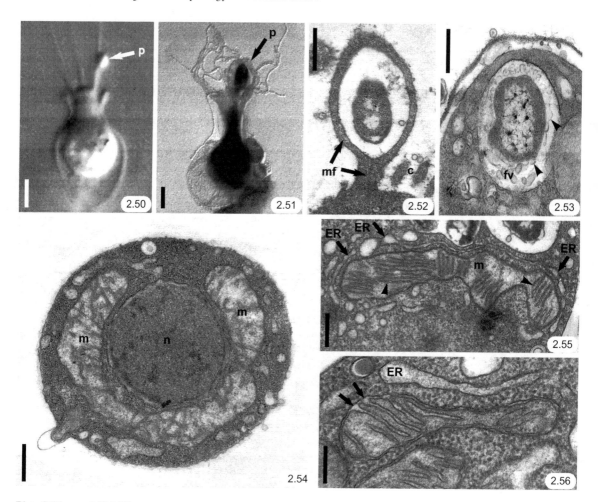

Plate 5 (Figures 2.50–2.56)

Figs 2.50–2.56 Pseudopod formation and mitochondrial reticulum. **Figs 2.50–2.51** *Salpingoeca amphoridium* (living) and *Choanoeca perplexa* (fixed), respectively. Cells with ingested particles in linguiform pseudopodia (arrows p). Bars = 2.5 μm and 1.0 μm, respectively.
Fig. 2.52 *C. perplexa*. Pseudopodium containing ingested bacterium. Microfilaments (arrows mf) and collar microvilli (c). Bar = 0.5 μm. Reproduced from Leadbeater (1983b).
Fig. 2.53 *C. perplexa*. Partially digested bacterium within food vacuole (fv). Note blebbing of bacterial capsule (arrowheads). Bar = 0.5 μm. Reproduced from Leadbeater (1977).
Fig. 2.54 *Stephanoeca diplocostata*. Transverse section of cell showing crescent-shaped mitochondrial profile (m) surrounding nucleus (n). Bar = 0.5 μm.
Fig. 2.55 *S. diplocostata*. Elongated mitochondrial profile (m) with stacks of flattened cristae (arrowheads). ER cisternal profiles (arrows ER) surrounding mitochondrion. Bar = 0.5 μm.
Fig. 2.56 *S. diplocostata*. Constricted mitochondrion in dividing cell showing flattened cristae continuous with inner mitochondrial membrane (arrows) and a cisterna of ER in close proximity (ER). Bar = 0.25 μm.

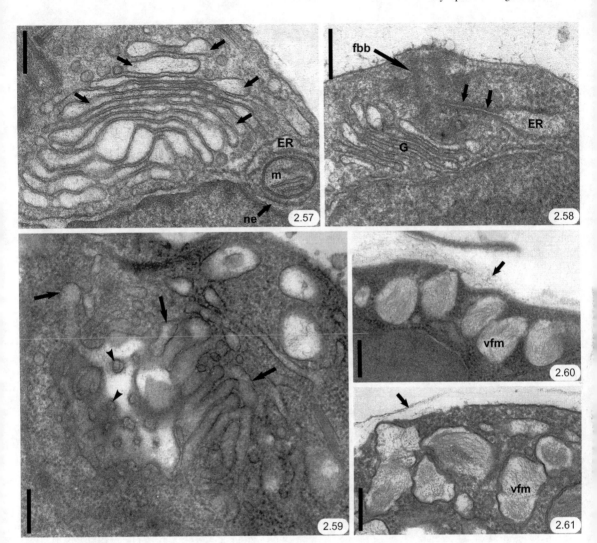

Plate 6 (Figures 2.57–2.61)

Figs 2.57–2.61 Golgi apparatus and associated vesicles. **Fig. 2.57** *Didymoeca costata*. Stack of Golgi cisterna in cell approaching division. Note the relatively large number of cisternae with inflated loculi containing fibrillar material (arrows). Mitochondrial profile (m) surrounded on lower surface by nuclear envelope (ne) and on outer surface by ER. Bar = 0.25 μm.

Fig. 2.58 *Salpingoeca infusionum*. Anterior end of cell showing location of Golgi apparatus (G), flagellar basal body (fbb) and a narrow cisterna with dense-staining surface (arrows) confluent with a cisterna of ER. Bar = 0.25 μm.

Fig. 2.59 *Stephanoeca diplocostata*. Tangential section of Golgi apparatus showing reticulate appearance of cisternae (large arrows) with lacunae in central region (arrowheads). Bar = 0.25 μm.

Figs 2.60–2.61 *Proterospongia choanojuncta* and *Stephanoeca diplocostata*, respectively. Vesicles of Golgi origin containing densely staining fibrillar material (vfm). In both examples vesicle contents are similar in appearance to material on the cell surface (arrows). Bar = 0.25 μm.

of manufacturing the new theca or lorica. In *Proterospongia choanojuncta* the distribution of these vesicles just beneath the plasma membrane of the emerging (upper) daughter cell is obvious (Figs 2.60, 3.36). Closer inspection reveals fibrillar material on the outer surface of the plasma membrane and it is likely that the contents of these vesicles will contribute towards the future cell covering (Fig. 2.60). A similar scenario applies to loricate species such as *Stephanoeca diplocostata*. The vesicles are bequeathed to the daughter cell (juvenile) that is pushed out of the parent lorica and which subsequently assembles its own lorica from accumulated costal strips (Fig. 7.34). Soon after cell division the vesicles with their fibrillar contents disappear. The most likely explanation of this phenomenon is that the contents have been exocytosed and will contribute to the organic layer present within the base of the lorica.

An unusual but distinctive narrow cisterna with dense-staining material on its outer surface has been observed close to the basal bodies and the anterior surface of the Golgi apparatus in some craspedid choanoflagellates (Fig. 2.58) (Leadbeater and Morton, 1974a). This cisterna is confluent with what appears to be an extension of the ER. No known function has been ascribed to this structure, but it has been found in both craspedid and acanthoecid species.

2.9 NUCLEUS AND NUCLEAR DIVISION (MITOSIS)

The spherical to ellipsoidal nucleus is typically located just above the centre of the cell, below the Golgi apparatus. Within the interphase nucleus the nucleoplasm comprises a mixture of moderate-staining background material and denser-staining patches of heterochromatin (Fig. 2.54). The latter are usually associated with the inner surface of the nuclear envelope, but may extend throughout the entire nucleus. In some nuclei a distinctive spherical to oval dense-staining nucleolus is apparent (Fig. 2.15, 2.38). The nuclear envelope is entire and contains pores.

Nuclear division, as determined by light microscopy, is similar for all those species that have been observed. The nuclear spindle is usually located perpendicular to the long axis of the cell, or diagonally if space is limited (Figs 2.67, 2.77). However, cell division (cytokinesis) is much more variable, being either longitudinal, as in cells within thin non-constraining coverings (Figs 2.62–2.64, 3.2–3.4, 3.11); diagonal in cells with constraining thecae (Figs 3.14,

3.24–3.27) and in nudiform loricate cells (Figs 6.12, 6.18, 6.40, 6.41); and inverted in tectiform loricate species (Figs 7.5, 7.6, 7.53a–e) (Table 1.3). The morphology of cell division and its evolutionary significance in craspedid (thecate) species is discussed in Chapter 3 and in acanthoecid (loricate) species in Chapters 6 and 7.

The ultrastructure of nuclear division has been analysed in detail in *Monosiga ovata* (Craspedida) and *Stephanoeca diplocostata* (Acanthoecida) (Karpov and Mylnikov, 1993; Leadbeater, 1994b; Karpov and Leadbeater, 1997). While two species can give only a limited insight into the process, nevertheless the comparison of a craspedid with an acanthoecid species is instructive. The general sequence of events is similar for both species. Prior to the onset of nuclear division the flagellum is withdrawn into the cell (Fig. 2.71a–e). This involves the axoneme of the shaft and transitional region, which becomes dislocated from the basal body, penetrating deeply into the cell, usually in the region of the Golgi apparatus and ER cisternae (Figs 2.72–2.74). Since so few sections of individuals at this stage in the cell cycle contain withdrawn flagella, it would appear that depolymerisation of the axoneme is rapid. The flagellar and non-flagellar basal bodies separate, the latter moving towards the cell surface just beneath the plasma membrane (Figs 2.66, 2.75). At this stage both basal bodies possess a ring of microtubules, which may in the case of *Stephanoeca diplocostata* be rather limited in length (Fig. 2.75). The two basal bodies move apart and become foci (poles) for the developing mitotic spindle (Figs 2.65, 2.77). At this stage, prophase, the nuclear envelope breaks down almost completely in *S. diplocostata* (Fig. 2.76), but in *Monosiga ovata* the envelope remains intact although many pores appear at the poles (Figs 2.67–2.69). The appearance of spindle microtubules radiating from the two poles and the appearance of a central band of dense-staining chromatin mark the beginning of metaphase (Figs 2.67–2.69, 2.77) (see also Karpov and Mylnikov, 1993). In both species the basal bodies act as foci for the converging spindle microtubules but there is usually a small gap between the spindle pole and its adjacent basal body (Figs 2.68, 2.77). A second basal body develops close to each of the polar basal bodies (Fig. 2.70). In *Monosiga ovata* the nucleus at metaphase still has an intact envelope, whereas at the same stage in *Stephanoeca* the nuclear envelope has completely disintegrated. However, in *Monosiga*, when the spindle elongates during anaphase the nuclear envelope also disintegrates (Fig. 2.70) (Karpov and Mylnikov 1993).

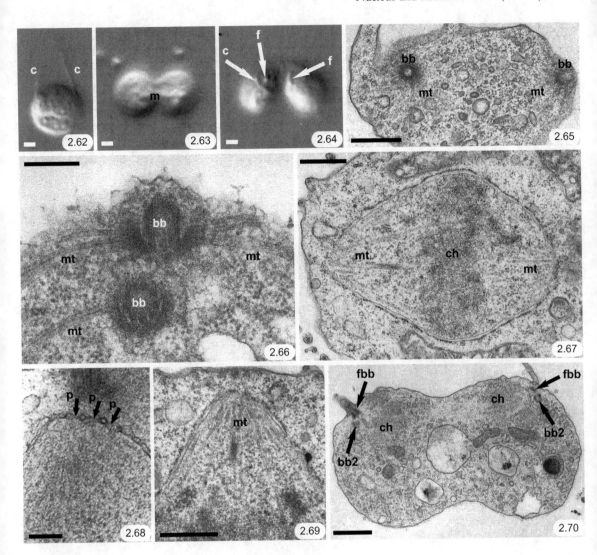

Plate 7 (Figures 2.62–2.70)

Figs 2.62–2.70 Mitosis in *Monosiga ovata* and *Salpingoeca rosetta* (Craspedida). Figs 2.62–2.64 *M. ovata*. Three stages in division: early anaphase, later anaphase and cytokinesis, respectively. Collar (c), flagellum (f), elongated mitochondrion (m). Bar = 1 μm.

Figs 2.65–2.66 *M. ovata* and *Salpingoeca rosetta*, respectively. Separation of two basal bodies (bb), both with surrounding cytoskeletal microtubules (mt), at beginning of nuclear division. Bars = 0.5 μm and 0.25 μm, respectively.

Fig. 2.67 *Monosiga ovata*. Metaphase nucleus showing nuclear envelope still more-or-less intact, with internal spindle microtubules (mt) and central dark-staining band of chromatin (ch) comprising metaphase plate. Bar = 0.5 μm.

Figs 2.68–2.69 *M. ovata*. Poles of two metaphase nuclei showing many pores in nuclear envelope (arrows p in Fig. 2.68) and convergence of spindle microtubules (mt). Bars = 0.25 μm and 0.5 μm, respectively. Fig. 2.68 reproduced from Karpov and Leadbeater (1997).

Fig. 2.70 *M. ovata*. Late anaphase; nuclear envelope has completely disintegrated and dark chromatin (ch) has moved towards the flagellar basal bodies (fbb), each with a second basal body (bb2), which represents the poles of the spindle. Bar = 1.0 μm. Reproduced from Karpov and Leadbeater (1997).

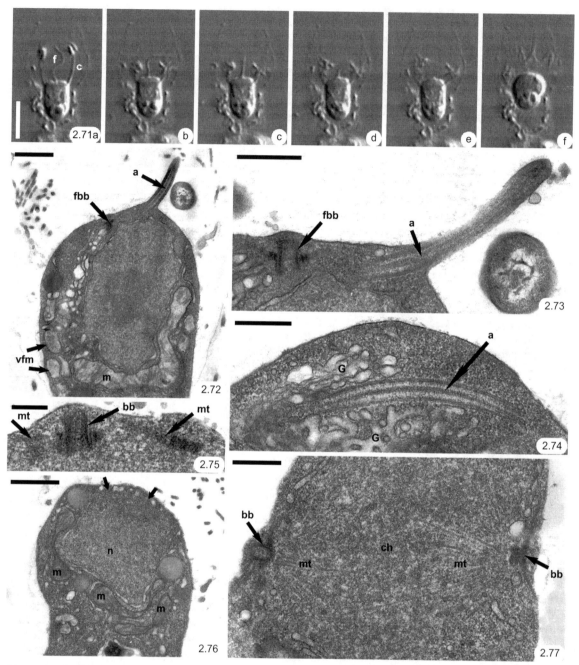

Plate 8 (Figures 2.71–2.77)

Figs 2.71–2.77 Mitosis in *Stephanoeca diplocostata* (Acanthoecida). **Figs 2.71a–f.** Sequence of images illustrating flagellar withdrawal and nuclear division. As the flagellum (f) is withdrawn (2.71a–d), the collar (c) bearing an accumulation of costal strips shortens and the anterior end of the cell enlarges. As nuclear division proceeds the cell becomes 8-shaped (2.71e–f). Bar = 5 μm.

Anaphase is characterised by the migration of chromatin to the two poles. In craspedid species the two polar basal bodies initiate the development of new flagellar axonemes and their elongation is rapid as the cells undergo cytokinesis (Figs 2.62–2.64). In *Stephanoeca diplocostata*, telophase is immediately followed by cytokinesis, which results in the inversion of the daughter cell (juvenile) that will be pushed backwards out of the lorica (see Section 7.2.2). Once the two discrete daughter cells are visible, with their flagellar poles facing each other, the polar basal bodies begin regenerating flagella (see Chapter 7).

Figs 2.72–2.73 Flagellar withdrawal prior to nuclear division. The proximal end of the 9 + 2 axoneme becomes detached from the flagellar basal body (fbb) and is withdrawn into the cell (arrow a), which at this time has an enlarged anterior end. The cell contains a prominent mitochondrial reticulum (m) and vesicles of fibrillar material (arrows vfm). Bars = 1 μm and 0.5 μm, respectively.

Fig. 2.74 Withdrawn 9 + 2 axoneme (arrow a) deep in cell surrounded by cisternae of the Golgi apparatus (G). Bar = 0.5 μm.

Fig. 2.75 Two rings of reduced cytoskeletal microtubules sectioned tangentially (arrows mt) associated with basal bodies, only one of which is visible (bb), during prophase. Bar = 0.25 μm.

Fig. 2.76 '8-shaped' cell containing prophase nucleus (n) with partial disintegration of the nuclear envelope at anterior end (arrows). Note mitochondrial profiles (m) with central constrictions. Bar = 1 μm.

Fig. 2.77 Metaphase nucleus with basal bodies (bb) at the two poles which act as foci for the microtubular spindle (mt). In the central region denser staining chromatin (ch) of the metaphase plate is visible. Bar = 0.5 μm. Figs 2.72–2.75, 2.77 reproduced from Leadbeater (1994b).

3 · Craspedida: choanoflagellates with exclusively organic coverings

Variations on a sedentary theme.

3.1 INTRODUCTION

Molecular phylogenetic analyses identify two major, well-supported monophyletic groupings within the Choanoflagellatea (Fig. 10.3) (Cavalier-Smith, 1997; Carr et al., 2008a; Nitsche et al., 2011). One grouping, order Craspedida Cavalier-Smith 1997, which subsumes Kent's (1880–2) original families Codonosigidae and Salpingoecidae, contains species with exclusively organic coverings, whereas the other, order Acanthoecida Cavalier-Smith 1997, contains taxa with siliceous loricae (see Tables 1.4 and 10.2).

Use of terminology relating to choanoflagellate organic cell coverings is discussed in Section 1.3.2. The term 'glycocalyx' or 'sheath' is used to describe craspedids with thin non-restrictive coverings (sensu Codonosigidae Kent) that permit unhindered longitudinal division of cells. The term 'theca' is used to describe craspedids with continuous restrictive organic coverings (sensu Salpingoecidae Kent) that usually inhibit longitudinal cell division and result in a parent cell partially emerging from the thecate covering in order to divide.

Choanoflagellates with organic coverings are cosmopolitan in distribution, being found in a wide range of freshwater, brackish and marine habitats (Kent, 1880–2; Zhukov and Karpov, 1985). They also contribute to the protozoan populations of soils and aquifers. Species with all the major forms of cell coverings, including cup-, tube- and flask-shaped investments, are represented in freshwater and marine biota. Because of their predominantly sedentary nature they are not a common component of the plankton, although their life cycles may include one or more colonial stages that are planktonic (Dayel et al., 2011). Sedentary cells may also enjoy a temporary planktonic lifestyle by settling as epibionts onto suspended algae, particularly diatoms, or particles of flocculated debris (see Section 9.7.1).

3.2 CELL CYCLE OF THE CRASPEDIDA

The basic cell cycle of craspedid choanoflagellates comprises a sedentary interphase cell (Figs 3.7, 3.24), whose major function is feeding, followed by cell division that gives rise to a transitory motile phase, whose function is dispersal (Figs 3.8, 3.27, 3.37). The flagellar apparatus is ideally suited for this dual functional role (see Chapter 2). In thecate species, cell division is equal in that two similar daughters result from mitosis, but unequal in that one inherits the parent theca while the other, known as the juvenile, must produce a new theca (Figs 3.24–3.27). This unequal aspect of cell division is not exclusive to cell coverings; it also applies to the segregation of basal bodies during mitosis, where one daughter acquires the flagellar basal body of the parent whereas the other receives the second basal body (see Section 2.9). Maturation of flagellar basal bodies over two or more cell cycles is a well-established phenomenon in many flagellates (Melkonian et al., 1987; Nohýnková et al., 2006). Other features, such as the collar microvilli of choanoflagellates, are probably shared equally between the two daughters, but nevertheless there may be generational differences between neighbouring microvilli within a collar (see also Section 7.2.3).

3.3 SPECIES WITH NON-RESTRICTIVE CELL COVERINGS

While the alternation of a sedentary interphase cell undergoing division to produce a motile daughter is common to many craspedid species, nevertheless there are numerous variations on this theme (Leadbeater, 1983a; Dayel et al., 2011). One feature of major importance is the degree to which the organic covering constrains cell division. If the covering is thin and flexible then there is no restriction to the mitotic spindle being oriented in the transverse plane and the cleavage plane can be vertical. The resulting division is termed longitudinal and the two daughter cells are located side by side. This situation is exemplified by *Monosiga*, where ultimately the daughters separate and swim away (Figs 2.62–2.64). However, there are many different ways in which daughter cells do not separate completely, leading to a variety of colonial forms (see Section 3.5).

Codosiga is a genus of stalked colonial species where the daughter cells resulting from division remain attached to the parent stalk, thereby forming a head of cells (Fig. 3.1). Figs 3.2–3.4 illustrate *Codosiga gracilis* during the later stages of division, including anaphase and telophase, and demonstrate how the daughter cells remain attached to the parent stalk. The developing flagella of the two daughter cells are visible, as is the elliptical outline of the collar (Figs 3.2, 3.3). During anaphase, the separating chromosomes (Fig. 3.2 arrows ch) move from the metaphase plate towards the poles represented by the basal bodies of the developing flagella. Immediately after this has been achieved the vertical cleavage furrow appears midway between the cells and deepens until the two cells separate. At the base of each cell a short pedicel maintains continuity with the parent stalk (Fig. 3.4). James-Clark's (1867b) original illustrations of *Codosiga* (*pulcherrima*) *botrytis* undergoing division are so accurate and elegant that they are reproduced here as Fig. 3.11. Colonies of *C. gracilis* are derived by the repeated division of daughter cells, the products of which remain attached to the parent stalk. A mature head may contain 10–20 cells (Fig. 3.1). Eventually the colony becomes so large that it disintegrates, releasing swimming cells that subsequently settle down and repeat the cycle. Occasionally an entire head of cells may become dislodged and the colony is then indistinguishable from that of a species of *Sphaeroeca* or *Astrosiga* (see Fig. 3.47) (Kent, 1880–2).

Four species of *Codosiga* – *C. gracilis*, *C. botrytis*, *C. balthica* and *C. minima* – have been observed in detail with electron microscopy (EM) (Leadbeater and Morton, 1974a; Hibberd, 1975; Wylezich *et al.*, 2012). *C. gracilis* is surrounded by a diaphanously thin extracellular matrix that is just visible in whole mounts (Figs 3.5, 3.7) (Leadbeater and Morton, 1974a). The extreme delicacy of the covering belies its strength and resilience in holding together a head of flagellated cells on the parent stalk. The latter comprises a longitudinal array of microfibrils coated with material that displays a pattern of diagonal striations with a spacing of 25 nm (Figs 3.9, 3.10). Where the stalk attaches to a substratum, the microfibrils splay outwards, thereby providing a tenacious holdfast. The stalk is extremely robust

and can withstand treatment in boiling 1.0 M alkali (NaOH) or acid (H$_2$SO$_4$). During such treatment the surface material is removed and the underlying microfibrils can be stained with calcofluor white or FITC-labelled wheat germ agglutinin (WGA), demonstrating a carbohydrate substructure. Sedentary cells typically have relatively long collar microvilli and a flagellum of moderate length, whereas motile cells for dispersal have short collars and relatively long flagella (Figs 3.7 and 3.8, respectively; see also Mah *et al.*, 2014). Settlement of motile cells is assisted by an array of posterior filopodia which secure the individual to a substratum and permit the secretion of the microfibrillar stalk (Fig. 3.8). *C. botrytis*, a freshwater colonial species, is surrounded by a more substantial covering than *C. gracilis*. Although this covering is two-layered and fibrous in texture, the cell is still able to undergo longitudinal division (de Saedeleer, 1929). A major difference between the two *Codosiga* species is the existence in *C. botrytis* of a narrow cytoplasmic bridge joining adjacent cells a short distance above the attachment of the stalk (Hibberd, 1975). The connecting strand contains a block of dense material which appears as a partition midway along the bridge (Hibberd, 1975). This connection is similar but not identical to the cytoplasmic bridges observed between adjacent cells in *Salpingoeca rosetta* (Dayel *et al.*, 2011) and some *Desmarella* colonies (Fig. 3.52) (Karpov and Coupe, 1998). The stalk of *C. botrytis* is more complex than that of *C. gracilis*. Narrow fingers of cytoplasm penetrate for about one-third of its length and give it a lacunar appearance in section. The outer surface of the stalk bears a regular pattern of diagonal ridges similar to those of *C. gracilis* but with a slightly larger spacing of 32 nm (Hibberd, 1975).

Another example of a colonial species where cells are held together by posterior connections, albeit in a completely different fashion, is shown in Figs 3.12 and 3.13. Here, individual cells are connected to each other by elongated posterior threads within tubular stalks which branch in an irregular dichotomous fashion.

While it is usual for daughter cells to remain motile for only a short time, there are occasions when they may divide in the swimming state. This occurs in nature, in some cultures during the mid-exponential phase of growth or after the introduction of a colony (rosette) inducing factor produced by certain bacteria (Fairclough *et al.*, 2010; Dayel *et al.*, 2011; Alegado *et al.*, 2012). Under these circumstances motile colonies may remain suspended in the medium for extended periods of time. Such colonies

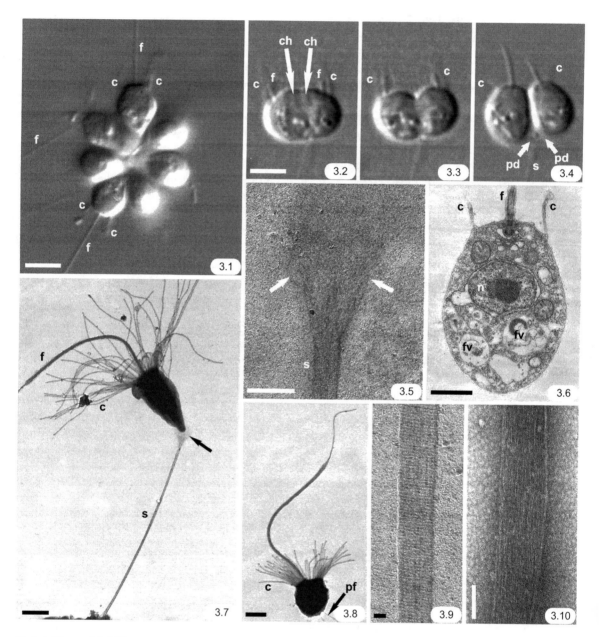

Plate 1 (Figures 3.1–3.10)

Figs 3.1–3.10 *Codosiga gracilis*, stalked colonial species with non-restrictive cell covering. Figs 3.1–3.4 Interference contrast images of living cells. **Fig. 3.1** Head of cells, each with a collar (c) and long flagellum (f). Bar = 5 μm.

Figs 3.2–3.4 Three consecutive images of cell division. Bar = 2.5 μm (shown on Fig. 3.2). **Fig. 3.2** Anaphase – the forming daughter cells each have a short flagellum (f) and share the parent collar (c). Separating chromosomes are visible (arrows ch).

Figs 3.3–3.4 Telophase and cytokinesis, respectively. **Fig. 3.4** Pedicel (arrow pd), stalk (s).

Fig. 3.5 Top of stalk (s) showing diaphanously thin covering that surrounds cell (arrows). Bar = 0.5 μm.

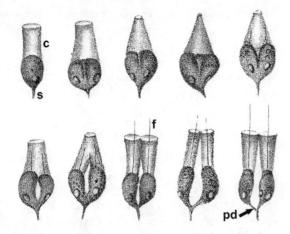

Fig. 3.11 Sequence of drawings reproduced from James-Clark (1867b) showing cell division in *Codosiga pulcherrima* James-Clark (= *Codosiga botrytis* (Ehrenberg) Bütschli). Collar (c), flagellum (f), stalk (s), pedicel (pd).

have been identified in the literature as species of *Proterospongia*, *Sphaeroeca*, *Astrosiga* and *Desmarella* according to morphology (Fig. 3.47). The development of colonies and details of their morphology and life cycles are discussed in Section 3.5.

3.4 SPECIES WITH RESTRICTIVE CELL COVERINGS

The defining character of species with restrictive coverings (thecate species) is that the theca is sufficiently close-fitting and inflexible as not to permit longitudinal cell division *in situ*. Instead the parent cell becomes amoeboid and partially emerges from the theca; in this position it undergoes nuclear division followed by cytokinesis. The daughter cell that remains partly within the theca returns to its covering, while the other daughter (juvenile), which is at an angle to the long axis of the parent cell, is pushed out by an elongating thread until it separates and swims

away (Figs 3.14c, 3.24–3.27). This type of division is, for convenience, known as 'emergent' division.

Stein (1878) was the first to illustrate emergent division in *Salpingoeca oblonga* (Fig. 3.14a). He referred to it as being a "very peculiar type of longitudinal fission" (Stein 1878, Plate X, Fig. 6). Kent (1880–2) recorded a similar morphology in *Salpingoeca inquillata* (Fig. 3.14b) but thought it was an example of 'transverse' division. Neither of these authors fully appreciated the association of this pattern of division with the limitations of a constraining theca. Griessmann (1913) was the first to observe emergent division correctly in all its stages in *Salpingoeca infusionum* (Fig. 3.14c). Comparable results were subsequently published by de Saedeleer (1929), working with *Salpingoeca amphoridium*, and Ellis (1929) with *Choanoeca perplexa*. Since the daughter cells of species with restrictive thecae do not remain side by side after division, there is not the same opportunity for immediate colony formation that there is for craspedids with flexible coverings (see Section 3.3). Occasionally large numbers of thecate cells may be observed colonising a surface or a fragment of suspended detritus; these are not 'true' colonies but rather 'aggregates' of juveniles settling close to parents. However, this does not preclude species with constraining thecae from producing colonies; it is common for 'naked' juveniles to divide in the motile state thereby forming 'Proterospongia' stages (see Section 3.5).

3.4.1 Thecal morphology

The functional role of the theca determines the range of morphology encountered in nature. The purpose of a theca is to contain the cell body and attach it to a substratum. Since the primary role of sedentary cells is to capture suspended food particles from water currents generated by flagellar activity, it is essential that the theca should not interfere in any way with the collar or flagellum of the cell. Complete enclosure of the feeding apparatus is exclusively a feature of some loricate species (see Chapter 4).

Fig. 3.6 LS cell with flagellum (f), collar (c), nucleus (n) food vacuoles (fv). Bar = 1 μm. Reproduced from Leadbeater and Morton (1974a).

Fig. 3.7 Stalked cell with long collar (c), flagellum (f), showing stalk (s) with diaphanously thin covering at base of cell (arrow). Bar = 2 μm.

Fig. 3.8 Motile cell with short collar (c) and posterior filopodia (pf). Bar = 2 μm.

Fig. 3.9 Shadowcast stalk showing pattern of diagonal ridges. Bar = 0.1 μm.

Fig. 3.10 Negatively stained (uranyl acetate) stalk showing parallel band of longitudinal microfibrils. Bar = 0.1 μm.

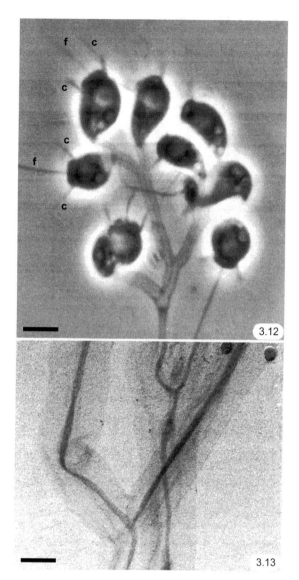

3.12

3.13

Plate 2 (Figures 3.12 and 3.13)

Figs 3.12–3.13 An unidentified colonial choanoflagellate showing irregular dichotomous branching.
Fig. 3.12 Each cell with collar (c) and flagellum (f) is extended posteriorly into a narrow thread surrounded by a parallel-sided translucent tube which holds the colony together. Phase contrast light microscopy. Bar = 5 μm.
Fig. 3.13 TEM whole mount of dichotomously branched posterior thread surrounded by translucent tube. Bar = 1 μm.

Many drawings and photographs of fixed thecate choanoflagellates are misleading for they show a collared cell entirely surrounded by the theca, whether it be a tube or cup. This appearance is artefactual, unless the cell is late in stationary phase, otherwise it results from cell shrinkage and post-mortem loss of collar rigidity.

3.4.2 Cup- and tube-shaped thecae

The most common forms of thecae encountered are stalked cups, tubes and flasks. The type species of *Salpingoeca*, namely *S. gracilis*, is a classic example of a species with a tubular theca (James-Clark, 1867b) and *S. infusionum* is a good example of a species with a cup-shaped theca (Figs 3.15–3.21). *S. infusionum* has a relatively long stalk comprising a substantial column of parallel microfibrils, similar in appearance to those encountered in *Codosiga gracilis* (compare Figs 3.21 and 3.10). As with the latter species, the microfibrils are extremely robust, withstanding treatment for several hours in boiling 1 M NaOH or 1 M H$_2$SO$_4$. Empty thecae of *S. infusionum* can be stained with a variety of non-specific dyes, such as cresyl violet (Fig. 3.15) and toluidine blue, as well as treatments specific for carbohydrate content. Thecae stain pale magenta with periodic acid–Schiff (PAS) reagent; controls involving the omission of periodic acid or acetylation showed no staining. Treatment with calcofluor white or FITC-labelled WGA resulted in marked fluorescence (Fig. 3.16), confirming a carbohydrate component in the theca.

Thecae of *S. infusionum* untreated except for washing with distilled water have a relatively amorphous appearance when viewed as whole mounts for TEM (Fig. 3.17). However, when heated to 80 °C in distilled water or 0.1–1.0 M sodium hydroxide, a framework of microfibrils becomes apparent in the cup as well as in the stalk (Figs 3.19 and 3.20). A similar pattern of microfibrils is observed in the cup-shaped thecae of *Salpingoeca rosetta* (Fig. 3.22). In particularly favourable specimens continuity between the microfibrils of the stalk and cup is apparent and it is likely that the microfibrils are deposited as continuous entities (Figs 3.20, 3.22). In section, the thecae of *Salpingoeca infusionum* and *S. rosetta* appear as a continuous single layer, the microfibrillar substructure is apparent in the latter as small dark dots on either side of a dark continuous line (Figs 3.18, 3.23).

Helpful comparisons can be made between the cup-shaped thecae of choanoflagellates and the microfibrillar

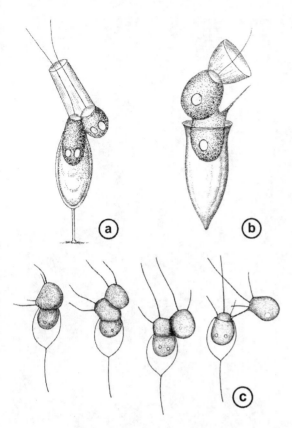

Fig. 3.14 Emergent cell division in: a. *Salpingoeca oblonga*.
Reproduced from Stein (1878); b. *Salpingoeca inquillata*.
Reproduced from Kent (1880–82); c. *Salpingoeca infusionum*.
Reproduced from Griessmann (1913).

coverings (loricae) of chrysophytes such as *Poterioochromonas stipitata*, for which detailed information on the ultrastructure and chemistry is available (Schnepf *et al.*, 1975; Herth *et al.*, 1977; Herth, 1980; Peck, 2010). *P. stipitata* secretes a cup-shaped microfibrillar lorica comparable to the theca of *Salpingoeca rosetta* (Fig. 3.22). The stalk of *P. stipitata* differs from those of choanoflagellates in that 12–24 bands (fibrils) of microfibrils run parallel to each other "in a steep helix, which is in nearly every case left-handed (Z-type)" (Schnepf *et al.*, 1975, p. 53). The outer layer of 'primary' fibrils comprising the cup of *P. stipitata* also describes a left-handed helix and is a continuous extension of the fibrils of the stalk. The inner secondary fibrils form an irregular meshwork.

Development of the lorica as well as individual microfibrils have been observed in *Poterioochromonas stipitata*.

Following settlement of a motile cell on a substratum, a cytoplasmic 'tail' develops and the first microfibrils of the stalk and foot are deposited. During the following two hours the stalk continues to increase in length and the cell rotates in a clockwise direction when viewed from the flagellar (anterior) pole (Schnepf *et al.*, 1975, p. 53). The final 30 min is occupied with the development of the cup. Individual microfibrils of the stalk appear on the outer surface of the plasma membrane. There are no indications of vesicle extrusion, but rather there is intimate contact between the developing microfibrils and the plasma membrane. Microfibrils elongate unidirectionally and so there is a polarity of growth (Schnepf *et al.*, 1975).

There are two published accounts of stalk development in choanoflagellates, namely *Codosiga botrytis* and *Salpingoeca rosetta*, and the pattern of growth in both is similar to that of *Poterioochromonas stipitata* (de Saedeleer, 1929; Dayel *et al.*, 2011). In cultures of *Salpingoeca rosetta*, cups of different heights were observed on mature stalks, but it is not clear whether these represent stages in development. The pattern of microfibrils in *S. rosetta* is superficially similar to that in *Poterioochromonas* and in favourable shadowcast whole mounts it is possible to discern a diagonal orientation to the microfibrils which might represent a flattened left-handed helix (Fig. 3.22, arrows). If this proves to be correct then this would be of great significance with respect to the indisputable left-handed rotation required for the costal patterns of choanoflagellate loricae (Acanthoecida) (see Chapters 4, 6 and 7).

The chemical composition of microfibrils comprising *Poterioochromonas* loricae has also been studied in detail (Herth *et al.*, 1977; Herth 1980). Unequivocal evidence has been obtained with respect to chitin being the structural polysaccharide of microfibrils. This conclusion is based on hydrolysis of the microfibrils, separation of the monomers by thin layer chromatography and their subsequent analysis by gas chromatography. Confirmation of this finding was achieved by X-ray diffraction analysis (Herth *et al.*, 1977). With respect to the chemical composition of microfibrils in thecate choanoflagellates, the results are equivocal in spite of considerable effort having been expended on carrying out similar analyses to those followed for *P. stipitata*. The fluorescent staining of thecae with FITC-labelled wheat germ agglutinin, while carried out with strict controls, remains ambivalent. Attempts to extract sugar

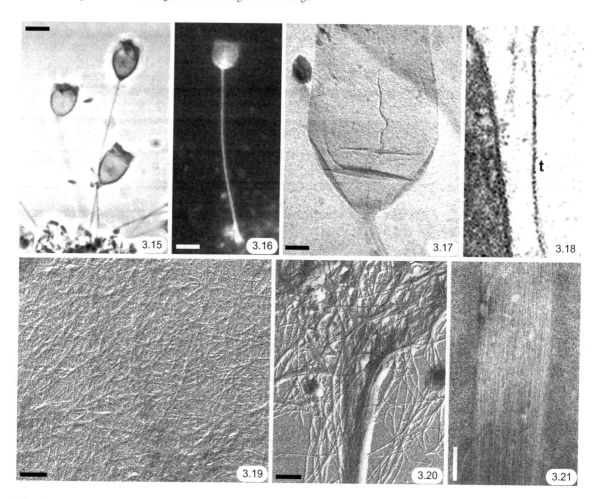

Plate 3 (Figures 3.15–3.21)

Figs 3.15–3.21 *Salpingoeca infusionum.* Light and electron microscopy of theca. **Figs 3.15–3.16** Phase contrast and fluorescence microscopy (FITC-WGA stain), respectively of long-stalked cup thecae. Bar = 5 μm.

Fig. 3.17 Shadowcast whole mount of distilled water-washed empty theca. Bar = 1 μm.

Fig. 3.18 Vertical section of cell showing theca (t) comprising a longitudinal array of dots, equivalent to sectioned microfibrils. Bar = 0.1 μm.

Figs 3.19–3.20 Shadowcast whole mounts of thecae after boiling in distilled water. **Fig. 3.19** Surface of cup. **Fig. 3.20** Attachment of stalk to cup. Bars = 0.25 μm.

Fig. 3.21 Negatively stained (uranyl acetate) stalk showing parallel band of microfibrils. Bar = 0.1 μm.

monomers from thecae also generated ambivalent results (unpublished information).

3.4.3 Flask-shaped thecae

Choanoflagellates with flask-shaped thecae are sedentary and are commonly found on immersed aquatic surfaces, including those of animals and plants where they are regular components of the epibiota (Fig. 9.22). In particle-rich water, groups of flask-shaped cells may attach to suspended clumps of microbial detritus giving them a colonial appearance. They are cosmopolitan in marine, brackish and freshwater habitats.

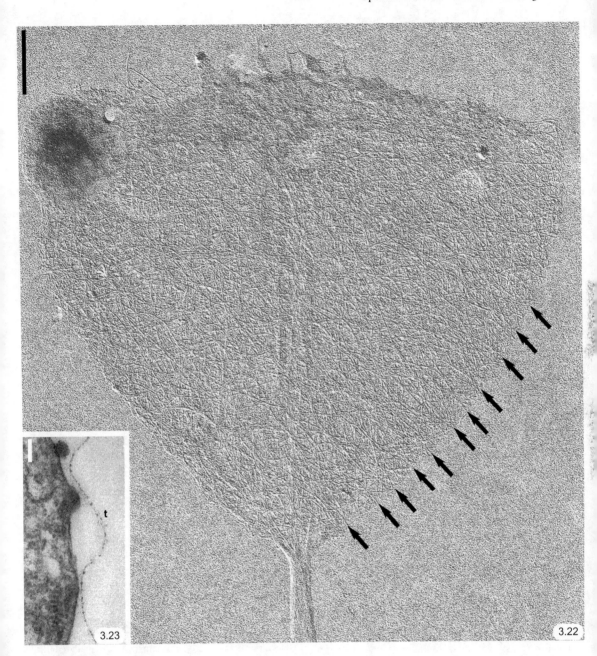

3.23

3.22

Plate 4 (Figures 3.22 and 3.23)

Figs 3.22–3.23 *Salpingoeca rosetta*. **Fig. 3.22** Shadowcast cup-shaped theca after boiling with distilled water for one hour. Note arrangement of microfibrils with a possible lower-right to upper-left (left-handed) alignment (arrows). Bar = 0.5 μm.
Fig. 3.23 Vertical section of theca (t) that appears as a thin line with dots representing microfibrils. Bar = 0.1 μm.

Flask-shaped thecae are usually round-bottomed (Figs 3.28, 3.30), although the chamber may be flattened horizontally as in *Salpingoeca angulosa* (de Saedeleer, 1927). They have a distinctive ultrastructure; the neck is parallel-sided when seen in empty specimens (Fig. 3.28). At the top of the neck the theca is recurved, forming a narrow rim (Fig. 3.33 arrowhead). A ribbed inner flange extends from the top of the chamber into the bottom part of the neck and serves as a connection between the theca and the enclosed cell (Figs 3.29, 3.33 arrow). In some species, such as the *Choanoeca* stage of *Proterospongia choanojuncta*, the outer surface of the neck is decorated with a patchwork of fine ridges (Fig. 3.32) (Leadbeater, 1983a). In section, the theca comprises a sandwich of three layers. The outer layers contain dense staining particles; the inner layer is lighter and amorphous (Fig. 3.33). Standard cytochemical tests for carbohydrate, including PAS, calcofluor white and FITC-labelled WGA, gave positive results. Unlike cup- and tube-shaped thecae, there is no obvious microfibrillar framework even after boiling or treatment with acid or alkali. However, with careful negative staining using uranyl actetate a pattern of parallel microfibrils can be seen within the neck of a *Salpingoeca urceolata* flask (Fig. 3.34). No such microfibrillar arrangement could be detected in the chamber. Attachment of the flask to a substratum is by one or more microfibrillar stalks that contain lacunae (Leadbeater, 1977).

Components of the theca appear within Golgi cisternae and the surrounding vesicles towards the end of interphase (Leadbeater, 1977). In particular, dense-staining particles, similar to those comprising the outer layers of the theca, are often observed within Golgi-associated vesicles (Fig. 3.35). Prior to cell division large numbers of vesicles filled with fibrillar material appear around the Golgi apparatus and collect within the peripheral cytoplasm. As the cell becomes amoeboid and emerges from the neck of the theca, it is the anterior daughter cell with the flagellum (the juvenile) that contains these vesicles in the cortical cytoplasm (Figs 3.36, 2.60). Some appear to discharge their contents to the outside before nuclear division is complete (Fig. 2.60). The juvenile cell, after separation from the daughter that remains in the parent theca, produces an array of posterior filopodia and settles within a short time (Figs 3.27, 3.37). At this stage the fibrillar contents of the vesicles have discharged to the cell exterior and presumably contribute to the developing theca (Fig. 3.37 arrows).

3.4.4 Cell division in *Choanoeca perplexa*

The inflexibility of flask-shaped thecae means that the enclosed cell must emerge from this straightjacket in order to divide. Figs 3.24–3.27 illustrate division in *Choanoeca perplexa*, which during interphase lacks a flagellum, although throughout the cell cycle it has the normal complement of two basal bodies (Figs 3.31, 3.35). At the onset of division, the apical basal body develops a long flagellum which will subsequently be bequeathed to the daughter cell (juvenile) that develops outside the theca (Figs 3.24, 3.36). The parent cell becomes amoeboid and the anterior end emerges from the flask (Fig. 3.25). Nuclear division, which begins while the cell is still within the parent theca (Fig. 3.36), is completed in the bulging cytoplasm (Figs 3.25, 3.26). The base of the flagellum acts as the mitotic pole that will pass with one telophase nucleus to the juvenile (Figs 3.24–3.27). Cytokinesis is longitudinal within the cytoplasmic bulge. The daughter cell that is still partly within the parent theca, and lacks a flagellum, returns to the covering while the other daughter (the juvenile), which is without a covering, is positioned at an angle to the long axis of the parent theca (Fig. 3.27). The flagellum of the juvenile by this time is long with respect to the radius of the cell (ratio length/radius $L/A = 13$ (see Section 2.3.2)). Linkage between the two daughters appears to involve at least one narrow thread associated with the respective collars (Fig. 3.27). This lengthens rapidly and the juvenile separates and swims away. It is instructive to compare emergent division in *Choanoeca perplexa* (Figs 3.24–3.27) with that illustrated by Griessmann (1913) in *Salpingoeca infusionum* (Fig. 3.14c).

Two points of significance arise from the pattern of division in *Choanoeca perplexa*. First, although *Choanoeca* is unusual in that the sedentary cell lacks a flagellum during interphase, nevertheless it possesses the usual complement of two basal bodies (Figs 3.31, 3.35). One of these, situated at the apex of the neck, is surrounded by a ring of cytoskeletal microtubules and is equivalent to the flagellar basal body of normally flagellated cells (Fig. 3.31). The flagellum that develops prior to division (Fig. 3.24) is derived from this basal body and, since it is bequeathed to the daughter cell (juvenile) destined to leave the parent theca, it serves as a 'marker' for interpreting basal body maturation. As nuclear division proceeds, a second basal body develops beneath the flagellar basal body of the juvenile, thereby restoring the usual complement of two basal bodies in a mature cell

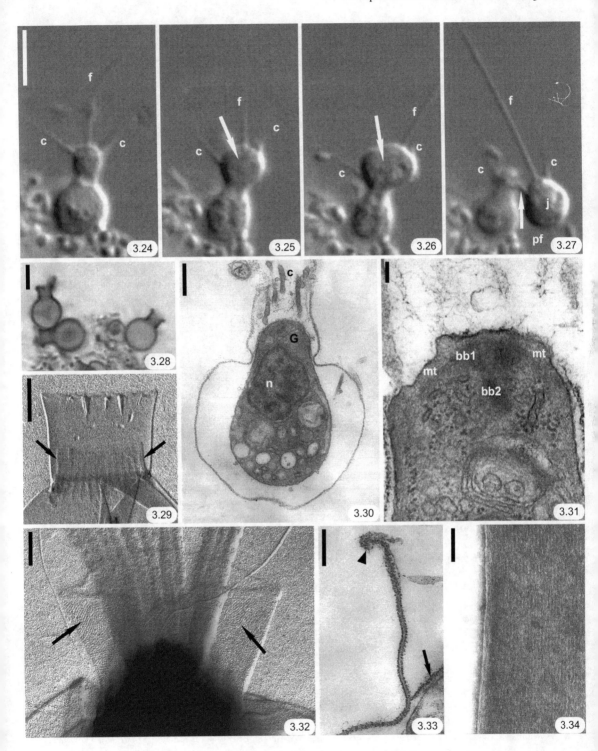

Plate 5 (Figures 3.24–3.34)

(Fig. 3.37). After separation, the flagellated cell locates a surface, secretes a theca and withdraws its flagellum (Fig. 3.37). When this sedentary cell subsequently divides the apical basal body develops a flagellum and the second flagellar base remains with the daughter cell associated with the parent theca. Thus, under normal conditions each basal body must experience two cycles of interphase to mature from the moment of its first appearance until the time that it first bears a flagellar shaft. From that time onwards it will always be the basal body to develop a flagellum during cell division.

The second point of significance is the use of an extending thread or threads emanating from the region of the collars of the two cells to effect the separation of the juvenile from the daughter that remains within the theca (Fig. 3.27). This method of separation is reminiscent of that in tectiform acanthoecids, where separation of the two daughter cells is achieved by the extension of two overlapping threads that slide over one another; one thread emanating from each of the daughter cells (see Section 7.2.2).

3.5 COLONIAL CHOANOFLAGELLATES

Coloniality within the choanoflagellates has evolved many times and in a variety of different ways (see also Chapters 6 and 9). In most instances, coloniality within craspedids is a phase in the life history of otherwise solitary species (Dayel

et al., 2011). This particularly applies to freely suspended 'pelagic' colonies, such as those attributed to *Proterospongia*, *Astrosiga*, *Sphaeroeca* and *Desmarella*, which are probably temporary stages in the life cycles of otherwise sedentary species. On the other hand, species of *Codosiga* represent permanently colonial stalked forms, except when the stalk carries only a single cell. Clusters of *Codosiga* cells can be dislodged from their supporting stalks and may become self-perpetuating colonies in that the head may break into smaller groups of cells that continue to divide in the motile state.

Before proceeding with examples it is necessary to consider what is meant by the term 'colony'. The standard dictionary definition refers to 'a group of identical cells derived from a single progenitor cell'. With respect to choanoflagellates it is necessary to qualify this definition by adding that the daughters resulting from division of the original progenitor cell do not fully separate but remain in unbroken continuity. Failure to separate can take a variety of forms; cells may remain attached by a common stalk, cytokinesis may be incomplete or the outer surfaces of daughter cells may remain adherent to each other. In the case of loricate species, colonies can be formed by a juvenile cell developing a lorica while still attached to the parent, as in the nudiform *Polyoeca dichotoma* (Figs 6.1b arrow, 6.25; Section 6.6.1) or by the incomplete separation of daughter loricae

---

Figs 3.24–3.34 Emergent cell division and thecal ultrastructure of flask-shaped choanoflagellates.

Figs 3.24–3.27 *Choanoeca perplexa*. Four consecutive images of emergent cell division. A flagellum (f) develops at the apex of the neck prior to division (Fig. 3.24) and specifies the flagellar axis of the daughter cell (juvenile) that will eventually swim away (Figs 3.25–3.27). The collar (c) is shared between the two cells. Nuclear division occurs within the emergent bulge of cytoplasm (Figs 3.25 and 3.26 arrows). Juvenile cell (j), with long flagellum (f) and posterior filopodia (pf), is pushed away by one or more cytoplasmic threads that link the necks of the two daughter cells (Fig. 3.27 arrow). Bar = 5 μm (shown on Fig. 3.24).

Fig. 3.28 *C. perplexa*. Toluidine blue-stained empty flasks. Bar = 2 μm. Reproduced from Leadbeater (1977).

Fig. 3.29 *Salpingoeca urceolata*. Whole mount of theca showing the neck with internal ribbed flange that serves to secure cell to theca (arrows). Bar = 1 μm.

Fig. 3.30 *Choanoeca perplexa*. LS of stationary-phase cell showing profile of flask-shaped theca and general disposition of organelles. Collar (c), Golgi apparatus (G), nucleus (n). Bar = 0.5 μm. Reproduced from Leadbeater (1977).

Fig. 3.31 Anterior end of cell showing position of two basal bodies (bb1 and bb2) and cytoskeletal microtubules (mt) associated with the apical basal body (bb1). Bar = 0.1 μm.

Fig. 3.32 *Proterospongia choanojuncta*. Neck of *Choanoeca* stage showing theca with a patchwork pattern of fine ridges (arrows). Bar = 1 μm.

Fig. 3.33 Vertical section through theca neck showing a three-layered substructure comprising a central light-staining layer covered on both sides by a surface layer of dense-staining material. Anterior end of neck is recurved (arrowhead). The flange inside the base of the neck can be seen extending to the cell surface (arrow). Bar = 0.25 μm. Reproduced from Leadbeater (1977).

Fig. 3.34 *Salpingoeca urceolata*. Negatively stained neck showing a pattern of parallel vertical microfibrils. Bar = 0.1 μm.

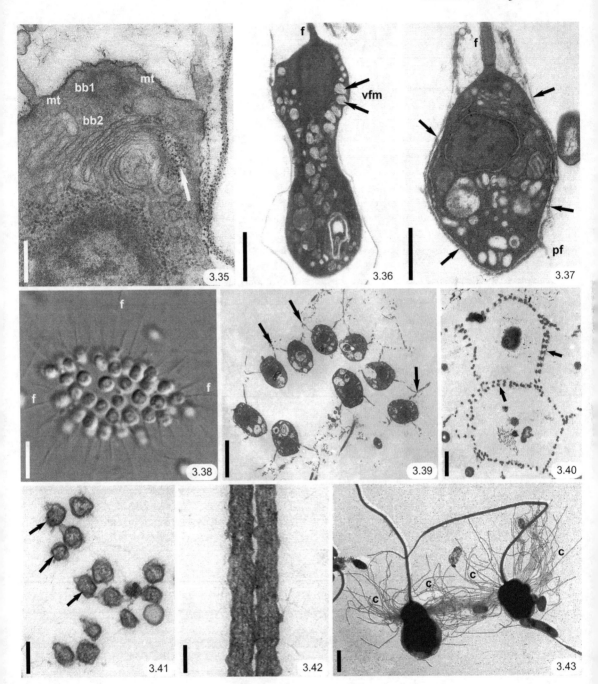

Plate 6 (Figures 3.35–3.43)

Figs 3.35–3.43 *Choanoeca perplexa* and *Proterospongia choanojuncta*. **Fig. 3.35** *Choanoeca perplexa*. Apical basal body (bb1) with surrounding cytoskeletal microtubules (mt) is subtended by second basal body (bb2). Golgi cisternae with dark-staining granules, similar to those on the surface of the theca, in a peripheral vesicle (arrow). Bar = 0.25 μm. Reproduced from Leadbeater (1975).

immediately after tectiform division, as in *Parvicorbicula socialis* (Figs 9.8, 9.9). The reason for including 'incomplete separation' in the definition is to distinguish colonies from aggregates. In the latter condition there is complete separation between daughter cells but even so the progeny may remain sufficiently close together to generate a more-or-less exclusive cluster. The definition of colony offered here is not without its limitations because it is possible that the juvenile of a flask-shaped thecate cell may settle and secrete its theca before the thread or threads responsible for dispersal have separated. Another possibility is that the juvenile may be released but the posterior filopodia may allow the cell to move over the surface of the parent theca and thereby secure settlement in its immediate proximity.

3.5.1 Colonial choanoflagellates in culture

Colony formation has been recorded in clonal cultures of several craspedids. The first cultured species for which colonies were described was *Proterospongia choanojuncta* (Leadbeater, 1983a). The sedentary phase of this species is morphologically and genetically indistinguishable from *Choanoeca perplexa* (Leadbeater, 1977; Medina *et al.*, 2003; Carr *et al.*, 2008a). Emergent division of individual cells predominates during the early and late stages of exponential growth and accounts for single motile cells, but during the middle phase colonies appear. Exactly how and from which cells such colonies arise has not been established. Colonial cells are larger than the usual single juvenile generated by emergent division. They are pyriform and have a long collar and flagellum (Fig. 3.43). A colony comprises a single layer of uniformly oriented 'naked' cells, with the flagella emerging from one surface. Except at the edge of the colony each cell is usually surrounded by five or six other cells in pentagonal or hexagonal close packing (Fig. 3.38). Cells are held together by the distal portion of their collars, the collar of each cell adhering laterally to the adjacent parts of the five or six neighbouring collars (Figs 3.39–3.42). Individual collars appear barrel-shaped, causing the plate-like colony to curve into a large hollow sphere or ovoid with the flagella radiating outwards (Figs 3.38, 3.39). The effect of combined flagellar movement is to propel a colony through the medium, usually with a slow rotation. Propulsion of large colony fragments can give rise to a temporary inversion of curvature of the colony surface, leaving the flagella on the concave rather than the convex outwardly directed surface. Electron microscopy reveals that attachment of the collars of adjacent cells is achieved by a one-to-one lengthwise adherence of the respective microvilli of neighbouring cells (Figs 3.39–3.41). In a transverse section of a colony in the collar region, the boundary of each collar appears approximately pentagonal and the precise one-to-one alignment between the collar microvilli of adjacent cells can be seen (Figs 3.40, 3.41 arrows). In whole mounts of cells it is apparent that the individual cell bodies are separate and that they are only held together by the distal portion of their collars (Fig. 3.43). Towards the end of a culture cycle the regular arrangement of colonies breaks down and small groups of cells may become non-motile. In late stationary phase cultures no colonies are observed (Leadbeater, 1977).

A recent surprise occurred when *Salpingoeca amphoridium*, which had been kept in clonal culture for many years

---

Figs 3.36–3.43 *Proterospongia choanojuncta*. **Fig. 3.36** Figure-of-eight-shaped dividing cell. Anterior emergent cell possesses an apical flagellum (f). Large number of vesicles containing fibrillar material (vfm) located in peripheral cytoplasm of anterior cell (see also Fig. 2.60). Bar = 0.5 μm.

Fig. 3.37 Motile cell with flagellum (f) and posterior filopodia (pf) with partially secreted wall material (arrows). Bar = 1 μm.

Fig. 3.38 Interference contrast light microscopy of *Proterospongia* colony comprising cells with long flagella (f). Bar = 10 μm.

Fig. 3.39 Section through colony showing outwardly directed cells held together by adherence of adjacent collar microvilli (arrows). Bar = 5 μm.

Fig. 3.40 Transverse section through collar region of colony showing the alignment of collar microvilli of adjacent cells (arrows). Bar = 1 μm.

Fig. 3.41 Higher magnification showing the pairing of collar microvilli between adjacent cells (arrows). Bar = 0.25 μm.

Fig. 3.42 LS of a pair of adjacent collar microvilli. Bar = 0.1 μm.

Fig. 3.43 Whole mount of two colony cells held together by anterior ends of their collar microvilli (c). Bar = 2 μm. Figs 3.37–3.40, 3.43 reproduced from Leadbeater (1983a).

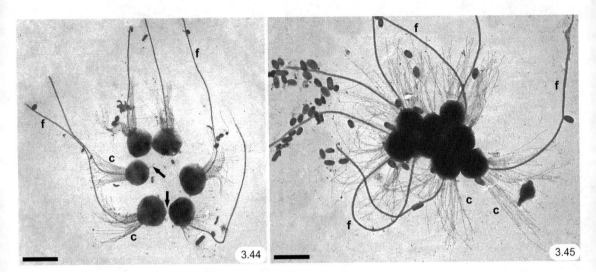

Plate 7 (Figures 3.44 and 3.45)

Figs 3.44–3.45 *'Proterospongia'* colonies.
Fig. 3.44 *Salpingoeca amphoridium* (ATCC 10153). Plate colony comprising six small cells with relatively long collars (c) and flagella (f). Cells attached to each other by narrow posterior filopodia (arrows). Bar = 10 µm.
Fig. 3.45 *Salpingoeca rosetta*. Close-packed colony of cells with long collars (c) and flagella (f). Bar = 5 µm.

and was the species extensively used by Pettitt *et al.* (2002) for feeding studies (see Section 2.3.3), suddenly produced distinctive colonies comprising four, six or eight naked cells arranged in a geometric array (Fig. 3.44). Colony cells, which are smaller than those enclosed within flask-shaped thecae, are held together by fine threads emerging from the posterior third of the rounded cell. A translucent plate of material envelops the basal portion of the cells. Individual collars and flagella are relatively long, with a ratio of flagellar length/cell radius (L/A) of 12 (see Section 2.3.2). Colonies occur during the mid-exponential phase of the culture cycle and disintegrate with the onset of the stationary phase.

Salpingoeca rosetta is an example of a sedentary craspedid with a cup-shaped theca (Fig. 3.22) that regularly produces motile colonies (Fairclough *et al.*, 2010; Dayel *et al.*, 2011). The life history of *S. rosetta* is complex and includes at least five different cell types. The sedentary stage, represented by a stalked thecate cell, undergoes mitosis and cell division to produce either a 'fast swimmer' daughter cell with a short collar, which serves for dispersal and subsequent settlement, or a 'slow swimmer' daughter cell which has a longer collar. Slow swimmers may

subsequently divide to generate either linear *Desmarella*-like colonies or rosette-shaped colonies. The latter comprise closely packed spheres of 4–50 cells with long collars and flagella (Fig. 3.45). Cells within rosette colonies are held together by a combination of extracellular material, posterior filopodia and lateral intercellular cytoplasmic bridges (Dayel *et al.*, 2011). Sedentary thecate cells can also become motile by vacating their thecae. This is achieved by the production of filopodia from the posterior portion of the cell, which serve to push the cell out of the surrounding theca. At the same time the collar microvilli shorten.

Alegado *et al.* (2012) observed that a clonal culture of *Salpingoeca rosetta* (ATCC 50818) with a mixture of co-cultured bacteria ceased to produce rosette colonies when treated with an antibiotic cocktail, even though the culture containing solitary cells proliferated by feeding on the remaining antibiotic-resistant bacteria. However, when the original bacterial population was reintroduced into this treated culture (designated RCA (rosette colonies absent)), rosette colony production was restored. Screening of the co-cultured bacterial population revealed that only one bacterial species, *Algoriphagus machipongonensis*, was

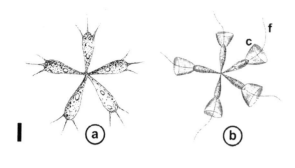

Fig. 3.46 *Astrosiga (Uvella) disjuncta.* a. Kent's (1880–2) copy of Stein's (1878) original illustration. b. Kent's embellished illustration of Stein's drawing with the addition of funnel-shaped collars. Bar = 5 μm. Reproduced from Kent (1880–2).

capable of inducing rosette colony formation in RCA cultures. *A. machipongonensis* is a Gram-negative, non-motile, non-spore-forming bacterial strain belonging to the phylum Bacteroidetes. The active ingredient was located within the medium of *A. machipongonensis* cultures as well as in the bacterial cell envelope. Chemical analysis revealed that the inducing molecule (designated RIF-1 (rosette inducing factor)) is a sulphonolipid that is active at concentrations between 10^{-5} and 10^{8} fM. *A. machipongonensis* conditioned medium contains approximately 10^{4} fM RIF-1. While nothing is known at present about the mode of action of RIF-1, nevertheless the shape of the dose–response curve and the potency of RIF-1 suggest that *Salpingoeca rosetta* perceives RIF-1 in a manner consistent with a receptor–ligand interaction, "albeit a receptor of exquisite sensitivity and remarkable dynamic range" (Alegado *et al.*, 2012, p. 5).

Roper *et al.* (2013) discussed the relative functional merits of the colonial versus flagellated unicellular states in *Salpingoeca rosetta*. Unlike some colonial organisms, such as *Volvox* species, there is no obvious coordination of flagellar movement between the cells of an *S. rosetta* colony. Since the flagella of individual cells within a colony 'push' in different directions, locomotory forces generated from different parts of a colony tend to cancel each other out with the result that colonies swim more slowly than single cells. However, asynchronous flagellar movement by colony cells does generate long-range feeding flows with a larger volume movement of fluid per cell than can be achieved by individual cells. Thus the relatively slow swimming

behaviour of a colony augments, rather than diminishes, the rate at which colonies draw in fluid, which, in turn, enhances the feeding rate. This, together with the fact that colony formation is usually triggered by high prey densities, probably favours the colonial habit in prey-enriched habitats.

3.5.2 Colonial choanoflagellates in nature

The first report of a colonial choanoflagellate would appear to be the freshwater species *Astrosiga (Uvella) disjuncta* which, according to Fromentel's (1874) original description, consisted of a colony of green fusiform cells each of which bore three 'short flagella' (Fig. 3.46a). Kent (1880–2) provided a robust re-interpretation of Fromentel's (1874) description in which he interpreted the 'position and direction' of the three flagella as representing "a central flagellate appendage and the two lateral margins of the hyaline infundibulate collar of a typical collared animalcule". He went so far as to embellish Fromentel's (1874) original illustration by giving it an 'improved' choanoflagellate appearance (Fig. 3.46b) and created the genus *Astrosiga* specifically for this species. In his discussion, Kent (1880–2, p. 341) commented on the possibility that the "colony stocks of this specific type (*Astrosiga*) present a close resemblance to the monad-clusters of *Codosiga botrytis* separated from their common footstalk and floating freely in the water". Kent (1880–2) was obviously aware of the possibility that choanoflagellate colonies could be the detached heads of sedentary species or a stage in the life cycle of another species.

Desmarella moniliformis was another distinctive colonial species that was recognised early by Kent (1878c) from a seawater sample. A freshwater strain of *D. moniliformis* comprises a single rank of cells held together by lateral cytoplasmic connections which contain a central dense-staining plug (Figs 3.51, 3.52), reminiscent of that seen in *Codosiga botrytis* and *Salpingoeca rosetta* (Hibberd, 1975; Karpov and Coupe, 1998; Dayel *et al.*, 2011). Colonies with many cells may be crescentic in shape with the outwardly directed flagella on the convex surface. Flagellar movement causes the colony to undulate in a graceful manner. Stein (1878) contemporaneously recorded an almost identical freshwater form that he called *Codonodesmus phalanx* (= *Desmarella phalanx* Kent), but he was doubtful whether this represented a permanent and independent entity and therefore his naming of this species was provisional (Kent, 1880–2).

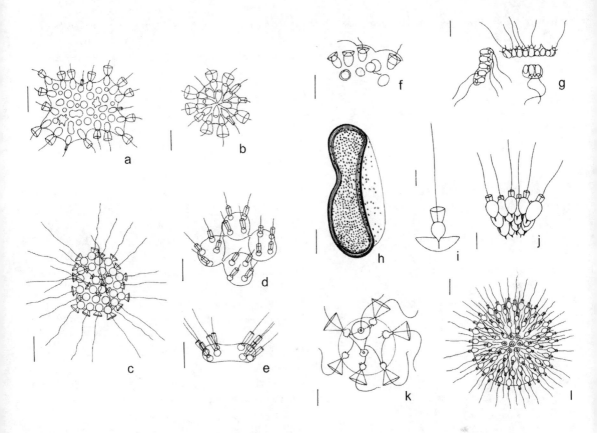

Fig. 3.47 Illustrations of *Proterospongia* spp. redrawn from the literature. a–b. *Proterospongia haeckeli* Kent. Bars = 10 and 20 μm, respectively. Reproduced from Kent (1880–2). c. *P. haeckeli* var. *clarki* Schiller. Bar = 40 μm. Reproduced from Schiller (1953). d. *P. skujae* (Skuja) Bourrelly. Bar = 40 μm. Reproduced from Skuja (1956). e. *P. skujae* var. *gracilis* (Skuja) Bourrelly. Bar = 20 μm. Reproduced from Skuja (1956). f. *P. nana* (Braarud) Throndsen. Bar = 10 μm. Reproduced from Braarud (1935). (g) *P. dybsoeënsis* (Grøntved) Loeblich. Bar = 10 μm. Reproduced from Grøntved (1956). h–i. *P. pedicellata* (Oxley) Ertl. h. Whole colony. Bar = 200 μm. i. Single cell. Bar = 10 μm. Reproduced from Oxley (1884). j. *P. pedicellata* (Oxley) Ertl, group of cells. Bar = 10 μm. Reproduced from Skuja (1932). k. *P. lackeyi* Bourrelly. Bar = 20 μm. Reproduced from Lackey (1959). l. *Sphaeroeca volvox* Lauterborn. Bar = 20 μm. Reproduced from Lauterborn (1894).

Since these early records there have been many sightings of colonial species. According to their morphology, as determined by light microscopy, they have been allocated to *Proterospongia*, *Astrosiga*, *Desmarella* or one of several more recent colonial genera including *Sphaeroeca* and *Cladospongia* (Fig. 3.47; Table 3.1).

3.5.3 Nomenclatural considerations

It is now apparent that some, if not the majority of colonial choanoflagellates described from field and aquaria samples are but ephemeral stages in the life cycles of other more stable, sedentary species. The allocation of colonial forms to specially nominated genera, such as *Proterospongia* and *Sphaeroeca*, without knowledge of the respective alternative sedentary phases, raises major nomenclatural problems. Recent work with cultured material has shown that even within one clonal isolate a variety of colonial forms may occur (Dayel *et al.*, 2011). There is no easy resolution of this matter. Without knowing the alternative phases of named species it is not possible to re-allocate them to alternative taxa. From this

Table 3.1 *Species of* Proterospongia *and related taxa reported in the literature summarising details of colony morphology and ecology. See also Fig. 3.47.*

Species	Shape of colony	Number of cells per colony	Cells held together by	Habitat	References
Proterospongia haeckeli	Irregular or circular film	6–60	Mucilaginous matrix	Freshwater, attached to water plants or water surface	Kent (1880–2)
P. haeckeli var. *clarki*	Circular plate	*c.*30	Plate of jelly	Freshwater, free-swimming colonies	Schiller (1953)
P. choanojuncta	Spherical or ovoid	9–120	Collars	Marine	Leadbeater (1983a)
P. dybsoeënsis	Single row of cells	3–10	Gelatinous, hyaline substance	Marine in coastal waters	Grøntved (1956)
P. lackeyi	Irregular or spherical	*c.*24	Jelly and interconnecting posterior threads	Freshwater in rivers	Lackey (1959)
P. nana	Irregular	*c.*40	Mucilaginous substance	Marine in coastal waters	Braarud (1935)
P. pedicellata	Elongate, rounded or subspherical	10 000–20 000	Exceedingly transparent, hyaline substance	Freshwater, free-swimming and attached	Oxley (1884), Skuja (1932)
P. skujae	Square or rectangular plate	4–16	Plate of jelly	Freshwater, planktonic	Skuja (1956), Ertl (1968)
P. skujae var. *gracilis*	Rectangular plate	4–16	Plate of jelly	Freshwater, planktonic	Skuja (1956)
Cladospongia elegans	Cylindrical, branched	>1000	Mucilaginous matrix and interconnecting protoplasmic strands	Freshwater in ponds	Iyengar and Ramanathan (1940)
Sphaeroeca volvox	Spherical	Several hundred	Sphere of jelly	Freshwater, free-swimming	Lauterborn (1894)
S. salina	Spherical or ellipsoidal	*c.*50	Sphere of jelly and interconnecting posterior strands	Brackish water, planktonic	Bourrelly (1957)

Reproduced from Leadbeater (1983a).

time forward it would be desirable not to add further species to colonial genera but to isolate species in clonal culture and determine their more stable sedentary phase and name the species accordingly.

A second matter arises with particular reference to *Proterospongia*. As discussed in Section 1.4.2, Kent (1880–2) considered that *Proterospongia haeckeli*, the type species of the genus, was a missing link ("the nearest concatenating form") between the flagellate infusoria and the sponges. On careful reading of Kent's numerous publications in which he discusses the link between *Proterospongia* and the sponges, it is the apparent co-existence of conventional collar-bearing cells with other cell types, including "amoebiform cytoblasts" and "sporular elements", within a "common gelatinous matrix" (see Fig. 1.8) that persuaded Kent (1880–2, pp. 364, 365) to place this enigmatic species in the Phalansteriidae. Unfortunately, by today's standards, the description of *Proterospongia* is so confused and the terminology is so archaic that both the name *Proterospongia* and the species *P. haeckeli* should be avoided until either the type is re-identified and re-described or the genus is rendered invalid (see also Section 1.4.2).

3.5.4 Ecological niche of colonial choanoflagellates

Since there is a functional requirement for a feeding cell to be more-or-less sedentary, the majority of craspedid species are attached to substrata (see Chapter 2). The specific function of the stalk is to secure a cell to the substratum and hold the cell at a distance above the surface. Coloniality has extended the range of microhabitats that species can inhabit. With respect to the sedentary habit, species of *Codosiga* have developed long stalks and in some cases are highly branched, thereby adding to the efficiency with which cells can filter the surrounding medium (Orme *et al.*, 2001).

Free-swimming colonial craspedids occur within freshwater, brackish and marine habitats, although in the latter they are usually only observed in the stagnant water of salt marshes, in rock pools and occasionally more generally within inshore waters. This is probably because they do not travel far from their related sedentary phase and they require relatively high densities of bacteria to maintain their active behaviour. They are common in freshwater bodies and may be abundant, particularly during the summer months. However, as in cultures, when prey concentrations decline individual cells decrease in size and eventually colonies disintegrate, probably re-forming the sedentary stage. In some instances, for example the freshwater *Desmarella moniliformis*, individual cells within a colony may encyst (see Section 3.6) (Leadbeater and Karpov, 2000). However, this eventually leads to fracturing of the moniliform thread of cells.

3.6 ENCYSTMENT

There are numerous accounts of encystment in freshwater choanoflagellates, including *Salpingoeca oblonga*

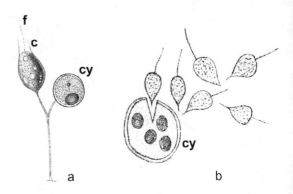

Fig 3.48 *Codosiga botrytis.* a. Stalked colony of two cells, one of which has encysted with a spherical cyst wall (cy). Flagellum (f), collar (c). b. Excystment apparently consisted of division of the cyst contents (cy) and release of many small flagellated cells. Reproduced from Fisch (1885).

(Stein, 1878), *S. amphoridium*, *S. fusiformis*, *S. infusionum*, *S. ringens* and *S. tintinnabulum* (Kent, 1880–2), *Codosiga botrytis* (Fisch, 1885; Griessmann, 1913) and *S. napiformis* (Francé, 1897). Unfortunately, with the exceptions of Fisch (1885) and Stoupin *et al.* (2012), most of the cells illustrated do not appear to be 'encysted' but rather moribund or in late stationary phase, having withdrawn their appendages (de Saedeleer, 1929). Stein (1878, Plate 10, Section IV, Figs 4, 5) mentions the existence in *Salpingoeca oblonga* of "a shell whose mouth is closed with a lid; the underlying spherical body is probably the animal that has passed into a quiescent state". Fisch (1885) illustrated a cyst of *Codosiga botrytis* with a thickened wall (Fig. 3.48a) and also described excystment during which many small flagellated cells were released (Fig. 3.48b). Stoupin *et al.* (2012) observed cyst formation in clonal cultures of a freshwater *Codosiga* sp. Cysts usually appeared 3–5 days after sub-culture into fresh medium and reached a peak in production during the late exponential phase of growth. Stoupin *et al.* (2012) also recovered viable *Codosiga* cysts from Siberian permafrost up to 43 000 years old. Jeuck *et al.* (2014) recently observed cysts in two cultured freshwater species of *Salpingoeca*.

There is only one ultrastructural study of cyst formation and that is for *Desmarella moniliformis* Kent (Leadbeater and Karpov, 2000). The freshwater *Desmarella* strain was isolated from the Rybinsk Reservoir in north-west Russia in 1996 and cells routinely produced thick-walled cysts (Figs 3.57 and 3.58 inset). An inoculum of swimming cells, cysts and a resident bacterial

a. Encystment

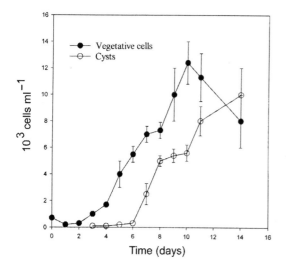

b. Excystment

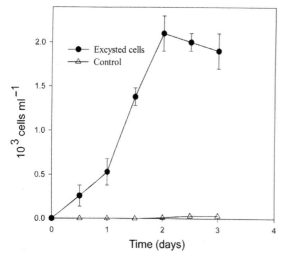

Fig. 3.49 Encystment and excystment in cultures of *Desmarella moniliformis*. a. Growth and cyst production in batch culture starting with an inoculum of vegetative (motile) cells. b. Production of vegetative cells from cysts when placed in fresh Pratt's medium without organic enrichment. The control consists of a lawn of bacteria covered with fresh Pratt's medium without organic enrichment which has been inoculated with 20 vegetative cells ml^{-1}. All points are the means ± SD of counts from three replicate dishes. Reproduced from Leadbeater and Karpov (2000).

flora was cultured in Pratt's medium enriched with cereal leaf infusion as an organic enrichment. Experiments were carried out using Petri dishes to provide a large surface area on which cysts could settle and develop. Fig. 3.49a illustrates a typical growth curve obtained from three replicate Petri dishes of culture. After a lag phase lasting approximately two days, the number of motile cells increased and typical moniliform *Desmarella* colonies comprising 4–15 cells developed. By the third day of growth cysts appeared and increased until the fourteenth day of growth, when there were approximately equal numbers of swimming and encysted cells. Cysts were usually associated with the bottom of the culture dish or the meniscus of the medium, but there were also many instances where one or more cells in a linear colony would encyst. Attempts to induce encystment by cold treatment, changes in salinity or the use of 'stale' medium did not appreciably alter the rate at which cysts appeared in culture.

Excystment could be induced by addition of fresh medium, without the cereal infusion, to cysts that had adhered to the bottom of a culture dish. Since swimming cells could never be eliminated, care was taken to ensure that the increase of motile cells was due to excystment rather than growth of the small background population of non-encysted cells. To achieve this, medium without organic enrichment was used and a control experiment was carried out containing a low density of motile cells, equivalent to that present in the experiment. The results are shown in Fig. 3.49b, where the increase in motile cells in the culture with cysts far exceeds that in the control culture. At the end of the experiment many empty cysts were observed, confirming that excystment had occurred. No evidence for division of an encysted cell followed by release of small flagellated cells, similar to that described by Fisch (1885) for *Codosiga botrytis* (Fig. 3.48b), was observed.

Whereas motile *Desmarella* cells are pear-shaped with a single flagellum and prominent collar, encysted cells are spherical and have thick, glistening walls (Figs 3.57 and 3.58 inset). A mature cyst wall is angular in outline and has a neck, but in the mature cyst the base of this is sealed (Fig. 3.57). Since cysts could be generated at will, it was possible to make a detailed ultrastructural study of cyst production. Early indication of encystment is marked by the accumulation of dense-staining extracellular material around the mid-region of a cell (Fig. 3.53). At this stage the cell changes from being pyriform (Fig. 3.50) to flask-shaped (Figs 3.53–3.55,) and retains its flagellum and collar. The Golgi apparatus within the neck rotates through 90° so that the long axis of the cisternal stack is vertical (Figs 3.53–3.55). Surrounding the cisternae are many small vesicles with dense-staining contents indicative of increased Golgi activity (Fig. 3.54). The number of mitochondrial profiles within the cell increases, suggesting elevated metabolic activity (compare Fig. 3.50 with Figs 3.53, 3.55, 3.56). As the cell covering thickens, the flagellum and collar are withdrawn and the neck shortens (Fig. 3.55). The cell then becomes spherical and the thickened wall forms a complete enclosure (Figs 3.56, 3.57). Cells that have been encysted for several weeks contain a high density of ribosomes, but there is a reduced Golgi apparatus (Fig. 3.57). Excystment involves the cell leaving through the neck of the cyst, which must involve breakdown of the continuous wall at the base of the neck (Fig. 3.58).

3.7 DISCUSSION

The distinction between non-restrictive and restrictive coverings among craspedids, while appearing fundamental at the morphological level because of its effect on cell division, does not appear to be recognised in molecular phylogenies (Fig. 10.3) (Carr et al., 2008a; Nitsche et al., 2011). There is variety in the ultrastructure of non-restrictive organic coverings. Whereas Codosiga gracilis has a diaphanously thin covering with little apparent substructure (Figs 3.5, 3.7), the two-layered fibrous 'sheath' covering C. botrytis is significantly thicker and more robust (Hibberd, 1975). However, since in both species there is no apparent restriction to longitudinal division in situ, the organic covering cannot be based on a framework of continuous microfibrils comparable to those of Salpingoeca infusionum and S. rosetta (Figs 3.19 and 3.22, respectively). A common feature shared by species with non-restrictive and restrictive coverings is the microfibrillar substructure of the stalk, which demonstrates that cells of both categories possess the necessary biochemical machinery for microfibril production.

Thecate structure is also varied, with a distinction between cup- and tube-shaped thecae, on the one hand, and flask-shaped thecae on the other. Cup- and tube-shaped thecae possess an obviously microfibrillar framework, although on specimens not treated with boiling water a coating of surface material usually masks the underlying substructure. Superficial patterns of ridges with a precise periodicity are common: sometimes these ridges are longitudinal, as in Salpingoeca gracilis (Thomsen, 1977b) and sometimes they are diagonal as in Proterospongia choanojuncta (Leadbeater, 1983a). In the treated specimens of cup-shaped thecae illustrated here, the underlying framework of microfibrils revealed is striking. Particularly in Salpingoeca rosetta, the microfibrils are seen to be continuous threads, and in favourable specimens such as that shown in Fig. 3.22, there is an indication that the microfibrils follow a helical path from lower right to upper left which would accord with a left-handed helix. The significance of this, should it be correct, is important when compared with the left-handed helical pattern of costae comprising the lorica of acanthoecids (see Chapter 4). The conventional wisdom regarding synthesis of continuous extracellular carbohydrate microfibrils in chrysophytes, such as Poterioochromonas (Schnepf et al., 1975) and Dinobryon (Franke and Herth, 1973; Herth, 1979) is that the cell rotates as the threads are polymerised on the surface of the plasma membrane. Certainly, if a microfibril is produced at a point source by a polymerisation (synthase) complex then some aspect of the cell will need to move to produce a continuous thread. Either the newly polymerised microfibril remains stationary and the synthase complex moves, or the synthase complex is anchored to a static plasma membrane and the thread is drawn out by means of the cell rotating.

Flask-shaped thecae are distinct from cup- and tube-shaped thecae in two respects. First, EM reveals a tri-laminate substructure comprising a translucent layer sandwiched between two denser layers. Second, a framework of microfibrils was not revealed after harsh treatment with boiling water, hot acid or alkali. Microfibrils in the thecal neck were visualised with negative staining and their apparent absence from the chamber may be because the matrix in which microfibrils are embedded is particularly resistant to extraction and completely masks a microfibrillar substructure.

There are similarities between the organic coverings of the Craspedida and Acanthoecida. As discussed in Chapter 4 (Section 4.5), most loricate choanoflagellates possess an organic investment of the lorica that is close to the cell surface, for example Stephanoeca diplocostata and Bicosta spinifera (Fig. 4.44), but in addition some species have an organic lining (veil) to the lorica that is at some distance from the cell, for example Diaphanoeca grandis and Didymoeca costata (Figs 4.37–4.42). Particularly in the veil, but also elsewhere, for instance the organic covering at the bases of the spines of Acanthoeca

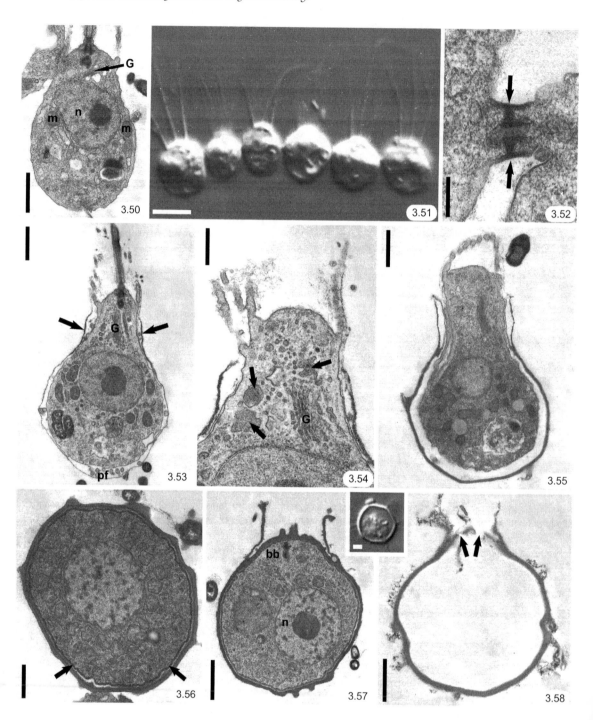

Plate 8 (Figures 3.50–3.58)

spectabilis, constituent microfibrils are obvious (Fig. 4.43). In *Diaphanoeca grandis*, as in the craspedid *Salpingoeca rosetta*, the microfibrils are embedded in an amorphous matrix which can be removed by washing or treatment with 1.5% sodium dodecyl sulphate (SDS) (Fig. 4.39; see Section 4.5). In some loricate species, such as *Bicosta spinifera*, a surface pattern of parallel ridges is evident (Fig. 4.44) similar to those observed in the Craspedida.

In conclusion, the ability of cells to polymerise microfibrils based on structural carbohydrates is widespread within the Craspedida, irrespective of whether the covering of the cell body is restrictive or non-restrictive, and can therefore be considered to be a plesiomorphic character that is also shared with loricate species. Whether non-restrictive coverings represent an evolutionary loss of constraining microfibrils, and are therefore an apomorphic character, or whether they are a modification of a universal glycocalyx-like covering and are therefore a plesiomorphic character cannot be decided without further ultrastructural and biochemical studies.

Figs 3.50–3.58 Encystment in a freshwater culture of *Desmarella moniliformis*. **Fig.** 3.50 Vegetative cell showing general disposition of organelles. The Golgi apparatus is horizontal (arrow G) and there is a single mitochondrial reticulum profile (m) on either side of the nucleus (n). Bar = 1 μm.
Fig. 3.51 Interference contrast microscopy of linear colony comprising six cells. Bar = 3 μm.
Fig. 3.52 Cytoplasmic junction between two adjacent colony cells. Note the dense-staining ring within the connection (arrows). Bar = 0.25 μm.
Fig. 3.53 Early stage in cyst formation with the wall extending around the neck and cell body (arrows). The Golgi apparatus (G) has rotated into the vertical plane and the posterior filopodia (pf) are located between the base of the cell and the surrounding wall. Bar = 1 μm.
Fig. 3.54 Higher magnification of a serial section of the cell neck shown in Fig. 3.53. The active Golgi apparatus (G) is surrounded by many vesicles containing granular material (arrows). Bar = 0.5 μm.
Fig. 3.55 Cyst development: the flagellum has been withdrawn (as determined by observation of serial sections) and the flask-shaped wall with projections is now complete. Bar = 1 μm.
Fig. 3.56 TS of cyst showing the thick wall surrounding a cell containing many mitochondrial profiles. Just beneath the plasma membrane is located a dense-staining layer of material (arrows). Bar = 1 μm.
Fig. 3.57 LS of cyst showing flask-shaped wall; anterior end of the cell at the base of the neck is covered by a thick ornamented cap of wall material. The enclosed cell has a prominent nucleus (n) with nucleolus, two basal bodies (bb), and the cytoplasm is filled with a dense array of ribosomes. Bar = 1 μm.
Fig. 3.58 LS of an empty cyst showing the flask-shaped wall and perforations in the cap at the base of the neck (arrows). Bar = 1 μm.
Figs 3.57 and 3.58 inset. Interference contrast microscopy of a cyst with a glistening wall. Bar = 1 μm. Figs 3.50, 3.51, 3.53–3.58 reproduced from Leadbeater and Karpov (2000).

4 • Loricate choanoflagellates: Acanthoecida

Lorica construction – searching for clues, making the rules.

4.1 INTRODUCTION

The choanoflagellate lorica must rank as one of the most distinctive protistan coverings ever to have evolved, competing with such iconic structures as the frustules of diatoms, the calcified discs of coccolithophorids and the scales of chrysophytes (Preisig *et al.*, 1994). Currently there are over 150 named loricate species ubiquitously distributed in marine and brackish water habitats, with a few freshwater records (see Section 9.4.8). They all possess loricae comprising silicified rod-shaped units attached to each other end-to-end to form costae that combine in two layers to produce a rigid basket-like cage in which is located a collared flagellate. Evolution of the lorica has allowed these microscopical suspension feeders to inhabit niches not usually available to a cell whose feeding habits require it to be secured to a substratum (see Sections 2.3.3 and 9.6). Diversification of the lorica in terms of size, form and costal pattern has been the secret of success.

Kent (1878c, 1880–2) was the first author to illustrate and describe a loricate choanoflagellate, *Salpingoeca ampulla*, which he found growing on algae and small zoophytes in seawater tanks at the Manchester Aquarium in May 1874 and subsequently in seawater from Brighton in February 1877 (Fig. 4.1a). He described the lorica of *S. ampulla* as being "sessile, narrow and ovate beneath, expanding superiorly in an inflated balloon-like manner, the external surface exhibiting even longitudinal sulci or striations" (Kent, 1880–2, p. 349) (Fig. 4.1a–c). He further commented that this "very beautiful variety is readily distinguished from all other representatives of the genus *Salpingoeca* hitherto described, both on account of the remarkable shape of its lorica, and the fact that the whole of the animalcule, including even flagellum and the hyaline collar, is completely enclosed

within that structure" (Kent, 1880–2, p. 349). While he did not actually observe siliceous costae, nevertheless he did appreciate that the lorica involved a superstructure that completely enclosed the cell and that this was an exceptional character for which he considered creating a separate genus.

Ellis (1929) must take the credit for appreciating unequivocally that a pattern of longitudinal and transverse costae provided the supporting framework of the lorica. Based on material collected from a saltmarsh in the south of England, he named three new genera, *Stephanoeca*, *Diaphanoeca* and *Acanthoeca*, and seven new species, including *Stephanoeca diplocostata*, *Diaphanoeca grandis* and *Acanthoeca spectabilis*, all well-established names and ubiquitous in inshore waters (Fig. 4.1d–g). Ellis' (1929) contribution was seminal for he not only recognised and photographed the lorica but he also correctly observed cell division and lorica assembly (Fig. 7.1).

Norris (1965), working with seawater samples collected from tide pools on the San Juan Islands, Washington State, United States, increased the number of described loricate species and produced the first electron micrograph demonstrating that silicified strips were the units of costal construction. This provided the basis for many subsequent publications by Throndsen (1969 *et seq.*), Leadbeater (1972a *et seq.*), Thomsen (1973 *et seq.*), Manton *et al.* (1975 *et seq.*) and Buck (1981 *et seq.*).

Unfortunately the functional role of the lorica is often overlooked. The elaborate pattern of siliceous costae provides a skeletal superstructure that surrounds the cell. The inner surface of the lorica is either partially or entirely covered by an organic microfibrillar investment not dissimilar in appearance from those seen in the Craspedida (see Chapter 3). Diversification, in terms of the different costal patterns and the position and extent of the organic covering, has facilitated the ecological radiation of loricate species in mostly marine and brackish water habitats (see also Section 9.4.8). At first sight the range of costal patterns may appear

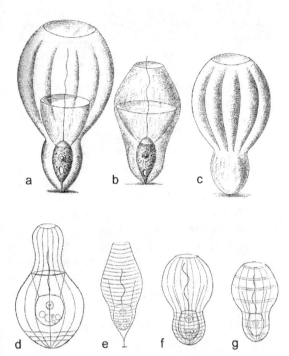

Fig. 4.1 Historical illustrations of loricate choanoflagellates. a–c. *Salpingoeca ampulla*. Reproduced from Kent (1880–2). Note that each cell is completely enclosed within a surrounding lorica. Two of the loricae (a, c) have longitudinal 'striations' (costae) on the anterior chamber. d–g. The first drawings of loricate choanoflagellates in which the unequivocal presence of siliceous costae was recorded. d. *Diaphanoeca grandis*; e. *Stephanoeca kentii* (= *S. norrisii*); f. *Stephanoeca ampulla*; g. *Stephanoeca diplocostata*. Reprinted from Ellis (1929).

bewildering and this is particularly the case if specimens are observed as dried whole mounts. Drying causes loricae to collapse and the stresses incurred distort the organisation of costae with respect to each other. Even with the most careful treatment to prevent flattening, some damage will inevitably be incurred. What can be particularly confusing is that the distorted patterns themselves may have a regularity that looks authentic (for example, Figs 4.51 and 4.52). Thus interpretation of results requires painstaking analysis of many specimens prepared by a variety of methods.

A logical system of principles, or 'rules', underpins every aspect of lorica construction, development and assembly. Apart from the occasional error, every stage from intracellular deposition of silica to make the first costal strip until the final assembly of the lorica is carried out with precision and accuracy. No detail relies on a chance event

and every stage must be viewed as part of an integrated whole. The challenge facing the observer is to deduce the system of 'rules' by analysing as many costal patterns as possible and by studying the mechanism of lorica assembly. Once a set of 'rules' has been established they can be translated into a graphical computer model and subsequently tested by constructing three-dimensional images (Figs 4.2, 4.4, 4.19, 4.22, 4.46, 4.52) (Leadbeater *et al.*, 2009). To a large extent this has now been achieved. Knowledge of the rules has also permitted insights into aspects of the evolutionary history of the lorica and, in particular, the possible relationship between the nudiform and tectiform conditions (see Chapters 6–8).

The organisation of this and the next four chapters provides a stage-by-stage explanation of the construction, assembly and possible evolution of the choanoflagellate lorica. This chapter will examine the substructure and diversity of loricae and list the 'rules' relating to lorica construction. In Chapter 5 the requirement of loricate species for silicon is discussed, together with the details of costal strip formation. In Chapter 6, the small group of nudiform species that divide to produce a motile dispersal stage is considered. In Chapter 7 the much larger, in terms of species, and more diverse tectiform group of loricate choanoflagellates is detailed. Chapter 8 explores the possible evolutionary relationship between nudiform and tectiform species.

For the benefit of clarity, two matters need to be discussed at the outset. The first is the use of terminology. The second is the relative merits of the various techniques that have been employed for preserving and treating loricae and the methods of analysing their three-dimensional structure.

4.2 TERMINOLOGY AND PRESERVATION OF CHOANOFLAGELLATE LORICAE

4.2.1 Terminology

Over the past 40 years a system of terminology has developed within the literature and, wherever possible, the standard terminology is used here. Definitions of the most important terms are presented below and nuances regarding their use in the following chapters are noted (see also the Glossary).

Cell axes – by convention the location of the flagellum, or flagellar basal body, denotes the anterior pole of the cell (see Section 2.1). The longitudinal (vertical) axis of the cell is a line extending through the long axis of the flagellar basal body and projected backwards through the centre of the cell to the opposite pole. The transverse (horizontal) axis of a cell is a line drawn in a plane at right angles (90°) to the long axis of the cell.

Lorica – (plural *loricae*) an extracellular basket-like structure comprising a system of silicified costae. Loricae range in size and morphology from a single horizontal costa on the surface of the cell body (*Monocosta fennica* – Thomsen, 1979), to a large, voluminous cage completely enclosing the cell. Basically, the lorica comprises two systems of costae. In the majority of species an outer arrangement of longitudinal costae overlies an inner pattern of helical and/or transverse costae. In one species grouping, characterised by *Parvicorbicula socialis*, both the transverse costae are outside the longitudinal costae. An organic investment lines part of the inner surface of the lorica, often in close proximity to the cell body. In some species an organic investment may cover the inner surface of the lorica at a distance from the cell, in which case it is called the *veil*.

Lorica chamber – a general term used to describe the enclosure of a volume by the lorica, usually distinguished by a convergence of the costae of the lorica anteriorly and posteriorly. For example, species of *Stephanoeca* possess a relatively small posterior chamber closely surrounding the cell and a larger anterior chamber (Figs 4.45–4.53). The two chambers are separated by a 'waist' which denotes the position at which the cell is secured to the lorica.

Costa – (plural *costae*) a linear series of costal strips attached to each other end to end with a short overlap. Costae are categorised according to their orientation in relation to the longitudinal axis of the cell. There are three categories. *Longitudinal costae* are oriented parallel to the long axis of the cell. *Transverse costae* (rings) are oriented perpendicular (horizontal) to the long axis of the cell. *Helical costae* are arranged in a helix around the long axis of the cell.

Costal strip – a rod-shaped unit of costal construction. The majority of loricate choanoflagellates possess narrow costal strips that are usually slightly crescentic in form. In a minority of species the costal strips may be flattened, spatulate (spoon-like) or spine shaped. Costal strips comprising a costa are attached to each other end to end with an overlap that is usually short, approximately one-tenth the length of a strip, but in some species may be longer.

Since the standard lorica comprises two layers of costae, the costal strips of one costal layer adhere to the costal strips of the other layer where they pass over each other.

The position of this adherence on a costal strip may be intermediate between the two ends of the strip. However, in many instances the extreme tip of a strip of one costa adheres, by means of a 'point' attachment, to an intermediate position of a strip from another costa. Such a connection is called an abutment (see below).

The *leading end* of a costal strip (opposite *trailing end*) is the end that points in the direction the costa was extended during lorica assembly. For longitudinal and helical costae this is in an anterior direction and for transverse costae this is to the left (clockwise) when seen from the outer surface (i.e. clockwise when viewed from the anterior end).

Imbrication – the regular pattern of overlaps between adjacent costal strips comprising a costa.

Transverse costa (ring) – an entire horizontal costa that is oriented perpendicular to the longitudinal axis of the cell. Transverse costae, according to species (see Table 4.3), may either be on the inside or outside of the longitudinal costae of the lorica.

Anterior ring – a transverse costa usually located at the extreme anterior end of the subtending longitudinal costae but sometimes with anteriorly projecting spines. According to species, the transverse costa may be on the outside or inside surface of the longitudinal costae (see Table 4.3).

Helical costa – a continuous costa that describes a helical path around the longitudinal axis of the lorica. Some helical costae describe a single turn of the helix (360°) and some describe several turns. Helical costae of choanoflagellates are left-handed (see also Section 6.2) and are on the inside of the lorica.

Left-handed helix – when viewed from below (posterior end) a left-handed helix rotates anti-clockwise away from the observer. When viewed from above (anterior end) a left-handed helix rotates clockwise towards the observer. When viewed from the outside with the anterior end uppermost, the helix rises from lower right to upper left. When a clockwise rotation is referred to in the text, it always refers to clockwise when viewed from above (see also Section 6.2).

Abutment between costal strips – in this context abutment refers to a 'point' attachment between the tip of a costal strip of one costa and the long axis of a strip from another costa. One strip usually abuts on another strip at an angle of 45° to 90°. Abutments occur between costal strips of transverse or helical costae against longitudinal costae or vice versa. In contrast, overlaps between adjacent costal strips *within* a costa are usually (but not always) longer and the strips are parallel to each other.

Proximal side of lorica (P) – the side of a lorica nearest to the observer; in flattened whole mounts the proximal side of the lorica will normally be seen from the outer surface.

Distal side of lorica (D) – the side of a lorica furthest away from the observer; in flattened whole mounts the distal side of the lorica will normally be seen from the inner surface.

Outer and inner surfaces of lorica – the outward and inward facing surfaces, respectively, of a lorica.

Note that the terms *proximal* and *distal* refer to the perspective in relation to the observer. *Outer* and *inner* are fixed properties of the lorica.

4.2.2 Numbering of costal strips within loricae

Since there is a basic uniformity in the construction of all choanoflagellate loricae and since there are two contrasting conditions, nudiform and tectiform, that are distinguished by the order in which different categories of strips are formed and stored, it is helpful to adopt from the outset a standard protocol for numbering the categories of costal strips. The value of this will become particularly apparent in Chapter 8. The numbering system that has been chosen is based on the organisation of strips in the assembled lorica and operates with the longitudinal costae first and the transverse or helical costae second. In all categories the numbering is from the posterior forwards (Fig. 4.2a, b). Three numbers – 1, 2 and 3 – are applied respectively to the posterior, intermediate and anterior portions of the longitudinal costae. Thus the basal strip of a longitudinal costa is 1; the middle strip or strips are 2; the anterior strip or strips are 3 (Fig. 4.2a, b). In some examples, such as *Didymoeca costata*, the three numbers refer to individual costal strips: petaloid (1), spatulate (2) and narrow (3) (Figs 4.15, 7.36). However, in other species, such as *Diaphanoeca grandis*, where the longitudinal costae comprise many strips, these numbers refer to several strips each (Fig. 4.27). In species such as the *micropora* form of *Savillea parva*, where there are only two strips in a longitudinal costa, these numbers are still used but are nominal. Helical and transverse costae are also numbered from the posterior forwards. For helical costae, again the numbering system is nominal with the basal portion of the costa being numbered 4, the intermediate strips 5 and the anterior strip or strips 6 (Fig. 4.2b). However, where there are transverse costae then each costa has a number (Fig. 4.2a). Thus, in *Didymoeca costata* the posterior transverse costa is 4, the intermediate costa 5 and the anterior ring is 6 (Fig. 7.36). Since the lorica of *Diaphanoeca grandis* has four transverse costae, these are

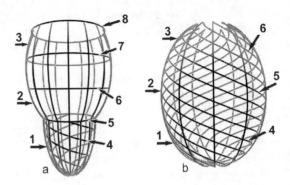

Fig. 4.2 Numbering of costal strips within loricae. Numbering proceeds from the longitudinal costae to other costae. Longitudinal costae are numbered: (1) basal section; (2) middle section; (3) anterior section. Helical and/or transverse costae are numbered as follows: in a. *Stephanoeca* sp. – a tectiform species (see also Figs 4.45, 4.46), helical costae in the posterior chamber are numbered (4) and the four transverse costae are numbered: (5) posterior (waist) costa, (6) posterior intermediate costa, (7) anterior intermediate costa, (8) anterior ring. In b. *Savillea parva* (*micropora* form) – a nudiform species (see also Figs 4.21, 4.22, 6.2–6.6), inner helical costae are numbered: (4) basal section, (5) middle section, (6) anterior section.

labelled as follows: posterior basal transverse costa 4; intermediate basal transverse costa 5; anterior basal transverse costa 6; anterior transverse costa 7 (Figs 4.26–4.28). Where the lorica comprises a series of helical costae within the posterior chamber and transverse rings in the anterior chamber, as for instance in *Stephanoeca* spp., the helical costa is numbered 4, and the transverse rings are numbered sequentially from the bottom 5, 6, 7 and 8, respectively (Fig. 4.2a). The numbering of the transverse costae is from the posterior forwards, irrespective of whether the costae are on the inside or outside of the longitudinal costae. This system of using three numbers (1, 2 and 3) for the longitudinal costae and the appropriate sequence of numbers for the transverse costae has been used consistently throughout this book.

4.2.3 Preservation of loricae for three-dimensional analysis

For detailed analysis of the costal arrangement in loricae the use of electron microscopy (EM) is essential. While most loricate species can be identified with phase or interference contrast light microscopy, the resolving power of the light microscope is not adequate for analysis of the junctions between costae and costal strips. The two most common methods of preparation for EM are dried whole mounts for transmission electron microscopy (TEM) and critical-point-dried preparations for scanning electron microscopy (SEM). Specimens that are air-dried inevitably flatten as a result of surface tension effects that cause distortion of the costal pattern and displacement of individual strips. Critical-point drying overcomes some of these damaging effects, but still there is a tendency for species with narrow costal strips to collapse.

Shadowcasting of fixed 'whole mounts' of cells with heavy metal for TEM has been particularly valuable for determining the three-dimensional relationship of costal strips at junctions. Shadowcasting enhances contrast and creates a three-dimensional effect, thereby making it possible to determine which costae or costal strips of a lorica overlie others. Thus in flattened specimens it is possible to distinguish the outer surface of the proximal side of a lorica from the inner surface of the distal side (see Section 4.3.1). Analysis of this level of detail has been essential for determining the relative positions of costal strips at junctions.

Determining the chirality (handedness) of helical costae and the relative positions of costal strips at junctions requires EM, photographic and scanned images to be maintained in their 'correct' orientation. Reversing an image creates a mirror reflection of the subject and this can never be matched with the original specimen unless viewed with a mirror. Therefore it is essential that whenever an image is manipulated optically or electronically it should be checked against a control image for correct orientation. A 'finder' grid with letters of the alphabet, recorded as a photographic image in an EM and subsequently scanned electronically, can be used as a control for actual images recorded and digitised in the same manner.

4.3 LORICA CONSTRUCTION

The costal pattern of every choanoflagellate lorica conforms to the basic rules governing its construction and assembly, although there are as many variations as there are species. Each variation, in turn, challenges the validity of the rules and at the same time provides an insight into their flexibility. To derive the rules requires a detailed understanding of how the different categories of costae – longitudinal, transverse and helical – interact with each other. Analysis of the junctions between costal strips is crucial since the pattern of overlaps within and between costae provides a historical record of how they were assembled. All species

contribute to an understanding of the rules but some, for reasons of special features, have been of seminal importance. Leading the list must be *Acanthoeca spectabilis* and *Savillea parva* (*micropora* and *parva* forms) (Figs 4.18, 4.19, 4.21, 4.22) for their prominent helical costae and *Didymoeca costata* for the morphological distinctions between its various categories of costal strips (Fig. 4.15) (see also Chapters 6 and 7). Fortunately, all these species were obtained in clonal culture thereby providing a long-term supply of high-quality material.

The following account opens with *Saepicula pulchra* as an exemplar because it shows to advantage features that occur in various combinations in many other loricae. The variety in costal strip morphology will be discussed. The evidence for a left-handed rotation of helical costae will be explored and the universality of a 'clockwise' organisation of costal strips comprising the transverse costae will be discussed. The substructure of longitudinal and transverse costae is described. The three-dimensional interactions of costal strips at junctions of longitudinal and transverse costae will be analysed and an elaboration found in *Diaphanoeca grandis* will be investigated. 'Variations on themes' explores the diversity of lorica construction in four distinctive groups of species. Following a discussion, the list of rules governing lorica construction is presented.

4.3.1 Saepicula pulchra

Saepicula pulchra has been selected as a model because of its distinctive shape, its reasonably undisturbed arrangement of costae when dried and a wide range of costal features not displayed in combination in most other species. Figs 4.3 and 4.5 show two shadowcast specimens and Fig. 4.4 is a computer-derived model of the lorica. Figs 4.6 and 4.7 provide higher magnifications of the anterior rings.

The lorica of *Saepicula* contains ten longitudinal costae which are seen to greatest advantage at the anterior end of the lorica (Figs 4.3, 4.5). These terminate in a transverse costa which, because it is located at the top of the lorica, is known as the anterior ring (Figs 4.6, 4.7). The pattern of costae in the bottom two-thirds of the lorica is more confused but comprises a series of helical costae within the outer layer of ten longitudinal costae. The computer model shown in Fig. 4.4 is an idealised reconstruction of the lorica and illustrates the relationship between the various categories of costae. Evidence for the helical pattern of the inner costal layer comes from the fact that there are about ten regularly spaced 'approximately diagonal' costae in the

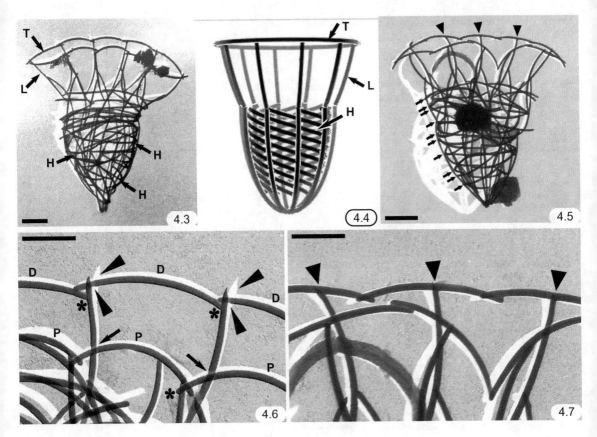

Plate 1 (Figures 4.3–4.7)

Figs 4.3–4.7 *Saepicula pulchra*. **Fig. 4.3** Empty lorica with anterior transverse costa (arrow T) containing ten costal strips that span the spaces between the ten longitudinal costae (arrow L). Dislocated helical costae (arrows H) are visible in the posterior chamber. Bar = 2 μm.

Fig. 4.4 Computer-generated image of *Saepicula* lorica illustrating the relative locations of transverse (arrow T), longitudinal (arrow L) and helical (arrow H) costae.

Fig. 4.5 Lorica with anterior transverse costa containing strips that form T-junctions with longitudinal costae (arrowheads). Ten arrows at side of posterior lorica chamber denote the gyres of ten helical costae. Bar = 2 μm.

Fig. 4.6 Part of a transverse ring illustrating the effects of metal shadowcasting. The white shadows of the proximal side (P) of the ring pass over the longitudinal costae (arrows) associated with distal side (D) of the ring. The tips of the spines cast long shadows (arrowheads) over the adjacent transverse strips showing that they are on the inner surface of the distal side of the ring. Note the abutment of each transverse strip with the respective longitudinal strip is at the right end of the transverse strip (asterisk) when seen on the inner surface of the distal side of the lorica. Bar = 1 μm.

Fig. 4.7 Part of the transverse ring of the lorica illustrated in Fig. 4.5. Note the T-junctions between the transverse costal strips and the longitudinal costae (arrowheads). Bar = 1 μm. Figs 4.3, 4.5 and 4.7 reproduced from Leadbeater (1980).

posterior chamber (see arrows in Fig. 4.5). Confirmation that they are not transverse but helical costae comes from the fact that they are not entire rings but have a one-to-one relationship at their anterior end with each of the longitudinal costae. This is seen more clearly in the closely related species *Acanthocorbis unguiculata* (arrowheads Fig. 4.23). Ten costal strips comprise the anterior ring but it is possible for them to interact with the longitudinal

Table 4.1 *Genera of loricate choanoflagellates listed according to whether the costal strips comprising the anterior ring form: (1) 'abutting' joins or (2) 'T-junctions' with the respective longitudinal costae.*

Abutting junctions between transverse costal strips of anterior ring and longitudinal costae	T-junctions between transverse costal strips of the anterior ring and longitudinal costae
Cosmoeca	*Parvicorbicula*
Conion	*Pleurasiga*
Nannoeca	*Polyfibula*
Saepicula	*Saepicula*
Stephanoeca	*Stephanacantha*
Platypleura	

costae in two distinctive conformations. In Figs 4.3 and 4.6 individual costal strips of the anterior ring span the gaps between neighbouring longitudinal costae and the junctions between the costal strips interact with the longitudinal costae. In Figs 4.5 and 4.7 individual costal strips of the anterior ring are pivoted centrally on the longitudinal costae and they span half the gap between neighbouring longitudinal costae on either side. This latter pattern is referred to as a 'T-junction' for obvious reasons. Both configurations are common in different genera (Table 4.1), but it is uncommon to find them as alternatives within the same species as shown here in *Saepicula pulchra*.

Fig. 4.6 also illustrates the value of metal shadowcasting in interpreting the three-dimensional relationships between costal strips at the junctions between longitudinal and transverse costae. Since the metal was deposited from above, the proximal layer (P) of costae casts a shadow (white shadow) over the inner surface of the distal costae (D) (Fig. 4.6 arrows). Shadowcasting also reveals that the anterior ring is on the outer surface of the longitudinal costae. The spine-shaped tips on the distal layer (D) of longitudinal costae have been pushed upwards (inwards) where they overlie the transverse ring and this causes them to cast a long white shadow over the anterior ring (Fig. 4.6 arrowheads).

4.3.2 Costal strip morphology

A quick glance at a lorica containing only narrow rod-shaped costal strips could give the impression that all costal strips are similar. In fact, this situation is rare and in most

species differences in length, thickness and radius of curvature, depending on where in the lorica individual strips are located, are usual. There is nothing casual about the form of costal strips, as is demonstrated by species such as *Didymoeca costata*, which has an obviously different morphology for each category of strip (Figs 4.11, 4.15, 7.35–7.37; Section 7.3.2). The nuancing of form can be subtle and minuscule. For instance, the tip of each anterior spine in *Diaphanoeca grandis* and *Acanthoeca spectabilis* is minutely forked (Figs 4.8, 6.27 inset, 6.30 inset, 7.55). The tip of each anterior longitudinal strip in *Saepicula pulchra* and *Acanthocorbis unguiculata* is nib-shaped (Figs 4.6, 4.9). Anterior spines are common but vary in form and size. They may be short and thorn-like at the end of a transverse costal strip, for example *Syndetophyllum pulchellum* and *Pleurasiga echinocostata* (Figs 4.14, 4.64). They may be long and attenuated as in species of *Calliacantha* and *Bicosta* and *Crucispina cruciformis* (Figs 4.70–4.72, 4.75, 7.58–7.65). The tips of strips are often flattened, spatulate or faceted where they attach to other costal strips (Figs 4.10, 4.13). This is particularly common at the leading end of a strip where it adheres (abuts) at right-angles to the respective strip of another costa. Flattened, broadened and sometimes perforated strips are common (Figs 4.11, 4.12, 4.14). In *Didymoeca costata* each category of strip is different; those around the cell (posterior chamber) are flattened with a thickened midrib and lateral perforations (Figs 4.11, 4.15, 7.35–7.37). Those forming the lower half of the anterior chamber are spatulate (Figs 4.15, 7.37) and those comprising the anterior ring are flattened and slightly wider (Fig. 4.15). The loricae of some species, for example *Syndetophyllum pulchellum*, contain entirely broad, flattened strips with a midrib and prominent perforations (Fig. 4.14). The end of each costal strip of the anterior ring is turned upwards.

4.3.3 Longitudinal costae

Longitudinal costae are usually outermost in a two-layered lorica. However, there are two important exceptions to this generalisation. The first is that in some genera the anterior ring is outside the longitudinal costae (see Table 4.3). The second is that in one group of species, characterised by *Parvicorbicula socialis*, all the transverse rings are outside the longitudinal costae (see Section 4.3.5). An explanation for these exceptions will emerge later (see Section 7.3.6). In well-preserved loricae (Fig. 4.15), and as illustrated in computer-generated models (Figs 4.4, 4.46, 4.52), the longitudinal costae are aligned parallel to the long axis

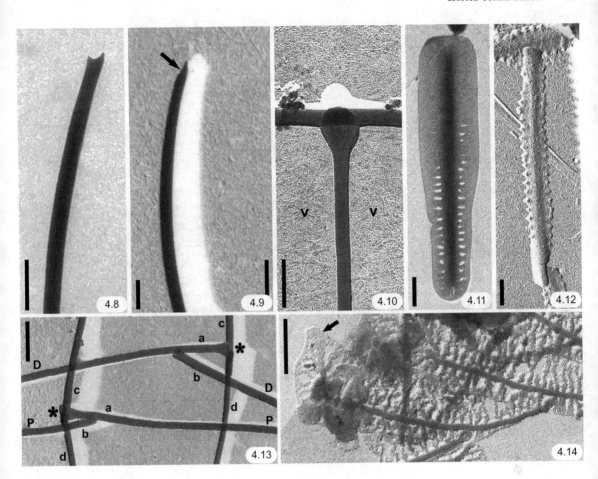

Plate 2 (Figures 4.8–4.14)

Figs 4.8–4.14 Costal strip morphology. **Fig. 4.8** *Diaphanoeca grandis*, anterior end of spine with forked tip. Bar = 0.5 μm.

Fig. 4.9 *Acanthocorbis unguiculata*, spine strip with nib-shaped point (arrow). Bar = 0.25 μm.

Fig. 4.10 *Stephanoeca elegans*, circular flattened anterior tip of longitudinal strip at junction with anterior ring. Microfibrils of veil (v) just visible. Bar = 0.25 μm.

Fig. 4.11 *Didymoeca costata*, petaloid longitudinal strip (see also Fig. 4.15). Bar = 0.5 μm.

Fig. 4.12 *Parvicorbicula serrulata*, broad longitudinal strip with serrated edges, attached at top to inner surface of anterior ring (inside view). Bar = 0.5 μm.

Fig. 4.13 *Cosmoeca ventricosa*, junctions of intermediate transverse costae with respective longitudinal costae (see also Fig. 4.66). Lower transverse costal strips are seen from the outer surface of proximal side (P) of the lorica – the spatulate left end of the transverse strip abuts onto the longitudinal strip (asterisk). Upper transverse costal strips are seen from the inner surface of distal side (D) of lorica – the right end of the transverse strip abuts onto the longitudinal strip (asterisk). (a)–(d) have been used in the same context as they are used in Figs 4.24 and 4.25. Bar = 0.5 μm.

Fig. 4.14 *Syndetophyllum pulchellum*, outer view of broad perforated strips comprising anterior ring. Upturned end is on the left (arrow). Bar = 0.5 μm.

of the cell. Any deflection from this orientation is either because of a self-imposed restriction, such as attachment by a stalk to a surface (Fig. 4.35) or a 'spiral twist' as in *Bicosta* spinifera (Fig. 4.71), or it is artefactual as a result of collapse (Figs. 4.50–4.52). The component strips of longitudinal costae adhere to each other end to end in linear array and

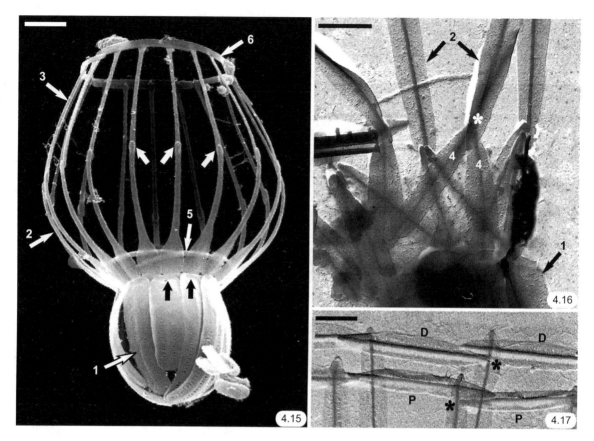

Plate 3 (Figures 4.15–4.17)

Figs 4.15–4.17 Imbrication of costal strips. **Fig. 4.15** *Didymoeca costata*. SEM of empty lorica showing longitudinal costae, consisting of basal petaloid strips (arrow 1), intermediate spatulate strips (arrow 2) and anterior narrow strips (arrow 3), and two transverse costae, namely the intermediate transverse costa (arrow 5) and anterior transverse costa (arrow 6). Imbrication of the three strips comprising a longitudinal costa is: petaloid outside spatulate (black arrows) outside narrow (white arrows). Bar = 1 μm.

Fig. 4.16 *Stephanacantha campaniformis*. Basal petaloid strips (arrow 1) overlap the inverted 'V' pattern of intermediate strips (4) which are themselves overlapped anteriorly (asterisk) by the longitudinal strips (arrow 2) of the anterior chamber. Since the pairs of intermediate strips are overlapped on the outside at both ends they are therefore part of an internal ring that has been modified to fill the gap between the basal petaloid strips (arrow 1) and the anterior longitudinal strips (arrow 2) in the longitudinal costae. Bar = 1 μm.

Fig. 4.17 *Platypleura infundibuliformis*. Flattened anterior ring showing imbrication of transverse strips at junctions with the longitudinal costae. The left end of each transverse strip (asterisk) abuts onto the respective longitudinal strip on the inner surface of the proximal side (P) and vice versa for the distal side (D). The transverse ring is inside the longitudinal costae. Bar = 1 μm. Reproduced from Thomsen and Boonruang (1983a).

their imbrication is posterior–outside–anterior (Fig. 4.15). This is excluding strips that contribute to pedicels or peduncles. In the majority of species, longitudinal costae converge at the posterior end, thereby closing the rear of the lorica (Figs 4.23, 4.26–4.28). However, there are exceptions; for example, *Crinolina aperta* and *Cosmoeca*

takahashii, where the longitudinal costae do not converge but project backwards as a ring of spines (Figs 4.57, 4.68, respectively). In *Crinolina* spp. the posterior end of the lorica is entirely open (Figs 4.56, 4.57).

At least one example exists in which costal strips of the inner layer have become integrated into the outer longitudinal layer. In *Stephanacantha campaniformis* the lorica comprises entirely broad costal strips (Fig. 4.16). The cell, which is located at the bottom of the lorica, is surrounded by 10–12 longitudinal strips (arrow 1 in Fig. 4.16). Each of these is attached anteriorly to a shorter strip (4 in Fig. 4.16) and these converge in pairs at their anterior end (asterisk in Fig. 4.16) and join with a single longitudinal strip (arrow 2 in Fig. 4.16) that then terminates in a T-junction with the anterior ring. The paired shorter strips are, in fact, strips from the inner layer. Evidence for this conclusion comes from two sources. First, the imbrication of each pair of strips with the anterior longitudinal strip is posterior–inside–anterior (Fig. 4.16 asterisk), whereas if the two strips were genuinely part of a longitudinal costa they would imbricate posterior–outside–anterior. Second, the strips that comprise the pairs are exocytosed and stored first in the strip development cycle, which is indicative of their being components of a separate (inner) layer (see Chapter 7; see also Fig. 3 in Thomsen and Boonruang, 1983a and Fig. 49 in Thomsen *et al.*, 1991).

There are many examples of longitudinal costae projecting forwards as spines beyond the anterior limit of transverse or helical costae. In some examples the spines are unspecialised strips, for example *Acanthocorbis apoda* (Fig. 4.48), but often they are long and attenuated (Figs 4.70–4.77) or they have pointed or forked tips (Figs 4.8, 4.9). *Acanthocorbis unguiculata* is a good example of a lorica in which the longitudinal costae project forwards as spines (Fig. 4.23). Superficially the lorica in this species appears to be based on an outer layer of longitudinal and an inner posterior layer of helical costae. However, when this pattern is compared with that of *Saepicula pulchra* (Figs 4.3, 4.5–4.7) it is clear that the two species are similar except for the absence of an anterior ring in *A. unguiculata*. This close relationship is further supported by the fact that the tips of the spines in both species have identical nib-shaped pointed tips (Leadbeater *et al.*, 2008a). In some species, a short projection of the longitudinal costae above the anterior ring appears to be optional and Thomsen and Boonruang (1984) concluded that projection of the longitudinal costae beyond the anterior ring by a distance of approximately one-eighth to one-half the length of a longitudinal strip should not be used as a specific character (Figs 4.63, 4.68). Long attentuated front and rear spines are characteristic of *Bicosta* and *Calliacantha* species, all of which have loricae with relatively few costal strips (Figs 4.70–4.77). Species of both genera are pelagic and the classic functional role of spines, as also found on the frustules of

planktonic diatoms, is to reduce sinking rate and thus to enhance suspension in the water column (see Section 9.6).

Acanthoeca species present another variation since in both *A. spectabilis* and *A. brevipoda* the chamber of the lorica comprises a closely wound helix of costae and only the anterior crown of spines resembles a recognisable array of longitudinal costae (Fig. 4.18). The crown is two-layered in that the spines are outside and can be considered as longitudinal costae with a ring of supporting strips inside. Since individual costae comprising the closely wound helix are continuous with the spines, these could be interpreted as a helical arrangement of longitudinal costae. However, since there is only one layer in this region and the costae are not longitudinal, the resolution of this structure is not immediately apparent. The significance of this helical arrangement and its importance in understanding the assembly mechanism are detailed in Section 4.3.4 and Chapter 6 (Leadbeater *et al.*, 2008b).

4.3.4 Helical costae

A helical costa is a linear array of strips that describes a helical path around the longitudinal axis of the lorica. Where a lorica comprises two costal layers, the helical costae are on the inside. *Acanthoeca spectabilis* and *A. brevipoda* are exceptional in that they present the most striking examples of helices so far observed among choanoflagellate loricae (Figs 4.18, 6.30). However, while the provenance of these helices is uncertain – the lorica chamber of *Acanthoeca* contains only a single layer of costae – nevertheless the detail they portray is fundamental to understanding all costal helices (see Chapter 6). The most important point to note is that helical costae are consistently left-handed (Leadbeater and Morton, 1974b; Leadbeater, 1979a; Leadbeater *et al.*, 2008b). This is clearly demonstrated because the costae rise from lower right to upper left when seen from the side. A line portraying the path of one costa has been superimposed on the actual SEM image shown in Fig. 4.18 and also the computer-generated image in Fig. 4.19.

A mathematical analysis was carried out on the helical costae of the specimen shown in Fig. 4.18 and allowance was made for slight flattening of the specimen (Leadbeater *et al.*, 2008b). The superimposed line delineates two turns of a non-flattened helix and, in addition, there is at least one-half of a further turn occurring within the stalk, making a total of 2.5 turns. On some long-stalked specimens the number of turns ranges from three to four. The number of costae contributing to one turn of the compound helix

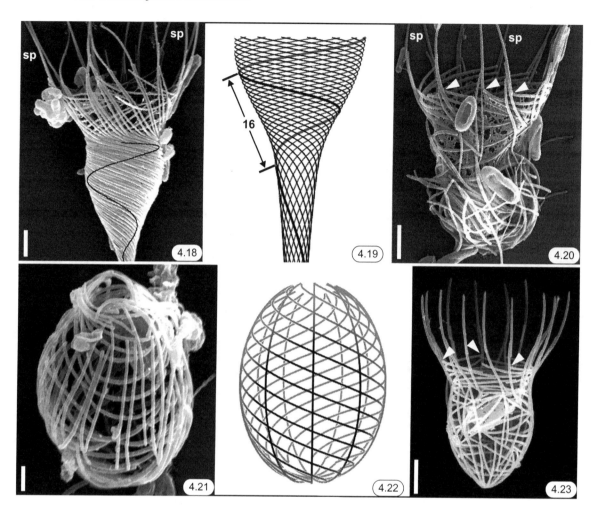

Plate 4 (Figures 4.18–4.23)

Figs 4.18–4.23 SEM and computer images of helical costae. **Fig. 4.18** *Acanthoeca spectabilis.* Chamber comprising a tightly wound system of helical costae and anterior ring of spines (sp). The superimposed black line derived mathematically illustrates the course of one continuous costa and depicts two turns of 360°. Each of the 17 spines is continuous with one helical costa and thus is continuous to the base of the chamber. The spines are supported internally by an arrangement of diagonal strips. Bar = 1 μm. Reproduced from Leadbeater *et al.* (2008b).

Fig. 4.19 *Acanthoeca spectabilis.* Computer-generated image approximating to the costal pattern illustrated in Fig. 4.18. There are 16 costae contributing to each turn of the helix.

Fig. 4.20 *Helgoeca nana.* Empty lorica showing general pattern of costae in spite of major dislocation. The anterior spines (sp) are continuations of the longitudinal costae. A helical costa abuts on to the base of each spine (arrowheads). The abutment is directed to the left and the costal strip is positioned to the right. Bar = 1 μm. Reproduced from Leadbeater *et al.* (2008a).

Figs 4.21–4.22 *Micropora* form of *Savillea parva.* **Fig. 4.21** Empty lorica viewed from the anterior end. The two-layered pattern of costae comprises an outer layer of longitudinal and inner layer of helical costae. Bar = 2 μm. Reproduced from Leadbeater (2008a).

Fig. 4.22 Computer image of lorica based on Fig. 4.21. In this reconstruction there is a one-to-one relationship between each of the ten longitudinal and helical costae and the latter undergo one turn.

Fig. 4.23 *Acanthocorbis unguiculata.* Lorica with same basic pattern of costae as *Saepicula pulchra* (see Figs 4.3–4.5) but lacking the anterior ring. The anterior end of each helical costa terminates in an abutment with the base of a spine (arrowheads). The abutment faces left and the costal strip is on the right when viewed from outside. Bar = 2 μm. Figure kindly provided by John van den Hoff.

ranges from 11 to 16 and this corresponds to the number of anterior spines (see Table 6.3). Careful inspection of the costae at the junction between the top of the helix, which is almost horizontal, and the spines, which are almost vertical, shows that they are continuous (Figs 4.18, 6.33). An inner layer of costal strips only appears within the crown of spines and these serve to stabilise the upright positioning of the spines. A more detailed analysis of the costal pattern in *Acanthoeca spectabilis* is presented in Chapter 6.

Contemporaneous with Leadbeater and Morton's (1974b) observation of helical costae in *Acanthoeca spectabilis* came another finding of left-handed helical costae, but this time as the inner lorica layer in the *micropora* form of *Savillea parva* (Fig. 4.21) (Leadbeater, 1975). The exact description used at the time was that "the inner or 'transverse' costae follow a spiral path rising from lower right to upper left when viewed in shadowcast whole mounts" (Leadbeater, 1975, p. 116). In hindsight, it is surprising that a connection was not made between the two instances of a left-handed helix in *Acanthoeca spectabilis* and the *micropora* form of *Savillea parva*. Instead it had to wait over 30 years until these two species were re-investigated in more detail with the benefit of high-quality SEM images and preliminary computer-generated models (Figs 4.19, 4.22) (Leadbeater *et al.*, 2009). The re-working of the *micropora* form of *Savillea parva* showed that the inner layer of helical costae underwent 1.5 turns and that there were two turns in the *parva* form (Fig. 6.10a–c). A search immediately followed for other examples of helical costae and this led to their discovery as the inner layer in the posterior part of the loricae of *Helgoeca nana* (Fig. 4.20; Table 4.2), *Acanthocorbis unguiculata* (Fig. 4.23), *Saepicula pulchra* (Figs 4.3, 4.4) and *Stephanoeca* species (Fig. 4.34). Whether helical costae would have been recognised in these species without the benefit of first witnessing the helices of *Acanthoeca* and *Savillea* is debatable. In these two latter genera the helical costae are reasonably stable and are not severely distorted during preparation. Where they occur only in the posterior part of the lorica and the turns are widely spaced, distortion readily renders the pattern confused (Figs 4.3, 4.5, 4.34). Nevertheless, there are two useful markers of helical costae. First, they are inclined and do not appear horizontal; second, the anterior end of each helical costa usually terminates in a point attachment (abutment) with the respective longitudinal costa (Figs 4.20, 4.23 arrowheads). Determining the number of helical costae present within a lorica can be problematical. In some species, for example *Saepicula pulchra*, there is a 1:1 ratio

Table 4.2 *Genera of loricate choanoflagellates listed according to whether non-longitudinal costae within the lorica comprise: (1) helical costae only; (2) helical costae and transverse rings; (3) transverse rings only.*

Helical costae only	Helical costae and transverse rings	Transverse rings only
Acanthoeca	*Saepicula*	*Conion*
Helgoeca	*Stephanoeca*	*Cosmoeca*
Savillea		*Didymoeca*
		Kakoeca
		Platypleura
		Polyfibula
		Stephanacantha
		Syndetophyllum

between helical and longitudinal costae (Figs 4.3, 4.4). In *Acanthoeca spectabilis* there is a one-to-one continuity between the number of spines and helical costae that is well illustrated in partially unwound loricae (Fig. 6.33). However, other ratios between helical and longitudinal costae are found, for instance in *Savillea parva* the ratio is 1:3 and in *Helgoeca nana* the ratio may be 2:1 (see Chapter 6).

4.3.5 Transverse costae (rings)

Transverse costae are entire horizontal rings of strips. They interact precisely with longitudinal costae with which they have clearly definable junctions (Figs 4.6, 4.13, 4.54). A transverse ring can be distinguished from a single turn of a helical costa in that it does not have an obvious leading and trailing end (see Section 7.7). In some instances this distinction may be difficult to demonstrate, particularly if loricae are flattened and damaged. However, there are some species, for instance *Saepicula pulchra* (Figs 4.3, 4.5), *Stephanoeca cupula sensu* Leadbeater (1972c) (Figs 4.24, 4.45) and species of *Crinolina*, *Cosmoeca* and *Parvicorbicula* (Figs 4.55–4.68), in which transverse costae are well preserved – a discussion of costal rings will begin with *Stephanoeca cupula*.

`STEPHANOECA CUPULA´ – JUNCTIONS BETWEEN TRANSVERSE AND LONGITUDINAL COSTAE
Stephanoeca cupula sensu Leadbeater (1972c) is a common inshore species that has a small cup-shaped lorica comprising an outer layer of eight longitudinal costae, an inner layer of helical costae in the posterior chamber and two transverse

costae in the anterior chamber (Fig. 4.45). Fig. 4.24 is a selected portion of a flattened anterior chamber showing the anterior ring and the intermediate transverse costa separated from each other by the distance of one longitudinal costal strip. Shadowcasting reveals that the left-hand, slightly lower pair of junctions belongs to the proximal side of the lorica (that is nearer the observer) and is therefore seen from the outside (P in Fig. 4.24). The right-hand, slightly higher pair of junctions belongs to the distal side (that is, further away from the observer) and is therefore seen from the inner surface (D in Fig. 4.24). The pattern of overlaps is illustrated diagrammatically beside the image. The interacting strips of the transverse costae are labelled *a* and *b* – where strip *a* is distinguished by the fact that its tip terminates with an abutment against a longitudinal costa (asterisks in Fig. 4.24). The anterior strips of the longitudinal costae are labelled *c* and the posterior strips *d* (Fig. 4.24).

With respect to the anterior costal ring, costal strip *b* overlaps strip *a* and is to the left of strip *a* when seen from the outside (proximal side P), and to the right when seen from the inside (distal side D) (Fig. 4.24 upper two diagrams). In *Stephanoeca cupula*, because the anterior ring is on the outer surface of the longitudinal costae, strips *a* and *b* are outside strip *c* when seen from the outer lorica surface (proximal side P) and vice versa. The organisation of transverse strips at junctions between the middle transverse and longitudinal costae is a mirror image with respect to the pattern for the anterior ring but otherwise the direction of the overlaps is the same (Fig. 4.24). The transverse costal strip *a*, which abuts the longitudinal costal strip *c*, overlaps strip *b*, which is to the left when seen in outer surface view (proximal side P), or to the right when seen in inner surface view (distal side D) (Fig. 4.24 lower two diagrams) and is on the inner surface of the longitudinal costae.

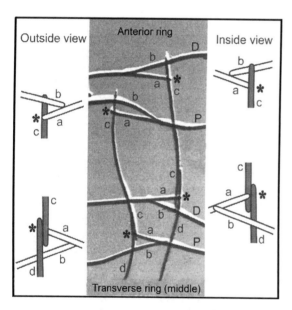

Fig. 4.24 *Stephanoeca cupula sensu* Leadbeater (1972c). In the centre panel is part of a shadowed preparation of a flattened lorica showing the interaction of the anterior ring and intermediate transverse costa with respective longitudinal costae. For both transverse costae, the upper junctions represent the inner surface of the distal side (D) and the lower junctions represent the outer surface of the proximal side (P). Diagrams of junctions illustrate the arrangement of strips seen from the outside and inside surfaces, respectively. Transverse costae that abut onto the longitudinal costae are labelled *a* and the position of the abutment is indicated by an asterisk. The overlapping strip is labelled *b*. Strips of the longitudinal costae are labelled *c* and *d*, respectively. For further illustration see Fig. 4.25.

GENERALISED ORGANISATION OF JUNCTIONS BETWEEN LONGITUDINAL AND TRANSVERSE COSTAE

Fig. 4.25 extends the analysis of costal strip junctions between transverse and longitudinal costae by illustrating the different conformations that occur when the anterior ring is on the outer or inner surface of the longitudinal costae, respectively (Fig. 4.25 A and B; Table 4.3; see Section 7.3.6). The imbrication of the overlaps is similar irrespective of whether the anterior ring is on the outer or inner surface of the longitudinal costae. For each junction illustrated, two versions are included: Fig. 4.25 A1, B1 and C1 represent the junctions as they would appear in pristine form without the displacement caused during preparation. Fig. 4.25 A2, B2 and C2 portray images that are commonly seen in flattened specimens (Figs 4.13, 4.24). Dislocation involves articulation and movement of the component strips with respect to each other; in most instances the two transverse strips are displaced to produce triangular gaps within the junction. The positions at which transverse strips *a* and *b* adhere to the longitudinal strips *c* and *d* are displaced and the greater the stress experienced the larger is the dislocation. Evidence in support of this interpretation comes from the fact that when transverse costae are held firmly in position, as for instance in the anterior transverse ring of *Diaphanoeca grandis*, the strips remain closely appressed to each other (Fig. 4.30). However, when the ring is subjected to stress by twisting as, for instance, when a stalked specimen of *D. grandis* is dried onto a plastic-coated grid, the junctions are prised open in a regular manner to generate triangular gaps (Fig. 4.36).

1. Anterior ring

Anterior ring **outside** longitudinal costae   Anterior ring **inside** longitudinal costae

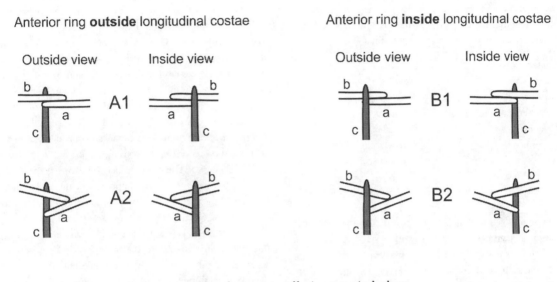

2. Intermediate costal ring

Transverse costa inside longitudinal costae

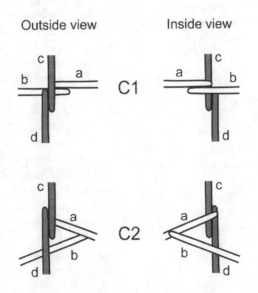

Fig. 4.25 Organization of junctions between transverse costal strips within (1) the anterior ring (A and B) and (2) the intermediate costal ring (C), where they respectively interact with longitudinal costae. For the anterior ring, the junctions are illustrated with respect to the ring being outside (A) and inside (B) the longitudinal costae. In each example the overlaps between transverse costal strips are shown in the 'pristine', closely appressed condition (A1, B1, C1, respectively) and the displaced condition (A2, B2, C2, respectively). In all examples, the transverse costal strip that abuts (terminates at) the longitudinal costa is labelled *a*, the overlapping strip *b* and the longitudinal strips *c* and *d* respectively.

Table 4.3 *Genera of loricate choanoflagellates listed according to whether the anterior ring is (1) inside or (2) outside longitudinal costae of the lorica.*

Anterior ring inside longitudinal costae	Anterior ring outside longitudinal costae
Conion	*Calotheca*
Cosmoeca	*Didymoeca*
Nannoeca	*Parvicorbicula*
Platypleura	*Pleurasiga*
Polyfibula	*Saepicula*
	Stephanacantha
	Stephanoeca
	Syndetophyllum

The ability of costal strip junctions to articulate in this manner indicates that the mechanism by which costal strips adhere to each other is not so rigid that they spring apart or break.

The pattern of costal strip overlaps documented here accounts for the majority, but not all, of loricae with internal lower transverse costae comprising narrow costal strips (Figs 4.13, 4.24). The diagrams illustrated in Figs 4.24 and 4.25 are not able to describe all the complex three-dimensional interactions that undoubtedly exist and, as illustrated in *Diaphanoeca grandis*, these may be considerable (Figs 4.29–4.33). There are variations to the standard pattern, for instance in *Didymoeca costata* and *Stephanoeca elegans* the tips of the 14–16 longitudinal costae adhere with regular spacing to the inside of the anterior ring comprising 5–6 costae (Fig. 4.15). In species with T-junctions comprising the anterior ring, including the 'T-junction' form of *Saepicula pulchra* (Figs 4.5, 4.7), imbrication of the strips comprising the ring is difficult to determine because there is a lack of rigidity and costae may tilt on drying. Other species with T-junctions and broadened strips, such as *Stephanacantha campaniformis* and *Syndetophyllum pulchellum* which have transverse costal strips with upright points on the left-hand side, are imbricated right over left when viewed from the outside (Fig. 4.14).

`DIAPHANOECA GRANDIS´ – A COMPOUND TRANSVERSE COSTA SECURED IN FORKED LONGITUDINAL JUNCTIONS

Diaphanoeca grandis is one of a relatively small group of species, also including *Crinolina isefiordensis* and *C. aperta* (Figs 4.56, 4.57), where the cell is suspended within a voluminous lorica chamber (see Section 4.6.2). Flagellar activity creates a flow of water that enters the lorica through its base and is expelled though the anterior end.

The lorica of *Diaphanoeca grandis* is barrel-shaped and the cell is suspended from the base of the neck by the collar microvilli (Fig. 4.26). Construction of the lorica is relatively straightforward, in that it contains 12 longitudinal costae and four transverse costae. For convenience, and by convention (see Section 4.2.2), the transverse costae are numbered 4–7, respectively, starting at the rear end (Figs 4.26–4.28). The anterior transverse costa (7) is located at the base of the neck and the three basal transverse costae, anterior (6), intermediate (5) and posterior (4), respectively, are located in the basal third of the lorica. The junctions between strips of the longitudinal and transverse costae comprising the three basal rings conform to the standard pattern illustrated for the intermediate transverse costa of *Stephanoeca cupula* (Fig. 4.24).

The junctions between the anterior transverse ring and the longitudinal costae are shown in Figs 4.29–4.33 with accompanying drawings. They are complex but, nevertheless, they conform to the standard pattern. In Fig. 4.30 the upper junctions represent the proximal side (P) seen from the outer surface and the lower junctions represent the distal side (D) seen from the inner surface. Within the anterior transverse costa, there is the same number of transverse strips as there are longitudinal costae, but because each transverse strip spans the distance between alternate longitudinal costae, each strip overlaps half the length respectively of its two neighbouring strips. The pattern of overlaps is similar to the now familiar arrangement illustrated for *Stephanoeca cupula* (Fig. 4.24). On the proximal side, outer surface view, the transverse strip *a*, which abuts the longitudinal costa *c*, approaches from the right and overlaps strip *b* by half its length on the left. An additional feature with respect to this junction is that the transverse ring is secured to the inner surface of the lower end of each upper longitudinal costal strip by an upwardly directed fork created by the inwardly angled upper end of the longitudinal strip below the region of overlap (arrows in Figs 4.32, 4.33). This extra support for the anterior transverse ring is necessary in order to secure the suspended cell and to resist the posteriorly directed propulsive force created by flagellar undulation.

SPECIES WITH EXTERNAL TRANSVERSE COSTAE

The majority of acanthoecid species that possess a two-layered lorica conform to the familiar pattern of longitudinal costae on the outside and helical or transverse costae on the inside (Section 4.3.1). A variation on this scenario is

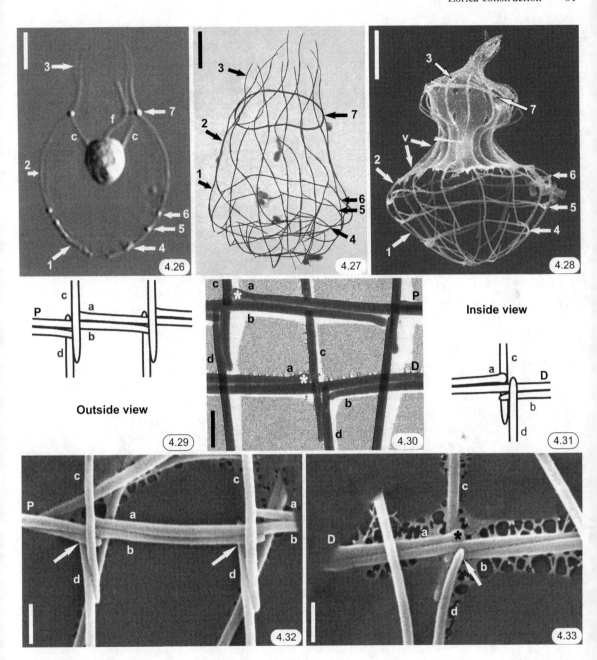

Outside view

Inside view

Plate 5 (Figures 4.26–4.33)

Figs 4.26–4.33 *Diaphanoeca grandis.* Transverse costal strip junctions with longitudinal costae. **Fig. 4.26** Interference contrast image of cell suspended within the lorica by collar microvilli (c) from anterior transverse costa. The overall form of the lorica is maintained by 12 longitudinal costae, labelled as follows: basal longitudinal strips (arrow 1), intermediate longitudinal strips (arrow 2), anterior longitudinal strips (arrow 3) and four transverse costae as follows: posterior basal ring (arrow 4), intermediate basal ring (arrow 5), anterior basal ring (arrow 6), anterior transverse ring (arrow 7). Flagellum (f). Bar = 5 μm.

Fig. 4.27 Whole mount of lorica illustrating the arrangement of 12 longitudinal costae and four transverse costae, labelled as for Fig. 4.26. Bar = 5 μm.

that some species have an outer anterior transverse ring (see Table 4.3). One group of species is characterised by all of their transverse costae being outside the longitudinal costae. At present these species are shared between *Parvicorbicula* and *Pleurasiga*. They include *Parvicorbicula socialis*, *Pa. quadricostata*, *Pa. circularis*, *Pleurasiga minima*, *Pl. reynoldsii*, *Pl. tricaudata* and *Pl. echinoscostata* (Figs 4.61, 4.64, 4.65, 4.67) (Manton *et al.*, 1976; Buck, 1981; Booth, 1990; Thomsen *et al.*, 1991, 1997) (see also Sections 7.6 and 8.4.2). The characters shared by this group of species include longitudinal costae comprising three costal strips (1–3) with the classic imbrication of strip 1 overlapping strip 2 which overlaps strip 3. The bottom of strip 2 projects inwards and connects with the organic covering of the cell. There are two transverse costae, both of which are outside the longitudinal costae, one comprising the anterior ring and the other located at or just below the join between costal strips 2 and 3 of the longitudinal costae (see Figs 9 and 11 in Buck, 1981). The anterior strip (3) of each longitudinal costa attaches by means of a flattened, slightly forked tip to the anterior ring midway along the respective transverse strip, thereby forming a T-junction. Strips comprising the posterior transverse costa in, for example, *Pleurasiga minima* (Fig. 4.61) and *Parvicorbicula quadricostata* (Fig. 4.67), form point attachments with respective strips of the longitudinal costae. *Pleurasiga echinocostata* is similar to *Pl. minima* but lacks the lower transverse costa (Fig. 4.64).

Two other species, *Parvicorbicula corynocostata* and *Pa. manubriata*, are similar to this grouping but the longitudinal costae around the cell contain only two strips (Thomsen *et al.*, 1997). The lower transverse costa of *Pa. corynocostata* is outside the longitudinal costae, whereas *Pa. manubriata* lacks a lower transverse costa. The current taxonomy of *Parvicorbicula* and *Pleurasiga* is confused and is in urgent need of revision.

4.4 EVIDENCE FOR A UNIVERSAL LEFT-HANDED ROTATION IN LORICA CONSTRUCTION

A detailed analysis of the junctions within and between costae provides valuable insights into both the construction (architecture) of loricae and the mode of costal assembly. Ultimately it is impossible to consider one without an understanding of the other. It is a well-established fact that lorica assembly is a single continuous event lasting a few minutes that involves costal strips sliding over each other to generate costae (Leadbeater, 1979c) (see Chapters 6 and 7). The question that must be addressed is what combination of movements can account for the variety of costal patterns and the range of junctions observed. The best way to answer this question is to consider each category of costae in turn, starting with helical costae.

To generate a left-handed helical costa from a vertical bundle of strips surrounding a cell requires a clockwise (as seen from above) rotational movement of the anterior end of the developing costa in relation to the posterior end which remains static (Fig. 6.21a–e arrowheads). Thus the two genera with obvious helices, *Savillea* and *Acanthoeca*, must be able to generate rotational movements ranging from 1.5 turns for the *micropora* form of *Savillea parva*, to 2 turns for the *parva* form and from 2.5 to 4 turns for *Acanthoeca spectabilis*. The evidence for this degree of turning and the possible mechanism by which it is achieved are discussed in Chapter 6. The rotational movement is usually accompanied by a forward movement of the anterior end of the developing lorica; for instance, in *Savillea parva* the cell is at the base of the lorica and much of the helix is above the cell body (Figs 6.2 inset, 6.8 inset, 6.11). However, this is not the situation in *Acanthoeca spectabilis*, where the helix is closely appressed to the cell surface and, although development of a posterior 'tail' does elongate the

---

Fig. 4.28 SEM image of critical-point dried lorica showing arrangement of transverse rings, labelled as in Fig. 4.26, and position of constricted veil (arrows v). Bar = 5 µm.

Figs 4.29–4.33 Complex junctions between transverse costal strips of anterior transverse costa and longitudinal costae. **Fig. 4.30** Flattened anterior transverse costa; upper junctions represent outer surface of proximal side (P), lower junction is inner surface of distal side (D). Transverse strip whose leading (left-hand) end is shown to abut onto a longitudinal costa is labelled *a* and the abutment is shown with asterisk. In the upper junction this abutting end is facing left while the remainder of the strip is directed right. In the lower junction of the distal surface the leading end abutment is directed to the right (asterisk). Bar = 0.5 µm. **Figs 4.29 and 4.31** illustrate diagrammatically outside and inside views, respectively, of the junctions shown in Fig. 4.30. **Figs 4.32 and 4.33** SEM images of junctions between strips of the anterior transverse costa and longitudinal costae. Note securement of the transverse costa within upwardly directed forks (arrows) that result from the overlap of longitudinal strips *c* and *d*, respectively, outside and inside the transverse strips. Labelling of strips and positioning of asterisks is in accordance with Figs 4.29–4.31. Bar = 0.5 µm. Figs 4.27, 4.28, 4.30 and 4.33 reproduced from Leadbeater and Cheng (2010).

helix, it is not projected beyond the front of the cell (Figs 6.45–6.50). For species such as *Saepicula pulchra*, *Helgoeca nana*, *Acanthocorbis unguiculata* and *Stephanoeca* spp. that have helical costae in the posterior part of their loricae in close proximity to the cell, a rotational movement is necessary to generate the helix and a forward movement to assemble the front part of the lorica.

The assembly of transverse costae also requires a rotational movement and the organisation of costal strip overlaps at junctions again suggests that the direction is clockwise as seen from above. A common feature of helical and transverse costae is that where abutments with longitudinal costae occur, the tip of an abutting strip often points towards the left of the lorica when viewed from the outside (Figs 4.13, 4.20, 4.23, 4.24). Thus the abutment is often at the leading end of the horizontal or diagonal strip in the direction of rotation (see Chapters 6 and 7). However, in species with compound transverse costae, such as *Diaphanoeca grandis*, and in species with external lower transverse costae, such as *Parvicorbicula quadricostata* and *Pleurasiga minima*, individual costal strips may abut against longitudinal costae at both ends (see Figs 4.29–4.33). In *Stephanoeca urnula* and *S. pyxidoides*, diagonal costal strips at the anterior end of the outer surface of the lorica terminate with an abutment against the respective longitudinal costae that is right-facing. However, while this might appear unusual, the inclination of costal strips (Fig. 4.51) or helical costae (Fig. 41 in Thomsen, 1979) within the remainder of the lorica indicate that a clockwise rotation has occurred (see Section 4.6.1).

Developing longitudinal costae require a vertical forward movement which is provided by extension of the lorica-assembling microvilli (see Chapters 6 and 7). However, if this movement is combined with the clockwise rotational movement necessary for the accompanying helical and transverse costae, the developing longitudinal costae would themselves describe a helical path. To overcome this possibility, the longitudinal costae must be able to rotate freely (carousel) around the long axis of the cell (see Sections 6.4.2, 7.2.2).

There is one way in which the existence of an inferred rotation can be tested and that is by reference to species that are normally freely suspended in a water column but which, on occasion, produce extra costal strips for a stalk that secures them to a surface. Under these circumstances the longitudinal costae are not free to carousel around the cell and instead they describe a helical path that provides a measure of the degree of turning. Two species, namely

Stephanoeca cauliculata and *Diaphanoeca spiralifurca*, illustrate this phenomenon well (Figs 4.35, 4.36). In *Stephanoeca cauliculata* the lorica comprises ten longitudinal costae, three transverse costae in the anterior chamber and a complement of helical costae in the posterior chamber. A long stalk consisting of longitudinal strips secures the cell to a surface. In empty loricae it is possible to trace backwards, for approximately one turn, the longitudinal costae in a left-handed helix around the posterior chamber (Fig. 4.35) (see also Fig. 11 in Nitsche *et al.*, 2011). In a similar manner the stalk of the lorica of *Diaphanoeca spiralifurca* shown in Fig. 4.36 has restricted the free rotation of the longitudinal costae and, instead, they have spiralled (hence the specific epithet) in a left-handed direction for about one turn. For this reason *D. spiralifurca* is, in fact, *D. grandis* with a restraining posterior stalk (compare Fig. 4.36 with Figs 4.27 and 4.37).

It is not possible with certainty to determine the extent of rotation in loricae with no trace of helical costae. One turn would appear to be the norm for those species with helical costae within the posterior chamber, such as *Saepicula pulchra* (Fig. 4.5) and *Diaphanoeca spiralifurca* (Fig. 4.36). This may well be the norm for all loricae, except for those that have considerably more, for example *Acanthoeca spectabilis* and *Savillea parva* (Figs 4.18, 4.21), or for those where there has been a reduction, for instance *Bicosta* species. In *B. spinifera* the two longitudinal costae undergo a left-handed cross-over of about half a turn (Fig. 4.71) (Hara and Takahashi, 1987a; Booth, 1990; McKenzie *et al.*, 1997) and in *B. antennigera* about the same but in a less obvious manner. *B. minor* is probably the only species that reveals no evidence of turning to the lorica (Fig. 4.72).

4.5 ORGANIC COMPONENTS OF LORICAE

Most acanthoecid species produce an organic investment that covers at least part of the internal lorica surface. Compared with the siliceous skeleton, which is so obvious in EM preparations, the organic component is often overlooked because it is either easily destroyed or rendered completely transparent, even in well-preserved shadowcast preparations. Thus our current knowledge of this aspect of lorica construction is limited, which is a significant shortcoming when searching for a functional role for the lorica (see Section 9.6).

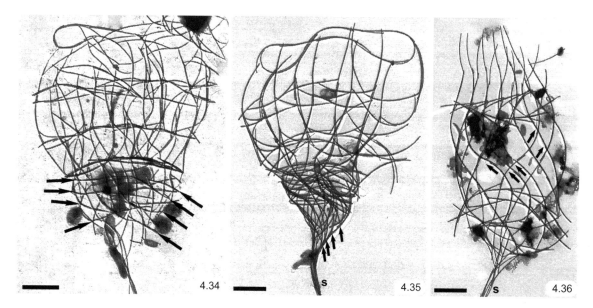

4.34   4.35   4.36

Plate 6 (Figures 4.34–4.36)

Figs 4.34–4.36 Evidence for left-handed (clockwise) rotation of costae. **Fig. 4.34** *Stephanoeca diplocostata*. Posterior lorica chamber comprising a slightly dislocated arrangement of longitudinal and helical (arrows) costae. Bar = 2 μm. Reproduced from Leadbeater (1979b). **Fig. 4.35** *Stephanoeca cauliculata*. Lorica attached to a substratum by long stalk (s) that has restricted the free rotational movement of longitudinal costae during assembly. The effect of clockwise rotation during lorica assembly is revealed by the helical deflection of the longitudinal costae in the posterior chamber (arrows). Bar = 2 μm. Reproduced from Leadbeater (1980). **Fig. 4.36** *Diaphanoeca spiralifurca*. Lorica is attached to substratum by stalk (s) and as with *Stephanoeca cauliculata* (Fig. 4.35) the longitudinal costae take on a left-handed helical appearance (arrows) because they were not able to rotate freely during assembly. Bar = 5 μm.

Buck *et al.* (1990) attempted a preliminary review of the organic investments of choanoflagellate loricae and, in summary, it is probably reasonable to distinguish between two categories of covering. The first, which is most easily visualised, is a covering closely associated with the cell. In many species this investment is attached to the anterior end of the cell just below the collar microvilli and lines the lorica in close proximity to the cell surface (Figs 4.44, 4.67, 4.75–4.77). It is present on species with completely enclosing loricae, such as *Stephanoeca*, *Savillea parva* and *Kakoeca antarctica*, as well as in species with minimalist costal coverings, such as *Bicosta* and *Calliacantha*. Sometimes this investment, which is not easily destroyed, bears a fine pattern of ridges on its outer surface (Fig. 4.44) similar to those on the thecae of the Craspedida (see Chapter 3). The function of this investment would appear to be attachment of the cell within the lorica. It obviously restricts access between the outside medium and the cell surface,

but does not interfere with the feeding functions of the outstretched collar microvilli on the living cell. Buck *et al.* (1990) referred to this type of investment as the 'suspensory membrane'. A similar investment surrounds the lower portion of the cell in *Diaphanoeca grandis* (asterisk in Fig. 4.37), although this is not in close proximity to the voluminous lorica.

The second category of organic investment is also associated with the inner surface of the lorica, but at a greater distance from the cell. In species such as *Acanthoeca spectabilis*, *Savillea parva* and *Stephanoeca cupula* this may be a forward continuation of the cell covering and takes the form of a precise rectangular meshwork of microfibrils (Figs 4.43, 6.6). In other species, such as *Cosmoeca* spp., the pattern is less regular but this could be due to damage during preparation (see *Cosmoeca norvegica* – Fig. 1 in Cavalier-Smith, 2006). Where the second category of organic investment covers the lorica at some considerable distance from the cell and is not an extension of the cell covering it is called the 'veil'. *Diaphanoeca grandis* is a particularly good example of a species with a veil lining the lorica. The extent of the veil is precise in that it extends from about one-third of the way down the anterior spines and terminates at the level of the anterior basal transverse costa (Figs 4.28, 4.37, 4.38). The veil in *D. grandis* comprises a layer of microfibrils embedded in a matrix which can be removed relatively easily by washing with detergent, 1.5% aqueous sodium dodecyl sulphate (SDS), or treatment with dilute acid. Its appearance can vary dramatically according to preparative treatment. For instance, in non-shadowcast whole mounts at low magnification only the siliceous costae are visible (Fig. 4.27). Even in shadowcast whole mounts of loricae prepared after washing in distilled water, the veil can appear transparent and is only visible because a series of small 'stretch' holes appear where it overlies the distal layer of costae (Figs 4.30, 4.33). However, in specimens critical-point dried for SEM, the veil is clearly visible and has shrunk, causing a constriction of the lorica above and below the anterior transverse ring (Fig. 4.28). Treatment of the lorica with crystal violet reveals the veil as if it were an old cloth (Fig. 4.38). Ruthenium red treatment gives a more delicate staining and also shows the cell covering to advantage (Fig. 4.37). In specimens prepared by washing with detergent (1.5% aqueous SDS), the matrix is removed and an irregular meshwork of fine microfibrils is evident (Fig. 4.39). Finally, negative staining reveals an impressive array of microfibrils (Fig. 4.40). The relatively enormous tubular lorica of *Diaphanoeca multiannulata* has a veil that extends from the anterior transverse costa for the depth of three transverse costae in the specimen illustrated by Thomsen *et al.* (1990) (see also Takahashi, 1981a). The cell is suspended within the lorica on a level with the second transverse costa from the anterior end. *Crinolina aperta*, another species in which the cell is suspended from the anterior transverse ring, also has a veil (Fig. 4.59).

The relatively small lorica of *Didymoeca costata* also has a veil, the extent of which is best seen in specimens critical-point dried for SEM (Fig. 4.41). The cobweb appearance is artefactual and probably represents a degraded version of the framework of microfibrils. The veil extends from the anterior transverse costa to just below the junction of the two longitudinal costae forming the anterior chamber (Fig. 4.41). This leaves a ring of spaces between the top of the posterior chamber and the mid-region of the lorica through which water can pass. In shadowcast preparations, the meshwork of microfibrils is just visible (Fig. 4.42).

It is tempting to think that all species should have a veil on part of the lorica, since this could provide a universal reason for the elaborate siliceous framework of the lorica. However, some species, for instance *Stephanoeca diplocostata*, have been investigated in great detail in both whole mounts and section and in this species there is no organic covering in the spaces between the anterior costal ring and the third costal ring down from the front end. It could well be that the variation in coverage of the lorica by an organic investment modifies the positions of inflow and outflow of water from the lorica and in this way each species may be adapted to a finely defined ecological niche (see Section 9.6).

4.6 VARIATIONS ON THEMES

Four groups of species have been selected to illustrate a variety of costal patterns based on four different themes. Overall these patterns demonstrate a mixture of small- and large-scale variations that must conform to the 'rules' of construction and assembly but must also be functionally and ecologically viable. Each costal pattern illustrated is distributed on a global scale. The order in which the four themes have been arranged reflects a morphological spectrum ranging from small loricae with many strips to larger loricae with fewer strips (see Table 4.4) and an ecological spectrum ranging from inshore, predominantly benthic (*Stephanoeca* and *Diaphanoeca*) to offshore, predominantly pelagic (*Crinolina, Cosmoeca, Parvicorbicula, Bicosta*) (see Table 9.8). Within the themes not all the species are attributable to the same genus; this highlights the fact that by subtle variations of costal pattern and lorica morphology a species of one genus may come to resemble those of another. Where appropriate, computer-generated three-dimensional images have been added to illustrate the patterns diagrammatically.

4.6.1 *Stephanoeca* theme (Figs 4.45–4.54)

Stephanoeca species are characterised by relatively small loricae comprising two chambers. The cell is located at the base of the lorica within the posterior chamber and the collar and flagellum are completely surrounded by the anterior chamber, which is ovoid in shape. The waist, which marks the boundary between the two chambers, is the position at which the cell is secured to the organic investment that lines the posterior chamber and extends forward to the first transverse costa of the anterior chamber. Kent's (1878c, 1880–2) original illustrations of *Salpingoeca ampulla*, as Ellis (1929)

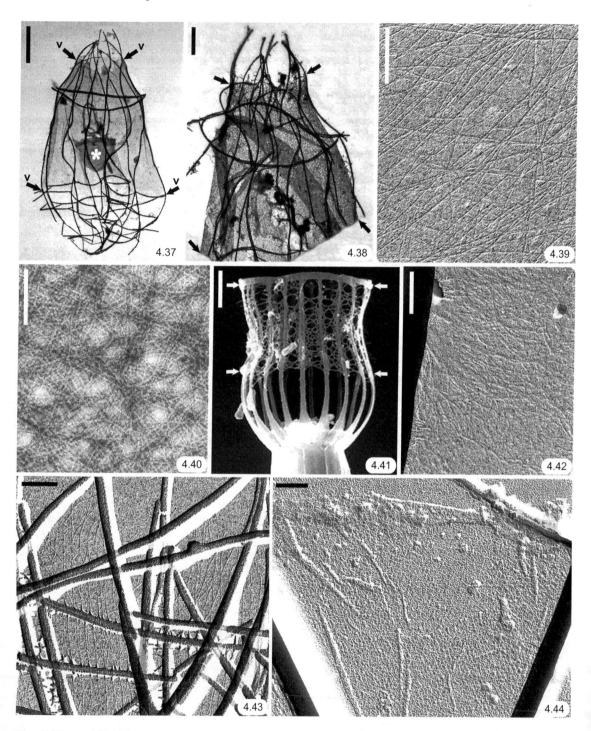

Plate 7 (Figures 4.37–4.44)

Figs 4.37–4.44 Organic component of loricae. Figs 4.37–4.40 *Diaphanoeca grandis*. Fig. 4.37 Lorica treated with ruthenium red stain reveals a delicately stained veil (arrows v) and the organic cup that surrounded cell prior to lysis (asterisk). Bar = 5 μm.

correctly surmised, have all the characteristics of a *Stephanoeca* species (Fig. 4.1a–c).

The *Stephanoeca* lorica is a classic two-layered structure based on an outer layer of regularly spaced longitudinal costae and an inner layer of costae that are helical in the posterior chamber and comprise a series of transverse rings in the anterior chamber. The three-dimensional computer image shown in Fig. 4.46 depicts an idealised pattern of *Stephanoeca* costae. *S. cupula sensu* Thomsen (1988) illustrates this costal pattern to greatest advantage (Fig. 4.45) (Thomsen, 1988). In the specimen shown in Fig. 4.45 there are eight longitudinal costae, which in the anterior chamber contain two strips. The four regularly spaced transverse costae are clearly visible. The anterior transverse costa comprises the anterior ring, which is outside the longitudinal costae; one ring is positioned at the waist and between these are the two intermediate rings. Details of the costal strip junctions between longitudinal and transverse costae have already been discussed in detail in Section 4.3.5 and Fig. 4.24. The helical nature of costae in the posterior chamber is difficult to discern because shrinkage and movement of the cell have dragged the organic investment away from its rightful place and disturbed the arrangement of costae. This is a common artefact exhibited by most *Stephanoeca* loricae. The posterior chamber of *S. norrisii* probably comes closest to revealing a helical pattern (Fig. 4.53).

Variations on the *Stephanoeca* theme include an increase in the number of longitudinal costae, each of which has an additional strip in the anterior chamber, making a total of three. In *S. diplocostata* the number of longitudinal costae varies between 15 and 24 (see Table 4.4). Within a clonal culture some loricae have strips projecting backwards in the form of a pedicel (Fig. 4.47). While this form was originally given the name *S. pedicellata* (Leadbeater, 1972c), this is not tenable since cells originating from the same clone cannot be given different taxonomic status. *Acanthocorbis apoda* has much in common with *Stephanoeca diplocostata* except that it lacks an anterior ring (Fig. 4.48). There are

fewer helical costae and only one transverse costa in the mid-region of the anterior chamber. This may be single or comprise several costae packed together. This costa marks the anterior margin of the organic investment extending from the posterior chamber.

Figs 4.49 and 4.50 are two specimens of *Stephanoeca pyxidoides*. There are between 10 and 12 longitudinal costae, each containing five strips. Within the posterior chamber the inner layer comprises an arrangement of helical costae that are considerably distorted. In the anterior chamber there are three or four transverse costae, one comprising the anterior ring, one at the waist and one or two intermediate costae (Figs 4.49, 4.50). The organic investment extends forward from the posterior chamber to the upper of the intermediate transverse costae and there are traces of organic material at the anterior end. Comparison of the anterior rings of the images illustrated in Figs 4.49 and 4.50 shows how confusing the effects of drying can be on the organisation of costae and costal strips. The anterior ring, which as in other *Stephanoeca* species is outside the longitudinal costae, contains the same number of costal strips as there are longitudinal costae, but each strip, instead of spanning the gap between neighbouring longitudinal costae, spans the distance between three or four longitudinal costae (Fig. 4.54). Thus the anterior ring at any location is three or four costal strips thick and each constituent strip abuts a longitudinal costa at either end and is overlapped for part of its length by neighbouring strips on either side. However, while this basic order can be identified, the pattern has been greatly distorted during preparation. The top of the lorica in Fig. 4.50, and that in *Stephanoeca urnula* (Fig. 4.51), has constricted probably due to shrinkage of the inner organic lining, and as a result the strips comprising the anterior ring have been prised apart and become partially dislodged from their respective longitudinal costae (Fig. 4.54b). In this distorted appearance the obvious abutments with longitudinal costae are confined to the right-hand end of each diagonal costal

Fig. 4.38 Lorica 'crudely' stained with cresyl violet showing veil (arrows). Bar = 10 μm.

Fig. 4.39 Shadowcast veil after treatment with 1.5% aq SDS detergent. Fine mesh of long, thin carbohydrate microfibrils evident. Bar = 0.25 μm.

Fig. 4.40 Negatively stained veil after treatment with detergent. Bar = 0.25 μm.

Fig. 4.41 *Didymoeca costata*. SEM of lorica showing location of veil (arrows). The open cobweb appearance of veil and constriction of the anterior chamber are artefactual. Bar = 2 μm.

Fig. 4.42 *Didymoeca costata*. Shadowcast whole mount of veil revealing microfibrillar substructure. Bar = 0.25 μm.

Fig. 4.43 *Acanthoeca spectabilis*. Regular meshwork of organic microfibrils at base of spines. Bar = 0.25 μm.

Fig. 4.44 *Bicosta spinifera*. Organic covering with diagonal array of parallel ridges. Bar = 0.25 μm. Reproduced from Manton *et al.* (1980).

Table 4.4 *Morphological details of the species illustrated in the four acanthoecid (loricate) 'themes' (Figs 4.45–4.77, see Section 4.6).*

	Cell width × length (µm)	Flagellar length (µm)	Flagellar length/cell radius ratio (L/A ratio)	Lorica length (µm)	Longitudinal costae	Transverse costae	Helical costae	Figure number	Reference
Stephanoeca theme									
Stephanoeca cupula	3 × 4	5	1.5	9–11	8–11	4	+	4.45	Thomsen (1988)
Stephanoeca diplocostata	6 × 7	8	2.5	15–20	15–24	4	+	4.47	Leadbeater (1994a)
Stephanoeca urnula	5 × 6	8	3	9–11	16	8	+	4.51	Thomsen (1973)
Stephanoeca pyxidoides	6 × 6	3–5	1.5	9–11	10–12	3	+	4.50	Leadbeater (1980)
Stephanoeca norrisii	4 × 5	9	3	13–18	c.60	>10	+	4.53	Tong (1997a)
Diaphanoeca theme									
Diaphanoeca grandis	5 × 5	13	4	37–39	12	4	–	4.55	Manton et al. (1981)
Crinolina isefiordensis	5 × 8	24	6	25–30	15–16	2	–	4.56	Thomsen (1976)
Crinolina aperta	5 × 8	40	10	50	12	2	–	4.57	Manton et al. (1975)
Cosmoeca/ Parvicorbicula theme									
Cosmoeca ventricosa	5 × 7	20	7	23–31	9–12	3	–	4.62	Thomsen and Boonruang (1984)
Cosmoeca norvegica	5 × 6	15–25	8	15–17	10	2–3	–	4.60	Thomsen and Boonruang (1984)
Cosmoeca phuketensis	4 × 2	9–11	5	7–9	9	2	–	4.63	Thomsen and Boonruang (1984)
Cosmoeca takahashii	7 × 4	25	6	16–22	12	3		4.68	Thomsen et al. (1990)
Pleurasiga minima	5 × 6	25	10	11	7	2		4.61	Thomsen et al. (1991)
Pleurasiga echinocostata	4 × 4	25	6	4.5	7	1		4.64	Thomsen et al. (1991)
Parvicorbicula socialis	5 × 7	14	4	13.8	10	2	–	4.65	Throndsen (1970b)
Parvicorbicula quadricostata	5 × 8	12	3	18	4	1	–	4.67	Throndsen (1970b)

Bicosta/
Calliacantha
theme

Bicosta spinifera	4 × 8	9	18	100	2	–	4.70	Manton et al. (1980)
Bicosta minor	3 × 5	8	12	14	2	–	4.72	Manton et al. (1980)
Calliacantha longicaudata	3 × 5	8	20	17	5	2	4.69	Manton et al. (1975)
Calliacantha simplex	5 × 8	10	25	20	4	2	4.75	Manton and Oates (1979a)
Calliacantha natans	6 × 9	9	27	19–34	4	2	4.77	Manton and Leadbeater (1978)
Calliacantha multispina	4 × 6	10	20	19–34	5	2	4.76	Manton and Oates (1979b)
Acanthocorbis haurakiana	3 × 4	2	8	20–25	6	2–4	4.74	Thomsen et al. (1991)

Most of the listed figures have been obtained from electron micrographs of whole mounts of cells. Since the majority of cell measurements were obtained from dried specimens they will considerably underestimate the dimensions of living cells. In all instances the upper range has been recorded but these may still be underestimated by up to 50%. Flagellar length and the lorica dimensions are not significantly affected by drying. The values for the ratio of flagellar length/cell radius will necessarily be approximations. The figure numbers of illustrations of individual taxa are listed, as are the references for the source of measurements.

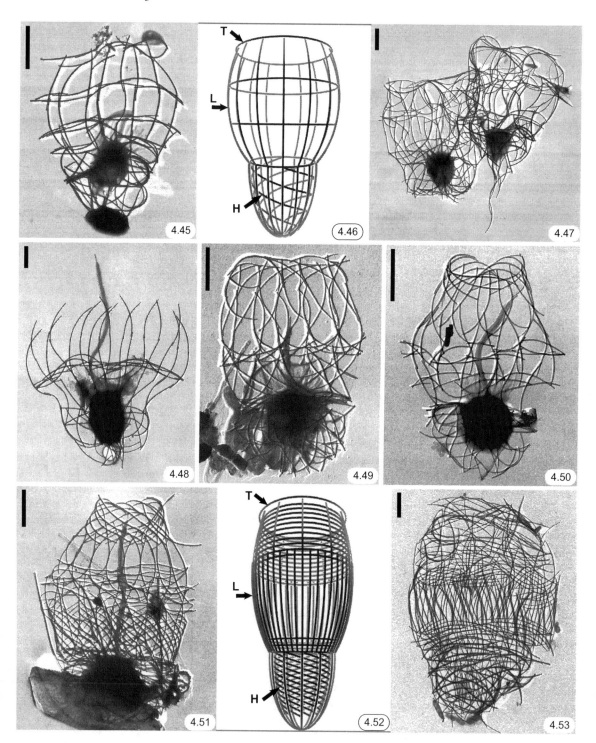

Plate 8 (Figures 4.45–4.53)

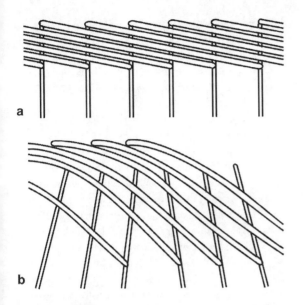

a

b

Fig. 4.54 *Stephanoeca urnula.* Diagram of outer surface of the lorica showing how the displacement of costal strips comprising the transverse ring can produce a misleading appearance. a. There are the same number of transverse costal strips as there are longitudinal costae but each strip spans the distance between at least four longitudinal costae. The result is that the ring comprises four or more closely appressed strips in depth. b. If, as a result of drying, the longitudinal costae converge at the anterior end, the transverse strips are displaced, giving a diagonal open appearance. Compare with Figs 4.50 and 4.51.

strip. The left-hand end of each strip has been dislodged from its respective longitudinal costa and has assumed a diagonal orientation (Thomsen, 1979; Leadbeater, 1980).

The lorica of *Stephanoeca norrisii* is exceptional for containing in the region of 400 costal strips (Fig. 4.53). This is a remarkably large number, particularly taking into account the relatively small size of the lorica (Table 4.4). The precise costal pattern is difficult to determine, although the general organisation is clear and it conforms to the *Stephanoeca* shape and overall plan. The outer layer of the lorica is based on regularly spaced longitudinal costae which extend from the base of the lorica to the anterior end. These appear to be supplemented by additional longitudinal strips in the mid-region of the anterior chamber, giving a total of 60 or more vertical strips. In the posterior chamber there is an inner layer of helical costae and at the base and top of the anterior chamber there are many rings of costae. The computer-generated image illustrated in Fig. 4.52 would probably be better described as an 'artist's impression' of the costal arrangement. However, it does indicate how such a pattern can be assembled according to the rules of lorica construction.

The size, shape and costal pattern of *Stephanoeca* loricae show a consistency that renders them easy to recognise. As presented here they show a series of increasing costal complexity within the confines of the general theme. There are many minor variations apart from the number of costae that have not been illustrated here. For instance, the number of costal strips comprising the anterior ring may be the same as the number of longitudinal costae, half the number or two-thirds. The number and positioning of the intermediate transverse rings in the anterior chamber may vary, as may the number of helical costae within the posterior chamber. The severe displacement of costae and costal strips that occurs in dried specimens renders this a difficult taxon to study for the principles of lorica

Figs 4.45–4.53 *Stephanoeca* theme. **Fig. 4.45** *Stephanoeca cupula sensu* Thomsen (1988). The basic *Stephanoeca* plan comprising longitudinal costae with four transverse rings in anterior chamber and helical costae in posterior chamber. Bar = 2 μm.

Fig. 4.46 Computer image of a *Stephanoeca* lorica showing positioning of transverse (arrow T), longitudinal (arrow L) and helical (arrow H) costae.

Fig. 4.47 *Stephanoeca diplocostata.* A round-bottomed (left) and a pedicellate (right) specimen. Lorica contains 14–18 longitudinal costae, four transverse costae in the anterior chamber and helical costae in the rear chamber. Bar = 2 μm.

Fig. 4.48 *Acanthocorbis apoda.* Lorica contains 16 longitudinal costae, a compound transverse ring around the middle of the anterior chamber and helical costae in posterior chamber. Bar = 2 μm.

Figs 4.49–4.50 *Stephanoeca pyxidoides.* Two loricae containing a similar arrangement of costae but the top of the lorica in Fig. 4.50 is constricted, probably due to shrinkage of inner organic investment, resulting in displacement of the strips of the anterior ring (see Fig. 4.54). Bar = 2 μm. Reproduced from Leadbeater (1980).

Fig. 4.51 *Stephanoeca urnula.* Lorica similar to that of *S. pyxidoides* (Fig. 4.50) but with more costae. Bar = 2 μm.

Figs 4.52–4.53 *Stephanoeca norrisii.* **Fig. 4.52** Computer image that approximates to the actual specimen in Fig. 4.53. Labelling as for Fig. 4.46. **Fig. 4.53** Lorica of *Stephanoeca* form but with greatly increased numbers of costae. Bar = 2 μm.

construction. This is certainly the case for helical costae in the posterior chamber, which almost certainly would have been overlooked if more convincing helices had not been seen in other genera. This presents a scenario whereby application of the rules gained from other taxa makes it possible to infer the presence of helical costae in the posterior chamber of *Stephanoeca*. With this in mind it is then possible to look for the indicators of such helical costae, such as the characteristic abutments with longitudinal costae.

Stephanoeca species, with small loricae and many costae, are mostly found in inshore benthic locations. Those with apertures between the costae of the anterior chamber, such as *S. cupula*, *S. diplocostata* and *S. pyxidoides*, are likely to draw in water laterally. However, *Stephanoeca norrisii* has too many costae to permit particle entry except through the anterior ring. In this respect, it resembles species of *Savillea* and *Kakoeca antarctica*.

4.6.2 *Diaphanoeca/Crinolina* theme (Figs 4.55–4.59)

Diaphanoeca grandis, with a large, voluminous lorica (Fig. 4.55), has already been discussed in relation to the compound layer of costal strips that comprise the anterior transverse ring (Section 4.3.5). Junctions between this ring and the longitudinal costae are additionally complicated by the presence of inwardly directed forks, each comprising a single prong deflected upwards (Figs 4.29–4.33). These forks provide a robustly secure lodgement for the anterior transverse ring, an essential requirement for a cell that is suspended by its collar microvillis from this costa. Flagellar undulation, which generates a base to anterior water flow through the lorica, also generates a propulsive motion of the cell in the opposite direction, which must be resisted. The deployment of an organic investment in the form of a veil, from the anterior basal transverse ring to two-thirds the distance up the lorica neck, is also an essential feature to ensure that water is effectively channelled through the lorica and not dissipated through the sides of the lorica (Section 4.5).

Three other species of this theme show a combination of all or most of the characteristic features of *Diaphanoeca grandis*. In particular, the two species of *Crinolina* – *C. isefiordensis* and *C. aperta* – possess loricae that are cylindrical with the cell suspended from the anterior costal ring (Figs 4.56, 4.57). In both, this ring shows some signs of reinforcement and is lodged within inwardly directed forks of the longitudinal costae in a manner similar to that in *Diaphanoeca grandis* (Figs 4.58, 4.59). In *Crinolina isefiordensis*, reinforcement of the anterior ring is achieved by relatively long overlaps between adjacent strips in the gaps

between longitudinal costae (Fig. 4.58). In *C. aperta* the junctions between strips of the transverse ring coincide with joins between the longitudinal costae. The end of the transverse costal strip that abuts onto the longitudinal costa is abruptly upturned (Fig. 4.59) (Buck, 1981). A veil can just be seen on the inner surface of *C. aperta* (Fig. 4.59) but is transparent in *C. isefiordensis*. One of Thomsen's (1976) original illustrations of this species shows what might be interpreted as fragments of a veil (Fig. 17 in Thomsen, 1976). *Diaphanoeca multiannulata* is the third species that has a cell suspended within a long, tubular lorica covered internally by an organic investment. Thomsen *et al.* (1990) summarised the costal details and dimensions of this relatively large species. The tubular lorica is 35–90 μm in length and 10–22 μm in diameter. There are 8–16 longitudinal costae and 3–6 transverse costae. The cell is usually suspended from the second transverse costa from the top and a prominent microfibrillar veil extends from the anterior transverse costa to the penultimate costa from the bottom (Takahashi, 1981a; Thomsen *et al.*, 1990). No reinforcement of a costal ring can be observed in published images (Takahashi, 1981a; Thomsen *et al.*, 1990).

The suspension of a cell in a voluminous lorica comprising narrow costae and lined with a veil permits a relatively small cell with flagellum of moderate length to process efficiently a relatively large volume of water. The large lorica comprising a lightweight framework also enhances form resistance, a valuable asset for species suspended in the water column since it reduces the sinking rate. *Crinolina isefiordensis*, *C. aperta* and *Diaphanoeca multiannulata* are commonly found in offshore pelagic communities (Leakey *et al.*, 2002) (Chapter 9). *Diaphanoeca grandis* is not a pelagic species but is commonly found in inshore waters, particularly in benthic collections (Table 9.8).

4.6.3 *Cosmoeca/Parvicorbicula–Pleurasiga* theme (Figs 4.60–4.68)

Cosmoeca and the *Parvicorbicula–Pleurasiga* species group (*sensu* Pa. *socialis* – see Section 4.3.5) are similar genera in that they are characterised by funnel- or barrel-shaped loricae comprising a relatively sparse framework of thin costal strips (Figs 4.60–4.68). For this reason they are common in pelagic environments and rarely found in benthic collections. Some species, such as *Cosmoeca norvegica*, *Pleurasiga minima* and *Parvicorbicula socialis*, have relatively long flagella and are probably capable of independent locomotion (Figs 4.60, 4.61, 4.65). However, the genera of this theme, *Cosmoeca* and *Parvicorbicula–Pleurasiga*, are easily

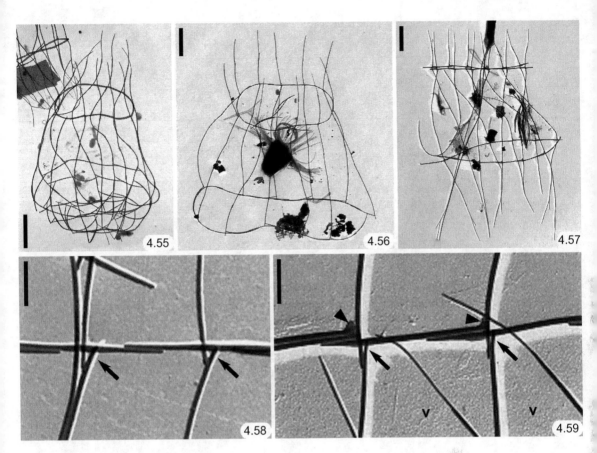

Plate 9 (Figures 4.55–4.59)

Figs 4.55–4.59 *Diaphanoeca/Crinolina* theme. **Fig. 4.55** *Diaphanoeca grandis.* Empty voluminous lorica containing 12 longitudinal and four transverse costae. Bar = 5 μm.

Fig. 4.56 *Crinolina isefiordensis.* Large, lightly silicified lorica comprising 11 longitudinal and two transverse costae. The cell is suspended from the anterior transverse costa; the rear of the lorica is completely open. Bar = 5 μm.

Fig. 4.57 *Crinolina aperta.* Empty lorica of similar appearance to *C. isefiordensis* except that the longitudinal costae are longer and project posteriorly. Bar = 5 μm. Reproduced from Manton *et al.* (1980).

Fig. 4.58 *Crinolina isefiordensis.* Inside view of anterior transverse costa, with long overlaps between strips, secured within the upwardly directed forked joints between strips of the longitudinal costae (arrows). Bar = 1 μm.

Fig. 4.59 *Crinolina aperta.* Inside view of anterior transverse costa, whose component strips have a 'buffer-like' upturned end (arrowheads), lodged within forked joints on longitudinal costae (arrows). Organic veil (v) is just visible. Bar = 1 μm.

distinguished from each other; all transverse costae of *Cosmoeca* are inside the longitudinal costae and the constituent strips span the gaps between neighbouring longitudinal costae (Fig. 4.60), whereas in the *Parvicorbicula–Pleurasiga* species group the anterior ring and lower transverse costa are outside the longitudinal costae and the component strips of the anterior ring form T-junctions with the longitudinal costae (Figs 4.61, 4.64, 4.65, 4.67). The similarity between these genera, emphasised here by illustrating them in Figs

4.60–4.68 according to pattern and form rather than taxonomically, would appear to be an example of convergent and parallel evolution, for there is no indication that a change in one or two characters could account for the overall differences that distinguish these two genera.

The basic lorica pattern of *Cosmoeca* consists of 9–12 longitudinal costae, each containing four linearly arranged costal strips (Fig. 4.66). There are two or three transverse costae. The anterior ring, which is inside the longitudinal

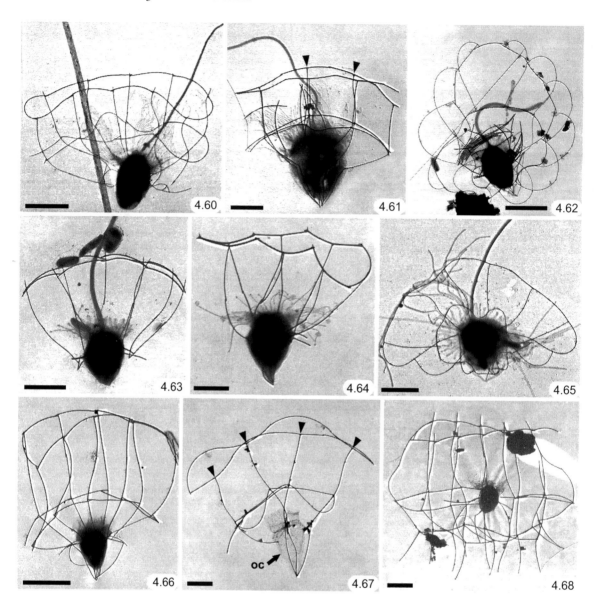

Plate 10 (Figures 4.60–4.68)

Figs 4.60–4.68 *Cosmoeca/Parvicorbicula* theme. **Fig. 4.60** *Cosmoeca norvegica*. Lorica with two transverse costae, the strips of both have conventional junctions (see Figs 4.24 and 4.25). Bar = 5 μm.

Fig. 4.61 *Pleurasiga minima*. Costal strips in anterior ring form 'T-joins' with respective longitudinal strips (arrowheads). Bar = 2.5 μm.

Fig. 4.62 *Cosmoeca ventricosa*. Barrel-shaped lorica with long, flexuous costae. Bar = 5 μm. Reproduced from Leadbeater (1973).

Fig. 4.63 *Cosmoeca phuketensis*. Small lorica without longitudinal strips around the bottom of the cell. Bar = 2.5 μm.

Fig. 4.64 *Pleurasiga echinocostata*. Lorica similar to that of *Pleurasiga minima* (Fig. 4.61) except that it lacks an intermediate transverse costa. Bar = 2.5 μm. Reproduced from Leadbeater (1973).

Fig. 4.65 *Parvicorbicula socialis*. Saucer-shaped lorica of similar construction to *Pleurasiga minima* (Fig. 4.61) except that it contains ten longitudinal costae (see also Fig. 9.9). Bar = 2.5 μm.

costae, and the intermediate ring both contain the same number of strips as there are longitudinal costae. The posterior ring contains 3–12 costal strips and is positioned at the top of the basal longitudinal strips. Junctions between strips of the transverse costae coincide with the longitudinal costae and are of the classic type illustrated in Figs 4.24 and 4.25. The combination of a straightforward arrangement of costae with strip junctions of the classic type makes *Cosmoeca* an ideal model species. However, as will be argued in Chapter 7, this 'apparent' straightforwardness is secondary and has only been achieved after a long and complex evolutionary history.

Cosmoeca species are initially distinguished according to the location of the intermediate costal ring. In *C. norvegica*, this ring is positioned near the tip of the second strip of the longitudinal costae, whereas in *C. ventricosa* it is located at the base of the second longitudinal strip from the top (Figs 4.60 and 4.66, respectively). Overall lorica shape in *C. ventricosa* can vary dramatically according to the length and thickness of the costal strips, particularly those comprising the longitudinal costae (compare Figs 4.62 and 4.66). The organic investment in the larger form of *C. ventricosa* (Fig. 4.62), and probably in other *Cosmoeca* species as well, extends to the intermediate transverse costa. In comparison with *C. norvegica* and *C. ventricosa*, *C. phuketensis* has a highly reduced lorica in which there are only two strips in each longitudinal costa, the basal strip and another are missing, and there is no intermediate ring (Fig. 4.63). *C. takahashii* is an unusual variation on the *Cosmoeca* theme in that it has a ring of posterior spines (Fig 4.68). However, the anterior end conforms in costal arrangement and in strip junctions with the standard pattern exhibited by other *Cosmoeca* species (Thomsen et al., 1990).

The basic lorica pattern of the *Parvicorbicula–Pleurasiga* species group has been discussed above (see Section 4.3.5). *Pleurasiga minima* possesses seven longitudinal costae and they extend throughout the length of the lorica. *Pleurasiga echinocostata* is similar except that there is no lower transverse costa (Fig. 4.64). *Parvicorbicula quadricostata* contains only four longitudinal costae although the anterior ring contains eight costae, four of which are unsupported by a longitudinal costa (Fig. 4.67). *Pa. socialis* has an open saucer-shaped lorica containing ten longitudinal costae (Fig. 4.65). This species, as its name implies, often occurs as large colonies of cells which are joined together by the outer lorica rim (Figs 9.4–9.9). *Parvicorbicula manubriata* is similar in morphology and colony formation to *Pa. socialis* (Thomsen et al., 1997).

The *Cosmoeca/Parvicorbicula* theme illustrates the convergence of two evolutionary patterns with the generation of several parallel forms. *Cosmoeca norvegica* closely resembles *Pleurasiga minima* (Figs 4.60, 4.61) and *Cosmoeca phuketensis* resembles *Pleurasiga echinocostata* (Figs 4.63, 4.64) almost to the point of confusion. Morphological similarity is matched by the frequency with which these species are found in association with each other in pelagic communities.

4.6.4 *Bicosta/Calliacantha* theme (Figs 4.69–4.77)

The *Bicosta/Calliacantha* theme is characterised by 'minimalist' loricae with long attenuated spines (Figs 4.70–4.77). The species involved are typically pelagic in the open ocean. Size is extremely variable both within and between species – for instance, the lorica of *Bicosta spinifera* can reach 100 μm or more in length (Fig. 9.19a) (see also Section 9.5). Interpretation of lorica construction focuses on how such minimalist costal patterns can be derived from the basic two-layered plan.

The loricae of *Calliacantha* spp. (Figs 4.75–4.77) are recognisably two-layered, with a pattern of longitudinal and transverse costae. *C. simplex* has the most straightforward arrangement, with four longitudinal costae that converge posteriorly and join a posterior spine (Fig. 4.75). Each longitudinal costa comprises three strips, two of which are positioned on the cell surface and the anterior strip projects forwards as a spine. Two transverse costae, each containing four strips that span the gaps between adjacent longitudinal costae, are located on the intermediate longitudinal strip (Fig. 4.75). An organic investment is present from the base of the lorica chamber to the lower

---

Fig. 4.66 *Cosmoeca ventricosa*. Lorica similar to *Cosmoeca norvegica* (Fig. 4.60) except that the intermediate transverse costa is lower on the anterior chamber. Part of this costa is shown in Fig. 4.13. Bar = 5 μm.

Fig. 4.67 *Parvicorbicula quadricostata*. Lorica with four longitudinal costae but eight strips in the anterior ring, four of which form T-joins with the longitudinal costae (arrowheads). Organic covering that encloses the cell (arrow oc). Bar = 2.5 μm.

Fig. 4.68 *Cosmoeca takahashii*. Lorica similar to *C. norvegica* (Fig. 4.60) but with ring of posterior spines. Bar = 2.5 μm. Reproduced from Thomsen et al. (1990).

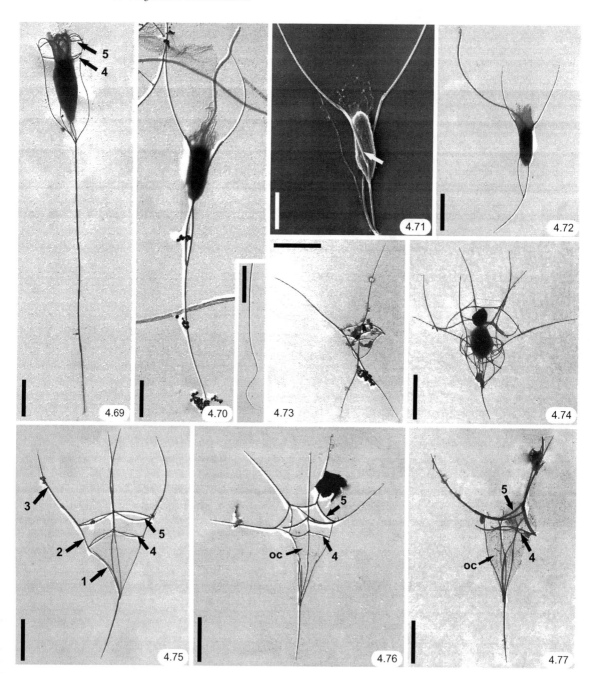

Plate 11 (Figures 4.69–4.77)

Figs 4.69–4.77 *Bicosta/Calliacantha* theme. **Fig. 4.69** *Calliacantha longicaudata*. Lorica chamber of four longitudinal costae, two transverse costae (arrows 4 and 5) and a long pedicel comprising a linear array of strips. Bar = 5 μm.
Fig. 4.70 *Bicosta spinifera*. Long, narrow lorica comprising two anterior spines and one posterior spine. The end of the posterior spine is often undulated (Fig. 4.70 inset). Bars = 5 μm.

of the two transverse costae and probably slightly beyond. The cell, which is not illustrated here, projects beyond the anterior transverse ring and bears a relatively long flagellum (Table 4.4).

The lorica of *Calliacantha multispina* is of similar construction to *C. simplex*, but has more anterior spines, usually between five and seven (Fig. 4.76). There is often a reduction in the number of longitudinal costae within the chamber – for instance, in Fig. 4.76 there are five spines but only three longitudinal costae. There are two transverse costae, the anterior of which contains the same number of strips as there are spines, and they are of similar thickness. The posterior transverse costa comprises a reduced number of thinner strips. The lorica terminates in a prominent posterior spine (Fig. 4.76).

Calliacantha natans, while resembling other related species, nevertheless differs in that the spines are not continuous with the longitudinal costae (Fig. 4.77). Typically there are three anterior spines which emerge from an anterior transverse ring comprising six thickened strips. There are between four and six longitudinal costae surrounding the cell, each costa comprising two strips. A second transverse costa is located below the anterior transverse ring. Posteriorly, the lorica terminates in a long spine (Fig. 4.77).

Acanthocorbis haurakiana presents a more elaborate arrangement of costae but, in common with *Calliacantha*, has anterior and posterior spines (Fig. 4.74). In the specimen shown in Fig. 4.74 there are six spines that are continuous with longitudinal costae. The anterior transverse ring consists of six strips, each bridging the gap between adjacent spines. The junctions are well displayed and are of the classic type illustrated in Fig. 4.24. The lower transverse costa comprises one or more rings of strips.

The loricae of *Bicosta* species contain seven costal strips in the form of two longitudinal costae that converge posteriorly on a prominent spine (Figs 4.70–4.72). Each longitudinal costa contains a linear array of three strips, the anterior of which forms a long attenuated spine. The two lower strips, the anterior being shorter than the posterior, are located close to the cell surface (Thomsen and Larsen, 1992). In *Bicosta spinifera* the longitudinal costae cross over in the mid-region of the cell. When seen in a SEM image the crossover is from lower-right to upper-left, which accords with the clockwise rotation essential to helical costae (Fig. 4.71) (see also Hara and Takahashi, 1987a; Booth, 1990; McKenzie *et al.*, 1997; Marchant, 2005). In *B. spinifera* one anterior spine is characteristically shorter than the other (Fig. 4.70) and the posterior spine often has an undulation at its distal end (Fig. 4.70 inset). The loricae of all *Bicosta* species vary greatly in size, for example: *B. minor* 12–45 µm; *B. spinifera* 38–120 µm; and *B. antennigera* 35–80 µm (Manton *et al.*, 1980; Thomsen and Larsen, 1992; see also Section 9.5). When species of *Bicosta* and *Calliacantha* co-occur in plankton samples, there is often a consensus in the category of size range found (see Tables 9.12, 9.13). This has led to the suggestion that there may be an environmental factor, such as temperature, affecting lorica size (Manton *et al.*, 1980). However, the results obtained by Thomsen and Larsen (1992) from a transect crossing the Scotia front in the Southern Ocean, where there is a sudden temperature drop between the Scotia and Weddell Seas, did not support a correlation between lorica length and temperature (see Section 9.5 for further discussion). The lorica chamber of *Calliacantha longicaudata* resembles that of *Bicosta multispina*, but there are no spines and the posterior pedicel comprises a linear array of costal strips (Fig. 4.69). The chamber of *Crucispina cruciformis* also resembles that of a *Calliacantha* species, but there are two posterior projecting spines (Fig. 4.73).

Fig. 4.71 *Bicosta spinifera*. SEM image of cell showing crossover of costae in a left-handed direction (lower right to upper left) (arrow). Bar = 5 µm. Figure kindly provided by Franco Novarino.

Fig. 4.72 *Bicosta minor*. Small lorica comprising three spines and containing seven strips. Bar = 5 µm.

Fig. 4.73 *Crucispina cruciformis*. Empty lorica with diagonal cross of longitudinal costae projecting forwards and backwards as spines, and a small chamber with two transverse costae. Bar = 5 µm.

Fig. 4.74 *Acanthocorbis haurakiana*. Lorica with chamber and ring of six anterior spines. Bar = 5 µm. Reproduced from Leadbeater (1973).

Fig. 4.75 *Calliacantha simplex*. Simple lorica comprising four longitudinal costae, each containing three strips (arrows 1, 2 and 3), one posterior spine and two transverse (arrows 4 and 5) costae. Bar = 5 µm.

Fig. 4.76 *Calliacantha multispina*. Lorica similar to *C. simplex* (Fig. 4.75), but there are five anterior spines. Two transverse costae – arrows 4 and 5. Organic covering (arrow oc). Bar = 5 µm.

Fig. 4.77 *Calliacantha natans*. Lorica comprising anterior transverse ring of six strips with three anterior spines and two transverse costae (arrows 4 and 5). Organic covering (arrow oc). Bar = 5 µm. Reproduced from Leadbeater (1972c).

The loricae of *Calliacantha* and *Bicosta* provide a logical sequence of costal reduction. *Calliacantha simplex* is central to understanding how this might have been achieved. The key features illustrated by *C. simplex* include: (1) a lorica comprising an outer layer of longitudinal costae and an inner layer of rings; (2) longitudinal costae comprising three costal strips, the anterior of which is a spine; (3) two transverse costae associated with the intermediate strip of the longitudinal costa; (4) an organic investment lining the chamber from base to lower costal ring. From this basic plan, the loricae of *C. multispina* and *C. natans* can be derived by additions or deletions to the numbers of longitudinal costae and the numbers of strips comprising the transverse costae. The relationship between *C. simplex* and *Bicosta* now appears obvious and is achieved by a reduction in the number of longitudinal costae and loss of transverse rings. The half-turn rotation in *Bicosta spinifera* is a residual clockwise rotation and provides the cell with a secure lodgement within the lorica. Only in *Bicosta minor* is there no evidence of a rotation (Fig 4.72).

4.7 DISCUSSION

It is obvious from a casual glance at EM images of choanoflagellate loricae that they are based on a universal substructure, namely costae comprising linear arrays of costal strips. What is less apparent is what universal principles might account for the variety of costal patterns observed. The physical properties of loricae have helped and hindered in this context. The good news is that loricae do not disintegrate or shatter when prepared for EM; the bad news is that during preparation many subtle costal strip displacements occur that can easily mislead the unwary observer. The physical properties of the organic component, in particular the difficulty with which it can be visualised with TEM, have further exacerbated the problem of assigning a functional role to the lorica. Overcoming these limitations has required the culturing of favourable species, the painstaking analysis of costal patterns and several strokes of good fortune.

The purpose of the lorica is to enclose a space around the cell and this is achieved by a two-layered arrangement of costae. Only in species of *Bicosta*, which has been discussed in Section 4.6.4, and *Monocosta fennica*, in which the lorica comprises a single ring (Thomsen, 1979), is one or other costal layer completely absent and this must be a secondary loss. Where two layers are present, the outer layer in the majority of species consists of longitudinal costae. However, in some species the anterior ring is outside the longitudinal costae and in the *Parvicorbicula–Pleurasiga* grouping all transverse costae are external (see Section 4.3.5).

For those species with an outer layer of longitudinal costae the inner lorica layer may contain helical or transverse costae or a mixture of the two. Helical costae resemble longitudinal costae in that they are continuous linear arrays of strips. In *Savillea parva* helical costae extend throughout the length of the lorica. In *Helgoeca nana* and *Acanthocorbis unguiculata* they extend from the bottom of the lorica to the base of spines. In loricae where there are also transverse rings, such as *Stephanoeca* species and *Saepicula pulchra*, the helical costae are confined to the lower part of the lorica below the transverse rings. *Acanthoeca* provides an exceptional variation on the helical theme for three reasons. First, it is the only known example of such a prominent tightly wound helix; second, there are at least two turns and maybe four to the helix; and third, it comprises a single layer on the surface of the cell. The helical costae of *Acanthoeca* are continuous with the anterior spines that are supported by an inner layer of costae. On this basis, the helical costae of *Acanthoeca* could be considered as modified longitudinal costae. However, the important point to note is that helical costae are always located adjacent to the cell surface, and only if a 'genuine' inner layer is absent can the outer layer of costae be helical. The reason for this is to be found in the assembly mechanism, which is discussed in Chapter 6.

Transverse costae differ from helical costae in that they comprise an entire ring and are not part of a drawn-out continuous curve. At their most obvious and straightforward, as illustrated by the anterior ring of *Saepicula pulchra* (Fig. 4.3) and the transverse rings of *Cosmoeca* species (Figs 4.60, 4.62, 4.63, 4.66, 4.68), the strips of a ring span the gaps between neighbouring longitudinal costae and form the classic junctions illustrated in Figs 4.24 and 4.25. However, where there is substantial overlapping between strips in a ring, as for instance in *Acanthocorbis apoda*, the distinction between a flattened spiral and a ring is difficult to discern (Fig. 4.48). Rings have logistical and mechanical advantages over helical costae. For example, rings make relatively economical use of costal strips by not being continuous threads, they can be of substantial diameter, as in *Diaphanoeca grandis* and *Crinolina* spp., and are mechanically robust. Conversely, helical costae are usually of smaller diameter and do not have the same mechanical robustness. It is no accident that helical costae are confined to smaller loricae, whereas transverse rings are present to the exclusion of helical costae in larger loricae. This distinction also has important ecological implications, for most species with helical costae are predominantly benthic and inshore, whereas larger loricae with thinner strips and transverse rings are predominantly pelagic and offshore. This gradation in size and ecology almost certainly reflects a trend in evolutionary development (see Chapters 6–9).

4.8 'RULES' OF LORICA CONSTRUCTION AND ASSEMBLY

On the basis of the information presented in this chapter it is now possible to compile a list of rules which define lorica construction. While the rules are universally applicable, they may be modified by an assortment of elaborations and variations. The list of rules is of fundamental importance to understanding the mechanism of lorica assembly discussed in Chapters 6 and 7.

4.8.1 List of rules

(1) The choanoflagellate lorica comprises a two-layered arrangement of costae with a thin organic investment covering part or all of the inner surface. In the majority of species the costae of the outer layer are aligned parallel to the vertical axis of the cell (longitudinal costae); the inner layer comprises either helical and/or transverse costae. The latter are arranged diagonally and perpendicular to the vertical axis of the cell, respectively. In a minority of species (characterised by the *Parvicorbicula–Pleurasiga* species grouping) the transverse costae form the outer layer and the longitudinal costae the inner layer.

(2) Costal strips comprising longitudinal and helical costae display a regular and consistent imbrication (overlapping). The direction of overlap between adjacent costal strips is posterior–outside–anterior. For a posteriorly projecting pedicel the direction of overlap is anterior–outside–posterior.

(3) Helical costae are left-handed in rotation. When the helix is viewed from below, the direction of rotation is anti-clockwise and away from the observer. When the helix is viewed from above, the direction of rotation is clockwise and towards the observer. When a lorica is observed in side view with the anterior end facing upwards, the helix rises from lower right to upper left with respect to the observer.

(4) Transverse costae are aligned perpendicular to the longitudinal costae (vertical axis of the cell) and each comprises an entire horizontal ring. The component costal strips display a regular overlapping imbrication.

(5) In species with internal transverse costae, where a component costal strip of a transverse or helical costa, comprising a single layer of strips, abuts onto a costal strip of a respective longitudinal costa, the abutting costal strip is usually located towards the right when seen from outside the lorica. The forward end of the abutment points in a clockwise direction when seen from above. Where the transverse costa is compound, comprising two or more layers of strips, and in the lower transverse costa of many species with external transverse costae, both ends of each component strip may abut onto respective longitudinal costae.

5 • Loricate choanoflagellates: requirement for silicon and its deposition in costal strips

Life in a glass cage.

5.1 INTRODUCTION

To enclose a collared flagellate cell within a surrounding chamber without interfering with its flagellar activity or feeding apparatus requires the mediation of a rigid skeleton and this is the mechanical function performed by the siliceous costal framework of the lorica. The combination of biogenic silica and an organic component within individual costal strips provides the support necessary to maintain the structural integrity of a lorica that is constantly subjected to a sub-ambient pressure resulting from flagellar activity.

Biogenic silica ($SiO_2 \cdot nH_2O$), sometimes known as biogenic opal, is relatively pure SiO_2 with a low water content. It is an amorphous material at the molecular level, usually without crystalline structure. Amorphous silica has been used extensively by lower animals, plants and microorganisms, usually for the provision of skeletal support (Leadbeater and Riding, 1986). In aquatic systems, silicon utilisation by protists, in particular diatoms, is so substantial that it has a major impact on the global geochemical partitioning and cycling of this element. The ubiquity of diatoms, often in high cell densities, and the ease with which some species can be cultured has resulted in an extensive literature on many aspects of silicon biogeochemistry, including uptake and dissolution kinetics, molecular and genetic control of intracellular deposition and global cycling (for reviews, see: Round et al., 1990; Martin-Jézéquel et al., 2000; Mann, 2001; Müller, 2003; Raven and Waite, 2004; Davis and Hildebrand, 2008). There are fewer studies on silicon utilisation by other protists, including loricate choanoflagellates, and those available have generally been influenced by diatom studies which have served as a role model (for reviews, see: Simpson and Volcani, 1981; Leadbeater and Riding, 1986; Leadbeater and Barker, 1995; Marron et al., 2013). Since loricate choanoflagellates are predominantly, but not exclusively (see Section 9.4.8), distributed in marine and brackish water habitats, their physiology with respect to silicon turnover is intimately affected by the saline environment.

In seawater the major chemical form of dissolved silicon available to living organisms is undissociated monomeric silicic acid, $Si(OH)_4$ (reactive silicate), with a minor quantity of $SiO(OH)_3^-$ (Martin-Jézéquel and Lopez, 2003). Diatoms have an obligate silicon requirement for growth and this is usually satisfied by monosilicic acid, with the exception of *Phaeodactylum tricornutum*, which can also utilise $SiO(OH)_3^-$ (Del Amo and Brzezinski, 1999). Monosilicic acid uptake and intracellular silica (frustule) polymerisation are coupled in diatoms and occur during one continuous stage of the cell cycle, beginning just before nuclear division and ending just prior to daughter cell separation (Sullivan, 1977). In diatoms, the rate of growth has a direct effect on the degree of frustule silicification; the longer a cell spends in the frustule-synthesis phase the more silicon is incorporated provided the element is non-limiting (Claquin et al., 2002). Kinetic parameters with respect to silicon uptake and specific Si-dependent rate of cell division follow Michaelis–Menten and Monod saturation functions, respectively (Martin-Jézéquel and Lopez, 2003).

The most important difference between loricate choanoflagellates and diatoms is that their requirement for silicon is facultative (Leadbeater, 1985). Loricate choanoflagellates make use of undissociated monosilicic acid throughout the cell cycle, but silicon depletion in *Stephanoeca diplocostata* does not halt cell division or the rate of growth (Leadbeater, 1985). Cells produced in the absence of dissolved silicon are 'naked' and lack loricae. However, there appears to be a coupling between the restoration of silicon and cell division since the latter is halted or reduced until the silicon deficit has been replenished, including the production of loricae for naked cells (Leadbeater, 1989).

In the majority of silica-depositing protists, silicon is polymerised intracellularly within specific silica deposition vesicles (SDVs) bounded by a membrane known as the 'silicalemma' (Reimann et al., 1966). The ultrastructure, physiology and biochemistry of intracellular silica polymerisation have been studied most thoroughly in diatoms (Pickett-Heaps et al., 1990; Hildebrand and Wetherbee, 2003; Martin-Jézéquel and Lopez, 2003; Raven and Waite, 2004), although inroads have now been made into other protistan groups, including loricate choanoflagellates and silica scale-bearing chrysophytes (Mignot and Brugerolle, 1982; Brugerolle and Bricheux, 1984; Leadbeater and Davies, 1984; Leadbeater 1985, 1987, 1989; Sandgren and Barlow, 1989; Leadbeater and Barker, 1995; Gong et al., 2010; Marron et al., 2013). In diatoms, silicon uptake is a low-energy requiring process (Raven, 1983). The energy for transport and biomineralisation appears to be provided by aerobic respiration and oxidative phosphorylation (Sullivan, 1977; Sullivan and Volcani, 1981) rather than ongoing photosynthesis, although there is some coupling with the light cycle (Chisholm et al., 1978). Silicon uptake across the plasma membrane in diatoms is mediated by specific silicic acid transporter proteins (SITs) and is concentrated 1000-fold within the cell, creating an intracellular pool of soluble silicate or organo-silicon compounds (Kroger, 2007; Curnow et al., 2012). In a recent transcriptome analysis of two loricate choanoflagellates, Marron et al. (2013) identified SIT-type genes in Stephanoeca diplocostata and Diaphanoeca grandis with a significant sequence similarity to the SIT genes of diatoms and chrysophytes (stramenopiles). The punctate taxonomic distribution of SIT homologues in such disparate eukaryotic groupings suggests that the capacity to metabolise silicon originated within these groups by lateral gene transfer. In marine diatoms silicon transport is dependent on Na^+ co-transport; thus the uptake of silicon is coupled with a favourable transport gradient for sodium (Bhattacharya and Volcani, 1980; Raven and Waite, 2004; Cunrow et al., 2012). The mechanism by which high silicon concentrations are maintained in intracellular pools is not clear (Davis and Hildebrand, 2008).

Diatom frustule synthesis within SDVs is a dynamic process involving the integration of several activities operating at different size scales. Currently at least two size categories are recognised (Davis and Hildebrand, 2008). First, microscale activity, previously called macromorphogenesis (Hildebrand and Wetherbee, 2003), which includes expansion and moulding of SDVs by organelles and cytoskeletal elements external to the SDV. Second, nanoscale

activity within SDVs, which involves the polymerisation of biogenic silica mediated by polypeptides called silaffins that have an affinity for silicon, long-chain polyamines (LCPAs) and silacidins (Brunner et al., 2009). The study of diatom biomineralisation, or nanofabrication as it is sometimes called, is a fast-developing topic that is being pursued by many groups with advanced technology because of its potential nano-biotechnological applications (Kroger, 2007). However, recalcitrant gaps remain in current knowledge, particularly with respect to the nature of the intracellular silicon pool and the precise mechanism that controls the elaborate patterning of frustules (Brunner et al., 2009).

5.2 BIOGEOCHEMICAL CYCLING OF SILICON IN SEAWATER

Silicon is universally present in seawater (pH 7.5–8.4), where it occurs mostly (~97%) as undissociated monomeric silicic acid $Si(OH)_4$. The total content of dissolved silicic acid in the world's oceans is approximately 10^{17} mol Si, which gives an average concentration of 70 μM globally (Tréguer et al., 1995). However, there are marked regional differences; for instance, in the central oceanic gyres the concentrations are usually less than 2 μM, whereas in the Antarctic concentrations may be as high as 80–100 μM in surface waters during winter. Deep and bottom waters are usually silicon-rich in comparison to surface waters. The distribution of silicic acid in different water masses is governed by complex interactions between physical, chemical, geological and biological processes.

Marine organisms such as diatoms, silicoflagellates, radiolarians and loricate choanoflagellates, the majority of which occur in surface waters (euphotic zone), utilise monosilicic acid to fabricate their biogenic silica frustules and skeletons. Once these siliceous structures make contact with seawater they have the potential to dissolve (Kamatani and Riley, 1979; Kamatani, 1982; Leadbeater and Davies, 1984). In diatoms there is an organic coating on the frustule which inhibits dissolution, but once this is removed, a process which occurs naturally by microbial action on dead cells and empty frustules, the rate of dissolution increases (Lewin, 1961; Bidle and Azam, 2001; Roubeix et al., 2008). Approximately 60% of the biogenic silica produced by diatoms in the euphotic zone dissolves in the upper 100 m of seawater (Tréguer et al., 1995). The remainder settles through the water column until it reaches the sea floor, where dissolution continues. Tréguer et al. (1995)

estimate that globally about 3% of gross biogenic silica production is eventually preserved within the sediments. However, this average reflects contrasting situations, with virtually no preservation in subtropical areas and preservation as high as 24% in the Southern Ocean.

5.3 TERMINOLOGY

Terminology with respect to silicon-containing compounds at various stages in the biogeochemical cycle is confusing, with a multiplicity of terms often used interchangeably. This is because silicon occurs in a variety of chemical combinations that are rarely in a single pure state that can be specified. Thus terms such as 'silicon', 'silica' and 'silicate' have taken on general as well as specific meanings. It is therefore worth qualifying at the outset the meaning of relevant terms used in this text.

Silicon refers to the element Si (atomic mass 28.1), which is classed as a metaloid or semi-metal with properties intermediate between those of typical metals and non-metals. Silicon is rarely, if ever, found naturally in the elemental state and within the biogeochemical cycle it is always combined. In this text the term silicon is used to refer to the element and in a more general sense to refer to 'silicon-containing' compounds.

Reactive silicate, sometimes known as molybdate-reactive silicate, is primarily undissociated monomeric orthosilicic acid $Si(OH)_4$ measurable by the ammonium molybdate colorimetric assay (Strickland and Parsons, 1968). Reactive silicate is the principal form of dissolved silicon available to living organisms (Burton et al., 1970). Concentrations are given as μM (1 μM Si = 28 μg Si l^{-1}). In this text reactive silicate specifically refers to soluble silicon-containing compounds assayed by the molybdate-blue test.

Biogenic silica, sometimes abbreviated to silica, is amorphous hydrated silica with the general formula $SiO_2 \cdot nH_2O$. It is based on a covalently linked polymeric network of randomly arranged tetrahedrally coordinated siloxane Si–O–Si centres with variable levels of hydroxylation (Mann, 2001). Traces of Al, Ca, K and other minor elements may also be present. In general usage, biogenic silica refers to a complete siliceous skeletal structure, which in diatoms contains an internal organic component and an external covering containing proteins and carbohydrates. Biogenic silica is not a stoichiometric mineral, its density, solubility, viscosity and composition may vary considerably from one organism to another.

Bound silicon refers to particulate silicon-containing compounds, including biogenic silica, that can be solubilised with NaOH or KOH, neutralised with dilute sulphuric acid and assayed using the molybdate-blue test (Strickland and Parsons, 1968).

5.4 *STEPHANOECA DIPLOCOSTATA*: THE EXPERIMENTAL SPECIES

Most of the experimental and ultrastructural work relating to growth and silicon utilisation by loricate choanoflagellates has been carried out on *Stephanoeca diplocostata* (Leadbeater and Davies, 1984; Leadbeater 1985, 1987, 1989; Geider and Leadbeater, 1988; Eccleston-Parry and Leadbeater, 1994a, b, 1995) (see Section 9.3). The lorica of *S. diplocostata* is barrel-shaped, comprising two chambers separated by a waist which is coincident with the position at which the lorica is attached to the cell (Figs 5.1–5.3). The anterior chamber is approximately twice the length (8–12 μm) of the posterior chamber (5–7 μm) and about 1.5–2.0 times the width. The bottom two-thirds of the cell are located within the posterior chamber, while the anterior portion of the cell, including the flagellum and surrounding collar, project into the anterior chamber (Fig. 5.3). Some loricae possess a long posterior pedicel (17–20 μm long), although this may be absent on sedentary cells (Fig. 4.47). The precise arrangement of costae is difficult to establish from dried specimens. However, with the aid of computer-generated images, such as that shown in Fig. 4.46, it is possible to estimate the overall costal arrangement. The lorica is two-layered, with the outer layer containing 15–24 (mode 19) longitudinal costae, each comprising five costal strips; three in the anterior chamber and two in the posterior. There are four transverse costae in the anterior chamber that are listed from the posterior forwards as follows: posterior transverse costa (waist) (category 5, see Fig. 4.2a); posterior intermediate transverse costa (category 6); anterior intermediate transverse costa (category 7); anterior ring (category 8) (Figs 5.1–5.3). All transverse costae except the anterior ring are within the longitudinal costae. The anterior ring is on the outside (see Section 4.6.1, Table 4.3). Table 5.1 lists the numbers of costal strips comprising the various costae. The inner layer of the posterior chamber contains between 4 and 13 helical costae (Fig. 5.2), although this category is the most difficult to count. The average total number of costal strips within a lorica is 172 ± SD 18.1 (mode 163) (Tables 5.1 and 5.6).

Table 5.1 Stephanoeca diplocostata. Numbers of longitudinal costae in anterior and posterior chambers of loricae. Numbers of costal strips comprising: (1) longitudinal costae, (2) helical costae of posterior chamber, (3) four transverse costae of the anterior chamber, (4) pedicel (if present); (n = sample number). See Fig. 4.2a and Section 4.2.2 for numbering of helical and transverse costae.

	Number of longitudinal costae in anterior chamber	Number of longitudinal costae in posterior chamber	Number of costal strips							
			Longitudinal costae	Helical costae (4)	Posterior transverse (waist) costa (5)	Posterior intermediate transverse costa (6)	Anterior intermediate transverse costa (7)	Anterior ring (8)	Pedicel	Total
Range	15–24	8–21	5	3–5	6–9	6–17	9–30	11–23	4–19	133–193
Mode	19	17	5	4	8	13	20	15	6	163
n	28	27	25	22	27	28	28	28	26	

While the lorica is consistent in shape and general costal pattern, nevertheless there are many variations, particularly with respect to the numbers of costae and the degree of silicification of individual strips (for example, compare Figs 5.2 and 5.3). *S. diplocostata* is a tectiform species and accumulates costal strips at the top of the collar throughout interphase (Fig. 5.3) (see Chapter 7). By the end of interphase a full complement of strips will have been accumulated and these are bequeathed to the daughter cell (juvenile) that leaves the parent lorica following cell division. Within minutes of separation, the juvenile assembles a lorica using the donated strips. Thus only at the beginning of interphase does a *Stephanoeca* cell have a biogenic silica complement equivalent to a single lorica. For the remainder of interphase each cell will have an additional quantity of biogenic silica equivalent to the number of strips undergoing production plus those stored at the top of the collar.

5.5 ULTRASTRUCTURE OF COSTAL STRIP PRODUCTION IN *STEPHANOECA DIPLOCOSTATA*

All loricate choanoflagellates produce costal strips intracellularly during interphase and, following exocytosis, store them on the surface of the cell. In nudiform species, such as *Acanthoeca spectabilis*, *Savillea parva* and *Helgoeca nana*, strips are produced immediately after cell division by the 'naked' juvenile daughter cell (see Chapter 6). Silica production continues until a complete set of strips has been accumulated on the surface of the juvenile, after which lorica assembly takes place (Chapter 6). Once the lorica has been assembled no further strips are produced until the

end of interphase, when the cell divides and the cycle is repeated in the juvenile of the next generation.

In tectiform species, cells that already possess a lorica produce new costal strips throughout interphase (see Chapter 7). Exocytosed strips are stored at the top of the inner surface of the collar (Fig. 5.3) and are subsequently bequeathed to the juvenile towards the end of cell division. Lorica assembly by the juvenile occurs within minutes of separation from the parent lorica. Percentages of cells at different identifiable stages of costal strip accumulation within the cell cycle can be obtained from samples taken during exponential growth (Table 5.2). Costal strip production would appear to extend for the majority of interphase and the onset of cell division coincides with the completion of strip production. In tectiform taxa, costal strip formation and accumulation do not interfere with the feeding activity of the flagellated cell, which can proceed continuously throughout interphase.

Costal strips are produced intracellularly within SDVs that are located parallel to the long axis of the cell within the peripheral cytoplasm (Fig. 5.5). Developing strips usually occur in groups at regular intervals around the cell. The origin of SDVs has not been ascertained in spite of thorough investigation. Although they may appear in close proximity to Golgi and endoplasmic reticulum (ER) cisternae, they have not been observed in continuity with either of these membrane systems. The bounding membrane of SDVs, known in diatoms as the 'silicalemma' (Reimann *et al.*, 1966), is considerably thicker than membranes attributable to either the Golgi apparatus or ER but is of similar thickness and appearance to the plasma membrane (Fig. 5.17). The earliest recognisable stages of SDV development comprise a long, thin cylindrical vesicle

Table 5.2 Stephanoeca diplocostata. *Mean percentages of cells at different stages in the cell cycle observed in asynchronous exponentially growing cultures at 20 °C. Mean values have been obtained from counts taken from three separate semi-continuous cultures.*

Non-dividing cells			Dividing cells		
Cells with no visible accumulations of costal strips	Cells with accumulations comprising:		Dumbbell-shaped cell	Sibling pairs	Recently liberated juveniles
	Intermediate number of strips	Many strips			
33	41	20	2	3	1

Reproduced from Leadbeater (1985)

subtended throughout its length by two, or occasionally three, parallel microtubules (Figs 5.6–5.10). Within the SDV is a light-staining thread, heterogeneous in appearance, which may be x-shaped in cross-section (Figs 5.6, 5.7, 5.9). The full length of an SDV and its enclosed strip is established early in development (Figs 5.8–5.10). Subsequent silica polymerisation results in a gradual increase in strip diameter and rapid darkening. As the strip darkens the surface stains densely and always two or more dense-staining central threads are visible (Fig. 5.29).

The microtubules subtending each SDV during the early stages of costal strip production are on the concave side, which usually faces towards the centre of the cell (Figs 5.6–5.8). Their function is to support and shape the SDV until the strips have acquired sufficient rigidity to support themselves. Once this has been achieved the microtubules depolymerise. Costal strip rigidity is closely related to cross-sectional diameter and the stage at which SDV-associated microtubules depolymerise can be defined more-or-less by strip diameter (Table 5.3). Microtubules are present on SDVs at strip diameters of less than 50 nm but are absent at strip diameters of 60 nm or more. In favourable cross-sections of SDVs containing immature strips, linking arms between the two microtubules and between the microtubules and the outer surface of the SDV can be seen (Fig. 5.7). In glancing longitudinal section the linkages appear to have a regular periodicity (Fig. 5.10). The supporting function of the microtubules can be demonstrated by the use of microtubule poisons, such as colchicine, which cause SDV-associated microtubules to depolymerise, resulting in the production

Table 5.3 *Percentage associations of microtubules with SDVs containing nascent costal strips of various diameters (sample number = 70).*

Cross-sectional diameter of nascent costal strip (nm)	% of SDVs with associated microtubules
0–29	100
30–39	94
40–49	94
50–59	15
60–69	3
70–100	0

Reproduced from Leadbeater (1987)

of misshapen strips of irregular length (Figs 5.14, 5.15, 5.18) (see Section 5.6.3). Similar effects are observed when silicon-deprived cells containing immature SDVs are resupplied with silicon. The first-formed strips are usually deformed and contain irregular amounts of silica.

Fully formed costal strips are exocytosed onto the outer surface of the cell, where they are stored until a complete set is available to produce a lorica. In nudiform species, such as *Acanthoeca spectabilis*, this involves a lateral movement through the plasma membrane of the cell body. However, for tectiform species, costal strips are exocytosed through the anterior surface of the cell between the flagellar apparatus and the ring of collar microvilli. Strips are then moved to the top of the collar where they are rotated through 90° and stored in bundles perpendicular to the long axis of the cell (see Chapter 7).

The chemical composition of costal strips has been subjected to analysis on several occasions (Leadbeater and Manton, 1974; Mann and Williams, 1982; Gong et al., 2010). The early study of Leadbeater and Manton (1974) on loricae of *Stephanoeca diplocostata*, involving X-ray analytical electron microscopy (EMMA), X-ray diffraction and hydrofluoric acid treatment, demonstrated that the principal element within costal strips was silicon and that it was present in a crystallographically amorphous form. Mann and Williams (1982), using ultra-high resolution EM, confirmed this observation and further refined this result by showing that no short-range order could be detected above 1.5 nm. Since mature strips appear electron dense when seen in whole mounts and sections they may give the impression of being homogeneous in substructure (Fig. 5.29). However, this is misleading since costal strips are flexible and rarely fracture when loricae are flattened by drying. Such flexibility would suggest that the silicon-containing component is intimately associated with a permeating organic matrix. EM images of costal strips produced by cells grown in low silicon-containing medium (see Section 5.7.3) and loricae undergoing dissolution (see Section 5.7.2) also reveal considerable structural heterogeneity (Figs 5.26–5.28, 5.34, 5.35, 5.38). The intense staining of the outer surface of strips may represent an organic coating or could be the product of cation binding as a result of fixation with osmium tetroxide followed by staining with uranyl actetate and lead citrate (Figs 5.29 - 5.32). Mann and Williams (1982) demonstrated that Co^{2+} and Fe^{3+} ions preferentially bind to costal strips.

Gong et al. (2010) subjected purified concentrates of costal strips extracted from *S. diplocostata* loricae to

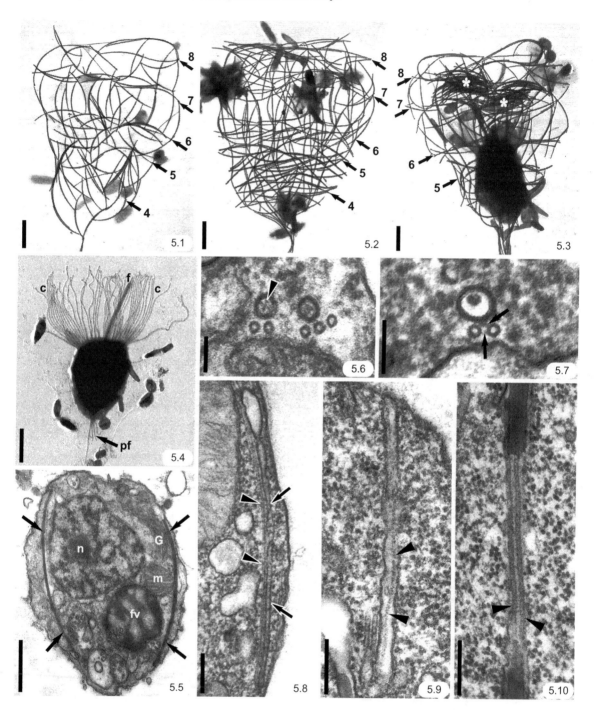

Plate 1 (Figures 5.1–5.10)

Figs 5.1–5.10 *Stephanoeca diplocostata.* Lorica structure and intracellular silica deposition. **Figs 5.1–5.2** Empty loricae illustrating the differences in costal numbers that can occur within loricae from a clonal culture. Helical costae in the posterior chamber indicated by arrow 4. The four transverse costae in the anterior chamber of each lorica are indicated by: (5) posterior transverse (waist) costa; (6) and (7) posterior and anterior intermediate transverse costae, respectively; (8) anterior ring. Bar = 2 μm.

intensive biochemical analysis. Scrupulous surface cleaning of strips with concentrated mineral acids followed by dissolution of biogenic silica with hydrofluoric acid left approximately 10% of the initial mass in an insoluble form. This insoluble material was analysed by means of electron dispersive X-ray spectrometry (EDX), which gave a spectrum containing a prominent peak for sulphur and a subsidiary peak for phosphorus, suggesting that there was an organic 'scaffold' within costal strips. In a subsequent experiment in which costal strips were partially demineralised with an alkaline phosphate buffer at pH 12, an insoluble 'tattered' organic matrix was revealed which was soluble in 8 M urea (pH 7.0). The supernatant solution was analysed by SDS-PAGE and the gel stained with Coomassie brilliant blue, revealing a protein band containing one major polypeptide with an apparent molecular mass of 14 kDa. Probing this band with wheatgerm agglutinin (WGA) gave a highly positive result indicating that the isolated protein was glycosylated. Polyclonal antibodies raised to the 14 kDa protein were conjugated with colloidal gold and applied to partially demineralised strips. TEM images of these strips revealed clusters of gold particles on the etched surfaces, again suggesting a possible organic scaffold within the silica matrix. Partially demineralised costal strips were also capable of nucleating silica deposition when placed in a solution of tetraethyl orthosilicate. Based on these results Gong *et al.* (2010) concluded that costal strips develop from an initial organic central filament of '(glyco)protein' composition. This arranges protein molecules around itself to form the organic scaffold which facilitates the polymerisation of biogenic silica. Further increase in circumference is achieved by appositional deposition of biogenic silica. It is envisaged that the protein scaffold remains entrapped within the deposited silica but that towards the end of silica deposition the central organic filament contracts. Both axial and lateral growth is guided by the organic template.

The mechanism by which costal strips adhere to each other to fabricate costae is one of the most important unresolved aspects of lorica construction. During lorica assembly, costal strips stored within close-packed bundles are pulled out by lorica-assembling microvilli to form costae (see Chapters 6 and 7). When developing costae are fully extended the accompanying microvilli are withdrawn, leaving the lorica complete in its final form. There is no obvious evidence of anything as gross as an adhesive at the junctions between costal strips. The connections are not rigid since the ends of overlapping strips can be displaced when loricae are dried for EM (see Sections 4.3.5 and 4.6.1, Fig. 4.54). Addition of concentrated sulphuric acid to loricae causes strips to separate, as does the more refined treatment with a general protease (1 mg ml^{-1} Tris buffer at pH 7.0) (Fig. 5.11). No firm conclusion can be drawn from these results except that it seems likely that an organic component, probably protein containing, facilitates adhesion. Costal strips in accumulated bundles at the top of the collar in *Stephanoeca diplocostata* and *Diaphanoeca grandis* usually scatter when prepared for EM. However, costal strip bundles of *D. grandis* remain intact when extracted from cells lysed with 0.13% sodium dodecyl sulphate (SDS) detergent in seawater and treated with 20% ethanol (unpublished results) (Fig. 5.12). This demonstrates that some physicochemical property of the strips, or material on their surface, is capable of holding a tightly packed bundle of strips together. Whatever is responsible for this effect must be able to operate over a short distance, probably of molecular dimensions, since loricae do not accrete strips and other particles onto their surface. A further feature of importance when considering possible mechanisms by which costal strips adhere to each other is that imperfectly formed and silica-depleted strips are still fully capable of adhering to each other to form loricae (Figs 5.15, 5.38) and during dissolution strip

---

Fig. 5.3 Cell nearing end of interphase with a substantial accumulation of costal strips at the top of the collar (asterisks). Transverse costae labelled as for Figs 5.1 and 5.2. Bar = 2 μm.

Fig. 5.4 Loricate cell treated with 1% hydrofluoric acid to dissolve siliceous costae leaving enclosed cell intact. Collar (c), flagellum (f), posterior filopodia (pf). Bar = 2 μm.

Fig. 5.5 Vertical section of cell showing developing costal strips (arrows) within SDVs in the peripheral cytoplasm. Nucleus (n), mitochondrial profile (m), Golgi apparatus (G), food vacuole (fv). Bar = 1 μm. Reproduced from Leadbeater (1979b).

Figs 5.6–5.7 Transverse sections of SDVs subtended by microtubules and containing immature costal strips. In Fig. 5.6, in the left-hand SDV an x-shaped costal strip template is visible (arrowhead). Reproduced from Leadbeater (1987). In Fig. 5.7 linkages between the two microtubules and between the microtubules and the silicalemma are visible (arrows). Bars = 0.1 μm.

Figs 5.8–5.9 SDVs in longitudinal section. Bars = 0.25 μm. Reproduced from Leadbeater (1985). Fig. 5.8 Elongated narrow SDV (arrows) with subtending microtubule (arrowheads). Fig. 5.9 Early stage in strip deposition (arrowheads).

Fig. 5.10 Longitudinal section of a pair of parallel SDV-associated microtubules. Fine periodic striations (arrowheads) probably represent linkages between microtubules and SDV (not seen). Bar = 0.25 μm. Reproduced from Leadbeater (1987).

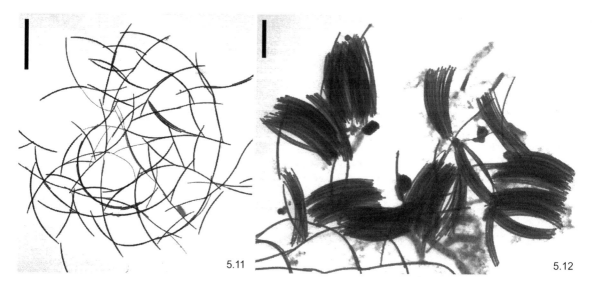

5.11

5.12

Plate 2 (Figures 5.11–5.12)

Figs 5.11–5.12 *Diaphanoeca grandis.* **Fig. 5.11** Partially disintegrated lorica base following treatment with a general protease (1 mg ml^{-1} at pH 7.0). Loose costal strips are evident. Bar = 5 μm.
Fig. 5.12 Collection of costal strip bundles extracted from cells lysed with 0.13% SDS detergent in seawater and treated with 20% ethanol. Bar = 2.5 μm.

adhesion remains intact until complete disintegration occurs (Figs 5.24, 5.25).

5.6 EFFECTS OF MICROTUBULE POISONS ON THE CYTOSKELETON AND COSTAL STRIP DEVELOPMENT

5.6.1 Background

There are four categories of microtubules within loricate choanoflagellate cells, namely those that: (1) contribute to the 9 + 2 axoneme of the flagellum and the two basal bodies (see Section 2.6.2); (2) form the microtubular cytoskeleton which comprises a radial array of cortical microtubules that converge to form a ring around the flagellar basal body (see Section 2.7.1); (3) comprise the mitotic spindle during cell division (see Section 2.9); (4) support and shape SDVs during the early stages of costal strip formation (Section 5.5).

Microtubule poisons, in particular colchicine and the vinca alkaloids, have been used extensively to investigate the cytological function of microtubules (Dustin, 1984). Since the relationship between SDVs and their subtending microtubules in loricate choanoflagellates is so specific and

unequivocal, it was anticipated that any disturbance in microtubule integrity or behaviour would have an immediately obvious effect on the morphology of costal strips. The experimental results obtained were everything that might have been wished for. However, additionally, more was learned about the mechanism involved in strip movement, storage and lorica assembly as well as the relative responses of the other different categories of microtubules to microtubule poisons.

Microtubules are highly dynamic polymers whose basic component is the α/β tubulin heterodimer. Under certain conditions tubulin dimers bind in a head-to-tail ($\alpha\beta\alpha\beta$) fashion to form linear protofilaments which associate in parallel array around a hollow cylindrical axis to form a microtubule approximately 25 nm in diameter. The polymer is initially formed from a microtubule nucleating complex. Thereafter, the microtubule can elongate slowly or shorten in a rapid fashion, which is achieved by the respective addition or removal of tubulin heterodimers. The repeated transition of tubulin between the dimeric and polymeric states is known as 'dynamic instability'. The head-to-tail organisation of tubulin heterodimers within microtubules creates an intrinsic polarity with α-tubulin at the slower-growing minus-end and β-tubulin at the faster growing plus-end (Wiese and Zheng, 2006). The relative stability of microtubules is controlled by microtubule-associated proteins (MAPs) and plus-end tipping proteins (TIPs) that consolidate or promote growth at the plus-end (Nogales, 2000).

Microtubule poisons bind to tubulin and suppress the dynamic behaviour of microtubules, in particular the mechanism of microtubule assembly (Dustin, 1984; Correia and Lobert, 2008). There are three primary modes by which poisons can disrupt microtubule assembly: (1) sub-stoichiometric poisoning (colchicine and podophyllotoxin); (2) alternative polymer formation (vinca alkaloids); (3) microtubule stabilisation (Taxol (paclitaxel)). During the course of work on *Stephanoeca diplocostata* colchicine, podophyllotoxin and the vinca alkaloids were used which are relevant to modes 1 and 2 listed above (Leadbeater, 1987).

Colchicine and podophyllotoxin inhibit the assembly of microtubules sub-stoichiometrically by selectively poisoning the ends of microtubules. It is likely that their mode of action is to bind to free tubulin heterodimers within the cell, which can then bind to the ends of microtubules, thereby inhibiting further heterodimer recruitment. Alternatively these drugs can bind to tubulin heterodimers stoichiometrically, creating a complex that has reduced or no affinity for the microtubule ends (Correia and Lobert, 2008).

At low concentrations the vinca alkaloids, including vincristine and vinblastine, decrease the rates of both growth and shortening at the plus-end of the microtubule, which, in effect, produces a 'kinetic cap' that suppresses function. Binding of the vinca alkaloids is responsible for the splaying of microtubules into spiral aggregates or spiral protofilaments, which leads to microtubule disintegration. Spiral protofilaments may then associate to form paracrystals, making them unavailable for polymerisation into microtubules (Correia and Lobert, 2008).

Relating the individual responses of choanoflagellate cells to the precise molecular effects of microtubule poisons is not possible at this time since there are too many unknown variables. Of the four categories of microtubules in loricate species, existing flagellar microtubules were relatively unaffected by treatment with microtubule poisons, although in cells dividing at the time of poison administration flagellar elongation was inhibited on daughter cells. Likewise cytoskeletal microtubules appeared to be relatively unaffected. However, both mitosis and SDV-associated microtubules were affected, the latter revealing rapid effects (see also Section 6.8.1).

5.6.2 Effects of colchicine on population growth of *Stephanoeca diplocostata*

Most of the experimental work involving treatment of choanoflagellates with microtubule poisons was carried out using colchicine on cultures of *Stephanoeca diplocostata* (see Section 5.5). Preliminary tests indicated that colchicine at a final concentration of 0.8–1.0 mM had an optimal effect on *S. diplocostata* in that growth was depressed without causing cell death and distorted costal strips were produced. Fig. 5.13 illustrates the effects of adding colchicine to give a final concentration of 0.8 mM on an exponentially growing culture of cells at 20 °C. Prior to colchicine addition the mean cell doubling time was 9.6 hours. For the following 15 hours after colchicine addition, growth rate, as measured by increase in cell number, was depressed by approximately 70%. Cell morphology was also affected in that there was a reduction in size and some cells divided into smaller units, usually grouped in a row one above another (Leadbeater, 1987).

After 15 hours most colchicine-treated cells can be returned to relatively normal vigour by transfer to fresh medium without the poison. To facilitate this transfer, it was necessary to wash cells several times which involved centrifugation to form a pellet followed by re-suspension in fresh medium. An unavoidable consequence of this procedure was loss of cells, which is illustrated in Fig. 5.13 by a discontinuity between the two counts at 65 hours. After a period of readjustment to fresh medium lasting approximately ten hours, during which there was a decline in cell numbers, exponential growth was once again achieved, albeit with a slower mean doubling time of 26 hours compared to that recorded before colchicine treatment.

5.6.3 Effects of colchicine on costal strip morphology and lorica assembly

Cultures of *S. diplocostata* were sampled at regular intervals after colchicine addition (0.8 mM final concentration) and the aliquots prepared as whole mounts for EM. In this way it was possible to monitor any disturbance to costal strip production and lorica assembly. Four hours after colchicine addition the first effects on strip morphology were observed. Deformed strips appeared at the top of the collar either singly, in small groups or in bundles mixed with normal crescentic strips (Fig. 5.14). Although the range of deformities produced was considerable (Figs 5.14, 5.18), the distortions can, nevertheless, be categorised in a relatively

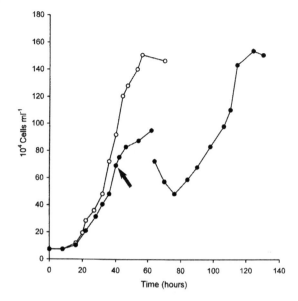

Fig. 5.13 *Stephanoeca diplocostata*. Effect of colchicine on population growth in stirred culture at 20 °C. Control culture (open circles) contains cells grown in Erdschreiber seawater with 80 μg ml^{-1} proteose peptone and 16 μg ml^{-1} yeast extract. Treated culture (closed circles) involved identical conditions to control but after 40 hours (arrow) 5 ml of an aqueous colchicine solution was added to give a final concentration of 0.8 mM colchicine. After 65 hours from the beginning of the experiment cells were pelleted by centrifugation, washed in fresh medium without colchicine and subsequently inoculated into fresh Erdschreiber seawater with organic enrichment. Centrifugation and washing resulted in a loss of cells noted in the graph by a discontinuity at 65 hours.

straightforward way. The majority of misshapen strips are narrow rods but, instead of being crescentic with the standard curvature (Fig. 5.26), they usually have irregular undulations or bends at intervals along their length. In extreme cases a strip may be branched or bent back on itself in the form of a hairpin (Fig. 5.18). At first the misshapen strips produced are only slightly shorter than normal, but most strips extruded after ten hours' treatment are one-half to one-third the standard length. The thickness of strips is also affected, with some shorter strips being wider and with an irregular gnarled appearance. In spite of the misshapen form of the strips, colchicine-treated cells are still able to exocytose and move them to the top of the collar. Accumulated bundles of misshapen strips on the collar lack the close-packed regularity characteristic of normal strips (compare Figs 5.14 and 5.3) and sometimes small numbers of long strips are replaced by larger numbers of shorter strips.

At 22 hours after colchicine addition some cells have loricae incorporating deformed strips. In a few instances (Fig. 5.15) the lorica may contain an entire set of deformed strips and yet still retain its distinctive *Stephanoeca* form. This result is somewhat unexpected since colchicine has well-established 'anti-mitotic' properties and yet lorica assembly in *S. diplocostata*, as in all tectiform species, is only achieved by a daughter cell (juvenile) after mitosis and division of the parent cell (see Chapter 7). Furthermore, for a parent cell to accumulate a complete set of misshapen strips to bequeath to a future juvenile cell would require it to have been subjected to colchicine treatment throughout the duration of interphase. In this example the effect of colchicine cannot have been severe enough to have prevented cell division, however irregular it may have been. Nevertheless, the absence of a flagellum on this specimen shows that flagellar regeneration, which normally would have occurred immediately after cell division, was inhibited (Fig. 5.15). Microtubule poisons do not affect the behaviour of the collar microvilli, which have a microfilamentous cytoskeleton (see Section 2.7.2).

The precise reason why misshapen strips are produced after treatment of cells with colchicine is uncertain. Various explanations are possible. Addition of colchicine may affect existing SDV-associated microtubules, thereby interfering with their supporting function. This could explain why early deformities include strips of normal length but with bends. Colchicine could also inhibit the polymerisation of microtubules for newly forming SDVs, which might explain the shorter, thicker misshapen strips. In sections of colchicine-treated cells, narrow SDVs containing the early stages of strip formation are observed without subtending microtubules (Figs 5.16, 5.17). Glancing sections of colchicine-treated cells with bent strips show that the cytoskeletal microtubules just beneath the plasma membrane have not been depolymerised (Fig. 5.19).

Less is known of the effects of podophyllotoxin, griseofulvin and the vinca alkaloids on the growth of *Stephanoeca*. Nevertheless, all resulted in the production of misshapen strips and all were effective in the range of 0.1–1.0 mM. They also inhibited cell division and the regeneration of flagella.

The results of experiments on *S. diplocostata* involving application of microtubule poisons are also informative in a more general context. The processes of costal strip exocytosis, accumulation at the top of the microvilli and subsequent assembly into a lorica by the juvenile cell once mitosis and cell division have taken place are not disturbed

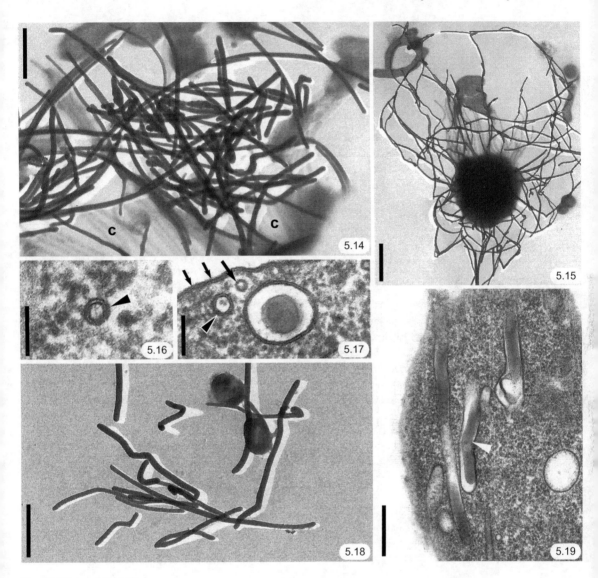

Plate 3 (Figures 5.14–5.19)

Figs 5.14–5.19 *Stephanoeca diplocostata*. Effects of colchicine (final concentration 0.8 mM) on costal strip morphology. **Fig. 5.14** Accumulation of misshapen costal strips at top of collar (c). Bar = 1 μm.

Fig. 5.15 Cell with collar, but lacking flagellum, within a recognizable lorica containing misshapen costal strips. Bar = 2 μm.

Figs 5.16–5.17 Transverse sections of SDVs within colchicine-treated cells. Note thickness of surrounding membrane (silicalemma). Bar = 0.1 μm. **Fig. 5.16** Single immature SDV (arrowhead) without subtending microtubules. **Fig. 5.17** Two SDVs adjacent to a cytoskeletal microtubule (arrow). Smaller SDV (arrowhead) should have two subtending microtubules but they are absent. Note similarity in thickness of SDV membrane and plasma membrane (small arrows).

Fig. 5.18 Group of misshapen costal strips of varying length. Bar = 1 μm.

Fig. 5.19 Misshapen developing costal strip (arrowhead) within SDV. Bar = 0.25 μm. Figs 5.14, 5.15, 5.17 and 5.18 reproduced from Leadbeater (1987).

by microtubule poisons in sub-lethal doses. None of these processes is dependent on costal strips being in pristine condition and each operates normally in spite of the strip being of imperfect shape. Many of the intermediate processes are microfilament based – for instance, movement of strips on the collar microvilli (see Chapter 7) and assembly of the lorica (see Chapters 6–8). The mechanism by which strips adhere to each other at the end of lorica assembly is also unaffected, in spite of strips not being tightly packed in bundles at the top of the collar. These conclusions are in accordance with those drawn from silicon depletion studies and are discussed in greater detail in Section 5.7.

5.7 GROWTH AND SILICON TURNOVER IN *STEPHANOECA DIPLOCOSTATA*

5.7.1 Background

The majority of experiments involving silicon turnover were accomplished using a specially designed apparatus that allowed cells to be grown in batch culture under controlled temperature conditions and samples to be taken at hourly intervals throughout a period of several days (Leadbeater and Davies, 1984). Initial experiments were concerned with growth and silicon turnover during an entire culture cycle under different starting conditions of organic enrichment and reactive silicate. Experiments were then extended to include growth under silicon-limiting conditions followed by the effects of silicon replenishment on growth of silicon-starved cells. The information presented here is based on the data published by Leadbeater and Davies (1984) and Leadbeater (1985, 1989). However, additional results have been included and the calculations relating to growth and silicon turnover kinetics have been thoroughly re-worked.

5.7.2 Batch culture growth of *Stephanoeca diplocostata* in silicon-enriched conditions

BACTERIAL PREY COMPONENT

Stock cultures of the clonal isolate of *Stephanoeca diplocostata* were maintained in Erdschreiber seawater with an assemblage of prey bacteria including at least two strains of *Pseudomonas* (Leadbeater and Davies, 1984). In the first phase of batch-culture experiments, growth of bacteria and, subsequently, *S. diplocostata* was controlled by the initial addition of organic enrichment comprising a mixture of proteose peptone and yeast extract. Despite neither the bacterial assemblage nor the chemical content of the medium being fully defined, the results in terms of prey and predator growth were consistent (Table 5.4) (Leadbeater and Davies, 1984). Increase in bacterial biomass in stirred culture at 20 °C, as measured by optical density, was rapid, reaching a maximum concentration within approximately eight hours of the start of an experiment for the lower concentration of organic enrichment and 16 hours for the higher. The maximum bacterial density in the lower concentration of organic enrichment was approximately half that in the higher, which reflects the ratio of organic addition. Reactive silicate and pH remained constant throughout the entire bacterial culture cycle, which lasted between 70 and 120 hours. Towards the end of batch culture experiments, in both unstirred and stirred conditions, bacteria eventually adhered to each other, forming clumps that were too large to be ingested by the predator and were therefore no longer available as food.

GROWTH OF `STEPHANOECA DIPLOCOSTATA´

Over a period of ten years (1978–88) approximately 35 experiments were carried out with respect to the growth and silicon requirement of *Stephanoeca diplocostata* in batch

Table 5.4 *Growth data, including duration of lag phase, mean specific growth rate* (μ), *mean doubling times* (G) *and mean maximum numbers of cells, for* Stephanoeca diplocostata *in stirred and unstirred cultures at 20 °C. Two concentrations of organic enrichment were used: concentration 1 refers to 40 µg ml^{-1} proteose peptone and 8 µg ml^{-1} yeast extract; concentration 2 refers to 80 µg ml^{-1} proteose peptone and 16 µg ml^{-1} yeast extract.*

Organic enrichment	Number of experiments	Lag phase duration range (h)	Mean specific growth rate $\mu \pm$ SE (h^{-1})	Mean doubling time $G \pm$ SE (h)	Mean maximum concentration of cells $\times$ 10^4 ml^{-1}
Concentration 1 stirred	8	8–22	0.071 ± 0.005	9.13 ± 0.44	54.0 ± 4.15
Concentration 2 stirred	15	4–24	0.074 ± 0.004	9.75 ± 0.54	139.5 ± 3.16
Concentration 2 unstirred	1	28	0.044	15.75	68.0

culture. The experiments selected for illustration here have been chosen either because the data sets are complete or because they demonstrate a specific feature particularly well (Figs 5.20–5.23).

All experiments involving *S. diplocostata* exhibited recognisable phases of lag, exponential and declining growth. For experiments involving stirred culture at 20 °C, a reasonable degree of uniformity was achieved with respect to the mean doubling time and maximum concentration of cells for a given concentration of organic enrichment (Table 5.4; Figs 5.20a, 5.21a, 5.22a). The lag phase varied between 8 and 24 h and once exponential increase was achieved the growth constant varied between 0.09 and 0.06 h^{-1}, corresponding to a doubling time of 7.3 to 11.4 h. The mean growth constant of all stirred experiments (0.071 h^{-1}) is close to the value of μ_{max} (0.079 h^{-1}) obtained by Geider and Leadbeater (1988) for *S. diplocostata* grown at 18 °C with a single pseudomonad strain of bacterium (Table 9.3). The effect of doubling the quantity of organic enrichment of the medium led to slightly more than doubling the maximum yield of *Stephanoeca* cells (Table 5.4). In a single experiment on an unstirred culture with the higher concentration of organic enrichment, growth was slower than in the comparable stirred culture, with a growth constant of 0.044 h^{-1} equivalent to a doubling time of 15.75 h. The maximum yield was 68×10^4 cells ml^{-1}, approximately half the concentration in the equivalent stirred culture (Table 5.4; Fig. 5.22b). In five stirred experiments in which cultures were sampled throughout stationary phase the duration varied from 32 to 60 hours and the average maximum concentration of cells was $154.0 \pm$ SD 25.8×10^4 ml^{-1}.

Cell morphology varies characteristically at different stages of growth (Leadbeater and Davies, 1984). With respect to the cell cycle, recently divided cells are relatively small and obconical in shape with a short flagellum and collar. As interphase progresses cells increase in volume and become ovoid, with bundles of costal strips appearing at the top of the collar. At the end of interphase, cells are rectangular in shape and possess large bundles of costal strips on the collar. Cells are larger in early exponential growth than in the later stage, and during stationary phase cells become rounded without accumulations of costal strips. As stationary phase progresses, cells remain spherical but gradually decrease in volume. Prior to death, the cell is so small that it becomes lens-shaped, with the long axis in the transverse plane held in position by a radial connection to the organic investment at the waist region between the two chambers of the lorica.

SILICON UPTAKE

For standard growth experiments, initial concentrations of reactive silicate in the medium ranged between 80 and 160 μM Si. Concentrations of this magnitude permitted a complete growth cycle to take place without silicon depletion (Figs 5.20a, 5.21a, 5.22a, b). The lowest concentrations of silicon attained were well above those recorded from the sea (2–3 μM Si) at the type locality in France. In these experiments silicon was neither a saturating (>1 mM Si at 20 °C) nor limiting nutrient.

During cell growth, silicon is transported from the medium, through the plasma membrane and cytosol to SDVs, where it is deposited as biogenic silica in costal strips. Once exocytosed and in direct contact with the medium, the bound silicon has the potential to dissolve, thereby recycling reactive silicate (see below). Thus both uptake and dissolution occur simultaneously in actively growing cultures with the result that all recorded silicon measurements are net values. This phenomenon can result in a 'cryptic' cycling of silicon, whereby minute quantities of reactive silicate are released from existing loricae only to be immediately utilised by other cells for costal strip production. In these circumstances, silicon cycling is unmeasurable by standard chemical means.

Within a culture of *S. diplocostata* silicon may be present in one of three forms: (1) in the medium as soluble reactive silicate; (2) intracellularly in a cytosolic pool, either as monosilicic acid or as soluble or bound organo-silicon compounds; (3) as biogenic silica within costal strips, which may be nascent within SDVs, in groupings at the top of the collar or assembled into loricae. In the experimental protocols reported here, which involved pelleting fixed cells and thereby separating particulate and soluble fractions, only two forms of silicon were measured: (1) reactive silicate, as determined by the molybdate-blue assay (as μM Si), which includes all soluble silicon in the medium and leachate from the cells; (2) bound silicon from the pelleted fraction, which includes intracellular particulate organo–silicon compounds from the cytosol as well as costal strips and loricae. Bound silicon was solubilised by addition of sodium hydroxide, neutralised with sulphuric acid and assayed as reactive silicate using the molybdate-blue assay. Since the cytosolic pool of silicon is probably minute in comparison with the other silicon components it was ignored. Thus in experiments such as those illustrated in Figs 5.20a, 5.21a, 5.22a, b only reactive silicate within the medium and bound silicon (costal strips and loricae) have been

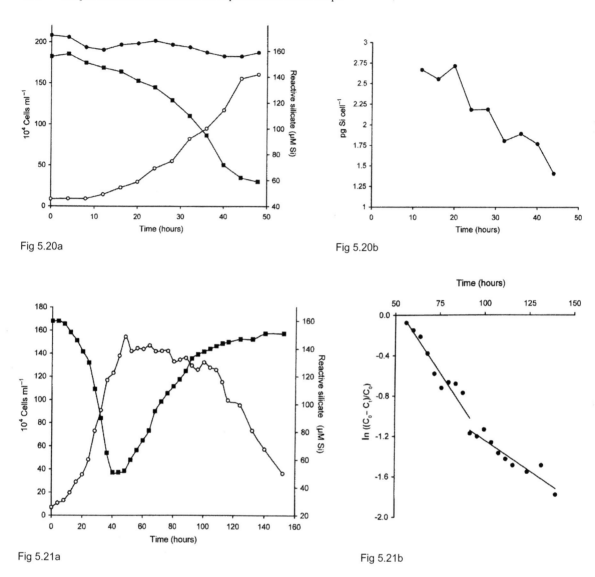

Fig 5.20a

Fig 5.20b

Fig 5.21a

Fig 5.21b

Figs 5.20–5.21 *Stephanoeca diplocostata*. Aspects of growth and silicon turnover during a culture cycle with stirring at 20 °C. Organic enrichment consists of a final concentration of 80 µg ml^{-1} proteose peptone and 16 µg ml^{-1} yeast extract. **Fig. 5.20a** Experiment 17 (see Table 5.5). Growth in cell numbers (open circles) and silicon uptake indicated by reduction of reactive silicate in the medium (closed squares) during the first 50 hours of culture. Closed circles denote total silicon (bound silicon + reactive silicate). **Fig. 5.20b** Experiment 17. Mass of bound silicon per cell throughout exponential phase of growth (see Fig. 5.20a). Range 2.67–1.41 pg Si cell^{-1}.
Fig. 5.21a Experiment 18 (see Table 5.5). Complete culture cycle illustrating growth (open circles) and reactive silicate (µM Si) in the medium (closed squares). **Fig. 5.21b** Plot of ln (($C_0 - C_t$) / C_0) as a function of time illustrating silicon dissolution based on data presented in Fig. 5.21a. The rate constants of dissolution, K_1 and K_2, are 0.0265 h^{-1} and 0.0119 h^{-1}, respectively.

recorded. Total silicon represents the combined values for soluble and bound silicon at a particular time.

Figs 5.20a, 5.21a, 5.22a, b summarise the characteristic patterns of growth and silicon turnover by *S. diplocostata*

in stirred and unstirred cultures, respectively, at 20 °C. During the lag phase of growth, the concentration of silicon in the medium is slightly reduced. Once exponential growth has been achieved, the curves for growth and net

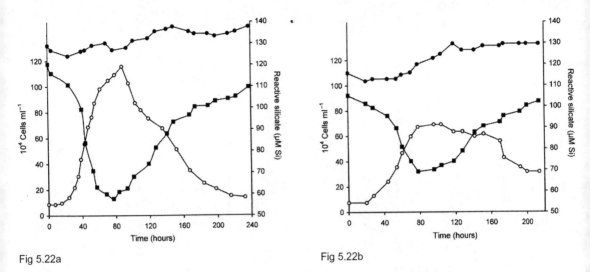

Fig 5.22a   Fig 5.22b

Fig. 5.22 *Stephanoeca diplocostata.* Aspects of growth and silicon turnover during a culture cycle at 20 °C (see Table 5.5). Organic enrichment consists of a final concentration of 80 µg ml⁻¹ proteose peptone and 16 µg ml⁻¹ yeast extract. Growth in cell numbers (open circles), reactive silicate in medium (closed squares), total silicon (bound silicon + reactive silicate) (closed circles). a. Stirred culture (Experiment 20). b. Unstirred culture.

silicon depletion are near mirror images of each other (Figs 5.20a, 5.21a, 5.22 a, b). The average mass of bound silicon per cell, calculated from the decrease in soluble silicon from the medium divided by the increase in cell number during the exponential phase, is $1.84 \pm SD\ 0.47$ pg Si cell⁻¹ ($n = 11$). However, if individual assays of bound silicon are expressed as a function of time throughout the exponential phase of growth, as illustrated in Fig. 5.20b for experiment 17 (see Fig. 5.20a), the value at the beginning of the exponential phase is higher (2.67 pg Si cell⁻¹) than at the end (1.41 pg Si cell⁻¹). Several reasons can be suggested for this variation. First, at the start of exponential growth there is usually a degree of synchrony as cells emerge from the lag phase, and during this time the majority will have a substantial complement of supernumerary costal strips on their collars; thus the value for bound silicon per cell will be in excess of that of a single parent lorica. Second, the loricae of cells at the beginning of exponential growth usually have more silicon per strip than at later stages. Third, at the beginning of exponential growth the net balance between silicon uptake and dissolution is weighted in favour of uptake because newly exocytosed costal strips are reasonably resistant to dissolution and there are fewer bacteria present that might enhance dissolution (see below). However, as growth proceeds, dissolution becomes more prevalent, particularly in older loricae. With the onset of stationary phase, the majority of cells lack accumulations

of costal strips and thus the value obtained for bound silicon probably comes closest to estimating the quantity of silicon within a single lorica. Table 5.5 lists the values of silicon per cell for five individual experiments as derived from soluble (mean $= 2.11 \pm SD\ 0.55$ pg Si cell⁻¹) and bound silicon (mean $= 1.95 \pm SD\ 0.59$ pg Si cell⁻¹) data, respectively, during exponential growth and from bound silicon data (mean $= 1.71 \pm SD\ 0.40$ pg Si cell⁻¹) at the beginning of stationary phase. Although deriving these values is rather like shooting at a moving target, nevertheless a value of between 3.5 pg Si cell⁻¹ for a loricate cell carrying a full complement of stored costal strips just prior to division is in line with the maximum values that have been obtained during this study. The silicon content of *S. diplocostata* cells is comparable to that of lightly silicified diatoms such as *Skeletonema costatum* (1–2 pg Si cell⁻¹) and *Thalassiosira pseudonana* (1.81 pg Si cell⁻¹) (Paasche, 1973).

It is possible to calculate theoretically an approximate value for the mass of silicon in an *S. diplocostata* lorica (Table 5.6). This is achieved by estimating the total volume of biogenic silica per lorica and multiplying this by an estimated density for biogenic silica, thereby obtaining a value for the mass of silica, and thus silicon, per lorica. Several assumptions have to be made, including the calculation of costal strip volume by treating each strip as a narrow cylinder of average diameter and length (Table 5.6). An estimate of the relative volume of the non-silica

Table 5.5 Growth and silicon turnover data for five selected Stephanoeca diplocostata experiments at 20 °C, four of which are illustrated graphically (Figs 5.20–5.22). Growth data include: (1) duration of exponential phase, (2) specific growth rate (μ), (3) duration of stationary phase, (4) maximum yield of cells. Silicon data include: (1) bound silicon per cell at the end of exponential growth, calculated from Δ Si (ng ml^{-1})/Δ cells ml^{-1} during this phase of growth, based on soluble and bound silicon data respectively, (2) bound silicon per cell at the start of stationary phase, (3) rate constants of silicon dissolution during and after stationary phase.

Experiment (stirred unless otherwise noted)	Duration of exponential phase (h)	Specific growth rate μ (h^{-1})	Bound silicon per cell at end of exponential growth (pg cell^{-1})		Bound silicon per cell at start of stationary phase (pg cell^{-1})	Duration of stationary phase (h)	Maximum concentration of cells (10^4 ml^{-1})	Rate constants of silicon dissolution (h^{-1})	
			From soluble Si data	From bound Si data				K_1	K_2
17 (Fig. 5.20)	24	0.088	1.73	1.70	1.41	–	155	–	–
18 (Fig. 5.21)	38	0.071	2.30	1.37	1.27	60	154	0.0265	0.012
19	32	0.072	2.89	2.89	2.30	52	165	0.0247	0.0039
20 (Fig. 5.22a)	31	0.071	2.14	2.14	1.79	0	116	0.0084	0.0029
Unstirred (Fig. 5.22b)	40	0.044	1.47	1.67	2.20	92	67	0.005	–

Table 5.6 Stephanoeca diplocostata. *Calculated mass of silicon per lorica comprising an average of 172 costal strips. Individual costal strips are treated as uniform narrow cylinders of biogenic silica with a mean diameter and length. The calculated mass of biogenic silica per lorica has been achieved by multiplying the total costal strip volume (minus 13.3% volume of protein) by the density, which is taken as 2.0 pg μm^{-3} (after Hurd, 1983). The silicon content of biogenic silica is taken as 42%, which accords with the biogenic silica having approximately 10% water content (SiO_2: $0.4\ H_2O$) (after Mortloch and Froelich, 1989). For comparison the range of silicon mass per lorica as determined experimentally is included (n = number of units sampled).*

Mean lorica length ($\mu m \pm$ SD)	Mean costal strip length ($\mu m \pm$ SD)	Mean costal strip diameter ($\mu m \pm$ SD)	Mean costal strip volume ($\pi r^2 l$) (μm^3)	Mean number of costal strips per lorica $\pm$ SD	Total volume of costal strips in lorica (μm^3)	Total volume of biogenic silica (μm^3)	Calculated mass of biogenic silica per lorica (density 2.0 pg μm^{-3}) (pg)	Calculated mass of silicon (Si) per lorica (pg)	Experimental determination of silicon (Si) mass per lorica (range pg)
15.6 ± 1.46 (62)	3.53 ± 0.43 (135)	0.077 ± 0.011 (100)	0.0164	172 ± 18.1 (60)	2.83	2.45	4.90	2.06	1.27–2.89

(protein) component of the costal strips also has to be made. This was calculated by dividing the percentage mass of protein (10%) which remains in costal strips after removal of the siliceous component with hydrofluoric acid (Gong et al., 2010), by the density of protein taken as 1.45 pg um^{-3} (see Fischer et al., 2004). The density of biogenic silica has been taken as 2.0 pg μm^{-3} (see Hurd, 1983) and the mass of silicon (Si) as a proportion of the biogenic silica mass has been taken as 42% (see Mortlock and Froelich, 1989). Despite these approximations, a value of 2.06 pg Si lorica^{-1} is obtained, which falls comfortably within the range of values obtained by experimental measurement (1.27–2.89 pg Si cell^{-1}) (see Table 5.6).

BIOGENIC SILICA DISSOLUTION

One of the most striking features of the *S. diplocostata* experiments is the rapidity with which biogenic silica dissolution occurs (Figs 5.20–5.22). Seawater is particularly favourable for the solubilisation of amorphous silica since it is highly undersaturated with respect to monosilicic acid, and has a relatively high pH and salt content. In physico-chemical terms dissolution of SiO_2 is due to the nucleophilic attack of water molecules causing the breakage of siloxane bonds (Si–O–Si) at the particle surface (Dove and Crerar, 1990). Water molecules orientate their electronegative oxygen atom towards the bound Si atom, which leads to a transfer of electron density to the Si–O (siloxane) bond, thereby increasing its length and eventually breaking it. The open linkage created then binds with the dissociating water molecule to form a Si–OH (silanol) group. A series of such reactions leads to the release of hydrated Si atoms in the form of monosilicic acid $Si(OH)_4$. Increasing pH leads to the deprotonation of surface silanol groups, thereby further facilitating the breaking of the bridging siloxane bonds (Dove and Elston, 1992). Salt solutions enhance the dissolution rates of quartz. Dove (1999) proposes that the cations of alkali metals (Na^+ and K^+) and alkaline earth elements (Ca^{2+} and Mg^{2+}) strengthen the nucleophilic properties of water molecules at the quartz surface. Hydrated cations adsorbed to the negatively charged SiO_2 surface improve the physical access of water to siloxane bonds by redirecting hydration water molecules to more favourable positions, thereby increasing the rate of hydrolysis of the bonds (Loucaides et al., 2008).

In most *Stephanoeca* experiments net silicon uptake ceases abruptly at the end of exponential growth and net dissolution becomes apparent throughout the phase of declining growth. Subsequent values for silicon dissolution during stationary phase must approximate to gross values since the rate of silicon uptake from the medium will be minimal. As illustrated in Figs 5.21a, 5.22a and 5.22b, within 5–7 days of the cessation of exponential growth the concentration of reactive silicate in the medium has almost returned to the level recorded at the beginning of the experiment.

According to Dixit et al. (2001) the solubility of biogenic silica at 20 °C in seawater at one atmosphere is in excess of 1 mM. Since the highest concentrations of dissolved silicon that occur in the *Stephanoeca* experiments are well below this value (<200 μM), the dissolution rate of loricae is limited by bound silicon depletion and not by solution saturation (Roubeix et al., 2008). In these conditions rate constants of net silicon dissolution can be obtained by plotting the fraction $l_n ((C_0 - C_t) / C_0)$ as a function of time, where C_0 = initial concentration of bound silicon; C_t = concentration of bound silicon that has dissolved in time t. The linear portions of the graph correspond to phases of exponential dissolution (Roubeix et al., 2008). Rate constants of dissolution (h^{-1}) can be determined graphically from the slopes of the linear portions of graphs. For each experiment one, and in some cases two, exponential phases of dissolution can be identified and characterised by rate constants K_1 and K_2, respectively. Fig. 5.23a illustrates the solubilisation of acid-cleaned *Stephanoeca* costal strips at 20 °C in seawater as measured by the increase of reactive silicate in the medium. Fig. 5.23b is the same data represented on a log–linear plot and from this graph a single rate constant of dissolution can be obtained ($K = 0.015$ h^{-1}). Fig. 5.21b illustrates a log–linear plot of the dissolution data obtained from experiment 18 illustrated in Fig. 5.21a. Here, two rates of dissolution are apparent with values for K_1 and K_2 being 0.0265 h^{-1} and 0.0119 h^{-1}, respectively. Table 5.5 lists K_1 and K_2 values obtained for other comparable experiments. In total seven experiments were carried out in which bound silicon was measured during its dissolution phase and the mean value for a single rate constant K_1 was 0.023 ± SD 0.008 h^{-1}.

The conditions under which *S. diplocostata* was grown during these experiments are particularly conducive to a rapid rate of silica dissolution. Already mentioned are the properties of seawater, with a pH value of 8.0–8.2 and salinity of 35‰. Silicon dissolution is also enhanced by an increase in temperature and active microbial activity. With respect to diatoms, bacteria accelerate silicon dissolution in the sea by colonising and enzymatically degrading the organic matrix of empty and detrital diatom frustules. In a series of experiments involving *Thalassiosira weissflogii* and *Chaetoceros simplex*, Bidle and Azam (2001)

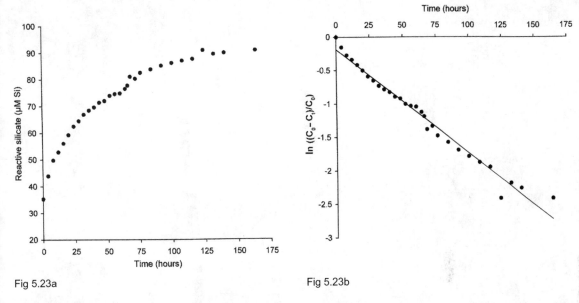

Fig 5.23a   Fig 5.23b

Fig. 5.23 *Stephanoeca diplocostata*. Dissolution of acid-cleaned costal strips stirred in Erdschreiber seawater at 20 °C. a. Standard plot of dissolution as measured by increase in reactive silicate in the medium. b. Plot of ln ((C_0 – C_t) / C_0) as a function of time based on data illustrated in part a. The rate constant of dissolution $K = 0.015$ h^{-1}.

demonstrated that protease activity was consistently attributable to the dominant bacterial ectohydrolase, and it strongly correlated with measured silica dissolution rates. Roubeix *et al.* (2008) further suggested that the general release of metabolic by-products by bacteria also appeared to facilitate frustule dissolution. The conditions within culture vessels containing *S. diplocostata* during stationary phase produce media that are rich in bacterial extracellular products and usually bacteria adhere to loricae in clumps. Such aggregations might shield the loricae from the medium, thereby reducing solubilisation, but would also bring costal strips into close contact with bacterial ectohydrolases thereby accelerating dissolution. It is quite probable that under stirred culture conditions at 20 °C the dissolution of *Stephanoeca* loricae is accelerated by one or more orders of magnitude.

The division of lorica dissolution kinetics into two exponential rates, K_1 and K_2, respectively (Fig. 5.21b; Table 5.5), is a common phenomenon also exhibited by frustule dissolution in diatoms (Kamatani and Riley, 1979; Kamatani, 1982; Roubeix *et al.*, 2008). Kamatani and Riley (1979) attributed this to the differential dissolution of two parts of diatom frustules. In *S. diplocostata* the central region of costal strips dissolves more rapidly than the surface (see below) and this could account for this phenomenon. The rate constant (K_1) of silica dissolution for *S. diplocostata* at

20 °C is higher by an order of magnitude than that for the majority of diatoms tested (Roubeix *et al.*, 2008). However, there are important caveats to making comparisons like this since there will be differences in the chemical composition of bound silicon and no allowance has been made for specific surface area (total surface area per unit of mass). The costal strips of *S. diplocostata* have a large surface area to volume ratio and once dissolution is underway the porous nature of costal strips with a central channel must increase the specific surface area considerably.

ULTRASTRUCTURE OF COSTAL STRIPS AND LORICAE UNDERGOING DISSOLUTION

Biogenic silica dissolution simultaneously affects both the surface and the centre of costal strips. The surface becomes progressively roughened and pitted while a central, longitudinal channel becomes increasingly apparent (Figs 5.26–5.32) (Leadbeater and Davies, 1984; Gong *et al.*, 2010). With progressive dissolution the central channel enlarges (Figs 5.30–5.32) and extends from one end of a strip to the other and the surface begins to fragment (Fig. 5.27). All these stages of dissolution are highly characteristic and cannot be confused with other conditions, such as silicon-depleted strips (Figs 5.34–5.38). Eventually strips lose their rigidity and disintegrate. Symptoms of dissolution

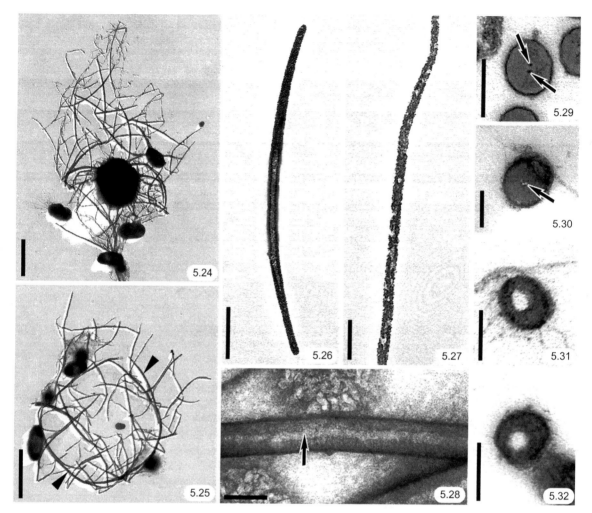

Plate 4 (Figures 5.24–5.32)

Figs 5.24–5.32 *Stephanoeca diplocostata*. Ultrastructural aspects of silicon dissolution. **Figs 5.24–5.25** Stages in the dissolution of loricae. Bars = 2 μm. **Fig. 5.24** Partially dissolved lorica, with disintegrated anterior chamber. **Fig. 5.25** Later stage of dissolution with most strips destroyed except posterior transverse (waist) costa (arrowheads).

Figs 5.26–5.27 Stages in costal strip dissolution. **Fig. 5.26** Partial dissolution; strip retains curved shape but has central hollow channel. Bar = 0.5 μm. **Fig. 5.27** Strip showing symptoms of advanced dissolution: loss of curved shape and severe pitting of outer surface. Bar = 0.25 μm.

Fig. 5.28 Negatively stained partially dissolved costal strip with prominent central channel (arrow). Bar = 0.1 μm.

Figs 5.29–5.32 Transverse sections of costal strips at successive stages of dissolution. Bars = 0.1 μm. **Fig. 5.29** Pristine strip with two central dots (arrows) and a densely stained surface. **Fig. 5.30** Strip with a minute central aperture (arrow). **Figs 5.31–5.32** Two strips each with a large central channel. Figs 5.24, 5.26 and 5.27 reproduced from Leadbeater and Davies (1984).

are exhibited equally throughout the length of an individual strip but not all strips in a lorica, even adjacent ones, show the same degree of corrosion at any one time (Figs 5.24, 5.25). Junctions between strips remain intact until strips dissolve away in these regions.

Dissolution of complete loricae in stirred cultures also follows a more-or-less consistent pattern. During late stationary phase the anterior chamber of most loricae shows signs of collapse as individual strips lose their rigidity (Fig. 5.24). Eventually the posterior chamber also disintegrates, leaving behind parts or all of the thickest costal strips that originally encircled the lorica waist (Fig. 5.25). The explanation for this sequence of events is that costal strips comprising the anterior chamber, above the posterior intermediate ring, are not protected on their inner surface by an organic investment and are therefore completely exposed to the medium. Costal strips in the posterior chamber are covered on their inner surface by an organic investment which partially limits immediate access to the surrounding medium.

5.7.3 Batch culture growth of *Stephanoeca diplocostata* in silicon depleted conditions

The *S. diplocostata* experiments that provided the data on growth and silicon turnover in silicon-enriched conditions (Section 5.7.2) made use of Erdschreiber seawater, which was based on natural seawater supplemented with sterile soil extract and additional NO_3 and PO_4. The advantages of using this medium were that it was easy to prepare and *S. diplocostata* grew well over a period of several days, giving relatively high cell yields. Silicon, at final concentrations of up to 200 μM, was supplied to the medium by addition of seawater that had been autoclaved in borosilicate glass containers. This overcame the necessity of having to add small quantities of sodium metasilicate, which could precipitate in the presence of soluble phosphate in seawater, and having to make further adjustments of pH. However, for experiments which require silicon-free medium, natural coastal seawater, which has a silicon content of 2–10 μM, was unsuitable. Silicon is a particularly difficult element to eliminate since even ultrapure chemicals including distilled (deionised) water contain minute traces of silicon as impurities. After a series of trial experiments, the artificial seawater recipe selected for these experiments was that of Harrison *et al.* (1980), omitting the metasilicate component. Silicon was supplied where appropriate by addition of a solution of sodium metasilicate ($Na_2SiO_3 \cdot 5H_2O$).

The first two experiments involving silicon restriction were carried out using Erdschreiber seawater with starting silicon concentrations of 20 and 35 μM, respectively (Table 5.7). In both experiments silicon was depleted to an unmeasurable concentration during the exponential phase and yet there was no apparent effect on growth (Fig. 5.33a, b). A similar growth pattern was obtained using silicon-free

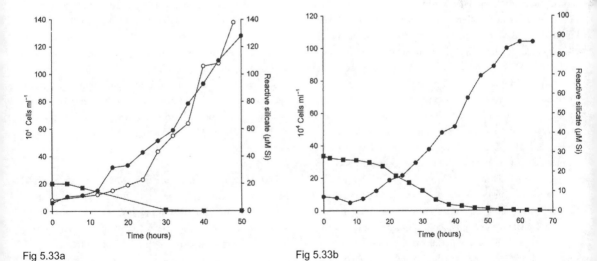

Fig 5.33a   Fig 5.33b

Fig. 5.33 *Stephanoeca diplocostata*. Growth of lorica-bearing cells in medium with limiting silicon at 20 °C. Organic enrichment consists of a final concentration of 80 μg ml^{-1} proteose peptone and 16 μg ml^{-1} yeast extract. In both experiments the silicon content (closed squares) of the medium is reduced to zero but growth is unaffected. a. Control culture with non-limiting silicon content (open circles); experimental culture with 20 μM initial concentration of reactive silicate (closed circles). b. Experiment with 35 μM initial concentration of reactive silicate.

Table 5.7 *Growth data, including duration of lag phase, mean specific growth rate (μ), mean doubling time (G) and maximum numbers of cells, for* Stephanoeca diplocostata *cultured at 20 °C in silicon-enriched and silicon-depleted Erdschreiber seawater and artificial seawater respectively. Concentration of organic enrichment used: 80 μg ml^{-1} proteose peptone and 16 μg ml^{-1} yeast extract.*

	Number of experiments	Lag phase duration range (h)	Mean specific growth rate $\mu \pm$ SD (h^{-1})	Mean doubling time $G \pm$ SD (h)	Maximum concentration of cells $\pm$ SD ($\times 10^4$ ml^{-1})
Silicon-rich Erdschreiber seawater (>150 μM Si)	15	4–24	0.074 ± 0.004	9.75 ± 0.54	139.5 ± 3.16
Silicon-limited Erdschreiber seawater (~30 μM Si)	2	4–16	0.068 ± 0.004	10.15 ± 0.05	118 ± 14.14
Silicon-rich artificial seawater (>50 μM Si)	2	16–32	0.089 ± 0.01	8.70 ± 1.04	47.7 ± 1.56
Silicon-impoverished artificial seawater (0–15 μM Si)	4	8–33	0.072 ± 0.009	9.78 ± 1.14	47 ± 9.0

synthetic seawater (Table 5.7). Silicon starvation had no visible effect on loricae already present in culture, but subsequent generations of cells showed increasingly severe symptoms of silica deprivation. These symptoms, which can be observed on living cells, included thin but complete loricae containing poorly silicified costae and incomplete loricae with reduced numbers of costae. Repeated subculturing of cells into silicon-free synthetic seawater eventually resulted in the production of entirely naked cells (Fig. 5.36). A complete absence of costal strips is extremely difficult to achieve since cells can metabolise even the minutest (immeasurable) concentrations of silicon that might have occurred as impurities in the medium or have been 'cryptically' recycled from the remnants of previous loricae. Silicon starvation of *Diaphanoeca grandis* had a similar effect to that on *S. diplocostata*. Thinner costal strips were produced which ultimately led to lorica collapse (Figs 5.37, 8.8). In extreme instances the only covering of the cell was the veil, which collapsed around the collar and flagellum.

The constantly changing conditions in batch culture make it difficult to pinpoint the precise concentration of soluble silicon at which normal costal strip deposition begins to be affected. The first detectable indication of silicon limitation is the occurrence of thinner and less opaque strips at the top of the collar (Fig. 5.34). Most incompletely silicified strips bear a regular array of constrictions, giving them a segmented appearance (Figs 5.34, 5.35, 5.38). Extremely thin strips bend or coil and may bear little resemblance to mature costal strips (Fig. 5.35 arrows). Unless silicon depletion is extremely rapid, symptoms of depletion occur uniformly along a strip and there is no reduction in the length (Fig. 5.38). Occasionally it is possible to detect a gradient of silica depletion within the time taken for a cell to produce a full complement of strips for a lorica. For instance, in *Diaphanoeca grandis* the first-produced strips are the thickest and thereafter strips become sequentially thinner. From particularly favourable specimens it is possible to determine the order in which strips have been produced (Fig. 8.8). Eventually silicon depletion is so severe that strips become shorter or cease to be produced (Fig. 5.35).

Weakly silicified strips are exocytosed from the cell and move to the top of the collar in the customary manner, where they are stored awaiting cell division (Figs 5.34, 5.35). In extreme instances of silicon deprivation, the number of strips stored on the collar is less than a full complement required for a lorica. Nevertheless, as illustrated by culture studies (Figs 5.33a, b), cell division proceeds at the normal rate in spite of silicon deprivation indicating that there is not a mandatory coupling between

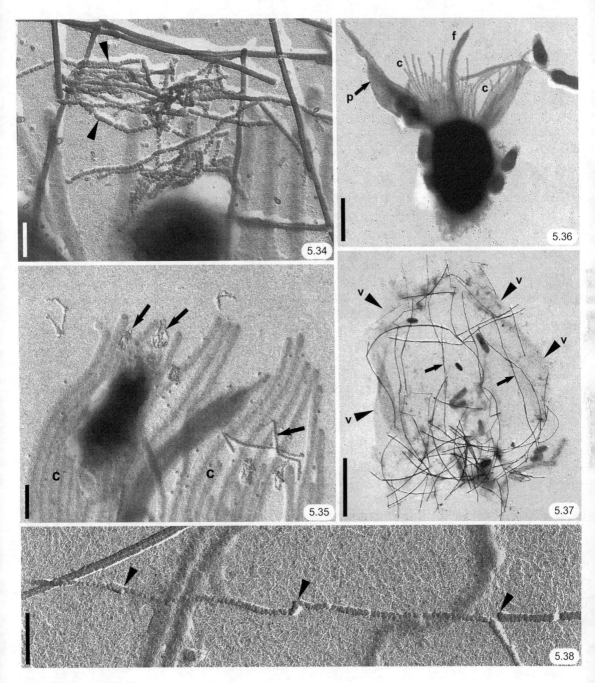

Plate 5 (Figures 5.34–5.38)

Figs 5.34–5.38 *Stephanoeca diplocostata* and *Diaphanoeca grandis*. Ultrastructure of silicon-depleted cells. **Fig. 5.34** *S. diplocostata*. Accumulation of thin, lightly silicified, costal strips at top of collar (arrowheads). Bar = 0.25 μm. Reproduced from Leadbeater (1985). **Fig. 5.35** *S. diplocostata*. Fragmentary contorted costal strips (arrows) associated with collar microvilli (c). Bar = 0.5 μm. **Fig. 5.36** *S. diplocostata*. Completely silicon-starved, naked cell with flagellum (f), collar (c) and associated pseudopodium (arrow p) containing an engulfed bacterium. Bar = 2 μm.

the full silicon requirement for lorica construction and growth. All normal processes associated with costal strip movement, including exocytosis, movement to the top of the collar and manipulation during cell division, continue normally in spite of strips being thinner or reduced in numbers. Furthermore, lorica assembly is still achieved in spite of costal strip deficiencies. It is quite remarkable how fragile silicon-deprived strips can be assembled into distinguishable costae (Figs 5.37, 5.38). If silicon depletion occurs rapidly within the span of one cell cycle it is possible, on the basis of costal strip thickness, to determine the order in which strips comprising a lorica have been formed and exocytosed (Figs 8.8, 8.10). Images of cells with costal strips graded by their silicon content have played an invaluable part in determining the order of strip production and movement outside the cell (see Chapter 8).

Since *Stephanoeca diplocostata* is a tectiform species, the costal strips that are accumulated at the top of the collar during interphase are bequeathed to the daughter cell (juvenile) that leaves the parent lorica after mitosis and division (see Chapter 7). This sequence of events continues as long as there are traces of silicon in the medium. Silicon-deprived juveniles, apparently without costal strips when seen with light microscopy, are still capable of extending lorica-assembling microvilli even though there are no costal strips to assemble (Leadbeater, 1985). However, when silicon has been completely depleted and there are no rudimentary strips, juvenile cells cease to undergo inversion during division and instead cells divide diagonally so that the flagellar poles of both daughters face forwards (Figs 5.41–5.44). A small percentage of cells undergo a partial form of division producing 'V-shaped' conjoined cells usually with two nuclei, collars and flagella (Figs 5.45, 8.23–8.26). The region of continuity between the partially divided cells is always at the posterior end (Figs 8.24–8.26) (see Section 8.6.2). Silicon-depleted cells can be subcultured into silicon-deficient medium and maintained in this state for several months. However, without loricae cells are susceptible to becoming surrounded and disabled by bacteria (Fig. 8.13) and therefore it is necessary to reduce the organic supplement so that bacterial, and therefore cell, densities are reduced.

5.7.4 Replenishment of silicon to severely silicon-depleted cells

Silicon replenishment of severely silicon-depleted cells causes a temporary cessation of growth. In the experiment illustrated in Figs 5.39a and b, 45 μM Si was added to a silicon-free culture containing cells without loricae. Over a period of eight hours cell numbers remained more-or-less constant and only after 12 hours was a small increase recorded. However, despite this apparent inactivity, silicon was immediately taken up, with a resulting decline in the silicon content of the medium from 45 to 39.5 μM Si eight hours after addition (Fig. 5.39b). In a separate experiment, not illustrated here, in which 45 μM Si was added to a low-silicon culture (~2 μM Si) of lorica-bearing cells, no disturbance in growth was recorded. Thus it appears that the cells themselves must be completely silicon deprived and not just immersed in a silicon-free medium.

To investigate this phenomenon further, a similar experiment was carried out in which an inoculum of completely silicon-starved cells was grown in silicon-free medium and during the exponential phase of growth was spiked with an aliquot of sodium metasilicate to give a concentration of 71 μM Si (Fig. 5.40a). The control for this experiment consisted of silicon-starved cells in silicon-free medium. In addition to observations on growth and silicon uptake, the percentage of cells with and without loricae was recorded (Table 5.7). Fig. 5.40a shows the pattern of growth in the control culture (open circles) and the test culture (closed circles) from the time of silicon addition. Whereas growth proceeded in the control culture with a doubling time of 18.1 hours, growth in the silicon-spiked culture was more-or-less static ($14.45 \pm 1.72 \times 10^4$ cells ml^{-1}) throughout the 40-hour period following silicon addition. However, during this period of apparent stasis, reactive silicate was taken up by cells for the first 16 hours, reducing the concentration from 71 to 67 μM Si, and thereafter the rate of uptake decreased such that over the next 32 hours the concentration dropped from 67 to 64.5 μM Si. Throughout the 48-hour period of this experiment the percentage of cells with loricae increased from 6% at time zero to 86% after 48 hours (Table 5.8). Since the

---

Fig. 5.37 *D. grandis*. Silicon-depleted lorica with thin costal strips lacking rigidity (arrows). Organic veil is clearly visible (arrowheads v). Bar = 2 μm.

Fig. 5.38 *S. diplocostata*. Thin, severely silicon-limited costal strips comprising anterior ring. Arrowheads denote junctions between neighbouring strips. Bar = 0.25 μm.

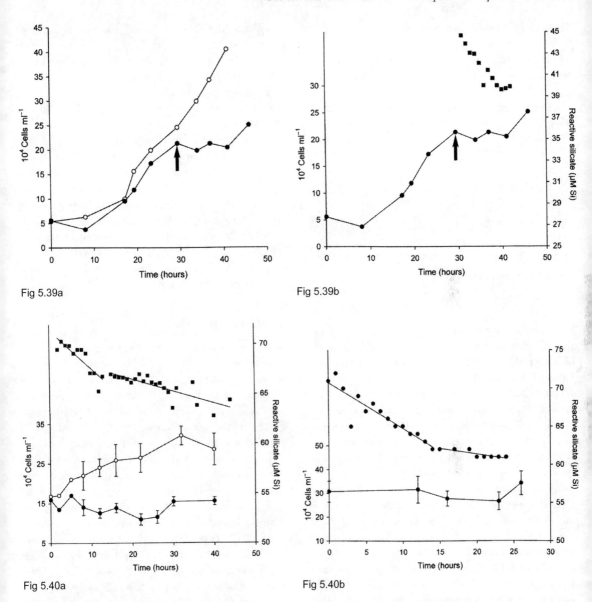

Fig 5.39a   Fig 5.39b

Fig 5.40a   Fig 5.40b

Figs 5.39–5.40 *Stephanoeca diplocostata*. Response of silicon-deprived cells without loricae grown in stirred silicon-free medium at 20 °C to a sudden addition of silicon. Organic enrichment consists of a final concentration of 80 µg ml⁻¹ proteose peptone and 16 µg ml⁻¹ yeast extract. Control cultures with cells grown in silicon-free medium (open circles). Test cultures received an instant addition of reactive silicate (closed circles) in late exponential phase. Uptake of silicon as indicated by decrease of reactive silicate in the medium (closed squares). **Fig. 5.39a** Experiment 4. After 30 h (arrow) the test culture received a silicon addition to give 48 µM Si. **Fig. 5.39b** Test culture from Fig. 5.39a illustrating reactive silicate uptake (closed squares) after silicon addition.
Fig. 5.40a Portion of graphs of a control and test culture showing immediately after addition of silicon to the test culture to give a concentration of 71 µM Si. Reactive silicate uptake by test culture (closed squares) can be resolved into two rates; an initial faster rate followed by a slower rate. Table 5.8 lists percentages of cells with and without loricae relevant to the test culture of this experiment.
Fig. 5.40b A similar experiment to that illustrated in Fig. 5.40a but only showing test culture from the point at which the silicon spike was added to give a concentration of 71 µM Si. Reactive silicate uptake (upper closed circles) can be resolved into two rates. Table 5.9 lists percentages of cells relevant to this experiment that are: (1) naked, (2) have exocytosed costal strips on surface of cells, (3) have exocytosed strips at top of collar and (4) have loricae.

Table 5.8 *Production of loricae by mid-exponential phase, naked cells after addition of silicon to culture medium to give a concentration of 71 µM Si. Cell numbers remained more-or-less constant during the period of sampling (see Fig. 5.40a).*

Time after addition of silicon (h)	Cells without loricae (%)	Cells with loricae (%)
0	94	6
8	74	26
12	56	44
22	35	65
30	27	73
48	14	86

Reproduced from Leadbeater (1989).

concentration of cells was reasonably static during this period, the increase in the percentage of loricate cells can only be explained by 'naked' cells assembling their own loricae. This scenario runs counter to the normal pattern of events for *S. diplocostata*, whereby a cell accumulates costal strips which are subsequently bequeathed to the juvenile during division, and subsequently the juvenile assembles a lorica. If this pattern of events were to be mandatory, then cells without loricae would not be expected to assemble their own loricae (see Section 8.6.2).

This apparently anomalous behaviour was investigated further by repeating the experiment whereby a culture in mid-exponential phase was spiked with silicon, but this time a more detailed analysis of cell morphology and costal strip production was undertaken. Fig. 5.40b illustrates the behaviour of a silicon-free culture spiked with silicon to give a final concentration of 71 µM Si. Immediately following this addition to the culture, cell numbers remained more-or-less static at $29.18 \pm 2.3 \times 10^4$ cells ml^{-1} for 24 hours. During the first 14 hours of this period the concentration of reactive silicate reduced by 9 µM Si (from 71 µM to 62 µM Si). After this period the rate of silicon uptake decreased. Light and electron microscopy revealed that at ten hours after silicon addition a substantial percentage (63%) of cells had deposited and stored their costal strips on the outer surface of the cell (Figs 5.47, 8.14–8.17) and only relatively few had transported them to the top of the collar (Figs 5.46, 5.48) (Table 5.8). Eventually the surface bundles of strips are assembled into loricae by the cells themselves, with the

result that after 47 hours 88% of cells have loricae (Figs 5.49, 8.18–8.22) (Table 5.9).

Towlson (1992) carried out an experiment in which a series of polypropylene vessels containing severely silicon-impoverished cells growing in silicon-free medium were spiked with silicon at concentrations of 1, 5, 10 and 40 µM Si, respectively. At all concentrations growth, in terms of cell numbers, was halted and this included 1 µM Si where there was no possibility that cells could deposit enough strips to assemble complete loricae. In a second experiment Towlson (1992) sub-cultured a concentrate of exponentially growing 'naked' cells into silicon-containing medium containing 10 µM Si and growth was immediately halted for 20 hours, during which time silicon was taken up. The control for this experiment involved the transfer of cells to fresh silicon-free medium and under these conditions exponential growth continued without change.

5.8 DISCUSSION

Stephanoeca diplocostata has proved to be an exemplary experimental species, not only because it grows well in laboratory conditions but also because the turnover of silicon during a culture cycle is easily measurable and falls within the optimal range of the standard molybdate-blue assay (Strickland and Parsons, 1968). Had this species possessed only a few lightly silicified costae, then the measurement of silicon turnover would have been more difficult. The most important limitation of *S. diplocostata* with respect to a full study of silicon deposition is that the majority of costal strips are of similar morphology and therefore it was not possible to determine the order in which they were produced or the logistics of strip movement once they had been exocytosed from the cell. This had to wait another 25 years for the availability of *Didymoeca (Diplotheca) costata* in culture (Leadbeater, 2010) (see Chapter 7).

The obvious function of biogenic silica within costal strips is to provide the rigidity necessary to support the framework of the lorica. Costal strips with reduced silicon content lack rigidity and buckle easily when subjected to compressive forces. In *Diaphanoeca grandis* silicon-impoverished strips are incapable of supporting the veil with the result that it collapses around the flagellar and collar apparatus. Unfortunately, nothing is known regarding the micromechanical properties of costal strips, individually or when assembled into costae and loricae, except that they possess some degree of flexibility and

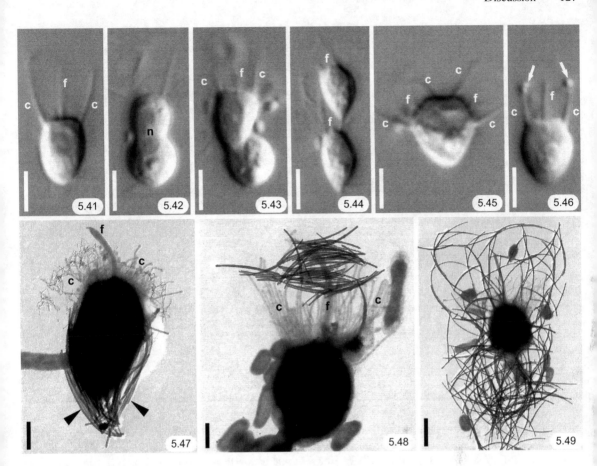

Plate 6 (Figures 5.41–5.49)

Figs 5.41–5.49 *Stephanoeca diplocostata.* **Figs 5.41–5.46** Interference contrast light microscopy of silicon-free cells. Flagellum (f), collar (c). Bar = 5 μm. **Fig. 5.41** Completely 'naked' cell.

Fig. 5.42 Naked cell undergoing mitosis, with elongated nucleus (n). Flagellum still present.

Fig. 5.43 Diagonal division of naked cell. Upper cell has a short flagellum and collar.

Fig. 5.44 Daughter cells resulting from diagonal division. Both have short flagella.

Fig. 5.45 'Monster' (conjoined) cell with two collars each surrounding a flagellum.

Fig. 5.46 Naked cell with accumulated costal strips at top of collar (arrows) 12 hours after addition of silicon.

Figs 5.47–5.49 Whole mounts of silicon-starved cells replenished with silicon. Flagellum (f), collar (c). **Fig. 5.47** Naked cell with recently exocytosed costal strips on sides of cell (arrowheads) in 'nudiform' mode. Bar = 1 μm.

Fig. 5.48 Naked cell with recently exocytosed costal strips at top of collar. Bar = 1 μm. Reproduced from Leadbeater (1985).

Fig. 5.49 Silicon-replenished cell with recently assembled lorica. Bar = 2 μm.

can be bent without breaking, as demonstrated when loricae are dried for EM preparation. It should be possible to explore the physical properties of costal strips, such as their elasticity and surface hardness, as well as their ability to withstand compressive, tensile and shear stresses. Measurements of physical criteria such as these have been achieved for diatom frustules using glass micro-needle tests and atomic force microscopy (AFM) (Almqvist *et al.*, 2001; Hamm *et al.*, 2003; Subhash *et al.*, 2005).

Ultrastructural aspects of costal strip formation in choanoflagellates conform to the general pattern of biogenic silica deposition in many other protists. The origin of silica

Table 5.9 *Production of costal strip accumulations and loricae by late exponential phase, naked cells after addition of silicon to culture medium to give a concentration of 71 μM Si. Cell numbers remained more-or-less static during the period of sampling (see Fig. 5.40b).*

| Time after addition of silicon (hours) | Naked cells without costal strips (%) | Naked cells | | Cells with incomplete loricae (%) | Cells with complete loricae (%) |
		Bundles of strips on surface of cells (%)	Accumulation of strips at top of collar (%)		
1	94	0	5	1	0
3	94	3	3	0	0
4	83	6	5	3	3
5	58	28	8	2	4
6	33	45	14	1	7
8	15	63	5	2	15
10	9	49	3	6	33
11	11	26	6	6	51
18	10	10	1	2	77
47	7	0	1	4	88

Reproduced from Leadbeater (1989).

deposition vesicles in choanoflagellates has not been resolved. While there is often a close association between SDVs and the Golgi apparatus and ER, there is no evidence that SDVs originate from either of these systems. It is possible that SDVs are an independent, self-generating membrane system within the cell. The function of the pairs of microtubules that subtend SDVs must be to support and shape the enclosed developing strips while they lack sufficient rigidity to maintain themselves. This conclusion is borne out by results from the microtubule poison experiments, where immature strips become deformed and misshapen as a result of a failure in the supporting role of SDV-associated microtubules. Subsequent silicon deposition, which is unaffected by this treatment, preserves the misshapen morphology. In *S. diplocostata*, normal exocytosis of mature strips occurs through the flattened anterior surface of the cell, between the flagellum and collar microvilli. However, in silicon-starved cells that have been replenished with silicon, costal strips are exocytosed laterally and accumulate on the sides of the cell. This change in location is now interpreted as part of a major developmental and evolutionary switch that occurs between the tectiform and nudiform conditions, respectively (see Chapter 8).

The *S. diplocostata* growth and silicon turnover experiments provide a number of important insights into the relationship between silicon metabolism and the cell cycle. In tectiform species, costal strip production appears to occupy most, if not all, of interphase. Under silicon-enriched conditions the rates of costal strip production and of growth are closely coupled, such that by the end of interphase there is exactly the correct number of fully silicified strips to produce a lorica. There is some flexibility in the silicon requirement/growth relationship in *Stephanoeca*. For instance, when silicon is moderately limiting a full complement of thinner strips is produced. Thus strip quantity takes precedence over quality. In extreme conditions of silicon deprivation, the coupling between silicon metabolism and growth proves to be facultative – growth proceeds more-or-less normally while only a few fragmentary, thinly silicified strips are exocytosed. This response is similar to that reported for chrysophytes such as *Synura* and *Paraphysomonas* (Sandgren and Barlow, 1989; Leadbeater and Barker, 1995; Sandgren *et al.*, 1996) but opposite to that of diatoms, where growth ceases under silica-limiting conditions and restarts when silicon is replenished. In *Synura petersenii*, although cells continue to grow without depositing silicified scales, trace quantities of silicon are essential for growth (Klaveness and Guillard, 1975).

The effect of silicon replenishment on silicon-deprived cells is to re-couple silica deposition in costal strips with the cell cycle and thus growth. It might have been expected that a cell that had been deprived of a lorica would be forever without a lorica. This would have been a logical conclusion in the context of the tectiform condition where a parent cell produces strips exclusively for the following generation. However, for the majority of *Stephanoeca* cells this scenario proved not to be the case; instead silicon replenishment immediately stimulated naked cells to produce strips for themselves and division was halted. In the most refined experiment 63% of naked cells exocytosed strips laterally and stored them on the sides of the cell whereas only, at most, 14% transferred strips to the top of the collar (Fig. 5.40b; Table 5.9). Storage of strips on the sides of the cell is a nudiform character and normally precedes self-assembly of the lorica (see Chapters 6 and 8). In the circumstances encountered here, to return a naked cell to the normal functioning tectiform condition requires the deposition of two sets of strips before division; the first for its own lorica followed by a second set for storage at the top of the collar and for subsequent use by the juvenile after division (see Section 8.6.1).

The rapidity with which costal strips dissolve in stirred culture was unexpected and means that all measurements of the uptake of soluble silicon are net values. As discussed above, the physicochemical conditions of culture are particularly favourable for dissolution and it is likely that under natural conditions dissolution is at least an order of magnitude slower. Nevertheless, in relative terms choanoflagellate loricae are less likely to survive sedimentation in the sea than many diatom frustules and even if they become incorporated into the sediments they are still subject to dissolution within pore water or may be crushed by compaction. The remnants of damaged choanoflagellate loricae have been observed within faecal pellets of marine zooplankton (see Section 9.4.9) (Marchant and Nash, 1986; Tanoue and Hara, 1986; Urban *et al.*, 1992). Incorporation of loricate fragments into faecal pellets protects strips from dissolution and accelerates their descent to the sediments. However, even if fragmented loricae were able to survive the subsequent compaction and preservation, the chances of recognising costal strip fragments among the mass of other siliceous structures would make reconstruction of a 'fossil' record extremely difficult.

6 • Loricate choanoflagellates: Acanthoecidae – nudiform species

Diversity based on spirals and mechanism of lorica assembly.

6.1 INTRODUCTION

The principles of choanoflagellate lorica construction have been discussed at length in Chapter 4. However, the basic microanatomy is so important to this and the next two chapters that it is worth repeating the essential details once again. The lorica comprises a basket-like framework of silicified costae that surrounds the cell. Typically it contains two layers of costae, a discrete layer of longitudinal costae and a layer of helical and/or transverse costae. The lower transverse costae in most species are on the inner surface of the lorica but in the *Parvicorbicula–Pleurasiga* species grouping they are on the outside. In addition, a thin delicate organic investment based on carbohydrate-containing microfibrils partially or entirely covers the internal lorica surface. Individual costae consist of linear arrays of rod-shaped costal strips arranged end-to-end.

The uniqueness of the choanoflagellate lorica and its universality of construction overwhelmingly support a single common origin of the Acanthoecida. This is also borne out by molecular phylogenies (see Section 10.4.1) (Carr *et al.*, 2008a; Nitsche *et al.*, 2011). Since no non-loricate species possesses anything remotely like a lorica, we must accept that with our current state of knowledge we have no morphological insight into the origin of this complex structure. However, this rather negative conclusion is not without a 'silver lining' for, although we have little insight into the evolution of the lorica *per se*, there is a remarkable evolutionary history related to its diversification once it evolved.

Since lorica construction in choanoflagellates is based on universally applicable rules (see Section 4.8), it is not surprising that the sequence of events involved in lorica production is also universally consistent. The overall control of

lorica production has much in common with the logistics of a 'supply chain' in manufacturing business. Logistics are at the heart of both processes and in business terms this means 'having the right thing, at the right place, at the right time'. Even the individual stages in both systems are analogous, namely: production, storage, transport and distribution. As will be demonstrated in this chapter and the next, lorica production comprises a series of coordinated stages in which costal strips are produced intracellularly, exocytosed, transported, stored, packaged and finally assembled into a lorica. Success in this process is dependent on costal strips, often of different morphologies, being initially produced in the 'correct' order and orientation since all subsequent stages proceed automatically with minimal provision for correction at a later stage. It might appear that such a highly coordinated sequence of events would leave little room for variation – if any stage were to alter radically the whole system would be expected to fail. However, remarkably, among extant species there are two scenarios which are characterised by their respective modes of division and associated procedures during lorica production.

The first and more straightforward scenario involves a loricate cell that divides to produce two daughter cells, one of which remains with the parent lorica, whereas the other, known as the juvenile, swims away, settles onto a substratum and subsequently produces its own lorica. This scenario is called *nudiform* because the juvenile is 'naked', that is, without a covering of costal strips (Manton *et al.*, 1981). At present there are four nudiform genera – *Polyoeca*, *Acanthoeca*, *Savillea* and *Helgoeca* – that between them contain five named species (see Table 6.2). Kent (1880–2, p. 360) originally described *Polyoeca dichotoma*, a stalked colonial species (Fig. 6.1a, b), as being "a collar-bearing flagellate that formed by serial conjunction of their daughter loricae a more or less extensive branching colony-stock". Although Kent (1880–2) did not record having seen naked juveniles, nevertheless he did appreciate that the pseudo-dichotomous colonies were generated by "each zoid giving rise by transverse fission to two new ones which attach themselves to the opposite sides of the parent lorica". The latter observation, which would have resulted in 'true' dichotomous branching, is with the benefit of hindsight incorrect. Dunkerly (1910, p. 190) completed the story correctly by illustrating 'diagonal' division (Fig. 6.1c) and concluding that "the daughter individual erects its lorica on the mouth of the mother individual's lorica instead of swimming away and settling elsewhere". Boucaud-Camou (1966) illustrated a *P. dichotoma* colony containing a juvenile cell surrounded by a partly formed lorica at the anterior end of the parent cell (Fig. 6.1b arrow). The occurrence of 'naked' juveniles has now been reported many times in nudiform species (Ellis, 1929; Leadbeater and Morton, 1974b; Leadbeater, 1979a, 2008a; Leadbeater *et al.*, 2008a, b).

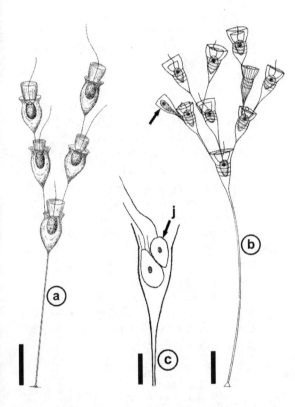

Fig. 6.1 *Polyoeca dichotoma*. a. Original illustration reproduced from Kent (1880–2). Bar = 10 μm. b. Stalked colony with recently divided daughter cell (juvenile) with newly formed lorica chamber (arrow). Bar = 10 μm. Reproduced from Boucaud-Camou (1966). c. Diagonal cell division showing two daughter cells each with a flagellum facing anteriorly. Bar = 2.5 μm. Reproduced from Dunkerly (1910).

The second scenario is in most respects more complex. Ellis (1929) twice observed 'inverted' division in *Stephanoeca ampulla* (Fig. 7.1). This type of division involved the juvenile cell being pushed backwards by a thin thread out of the parent lorica. He also recorded that "the gradual development of the lorica takes place whilst the young cell (juvenile) is floating away from the parent lorica" (Ellis, 1929, p. 72). Ellis (1929) had, in fact, noticed three key features regarding the tectiform mode of division, namely that the juvenile is inverted with respect to the parent lorica and it is neither motile nor naked. We now know that costal strips for the future juvenile are produced by the parent cell and stored at the top of the collar in anticipation of division (Figs 7.2, 7.3). During division the stored strips are bequeathed to the juvenile, enabling it to assemble its own lorica within minutes after separation from the parent lorica (see Chapter 7). Since costal strips stored at the top of the collar resemble a 'cover' or 'roof', this type of division is called *tectiform* (*tegere* = to cover) (Manton *et al.*, 1981). The overwhelming majority of loricate species (>150) are tectiform and details of their division have been described many times (Leadbeater, 1979b, c, 1994a, b, 2010; Leadbeater and Cheng, 2010). The difference in the morphology of division, although a relatively minor feature in itself, is but one of a number of interrelated characters that distinguish the nudiform and tectiform evolutionary groupings, now given family status as Acanthoecidae and Stephanoecidae, respectively (Table 6.1) (Nitsche *et al.*, 2011).

To appreciate the possible evolutionary relationship between the nudiform and tectiform conditions, it is necessary first to explore each group in detail. Thus the object of this chapter is to review nudiform species, and that of Chapter 7 is to consider the tectiform condition. In Chapter 8 the information from Chapters 6 and 7 will be brought together with the experimental conclusions from Chapter 5 and an evolutionary hypothesis will be presented. This proposes that the tectiform condition was derived from the nudiform condition by a single abrupt change involving: (1) the timing of costal strip development with respect to the cell cycle and division; (2) the order and orientation in which costal strips are produced within the cell and stored extracellularly; (3) the inversion of the juvenile during division. What is so surprising about this scenario is that these changes should have occurred concurrently and that instead of resulting in a catastrophic failure they were, in fact, of immediate evolutionary advantage.

Table 6.1 *Characters relating to costal strip production and the cell cycle that distinguish nudiform species (Acanthocidae) from tectiform species (Stephanoecidae).*

Acanthoecidae (nudiform)	Stephanoecidae (tectiform)
Cell division diagonal – flagellar poles of both daughter cells face forwards	Cell division inverted – flagellar poles of daughter cells face each other
Cell division produces a naked, motile juvenile that swims away from parent lorica	Cell division produces a non-motile juvenile that receives a full set of strips from the parent cell
Juvenile cannot assemble lorica until a set of strips has been produced	Juvenile, immediately after release from parent lorica, assembles a lorica from set of strips bequeathed by parent cell
Costal strips produced by naked juvenile at beginning of interphase for self use	Costal strips produced throughout interphase by lorica-bearing cell and passed on to juvenile during cell division
Costal strips are produced in correct orientation and stored on sides of cell	Costal strips are produced 'upside-down' and stored at top of inner surface of collar microvilli
Costal strips are produced: (1) outer layer first – base to anterior (2) inner layer second – base to anterior	Costal strips are produced: (1) inner layer first – base to anterior* (2) outer layer second – base to anterior*

* The *Parvicorbicula–Pleurasiga* species group possess transverse costae on the outer surface of the lorica (see Section 4.3.5). Costal strips for transverse costae are produced before those for the longitudinal costae.

6.2 TERMINOLOGY: SPIRALS AND HELICES

Since helical costae are a prominent feature of nudiform loricae it is helpful to define the terms relating to spirals and helices (three-dimensional spirals) that will be used in this text:

Spiral – a curve on a plane that winds around a fixed centre point at a continuously increasing or decreasing distance from the point. A spiral is a flat, often two-dimensional, planar curve that extends primarily in length and diameter, but not in height. Examples of spirals include a groove on a record and the arms of a spiral galaxy. A so-called 'spiral staircase' is a good example of a 'helix' (see below).

Helix – a three-dimensional curve that turns around an axis at a constant or continuously varying distance while moving parallel to the axis. A standard (*cylindrical*) helix is a curve that lies on the surface of a cylinder with a constant distance between adjacent gyres. Examples of helices include a metal spring and a strand of DNA. A *conical helix* is a helix on the surface of a cone and a *spherical helix* is a helix on the surface of a sphere.

Chirality or *handedness* – helices can be either right-handed or left-handed. With the line of sight along the axis of the helix, if a clockwise turning motion moves the helix towards the observer, then it is called a left-handed helix; if away from the observer then it is a right-handed helix. Handedness is a property of the helix, not of the perspective: a left-handed helix cannot be turned to look like a right-handed one unless it is viewed in a mirror, and vice versa. When the direction of a rotation is referred to in the text, it always refers to the direction when viewed from above (see also Section 4.2.1).

A spiral, conical helix and spherical helix may be *equiangular* – where the distances between successive turns increase in a geometric progression (= logarithmic spiral) – or *Archimedean* – where the distances between successive turns remain constant (= arithmetic spiral).

6.3 NUDIFORM CHOANOFLAGELLATES (ACANTHOECIDAE)

At present there are five named nudiform species attributable to four genera, namely *Polyoeca* Kent, *Acanthoeca* Ellis, *Savillea* Loeblich and *Helgoeca* Leadbeater (see Table 6.2). Of the five species, *Savillea parva* has occurred as two different forms in a single clonal culture and intermediate forms between *Acanthoeca spectabilis* and *Polyoeca dichotoma* have been observed. Thus the number of nudiform taxa could be as few as four. Whether or not there are other nudiform species yet to be found and named it is impossible to say at present.

While representatives of the four nudiform genera listed in Table 6.2 unequivocally display the key nudiform characters noted in Table 6.1, nevertheless, based on lorica morphology alone, the four genera neither appear as a coherent group, nor do they possess obviously distinctive features that separate them from tectiform species with loricae of similar morphology. For example, *Helgoeca nana* was originally attributed to the tectiform genus *Acanthocorbis* until culture studies showed it reproduced by means of naked flagellated cells (Thomsen *et al.*, 1997; Leadbeater *et al.*, 2008a). This is not surprising seeing that the same rules regarding lorica construction apply to all species (see Chapter 4). An accumulation of costal strips at the top of the collar can be taken as a defining character of the tectiform condition, but the absence of this character is inconclusive. Of the four nudiform genera, *Savillea* and *Helgoeca* have loricae that are of the standard two-layered construction and are discussed first. *Acanthoeca* possesses a highly modified lorica both in terms of its prominent spiral

and single layer of costae throughout the lorica chamber. However, a full understanding of *Acanthoeca* is seminal to understanding how lorica assembly is achieved by a forward and clockwise rotational movement of the lorica-assembling microvilli and will be discussed last in a wider context.

6.4 *SAVILLEA*

6.4.1 *Savillea*: lorica construction

Currently there are two described species of *Savillea*, namely *S. parva* (Ellis) Loeblich and *S. micropora* (Norris) Leadbeater. Both are common in stagnant coastal waters, particularly saltmarshes, and both have been obtained in mixed culture on a number of occasions (Leadbeater, 1975, 2008a). In 2000 a clonal culture of *S. micropora* was obtained and for four years this exclusively produced cells with the *S. micropora* form of lorica. In 2004 one subculture spontaneously produced cells with the *S. parva* type of lorica. Since that time several culture lines have been maintained containing exclusively *S. micropora*, but others contain a mixture of species. Since both forms occurred in a clonal culture originating from a single cell, it has been necessary to merge the two species into one taxon, and the

Table 6.2 *Cell and lorica dimensions of the five currently known extant nudiform species (n/a = not applicable).*

	Cell dimensions width × length (µm)	Flagellar length (µm)	Flagellar length/cell radius (*L/A* ratio)	Lorica chamber length (µm)	Lorica stalk length (µm)
Polyoeca					
Polyoeca dichotoma Kent	4 × 6	8	4	19	37
Acanthoeca					
Acanthoeca spectabilis Ellis	6 × 8	10	3.5	14	9
Acanthoeca brevipoda Ellis	4 × 6	10	5	10	None
Savillea					
Savillea parva (Ellis) Loeblich					
parva form	4 × 4	7	3.5	11	None
micropora form	4 × 4	None	n/a	5	None
Helgoeca					
Helgoeca nana (Thomsen) Leadbeater	3 × 5	5	3.5	7–10	None

Data taken from published information, in particular Kent (1880–2), Dunkerly (1910), Ellis (1929), Boucaud-Camou (1966), Norris (1965), Thomsen *et al.* (1997), Leadbeater and Morton (1974b), Leadbeater (1975, 1979a), Leadbeater *et al.* (2008a, b).

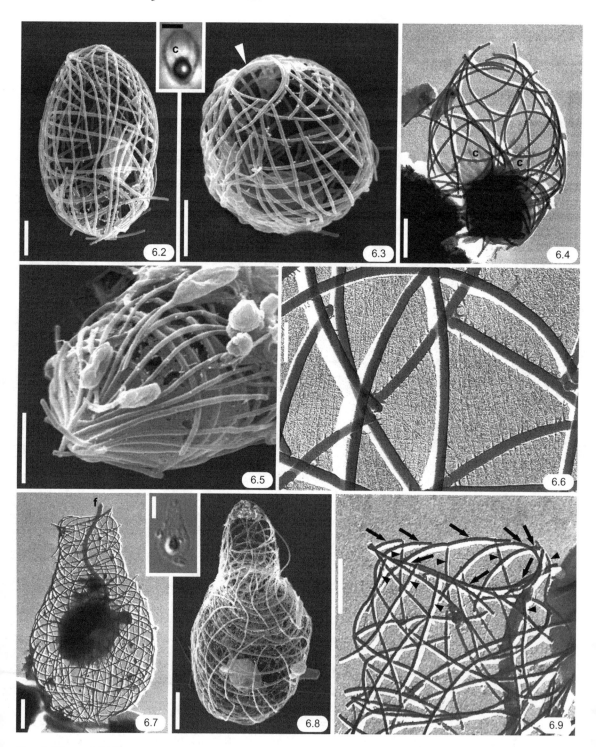

Plate 1 (Figures 6.2–6.9)

specific epithet must come from the first named, which is *S. parva*. However, it is still useful to distinguish between the two lorica types which are therefore known as the *parva* and *micropora* forms, respectively.

Savillea parva is a minute species with a cell approximately 2–3 μm in diameter located at the base of a pear-shaped (*parva* form) or ovoid (*micropora* form) lorica 5–11 μm long (Figs 6.2–6.4, 6.7, 6.8). The *parva* form possesses a flagellum slightly longer than the cell (Fig. 6.7) but the *micropora* form lacks an emergent flagellum (Figs 6.2 inset, 6.4). The collar is approximately 2 μm in length and contains between 15 and 20 collar microvilli. In ultrastructural terms, the basic costal construction of the lorica is similar in both forms except that in the *parva* form the anterior end of the lorica projects forwards in a chimney-like extension. The description given below is based on the *micropora* form and will be supplemented subsequently by detail of the *parva* form.

The *micropora* lorica is ovoid in shape and contains an arrangement of outer longitudinal and inner helical costae (Figs 6.2, 6.3). The longitudinal costae, which are on the outer surface of the lorica, extend from the extreme posterior end (Fig. 6.5) to the edge of the anterior pore (Fig. 6.3). Each longitudinal costa comprises two narrow costal strips with a considerable overlap; these are often displaced in dried specimens. The number of longitudinal costae per lorica varies from 14 to 25 in cultured material with a mode of 22. Some specimens collected from the field have as few as eight longitudinal costae (Fig. 6.4). The helical costae do not extend to the extreme posterior end of the lorica but arise about 1 μm from the bottom (Fig. 6.5). The direction (chirality) of the helix is left-handed; that is, the helix rises from lower right to upper left when observed in side view (Fig. 6.2). The spacing between gyres of the helical costae is more-or-less equal except around the apical pore, where the

costae flatten to form a rim (Fig. 6.3). The number of helical costae in the loricae of cultured cells varies from 9 to 13 with a mode of 11. At the posterior end of the cell each helical costa terminates adjacent to a longitudinal costa in a regular manner (Fig. 6.5).

To determine the numerical ratio of helical to longitudinal costae, it is necessary to generate a computer model of the lorica (Figs 6.10a–c, 6.22a–c). For this, not only do the respective numbers of costae have to be known but a value for the inclination of the helical turns with respect to the horizontal axis of the lorica is required. Using these parameters a computer image can be generated that resembles electron microscopy (EM) images (compare Figs 6.10a and b with 6.2 and 6.3, respectively; Fig. 6.22c with 6.22d). From such a model it is possible to determine both the number of helical costae present and the number of turns described by the helix (Leadbeater *et al.*, 2009). The values of these two features are 8 and 1.5, respectively. Thus a typical lorica of a cultured *micropora* cell contains 22 longitudinal costae, and eight helical costae, each of which undergoes 1.5 turns

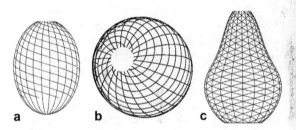

Fig. 6.10 Computer-generated images of loricae illustrating the organisation of vertical and helical costae in *Savillea parva*: a–b. *micropora* form – ratio 1:3 helical to longitudinal costae with 1.5 rotations (540°). c. The *parva* form – ratio 1:1 helical to longitudinal costae with two rotations (720°). Reproduced from Leadbeater (2008a).

Figs 6.2–6.9 *Savillea parva*. **Figs 6.2–6.3** The *micropora* form. SEM images showing two-layered construction of lorica, with an outer layer of longitudinal costae and inner layer of helical costae. Anterior pore (arrowhead) shown in Fig. 6.3. Bars = 1 μm. Fig. 6.2 inset. Living cell, with prominent collar (c) but no flagellum, located at bottom of lorica. Bar = 2 μm.

Fig. 6.4 The *micropora* form. Cell, with collar microvilli (c), at base of lorica with relatively few costae. Bar = 1 μm.

Fig. 6.5 The *micropora* form. Base of lorica showing the convergence of the outer longitudinal costae. Bar = 1 μm.

Fig. 6.6 Higher power of part of the lorica shown in Fig. 6.4 showing regular meshwork of organic microfibrils in spaces between costae. Bar = 0.25 μm.

Figs 6.7–6.9 The *parva* form. **Figs 6.7–6.8** TEM and SEM images, respectively, of cells located in pear-shaped loricae with chimney-shaped neck. Flagellum (f). Bars = 2 μm. **Fig. 6.8 inset.** Interference contrast micrograph of living cell. Bar = 1 μm.

Fig. 6.9 Anterior ring comprising eight almost horizontal overlapping helical costae (arrows) in a left-handed conformation. Each helical costa is associated with one longitudinal costa (arrowheads). Between the costae faint microfibrils of the organic investment can just be seen. Bar = 0.25 μm. Figs 6.5, 6.7, 6.8 and 6.9 reproduced from Leadbeater (2008a).

(540°). The ratio of helical to longitudinal costae is approximately 1:3. Since the anterior pore is formed by eight overlapping helical costae, on average each helical costa is associated at its anterior end with three longitudinal costae. Loricae with ratios approximating to 1:1 and 1:2 helical to longitudinal costae have been observed in material collected from the field.

In contrast to the *micropora* form, cells with the *parva* type of lorica possess a prominent anterior flagellum (Fig. 6.7) which usually undulates vigorously. The arrangement of costae comprising the *parva* lorica is similar to that of the *micropora* form, the major difference being the anterior extension of the lorica to give an overall pear-shape (Figs 6.7, 6.8). In common with the *micropora* form the costae of the *parva* form are arranged in two layers, the outer layer being longitudinal and the inner layer a series of helical costae (Figs 6.7, 6.8); the ratio of helical to longitudinal costae in the *parva* form is 1:1 (Fig. 6.10c). The number of longitudinal costae varies from eight to ten and the anterior ring comprises a similar number of strips overlapping each other in a clockwise direction when viewed from above (Fig. 6.9). The number of helical costae that can be discerned in lateral view (16–20) is approximately twice the number of longitudinal costae, indicating that each helical costa undergoes two turns of 360° (Fig. 6.10c). The helical costae are equally spaced throughout the length of the lorica, except for the anterior pore where they converge. Although the junctions between the helical and longitudinal costae at the anterior pore are difficult to resolve, the anterior strip of each helical costa approaches its abutment onto the respective longitudinal strip from the right when viewed from the outside and left when viewed from the inside (see direction of arrows in Fig. 6.9).

The inner lorica surface of both forms of *S. parva* is covered by an organic investment, which in shadowcast whole mounts consists of a precise rectangular pattern of microfibrils (Fig. 6.6). In some specimens the spaces within the meshwork are open and in others closed. It is probable that the meshwork provides support for a continuous matrix. Critical-point dried specimens prepared for scanning electron microscopy (SEM) often have a 'cobweb' pattern of microfibrils (Fig. 6.8), which is probably artefactual. The closely positioned costae, together with an inner organic investment, must prevent fluid passage through the side wall of the lorica. The only means of ingress or egress can be through the relatively small anterior pore. The absence of an emergent flagellum on the *micropora* form and the short collar cast doubt on whether *micropora* cells

actually ingest solid particles at all. The *micropora* form of *S. parva* typically occurs embedded within biofilms and it is possible that it makes use of microbial exudates, including extracellular polysaccharide substances (EPS) in its nutrition.

6.4.2 *Savillea*: cell division and lorica assembly

Prior to cell division of the *micropora* form, the hitherto non-flagellate cell elongates and produces a short flagellum (Fig. 6.11). Cell division is diagonal, so the two daughter cells are located one above the other. The flagellum passes to the upper (juvenile) cell which subsequently moves away (Fig. 6.12). When the parent lorica is embedded in a microbial biofilm, the juvenile rarely moves further than just outside the anterior pore. The juxtaposition of a juvenile in the process of accumulating costal strips adjacent to the apical pore of a parent lorica is common.

On settlement the spindle-shaped juvenile withdraws its flagellum and immediately produces costal strips which are exocytosed and accumulated in longitudinal bundles on the cell surface (Leadbeater, 1975). The strips destined for the longitudinal costae are exocytosed first and are therefore on the outside. These are followed by the strips destined for the helical costae which will be on the inside of the accumulation nearest to the plasma membrane. An insight into this organisation is gained from cells disrupted during preparation for EM (Figs 6.13, 6.14). Dried cells shrink and the strips splay out to reveal an outer layer of thinner strips in pairs or fours that are destined for future longitudinal costae (Figs 6.13 and 6.14 smaller arrows). The inner, thicker curved strips are arranged in larger groups and are titled at an angle to the cell. The direction of this tilt, lower right to upper left, reflects the direction that the left-handed helical costae will assume once they have been assembled (Figs 6.13 and 6.14 larger arrows). The posterior organic sheath is also visible when the accumulation of strips is complete.

Once a full complement of strips has been accumulated, the cell assembles them into a lorica in a single continuous action lasting two or three minutes. This is achieved by the extension of sub-groupings of strips on the surface of the cell into longitudinal and helical costae, respectively. While stages of lorica assembly in *S. parva* are not available from EM, it is nevertheless possible to reconstruct the sequence of events that must occur on the basis of what is known for other species. In *Acanthoeca spectabilis*, *Stephanoeca diplocostata* and *Didymoeca costata* a specific set of

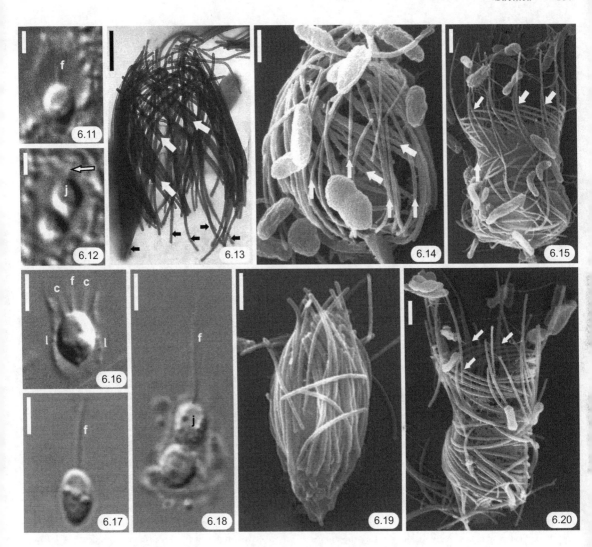

Plate 2 (Figures 6.11–6.20)

Figs 6.11–6.20 *Savillea parva – micropora* form and *Helgoeca nana*. **Figs 6.11–6.12** *S. parva*. Interference contrast micrographs of living cells. Bars = 1 μm. Reproduced from Leadbeater (2008a). **Fig. 6.11** Cell with flagellum (f) immediately prior to cell division. **Fig. 6.12** Two daughter cells following division; the upper, juvenile cell possesses the flagellum (arrow).

Figs 6.13–6.14 *S. parva*. Flattened juvenile cells showing the outer arrangement of vertical strips destined for the longitudinal costae (small arrows) and the inner sloping strips destined for the helical costae (large arrows). Bars = 0.5 μm. Fig. 6.14 reproduced from Leadbeater (2008a).

Fig. 6.15 *Helgoeca nana*. SEM of lorica showing outer longitudinal and inner helical costae. Smaller (left) arrow denotes one helical costa abutting onto a single longitudinal costa; the larger arrows denote two helical costae abutting onto a single longitudinal costa. Bar = 1 μm.

Figs 6.16–6.18 *Helgoeca nana*. Interference contrast micrographs of living cells. Bars = 0.5 μm. Reproduced from Leadbeater *et al.* (2008a). **Fig. 6.16** Interphase cell with collar (c) and flagellum (f) in small beaker-shaped lorica (l).

Fig. 6.17 Spindle-shaped 'naked' juvenile cell with flagellum (f).

Fig. 6.18 Recently divided cell; upper cell (juvenile) with long flagellum (f).

Fig. 6.19 *H. nana*. SEM of spindle-shaped juvenile cell surrounded by vertical bundles of costal strips. Bar = 1 μm. Reproduced from Leadbeater *et al.* (2008a).

Fig. 6.20 *H. nana*. SEM of lorica showing dislodged outer longitudinal and inner helical costae. Arrows denote anterior end of helical costae abutting onto respective longitudinal costae. Bar = 1 μm.

'lorica-assembling' microvilli, located just behind the collar microvilli, is responsible for providing the combination of a forward and clockwise rotational movement (Figs 6.55, 6.56, 7.16–7.21, 7.31) (Leadbeater, 1979c, 1994b; Leadbeater et al., 2008b). Individual lorica-assembling microvilli, like the collar microvilli (see Section 2.7.2), are capable of active changes in length. Thus they are capable of providing the forward movement. To achieve a clockwise rotational movement, the entire ring of microvilli is required to rotate around a central axis. In the case of the micropora form of Savillea this requires 1.5 complete turns amounting to 540°, whereas for the parva form two complete turns, amounting to 720°, are necessary (see Section 4.3.4).

A sequence of three-dimensional computer images shows how a combination of a forward and clockwise rotational movement can convert sub-groups of costal strips that are more-or-less parallel to the long axis of the juvenile into an ovoid lorica comprising an outer layer of longitudinal and inner layer of helical costae (Figs 6.23a–c). Essential to achieving this are two important criteria: (1) sub-groupings of strips destined to produce the longitudinal costae must remain vertical and rotate (carousel) horizontally in a clockwise direction (as seen from above) around the long axis of the cell; and (2) sub-groupings of strips destined to produce the helical costae must be attached at the front end to a respective longitudinal costa and at the rear end to a static position on the surface of the cell. The lorica-assembling microvilli that are on the inner surface of the developing longitudinal costae, and the helical costae where they cross the longitudinal costae, are responsible for extending the longitudinal costae forwards and the helical costae diagonally. In this context the microvilli have a dual role of moving and positioning the two sets of costae. Further circumstantial evidence for the mechanism of lorica assembly will be provided in the detailed discussion of Acanthoeca (see Section 6.6.4).

In summary, Savillea parva provides one of the best examples of a two-layered lorica. It also raises many fundamental questions regarding the ultrastructure and mechanism of lorica assembly. Most important among these concerns is the need for evidence that a clockwise turning motion is required for lorica assembly and a possible mechanism whereby this could be achieved. Example species presented in this and the following two chapters will build up an unequivocal case for the hypothesis outlined above. Savillea parva also illustrates the key features of a nudiform species. These include reproduction by

means of a naked flagellated cell; costal strip exocytosis onto the surface of the cell; the production of strips for the outer longitudinal costae first and the inner helical costae second; and the normal orientation of stored costal strips with the anterior end forwards and the posterior end rearwards. Savillea parva has loricate features in common with Helgoeca nana but is dissimilar from Acanthoeca and Polyoeca. However, the shared nudiform characters between these species are unequivocal.

6.5 HELGOECA

6.5.1 Helgoeca: lorica construction

Helgoeca is currently represented by one species, H. nana, which was originally allocated to the tectiform genus Acanthocorbis (Thomsen et al., 1997; Leadbeater et al., 2008a). However, when observed in culture the production of naked motile cells immediately demonstrated its nudiform condition. H. nana is a minute species; the cell is approximately 4 μm in diameter and is located in a beaker-shaped lorica 7–10 μm long by 4–5 μm wide (Fig. 6.16). Cell division is preceded by withdrawal of the parent flagellum. Division is diagonal and results in a juvenile cell with a long flagellum (Figs 6.17, 6.18) that swims away from the parent lorica, settles onto a surface, becomes spindle-shaped and produces a covering of costal strips (Fig. 6.19) (Leadbeater et al., 2008a).

The lorica is of standard two-layered construction, with an outer layer of 10–12 longitudinal costae that project forwards as spines (Figs 4.20, 6.15, 6.20, 6.22d). The inner layer consists of helical costae that anteriorly abut onto the respective longitudinal costae in an approximately 1:1 ratio (helical to longitudinal costae) (Fig. 6.22c). Unfortunately loricae are prone to collapse when critical-point dried but observation of many specimens allows a reasonable appraisal of the costal pattern. Each of the anterior spines is subtended by two supporting strips and it is in the region of this junction that the anterior end of the respective helical costa terminates (Fig. 6.20). Observation of many loricae reveals that the inner layer comprises 10–12 equally spaced helical costae (Fig. 6.15). In most specimens the pitch of the helix flattens anteriorly. The direction of coiling is left-handed so that the direction of tilt when seen from the side is lower-right to upper-left. It has not been possible to determine with certainty the number of turns for each helix, but it is approximately a single turn of 360° (Fig. 6.15). A computer-generated image illustrates an

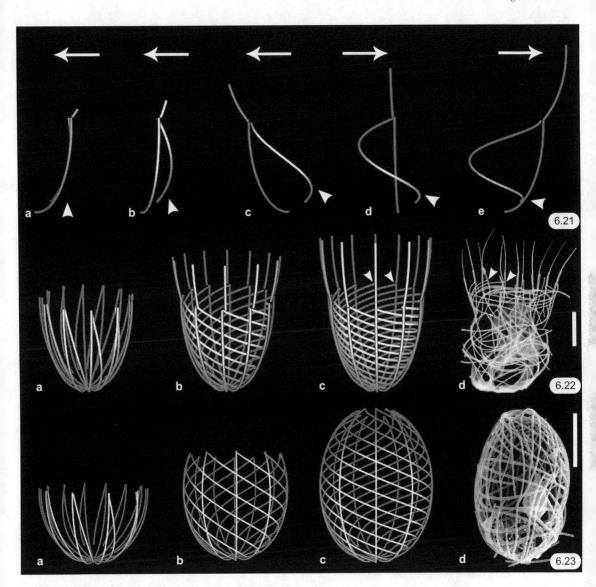

Plate 3 (Figures 6.21–6.23)

Figs 6.21–6.23 Computer-generated images of developing nudiform loricae (*Helgoeca nana* and *Savillea parva*) from vertical bundles of strips to completed loricae and actual examples from electron microscopy. Bars = 2 μm. Reproduced from Leadbeater *et al.* (2009).
Fig. 6.21 Formation, during one complete rotation (360°), of an outer longitudinal and inner helical costa from two vertically aligned strips (Fig. 6.21a). Arrows denote direction of movement of tip during rotation (clockwise as seen from above – starting left and finishing right). Arrowheads denote angle of inclination of helical costa from vertical (Fig. 6.21a) to almost horizontal (Fig. 6.21e).
Fig. 6.22 *Helgoeca nana*. Fig. 6.22a–c Assembly of the lorica as a result of a forward and 360°clockwise movement. Fig. 6.22d Actual lorica. Arrowheads in Fig. 6.22c–d point to junctions between anterior ends of the helical costae and their respective longitudinal costae.
Fig. 6.23 *Savillea parva – micropora* form. Fig. 6.23a–c Assembly of the lorica from approximately vertical bundles of strips as a result of a forward and 540° clockwise movement. Fig. 6.23d Actual lorica.

idealised pattern of costae for *Helgoeca nana*. Fig. 6.22a–c presents a set of computer-generated images illustrating how a *Helgoeca* lorica, comprising outer longitudinal and inner helical costae, can be generated from vertical sub-groupings of strips. The same criteria regarding the logistics of assembly as noted for *Savillea* in Section 6.4.1 are also relevant to *Helgoeca*. However, whereas in *Savillea* the helical costae extend to the top of the lorica, in *Helgoeca* they remain at more-or-less the same height as the originating vertical sub-groupings of strips (compare Figs 6.23a–c with 6.22a–c). The manner in which an initial vertical sub-grouping of strips can be rotated to form a flattened helix without any increase in height is shown in Fig. 6.21a–e. This sequence of computer-generated images illustrates how a single turn of 360° by a developing longi-tudinal costa can extend the tilted strips to produce a flattened helical costa. This phenomenon is also relevant to tectiform genera such as *Stephanoeca* and *Saepicula* that have a flattened helix in the base of their loricae (see Sections 4.3.1, 4.6.1 and Chapter 7) (Leadbeater *et al.*, 2009).

There are similarities between the costal patterns of *Savillea parva* and *Helgoeca nana*. The lorica chamber of *Helgoeca*, comprising helical and longitudinal costae, is not dissimilar to that of the *micropora* form of *Savillea parva* (compare Fig. 6.22d with 6.23d). This is particularly appar-ent if computer reconstructions are compared (compare Fig. 6.22c with 6.23c). The anterior ring of spines and the supporting strips in *Helgoeca nana* are reminiscent of the equivalent structures in *Polyoeca dichotoma* and *Acanthoeca spectabilis* (compare Figs 6.20 and 6.22d with Figs 6.27 and 6.30). The inner surface of the lorica cham-ber of *Helgoeca nana* is covered by a continuous organic investment (Figs 6.15, 6.20).

6.6 THE *POLYOECA–ACANTHOECA* CONTINUUM

6.6.1 *Polyoeca dichotoma*

Polyoeca dichotoma Kent was the first nudiform species to be recorded, although neither Kent (1880–2) nor Dunkerly (1910) could have been aware of its evolutionary signifi-cance at the time of their respective publications (see Section 6.1). When these authors observed *Polyoeca* there was no appreciation that the periplast could be anything other than organic and in many respects *Polyoeca* appeared to be an unexceptional member of the Salpingoecidae

(Craspedida). Despite Dunkerly (1910) reporting 'oblique' (diagonal) cell division giving rise to a naked flagellated juvenile (Fig. 6.1c) that failed to disperse, but instead assembled its own lorica while still attached to the parent, this, at the time, would not have appeared to be unusual behaviour for a member of the Craspedida.

The form of branching in *Polyoeca* is similar to that of the planktonic chrysophyte alga *Dinobryon* and for similar reasons (Herth, 1979). While superficially the form of branching might appear dichotomous (Figs 6.1a, b), this is not the case since each cell does not divide equally to produce two similar branches. Instead, division is diagonal and unequal – the lower daughter cell remains within the parent lorica while the upper daughter (juvenile) does not disperse but subsequently assembles its own stalked lorica using the parent as a substratum (Fig. 6.1b arrow). Succes-sive divisions of the lower cell can result in two or more attached daughters belonging to different generations (Fig. 6.1b) (Dunkerly, 1910; Boucaud-Camou, 1966). Dunkerly (1910) considered that his specimens of *Polyoeca* differed from Kent's (1880–2) original description because the cells fitted more tightly within their respective lorica chambers. He also took exception to the specific epithet *dichotoma* on the grounds that it was a misnomer. For these reasons he called his specimens *Polyoeca dumosa* (*dumosa* = bushy, tufted). However, Kent (1880–2) did acknowledge that the branching was 'sub-dichotomous' and for this reason Boucaud-Camou (1966) re-combined the two species, returning them to their original name *Polyoeca dichotoma*. Norris (1965) named a long-stalked species with anterior spines *Acanthoecopsis spiculifera*. However, his illus-tration (Fig. 14 in Norris, 1965) leaves little doubt that this is also *Polyoeca dichotoma*.

Polyoeca dichotoma is a sedentary species with a lorica that comprises a long stalk subtending a funnel-shaped chamber with an anterior ring of spines (Figs 6.24–6.27). A series of longitudinal costae extends from the base of the stalk through the chamber, eventually giving rise to the spines (Figs 6.26, 6.27). The narrow, parallel-sided stalk contains approximately ten longitudinal costae that form a steep left-handed helix of one or more turns. Within the chamber wall the number of longitudinal costae increases to between 22 and 43 (Table 6.3). Of these, 13–16 extend forwards to form a regular array of spines while the remain-der terminate at the level of the anterior transverse band of costal strips (Figs 6.26, 6.27). Thus, neighbouring spines are separated by one or two shorter longitudinal costae. Each spine contains two overlapping strips and terminates

Plate 4 (Figures 6.24–6.29)

Figs 6.24–6.29 *Polyoeca dichotoma, Acanthoeca spectabilis* intermediate form and *Acanthoeca brevipoda*. **Figs 6.24–6.25** *P. dichotoma*. TEM images of loricae showing long stalk (p) and lorica chamber comprising longitudinal costae with three bands of helical costae (arrowheads) and anterior ring of spines. Bars = 5 μm. **Fig. 6.26** Higher power of lower chamber in Fig. 6.25 showing the bands of helical costae (arrowheads); strips from the anterior band on the distal side (inner surface) abut base of spines from left-hand side (arrows). Base of stalk of a daughter cell (asterisk) attached to parent lorica at height of the intermediate helical band. Bar = 2 μm.
Fig. 6.27 *P. dichotoma*. Lorica chamber with fewer longitudinal costae showing free portion of spines comprising two costal strips. Bar = 2 μm. Fig. 6.27 inset. Anterior end of spine showing forked tip. Bar = 0.1 μm.
Fig. 6.28 *Acanthoeca spectabilis* intermediate form. Base of lorica chamber comprises a closely wound left-handed helix of costae that subtend a ring of longitudinal costae and a band of helical costae (arrowheads). Free ends of anterior spines contain two costal strips. Bar = 2 μm. Fig. 6.28 inset. Anterior end of spine showing forked tip. Bar = 0.1 μm.
Fig. 6.29 *A. brevipoda*. Lorica chamber comprises a left-handed helix of broad costal strips subtending a band of diagonal strips. Free end of each spine comprises one costal strip. Shrivelled cell (asterisk) with flagellum (f) can be seen inside lorica. Bar = 2 μm. Figs 6.25 and 6.26 reproduced from Leadbeater (1979a).

in a minute forked tip (Fig. 6.27 inset). On the inner surface of the longitudinal costae three approximately equally spaced bands of helical costae maintain the cross-sectional diameter of the chamber (Figs 6.24–6.27 arrowheads). The anterior band comprises two or three layers of costal strips, each strip abutting onto the respective longitudinal strip at the junction between the second and third strips from the tip of the spine (Figs 6.26, 6.27). When the chamber is viewed from outside the abutting transverse (helical) strip is on the right-hand side and when seen from the inside each abutting strip is on the left-hand side as shown by arrows in Fig. 6.26. The stalk and lorica chamber, up to the level of the intermediate transverse band of costae, are covered internally by a continuous organic investment. Between the intermediate and anterior transverse costal bands the organic investment takes the form of a regular meshwork of microfibrils similar to that observed on *Acanthoeca* loricae (Fig. 4.43). In colonial specimens the base of the daughter lorica stalk is located in the mid-region of the parent chamber, demonstrating that the juvenile cell does not migrate along the parent lorica after division (Figs 6.25, 6.26). This is contrary to the equivalent situation in the chrysophyte

alga *Dinobryon*, where after division the juvenile migrates to the rim of the parent lorica (Herth, 1979).

6.6.2 *Acanthoeca*

The genus *Acanthoeca* contains two species, *A. spectabilis* Ellis and *A. brevipoda* Ellis, which are both ubiquitously distributed and typically occur as components of marine microbial biofilms (Figs 6.29, 6.30). *A. spectabilis* is common on the surfaces of seaweeds, particularly *Ulva*, *Enteromorpha* and filamentous algae (Ellis, 1929; Leadbeater and Morton, 1974b; Leadbeater, 1979a), although it probably occurs on all types of suitable marine and brackish water surfaces. *Acanthoeca brevipoda*, while occurring in similar localities to *A. spectabilis*, is far less common. The *Acanthoeca* lorica is unmistakable and unique because of the closely wound helical pattern of costae shaping the chamber and an apparent lack of longitudinal costae. When specimens are dried and preserved for EM the helical organisation is so stable that it is rarely disturbed (Figs 4.18, 6.30, 6.51–6.53, 6.57–6.59) (see Section 4.3.4).

Table 6.3 *Lorica dimensions and spine and longitudinal costal numbers of* Polyoeca dichotoma, Acanthoeca spectabilis *(intermediate form),* A. spectabilis *(from field samples and clonal culture, respectively) and* Acanthoeca brevipoda *(mean ± SD unless stated otherwise; (*n*) = sample size) (n/a = not applicable).*

Lorica feature	*Polyoeca dichotoma*	*Acanthoeca spectabilis* (intermediate form)	*Acanthoeca spectabilis* (field specimens)	*Acanthoeca spectabilis* (cultured cells)	*Acanthoeca brevipoda*
Overall lorica length (μm)	56.9 ± 12.22 (5)	26.0 ± 4.13 (4)	23 ± 6.67 (21)	17.9 ± 1.81 (20)	10.4 ± 2.16 (5)
Length of chamber and spines (μm)	19.2 ± 4.22 (25)	15.6 ± 0.88 (9)	13.7 ± 2.41 (31)	10.5 ± 0.95 (20)	10.4 ± 2.16 (5)
Length of spines (μm)	9.9 ± 1.42 (25)	8.3 ± 0.36 (9)	6.6 ± 1.21 (31)	4.3 ± 0.47 (20)	4.1 ± 0.68 (5)
Length of stalk (μm)	37.4 ± 15.09 (5)	10.4 ± 3.14 (4)	9.2 ± 5.04 (21)	7.4 ± 1.33 (20)	None
Number of spines (range and mode)	13–16 mode 14 (28)	10–15 mode 13 (11)	11–16 mode 13 (33)	13–15 mode 15 (20)	10–12 mode 11 (6)
Number of longitudinal costae comprising lorica chamber (range and mode)	22–43 mode 36 (22)	28–41 mode 34 (11)	n/a	n/a	n/a

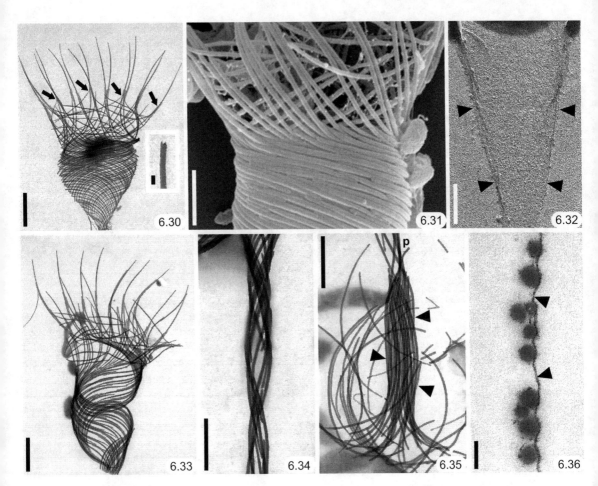

Plate 5 (Figures 6.30–6.36)

Figs 6.30–6.36 *Acanthoeca spectabilis*. SEM and TEM images of lorica. **Fig. 6.30** TEM of lorica showing closely wound costal helix and spines with diagonal strips that, on the distal surface, abut from the left (arrows). Free ends of spines comprise a single costal strip. Bar = 2 μm. Fig. 6.30 inset. Anterior end of spine showing forked tip. Bar = 0.1 μm.

Fig. 6.31 SEM of top of lorica chamber showing closely wound left-handed helix of costae that, at the anterior end, deflect upwards to give rise to spines that contain a single costal strip. At junction of spines and helix an inner layer of diagonal strips (lower left to upper right) serve to position and stabilise spines. Bar = 1 μm. Reproduced from Leadbeater *et al.* (2008b).

Fig. 6.32 Lorica, treated with hydrofluoric acid to remove strips, showing underlying microfibrillar organic investment (arrowheads). Bar = 0.5 μm.

Fig. 6.33 Partially 'unwound' lorica showing continuity of spines and (longitudinal) costae within helix. Bar = 2 μm.

Fig. 6.34 Intermediate region of stalk showing left-handed helical arrangement of costae. Bar = 2 μm.

Fig. 6.35 Base of stalk (p) surrounded by an outer cylinder of residual costal strips (arrowheads). Bar = 2 μm.

Fig. 6.36 Vertical section of lorica chamber showing costal strips of helix and thick organic lining (arrowheads). Bar = 0.1 μm. Reproduced from Leadbeater *et al.* (2008b).

The lorica of *Acanthoeca spectabilis*, like that of *Polyoeca dichotoma* (see above), is characterised by a relatively long stalk subtending a funnel-shaped chamber with an anterior ring of spines (Table 6.3). Costal strips contributing to the lower portion of the stalk are more-or-less parallel to the long axis of the lorica, although even here there may be evidence of turning (Fig. 6.34). The costae of the chamber describe a conical helix which flattens as the anterior end is approached (Figs 6.30, 6.31). The direction (chirality) of the helix is left-handed, ascending in a clockwise direction when viewed from the front end. When seen in side view with the flagellum directed upwards, the helix ascends from lower-right to upper-left (Figs 4.18, 4.19, 6.31). The anterior part of the lorica comprises a crown of 11–16 vertically positioned spines (Table 6.3). Each spine consists of a single anterior strip with a minute forked tip (Fig. 6.30 inset), supported by two or three strips that are continuous posteriorly with the top of a helical costa (Figs 6.30, 6.33). Two series of strips on the inner surface of the spines stabilise the crown (Figs 6.30, 6.31). One set of diagonal strips just above the anterior end of the helical costae, with upper-right to lower-left inclination, supports the lower end of the strips subtending the spines (Fig. 6.31). At the junction of the anterior spines with subtending strips there is a band comprising one or two layers of almost transverse strips (Fig. 6.30).

Careful analysis of the organisation of the helical costae and their component strips reveals that each spine is continuous with a single compound costa that probably runs the entire length of the lorica. There is an abrupt change of direction between the spine together with its supports, which are parallel to the long axis of the cell, and the top of the lorica chamber where the costae are more-or-less horizontal. The continuity of individual costae with the spines and the extent of overlap between costal strips can be seen in a lorica that has partially unwound during preparation (Fig. 6.33). In the mid-region of the lorica chamber gaps are often evident between overlapping costal strips. A mathematical analysis of the helical costae of the specimen of *A. spectabilis* illustrated in Fig. 4.18 demonstrated that they underwent two turns within the height of the chamber (see also Fig. 4.19) (Leadbeater *et al.*, 2008b; Frösler and Leadbeater, 2009). However, in some whole mounts it is apparent that these costae may undergo one or more additional turns within the stalk (Fig. 6.34). Thus assembly of the lorica may require costae to undergo up to four complete turns of 360° around the cell during assembly. A substantial organic investment covers the inner surface of the hollow stalk and the lorica chamber up to the top of the helix (Figs 6.32, 6.36). Between the top of the helix and the base of the anterior spine-strip the organic investment is continued as a regular meshwork of microfibrils (Fig. 4.43).

Occasionally in samples collected from the field a form of *Acanthoeca spectabilis* is found which possesses a ring of spines more similar to those of *Polyoeca dichotoma* than to the 'standard' version of *A. spectabilis* (Fig. 6.28). In Table 6.3 this version is designated '*Acanthoeca spectabilis* intermediate form'. The stalk and the chamber are exactly equivalent to *A. spectabilis*; the chamber contains the familiar helical costal arrangement. However, the crown of spines differs from standard *A. spectabilis* in two respects: (1) the spine bases are not confluent with the helical costae of the chamber and (2) the length of each spine above the helix comprises three costal strips, two of which project forwards as the free end of the spine (Fig. 6.28). This second point is at variance with the standard *A. spectabilis* form that has only two strips above the helix and only one of these contributes to the free end of the spine (compare Fig. 6.28 with 6.30 and note measurements of respective spine lengths in Table 6.3).

Acanthoeca brevipoda is an enigmatic species. In spite of several sightings in samples from European, Japanese and West Coast North American waters (Ellis, 1929; Norris, 1965; Leadbeater, 1972c; Thomsen, 1973; Hara and Takahashi, 1984; Tong, 1997a), it has never been observed in anything except small numbers. *A. brevipoda* is unmistakably an *Acanthoeca* species on account of its single layer of closely wound broad helical costae around the chamber and its anterior ring of 10–12 spines (Fig. 6.29). Between the spine bases are two or three longitudinal costal strips, resembling a similar organisation of longitudinal costae in *Polyoeca dichotoma* and the intermediate form of *A. spectabilis* (compare Figs 6.27 and 6.28 with 6.29). Occasionally specimens of *A. spectabilis* with some *A. brevipoda* characters have been recorded. In particular, the helical costae are broader and the anterior ring of spines resembles that of *A. brevipoda*, but the cells are stalked which is an *A. spectabilis* character.

6.6.3 Significance of *Polyoeca–Acanthoeca* lorica morphology

Without a doubt *Acanthoeca spectabilis* is the species that has contributed most to our understanding that a clockwise rotational movement is essential for lorica assembly. Without *Acanthoeca*, at best this conclusion would have been delayed and at worst overlooked completely. With this knowledge in hand, it is now relatively easy to appreciate the many 'tell-tale' signs that show that a rotational movement is essential for the construction of all choanoflagellate loricae, whether or not they possess helical costae (see Chapter 4).

However, in spite of the ubiquity of *Acanthoeca spectabilis* and its obvious lorica morphology, it is an unusual species for several reasons. First, it is closely related to another species, *Polyoeca dichotoma*, which has the more conventional longitudinal costae rather than a layer of tightly wound helical costae. Second, the helical costae of *A. spectabilis* do not conform to the normal definition of this costal category, which envisages them being derived from the inner layer of the lorica (see Section 4.3.4). It is probable that both of these 'unusual' features can be accounted for by a detailed comparison of the loricae of *Polyoeca* and *Acanthoeca*.

There are obvious morphological similarities between *Polyoeca* and *Acanthoeca* (see Table 6.3 for comparative measurements). Both have relatively long hollow stalks made up of a cylinder of 'helical' costae. Both have a funnel-shaped lorica chamber with a ring of spines that are supported by a transverse arrangement of overlapping strips. Both reveal ample evidence that a left-handed rotation has taken place during lorica assembly. However, there are obvious differences. In *Polyoeca* the lorica chamber comprises an outer layer of longitudinal costae supported by two transverse bands (posterior and intermediate) of helical costae. The crown of spines differs for the reasons outlined above, namely that in *Polyoeca* each spine comprises three costal strips, two of which contribute to the free spine. In *Acanthoeca* the chamber comprises a single layer of helical costae; there are no bands of transverse costae within the helix; the spines are only two strips in length, the free terminal portion consisting of one strip.

Bearing in mind these differences in lorica construction, what changes would be required to effect the transformation of *Polyoeca* to *Acanthoeca*? The key change necessary would be loss of the two lower bands of transverse costal strips (posterior and intermediate) from the chamber of *Polyoeca*. If this was achieved, then the longitudinal costae would come into intimate contact with the cell surface so any rotation that occurred would generate a helix from otherwise longitudinal costae. This scenario highlights the fact that the helical conformation is not a property of the costa *per se*, but rather a result of the direction in which the contributing costal strips are extended during assembly. Helical costae, just like longitudinal costae, are extended lengthwise, but whereas longitudinal costae are extended in the vertical plane, helical costae are drawn out diagonally by a rotational movement of the cell. If this scenario regarding *Polyoeca* is correct, then only a relatively minor change would be required to convert the longitudinal costae into

the helical costae of *Acanthoeca*. Another feature of more general importance is that the generation of a helix does not necessarily require an increase in height. This is illustrated by the chamber of *Acanthoeca*, where the anterior flattening of the helix results in a transition from almost horizontal costae to vertical spines. Generation of flattened helices is a particularly important feature of the posterior chambers of tectiform species, such as *Stephanoeca* and *Saepicula* (see Chapters 4 and 7). There are other features about the *Acanthoeca* lorica that are exceptional. In particular, the helical patterns observed require at least 2.5–4 turns of 360°. This must be a record for a choanoflagellate lorica; the *parva* form of *Savillea parva* is the nearest runner-up with just two turns (see Section 6.4.1).

Details relating to the logistics and mechanism of lorica production in *Acanthoeca spectabilis* have been gleaned from a combination of ultrastructural and experimental studies. Since lorica assembly is so intimately related to the cell cycle, this is described first, followed by details of an experimental programme involving cytoskeletal poisons.

6.6.4 *Acanthoeca spectabilis*: cell division and lorica assembly

Cell division in *Acanthoeca spectabilis* is of the nudiform type (Figs 6.37–6.41). Immediately prior to division, the flagellum is withdrawn and the nucleus undergoes mitosis (Figs 6.37, 6.38). The long collar microvilli of the parent appear to be transferred directly to the daughter cell that remains within the parent lorica. During cytokinesis the front end of the cell enlarges and gives rise to a small pyriform daughter (juvenile) that quickly acquires a long flagellum (Figs 6.41, 6.42). Developing costal strips can be seen within the juvenile before cytokinesis is complete so the initiation of silica deposition is immediate. Following cytokinesis, the juvenile swims away (Fig. 6.43), settles onto a surface, withdraws its flagellum and becomes spindle-shaped. The juvenile, prior to dispersal and as a free individual, lacks visible microvilli (Figs 6.42–6.44). Costal strips are exocytosed onto the surface of the juvenile in well-defined longitudinal groupings (Fig. 6.44 left cell – asterisks). From estimates based on SEM there are approximately 14 such groupings which is equivalent to the number of spines. The first strips to be deposited are those that will form the lorica stalk, followed by those for the chamber. Later strips are those that will form the crown of spines and the inner layer of strips that stabilise the spines.

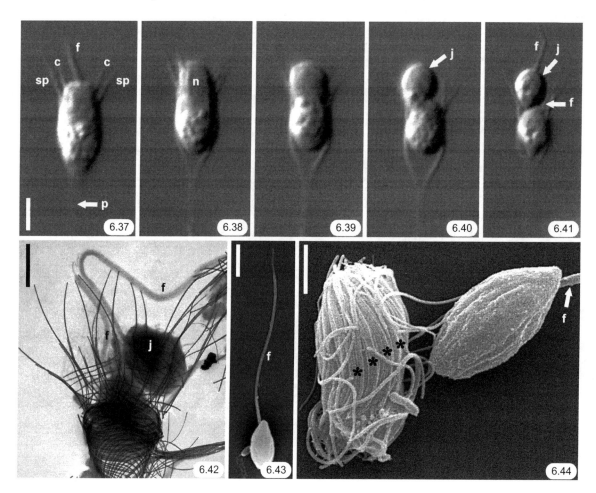

Plate 6 (Figures 6.37–6.44)

Figs 6.37–6.44 *Acanthoeca spectabilis.* Light and electron microscopy of cell division. **Figs 6.37–6.41** Sequence of stages during cell division; interference contrast light micrographs. Bar = 2 μm. **Fig. 6.37** Cell with anterior flagellum (f) that is in the process of being withdrawn prior to mitosis. Collar (c), anterior spines (sp), stalk (p).

Figs 6.38–6.39 Nuclear division (n) occurring within enlarged anterior end of cell.

Figs 6.40–6.41 Cytokinesis: development of pyriform juvenile cell (j) above the daughter cell that will remain within parent lorica. Flagella (f) of two daughter cells develop rapidly following completion of cytokinesis (Fig. 6.41).

Fig. 6.42 TEM image of recently divided cell showing both daughters with flagella (f). Juvenile (j). Bar = 2 μm.

Fig. 6.43 Recently released pear-shaped juvenile with long flagellum (f) but without a collar. Bar = 2.5 μm.

Fig. 6.44 Two juvenile cells; right-hand cell is a naked juvenile that still has a flagellum (f). Left-hand cell has a substantial accumulation of costal strips arranged in vertical sub-groupings (asterisks) on its surface. Bar = 1 μm. Figs 6.37–6.42 and 6.44 reproduced from Leadbeater *et al.* (2008b).

When a full complement of strips has been exocytosed, lorica assembly proceeds as a single, continuous event that takes up to 15 minutes. When viewed with light microscopy, juveniles in anticipation of lorica assembly are ellipsoid and are surrounded by a thick coat of costal strips (Fig. 6.45 arrowheads). Lorica assembly can be divided into four overlapping stages (Frösler and Leadbeater, 2009). The first entails a shimmering movement as the covering of costal strips moves around the cell body. The second involves the posterior extension of the cell to form a long stalk (Figs 6.46, 6.47). If the cell is attached to a surface, this pushes the cell body forwards. During the third stage the spines emerge at the anterior end and the collar and flagellum appear (Figs 6.48, 6.50). The fourth stage involves the relatively slow retraction of the posterior cytoplasmic protrusion from the surrounding stalk and dilation of the lorica chamber (Figs 6.49 arrowheads, 6.50). The latter is accompanied by a slight shortening of the chamber and a rounding of the enclosed cell.

The four stages of lorica assembly identified in living cells can also be recognised in cells prepared for SEM. The first stage reveals the costal strips in discrete bundles undergoing the beginnings of a left-handed rotation at the front end (Fig. 6.51 upper arrows). It is a measure of the flexibility of individual strips that the anterior end can turn almost 45° from the vertical while the posterior end remains approximately parallel to the longitudinal axis (Fig. 6.51 lower arrows). During the second stage, the turning of the helical costae continues with the strips at the anterior end becoming almost horizontal (Fig. 6.52 black arrows). At the same time the posterior end of the cell protrudes backwards, taking with it the outer costal strips from each bundle. Occasionally a ring of outer strips remains at the base of the stalk (Fig. 6.35 arrowheads), confirming that the forward movement of the stalk and chamber occurs from the inside. The third stage begins once the posterior movement is more-or-less complete. The bundles of strips forming the helical costae of the lorica chamber are still overlapping each other to a considerable extent (Fig. 6.53 lowest black arrow) and at the anterior end groups of strips that will form the spines and their supports are distinguishable but still as a continuation of the helix (Fig. 6.53 upper white arrows). The narrow, dome-shaped anterior end of the cell (Fig. 6.53 asterisk) is visible with the short lorica-assembling microvilli radiating like the spokes of a wheel from the cell surface (Fig. 6.53 small black arrows). During the third stage, the helical costae continue to turn and at the same time the strips forming the spines are moved forwards by the lorica-assembling microvilli (Fig. 6.55). Each microvillus is aligned with the long axis of a spine (Fig. 6.56), and also plays a key role in positioning the horizontal bands of strips on the inner surface of the spines. Once the lorica is complete, the microvilli are withdrawn and become part of the developing collar. Finally, during the fourth stage, the lorica chamber is dilated to form a single layer of costal strips and the organic layer is secreted on the inner surface of the lorica. The

posterior protuberance is withdrawn. The mature interphase cell has a flagellum and collar comprising approximately 30 microvilli.

Fig. 6.54 illustrates a whole mount of a cell at the third stage of lorica assembly treated with hydrofluoric acid to remove the siliceous costae. The elongated form of the cell, with a long posterior cylindrical 'tail' of cytoplasm and a ring of lorica-assembling microvilli (Fig. 6.54 arrowheads), is clearly visible. Within the prolonged tail, longitudinal components of the cytoskeletal system, almost certainly microtubules, are apparent. There is no indication whatsoever of a twisting of the cell which would seem to confirm that any rotational movement is not achieved by a 'screwing' motion generated by the cell. This is in contrast to the apparent situation of chrysophyte flagellates such as *Dinobryon*, where the whole cell is observed to rotate (Herth, 1979).

6.7 LOGISTICAL AND MECHANICAL ASPECTS OF LORICA ASSEMBLY

Lorica production can be divided into two components. The first is logistical, whereby costal strips are produced, exocytosed and stored in bundles in the correct order and location on the juvenile. The second is mechanical, and includes the movements necessary to transform costal strips within bundles on the juvenile into costae comprising linear arrays of strips on the adult. With respect to the logistical component in nudiform species, the order in which strips are produced and stored is outer lorica layer first – posterior to anterior; inner lorica layer second – posterior to anterior. Within a full complement of stored strips on the nudiform juvenile, each future costa is represented by one strip bundle. An important logistical achievement of the juvenile during assembly is to retain contact with every strip in every bundle to ensure that they all reach their respective destinations correctly. The mechanical component of lorica assembly includes the provision of spatial and temporal movements capable of converting bundles of strips into costae. The generation of traction forces must involve intracellular molecular motors working in close connection with the cytoskeleton.

6.8 ROLE OF THE CYTOSKELETON DURING LORICA ASSEMBLY

The mechanical component of lorica assembly is universal and can be inferred from a study of lorica construction, as discussed in Chapter 4. It must involve the simultaneous

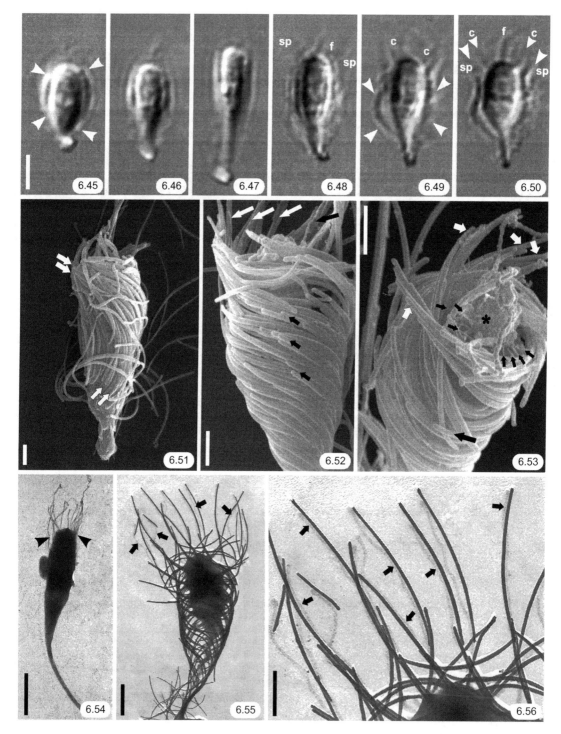

Plate 7 (Figures 6.45–6.56)

generation of two discrete movements, namely a forward vertical movement and a clockwise rotation. Interactions between these two motions can theoretically account for all costal patterns observed. The forward vertical movement is essential for extending longitudinal costae. A combination of the forward movement and clockwise rotation can create a left-handed diagonal motion that is required for extending helical costae. A rotational motion alone is essential for the extension of transverse costae in tectiform species and the forward movement serves to space these horizontal costae vertically (see Chapter 7). *Acanthoeca spectabilis* undergoes a further posterior lengthening to generate a stalk, but this is additional to the vertical and rotational motions mentioned above.

Light and electron microscopical images of lorica assembly in *A. spectabilis* (Figs 6.45–6.56) show that posterior elongation of the stalk occurs at the same time as a left-handed rotation generates the helix. Projection of the stalk backwards can be accounted for by the posterior elongation of the cell and in whole mounts treated with hydrofluoric acid to remove the silica covering, components of the cytoskeleton, probably microtubules, can be seen within the cylindrical protuberance (Leadbeater et al., 2008b). Thus it is likely that the posterior extension is a microtubule-generated activity. Helical turning is less easily explained since much of this movement occurs on the surface of the cell. However, at the anterior end during lorica assembly in *A. spectabilis* it is apparent that the lorica-assembling microvilli, which at this stage are relatively short, are intimately connected to the developing costae (Fig. 6.53). Thus it seems likely that the ring of microvilli, and possibly the surface of the cell, is rotated by cytoskeletal generated traction forces in the cortical cytoplasm. Once the full length of the stalk and chamber has been achieved, the spines are projected forwards by anterior extension of the lorica-assembling microvilli (Figs 6.55, 6.56). Each spine is accompanied by a lorica-assembling microvillus and the movement forwards is entirely vertical. Since the microvilli contain a core of actin microfilaments this aspect of lorica assembly must be actin-generated.

Much supporting information relating to the above scenario has been obtained from other nudiform species, in particular *Savillea parva* (Section 6.4.1) (Leadbeater, 2008a), and from tectiform species such as *Stephanoeca diplocostata* and *Didymoeca costata* (see Chapter 7) (Leadbeater, 1979b, c, 1994a, b, 2010). In all instances the forward movement is generated by the vertical extension of a ring of lorica-assembling microvilli. To achieve a rotational movement of the developing lorica, as displayed by species with a chamber projecting above the level of the enclosed cell, the ring of lorica-assembling microvilli must

Figs 6.45–6.56 *Acanthoeca spectabilis*. Lorica assembly. Figs 6.45–6.50 Sequence of stages in lorica assembly, based on two cells (Figs 6.45–6.47 and Figs 6.48–6.50, respectively). Bar = 2 µm. Fig. 6.45 Stage 1: ellipsoid cell with thick coat of costal strips (arrowheads). Figs 6.46–6.47 Stage 2: cell with posterior protrusion.
Fig. 6.48 Stage 3: anterior extension of spines (sp) and development of flagellum (f).
Figs 6.49–6.50 Stage 4: dilation of lorica chamber (Fig. 6.49 arrowheads) and development of collar (c) and flagellum (Fig. 6.50 (f)).
Figs 6.51–6.53 SEM images of lorica assembly. Bars = 1 µm. Fig. 6.51 Stage 1: juvenile immediately before elongation showing the beginnings of left-handed rotation of costal strips at top of cell (upper arrows) while rear end remains stationary (lower arrows).
Fig. 6.52 Lorica nearly complete showing a residual amount of overlapping of costal strips forming the lorica chamber (shorter black arrows). Anterior spines have pushed forwards with underlying lorica assembling microvilli (white arrows). Microvillus associated with developing spine (larger black arrow).
Fig. 6.53 Stage 3: anterior view of cell showing left-handed rotation of costal strips. Groups of strips that will form spines (white arrows) associated with short individual lorica-assembling microvilli (small black arrows) around anterior end of cell (asterisk). Strips around lorica chamber with overlaps (large black arrow).
Figs 6.54–6.56 TEM images of Stage 3 of lorica formation. Fig. 6.54 Greatly elongated cell at late stage of lorica assembly treated with hydrofluoric acid to remove costae, showing long tail and lorica assembling microvilli (arrowheads) emerging from cell at a short distance behind the anterior end. Bar = 2 µm.
Fig. 6.55 Cell with a long cytoplasmic tail within the lorica stalk. Lorica-assembling microvilli are present, associated with individual spines (arrows). The left-handed rotation of the helical costae in the lorica chamber is obvious. Bar = 2 µm.
Fig. 6.56 Higher magnification of the spines shown in Fig. 6.52, each spine is associated with a lorica-assembling microvillus (arrows). Bar = 1 µm. Figs 6.45–6.50 and 6.53 reproduced from Frösler and Leadbeater (2009). Figs 6.51, 6.52, and 6.54–6.56 reproduced from Leadbeater et al. (2008b).

be rotated in a manner similar to the upright columns on a carousel (merry-go-round). As already discussed in the context of *Savillea parva*, to generate a two-layered lorica comprising an inner layer of helical costae and outer layer of longitudinal costae requires that the longitudinal costae should carousel around the cell while the helical costae are attached to a respective longitudinal costa at the anterior end and the cell surface at the posterior end (see Section 6.4.1).

In summary, circumstantial evidence, gleaned from a study of lorica construction (see Chapter 4), an incomplete record of developmental observations (see above and Chapter 7) and from computer-generated images (Figs 6.21–6.23), supports the conclusion that lorica assembly requires a simultaneous combination of forward vertical and clockwise rotational movements. Without doubt the forward movement is generated by actin-based lorica-assembling microvilli. However, the rotational movement is less well understood, although the ultrastructural evidence currently available suggests that this movement is generated by the cortical cytoskeleton. It seems unlikely that the whole cell rotates, as is apparently the case for lorica deposition in *Dinobryon* (Herth, 1979). In an attempt to test this hypothesis a suite of experiments was carried out on *Acanthoeca spectabilis* using a range of microtubule (tubulin) and microfilament (actin) poisons (Frösler and Leadbeater, 2009).

6.8.1 Effects of microtubule poisons on lorica assembly

Acanthoeca spectabilis, in common with other loricate cells, possesses four categories of microtubules: flagellar, cytoskeletal, mitotic and those associated with SDVs and costal strip deposition (see Sections 2.6, 2.7, 2.9 and 5.5). In experiments on *A. spectabilis*, colchicine was the principal microtubule poison used but supplementary results were obtained using nocadozole (see Section 5.6.1 for mode of action of microtubule poisons). Colchicine at a final concentration of 0.8 mM affected all microtubule systems within *A. spectabilis* without killing cells (Table 6.4; Section 5.6.2). Division was halted, but since *A. spectabilis* is a nudiform species, existing juveniles continued to produce and accumulate costal strips. The first ultrastructural disturbance noted was the appearance of misshapen costal strips (Fig. 6.57). Initially, strip deformities were restricted to minor bends but subsequently the abnormalities became more severe and included major irregularities in length, thickness and shape. The appearance of misshapen costal strips is a good indicator that colchicine has effectively penetrated cells and that other cell abnormalities are likely to be a result of microtubule disturbance.

In cultures treated for 24 hours with 0.8 mM colchicine, many 'dwarf' cells, which lack extended stalks and possess truncated chambers of compacted helical costae with an anterior ring of spines, developed (Fig. 6.57). Their striking appearance makes them instantly distinguishable from all other cells, including juveniles. The helical costae of these cells undergo approximately one turn, always less than the 2–4 turns of normal untreated loricae, and the extent of overlap between adjacent strips of each costa is extensive (Fig. 6.57). The spines are approximately vertical and the transverse costal strips stabilising the spines are sometimes reduced in number or absent. Since the last strips to be produced and exocytosed by a juvenile prior to lorica assembly are the spines and their stabilising strips, these are the earliest misshapen strips to appear on dwarf cells (Fig. 6.57). Those dwarf cells that appear later have more deformed strips extending backwards within the chamber. In dwarf specimens, as in untreated cells, the final stage of lorica assembly involves the deposition at the base of the spines of an organic investment containing a regular mesh of fibrils (Fig. 4.43). The dwarf cell possesses a collar of microvilli (Fig. 6.57), but there is no flagellum, axoneme development being inhibited by the microtubule poison. Similar effects resulted from treatment of cells for 24 hours with 1 μM nocodazole, another microtubule poison (see Table 6.4).

The concentration of colchicine that is just inhibitory to growth of *A. spectabilis*, its deforming effects on costal strip morphology and the response time to drug application are in keeping with results reported for *Stephanoeca diplocostata* (see Sections 5.6.2 and 5.6.3). Dwarf *Acanthoeca* cells derive from an overall failure of juveniles to increase in height. Inability to produce a stalk is the most obvious symptom, but also the chamber is unusually squat. On the other hand, anterior spine extension is unaffected. These symptoms probably reflect a failure of the microtubular cytoskeleton to effect elongation of the cell body and posterior protuberance during the second stage of lorica assembly. This malfunction may well be caused by inhibition in the rearwards polymerisation of cytoskeletal microtubules that would normally occur during this phase of development.

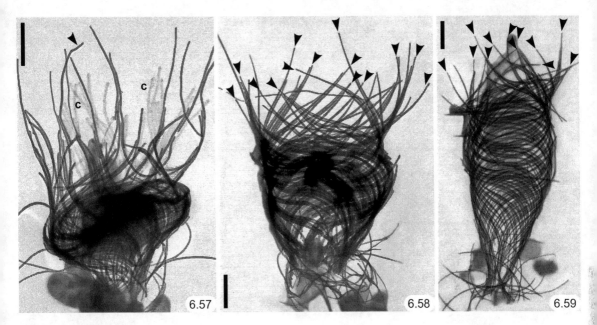

6.57   6.58   6.59

Plate 8 (Figures 6.57–6.59)

Figs 6.57–6.59 Effects of cytoskeletal poisons on *Acanthoeca spectabilis*. Whole mount preparations for TEM. Bars = 1 µm. Reproduced from Frösler and Leadbeater (2009). **Fig. 6.57** Cell treated with 0.8 mM colchicine for 24 h. The truncated 'dwarf' lorica has deformed spines (arrowhead). Cell has collar microvilli (c) but lacks a flagellum.
Fig. 6.58 The dwarf lorica of a cell treated for 24 h with 0.25 µM cytochalasin and 0.8 µM colchicine. The anterior array of tilted strips represents non-extended spines. Arrowheads point to anterior ends of spine strips distinguished by their forked tips.
Fig. 6.59 Lorica of cell treated for 24 h with 10 µM cytochalasin D. The lorica is not as long as on untreated cells but it is not dwarf. Groups of strips that would normally form spines are tilted and not extended. Arrowheads point to anterior ends of spine strips distinguished by their forked tips.

6.8.2 Effect of microfilament poisons on lorica assembly

Two actin microfilament poisons, cytochalasin D and latrunculin A, were used in experiments on *A. spectabilis*. Cytochalasin D causes disruption and destabilisation of microfilaments by preventing actin polymerisation. This is achieved by binding and blocking the barbed (fast-growing) ends of actin microfilaments so that polymerisation and depolymerisation are inhibited or at least slowed down (Brown and Spudich, 1981; Bonder and Mooseker, 1986). Cytochalasin D also binds to actin monomers and maybe dimers and accelerates the ATPase activity of actin.

Whereas the effect of microtubule poisons on *A. spectabilis* was to produce 'dwarf' loricae, there was no immediately obvious deformity associated with actin poisons

such as cytochalasin D. This made it difficult to distinguish between cells that had assembled their loricae before and after drug treatment. To overcome this problem it was decided to apply both colchicine and cytochalasin together (Frösler and Leadbeater, 2009). In this way, dwarf cells resulting from colchicine treatment served as an indicator that their loricae had been assembled subsequent to drug application. The task was then to look for additional effects that might be ascribed to cytochalasin treatment. This, of course, does not rule out potential interactions in effect between the two drugs, but subsequently it was possible to proceed to treatment with cytochalasin alone with some prior knowledge of what results might be expected.

Application of 0.25 µM cytochalasin D combined with 0.8 mM colchicine for 24 hours to *Acanthoeca* cultures

Table 6.4 *Effects of cytoskeletal inhibitors on* Acanthoeca spectabilis *cultures.*

	Adult cells – lorica assembled before drug treatment			Juvenile cells – lorica assembled after drug treatment				
	Arrested division	Flagella present	Microvilli present	Dwarf cells	Flagella present	Microvilli present	Deformed costal strips	Spines extended
Microtubule inhibitors								
Colchicine 0.8 mM	+	±	+	+	–	+	+	+
Nocodazole 1.0 µM	+	–	+	+	–	+	+	+
Microfilament and microtubule inhibitors								
Cytochalasin D 0.25 µM Colchicine 0.8 mM	+	±	±	+	–	–	+	–
Latrunculin A 0.1 µM Colchicine 0.8 mM	+	–	–	+	–	–	+	–
Microfilament inhibitor								
Cytochalasin D 0.5–10 µM	+	+	±	±	+	–	–	–
Control cells (without drug treatment)	–	+	+	–	+	+	–	+

Adult cells refers to cells with loricae assembled prior to treatment. Juvenile cells refers to cells with loricae assembled after application of the drug. Results were obtained after 24 hours of treatment. Symbols: + character observed; – character absent; ± character not observed on all cells.
Reproduced from Frösler and Leadbeater (2009).

resulted in the appearance of dwarf cells similar to those produced by colchicine alone (Fig. 6.58). However, the spines were not fully extended in this dual treatment nor were they oriented vertically. Instead, they remained as groups of strips tilted at an angle that was more-or-less continuous with the top of the helix (Fig. 6.58). The presence of the minute terminal forked tips served as a marker for potential spine strips (Fig. 6.58 arrowheads). Combined drug-treated dwarf cells lacked both collar microvilli and flagella since both actin and tubulin polymerisation had been inhibited. However, the organic investment was visible within the lorica. The results of treating cells with a combination of colchicine and latrunculin A were similar in most respects to those obtained with colchicine and cytochalasin D. In particular, the anterior spines were not extended and the collar microvilli were absent (Table 6.4).

To ensure that the morphology of dwarf cells resulting from dual treatment represents the completed product of lorica assembly and is not a delayed stage in development, cells treated with this drug combination for 24 hours were subsequently concentrated, washed and returned to normal culture. Within 24 hours dwarf cells had developed collar microvilli and flagella. However, the truncated loricae were not modified in any way, and the spines remained in a non-extended state, indicating that lorica assembly was complete and that no further adjustments had been made. *Acanthoeca* cultures treated with 0.5–10 µM cytochalasin D alone also contained cells with loricae that lacked extended spines (Fig. 6.59 arrowheads). However, they were not dwarf; some posterior extension had taken place although they lacked the narrow, elongated stalks typical of untreated cells. Cells that had formed their loricae during cytochalasin treatment resembled normal untreated cells during the third stage of assembly (compare Fig. 6.59 with 6.53), but differed in that the helical costae had completed sliding past each other and yet the spines had not extended. Thus treatment with actin microfilament poisons confirms the role of the lorica-assembling microvilli in mediating the forward vertical movement of the spines.

6.8.3 Contribution of experiments to understanding the mechanism of lorica assembly

The conclusion to emerge from microtubule poison experiments involving *Acanthoeca spectabilis* is that the rearwards extension of the stalk and chamber is inhibited during lorica assembly. This probably results from an inhibition of the posterior extension of the cytoskeletal microtubules during the second stage of lorica assembly. This movement is specific to *A. spectabilis* and will probably only be important to other stalked species, such as *Stephanoeca cauliculata* (Fig. 4.35). Judging from microtubule inhibitor experiments on *Stephanoeca diplocostata*, it is unlikely that cytoskeletal microtubules already in existence in juvenile cells prior to the drug application will subsequently be depolymerised by colchicine treatment (Leadbeater, 1987). In species without stalks, such as *Stephanoeca diplocostata*, colchicine treatment does not inhibit lorica assembly (Fig. 5.15). Disruption, if it does occur in this species, appears to be more associated with deformities of the costal strips. Treatment of *A. spectabilis* with microfilament poisons affects the vertical extension of the lorica-assembling microvilli and therefore interferes with one of the designated 'discrete' movements of lorica assembly. However, this is not unexpected since the behaviour of the microvilli, as determined by EM, anticipated such an effect (Leadbeater, 1994b; Leadbeater, 2010; Leadbeater and Cheng, 2010).

In none of the current cytoskeletal poison experiments was there inhibition of the second designated 'discrete' movement, namely the clockwise rotation anticipated from morphological evidence. However, this may be because these drugs are more effective at inhibiting processes that involve polymerisation of microtubules and microfilaments rather than affecting the operation of existing systems that are already in place. For the time being evidence in support of a rotational movement must remain circumstantial.

If, as suggested, the ring of lorica-assembling microvilli does rotate in the form of a carousel during lorica assembly, then it is necessary to enquire whether this is structurally feasible. As described in Section 2.7, the microfilamentous core of each collar microvillus penetrates the cortical cytoplasm of the cell for some considerable distance and overall the microvillar bases interdigitate with the radial cytoskeletal microtubules. Post-mortem displacement of cytoskeletal microtubules usually leads to a displacement of the microfilamentous bases of collar microvilli, which is indicative of there being a close structural attachment between the two components of the cytoskeleton. It was suggested in Section 2.7 that the cytoskeletal microtubules could provide a framework for anchoring the microvilli and ensuring their regular spacing within the collar. Since the cytoskeletal microtubules radiate from a ring around the emergent flagellar base, is it possible that the entire ring of microtubules with the regularly spaced microvilli in place may rotate within the cortical cytoplasm during lorica assembly? For the moment we do not have a definitive answer to this question but the topic will be considered further in the context of tectiform species (Chapter 7) and the possible evolutionary relationship between nudiform and tectiform species (Chapter 8).

6.9 EVOLUTIONARY SIGNIFICANCE OF NUDIFORM CHOANOFLAGELLATES

Nudiform choanoflagellates are an evolutionary paradox. On the one hand the five species attributed to this grouping (Table 6.2) are consistent in their mode of cell division; in the generation of naked motile juveniles; in the timing of costal strip deposition within the cell cycle; and in the order and orientation in which costal strips are deposited and exocytosed by the juvenile. These features are interlinked and cannot either individually or collectively be confused with the sequence of events in tectiform species (see Chapter 7). On the other hand, in terms of lorica morphology, nudiform species do not represent a coherent and well-circumscribed grouping. On the basis of the costal pattern alone, there is neither a reason for grouping them together nor do they possess characters that would convincingly distinguish them from the much larger group of tectiform species. From this conclusion follows a second paradox, namely why should two groupings of species, nudiform and tectiform, that so obviously share so many features in common, display such unequivocal differences? The answer to this question only becomes apparent after an in-depth study of tectiform species, which is the subject of Chapter 7. The conclusion that will subsequently emerge from Chapter 8 is that a single abrupt change occurred affecting concurrently the timing and sequence of costal strip production within the cell cycle and the orientation of developing strips and cytokinesis. Fortuitously, this drastic change must have been of immediate evolutionary benefit and led to the extensive ecological radiation of tectiform species that we witness today. Molecular phylogenetic studies distinguish nudiform acanthoecids as being a distinctive clade, but their relationship with tectiform acanthoecids is uncertain depending on the data used to generate a phylogeny and, in particular, whether the third codon positions are included or excluded (for further discussion see Section 10.4.1). If they are included the nudiform and tectiform groups appear as sister clades represented by the families Acanthoecidae Norris *sensu* Nitsche *et al.*, 2011 and Stephanoecidae Leadbeater, respectively, in Fig. 10.3 (see also Carr *et al.*, 2008a; Nitsche *et al.*, 2011). However, if the third codon positions are omitted the root of the nudiform clade appears within the tectiform clade, which, if correct, would indicate that the extant nudiform grouping evolved from a tectiform ancestor.

7 • Loricate choanoflagellates: Stephanoecidae – tectiform species

Diversity based on rings – logistics of costal strip production, storage and lorica assembly.

7.1 INTRODUCTION

The overwhelming majority of loricate choanoflagellates are tectiform and attributable to the Family Stephanoecidae. In contrast to the five presently known nudiform species (Family Acanthoecidae) discussed in Chapter 6, there are, at the present time of counting, approximately 150 named tectiform species. They display considerable morphological variety (see Chapter 4) but it is in terms of their ecological diversity that the distinction between the two families is particularly marked. Whereas the lorica chambers of the five currently known extant nudiform species are mostly minute, sedentary and distributed within biofilms or on surfaces, tectiform species have diversified to occupy many microhabitats, principally in marine and brackish water environments (see also Section 9.4.8). In particular, many are freely suspended within the water column and contribute universally to marine planktonic microbial foodwebs (see Chapter 9).

Table 6.1 lists the important morphological characters that distinguish nudiform and tectiform species. The most obvious differences centre on whether it is the juvenile or parent cell that produces and accumulates costal strips and where on the cell they are stored; at what position in the cell cycle the lorica is assembled; and whether cell division is diagonal or inverted. At first glance this might appear to be an unrelated list of characters without much meaning or purpose. However, on detailed analysis it turns out that the characters are intimately related and of major evolutionary significance (see Chapter 8). Tectiform species are in most respects more complex than their nudiform counterparts; a parent cell, which is in possession of a lorica, produces and exocytoses costal strips that are stored at the top of the collar. When a full complement has been produced the cell undergoes mitosis followed by an inverted form of division; the upper daughter cell (juvenile) is pushed backwards out of the parent lorica with the accumulated strips. Once free of the parent, the juvenile immediately assembles a lorica.

Ellis (1929) was the first to observe and illustrate tectiform division in *Stephanoeca ampulla* (Fig. 7.1), although the significance of 'inverted' division would not have been apparent to him at the time. Most of our current knowledge of the tectiform condition has come from three species, namely *Stephanoeca diplocostata*, *Didymoeca costata* and *Diaphanoeca grandis*. Since *Stephanoeca diplocostata* was the first to be obtained in culture, this species has supplied much of the basic information on costal strip formation (see Section 5.5), storage and cell division (Leadbeater, 1979b, c, 1994a, b). However, *S. diplocostata* has a major limitation in that all of its strips are of similar morphology and therefore cannot be distinguished from each other with certainty either during formation or subsequently on the cell surface. To overcome this shortcoming a species was required that had a lorica containing an intermediate number of strips that are clearly distinguishable into categories of different morphology. *Didymoeca costata* fulfilled this requirement to perfection. Its lorica contains between 60 and 75 costal strips separable into six easily identifiable categories that occupy specific locations within the lorica (Figs 4.15, 7.36) (Leadbeater, 2010). *Diaphanoeca grandis* has provided valuable supplementary information on how a cell manages to assemble a voluminous lorica at some distance from its surface (Leadbeater and Cheng, 2010). Since each of these species has contributed important but complementary insights into the tectiform condition and the chronological order of the respective studies coincides with a logical descriptive sequence, the three species are considered separately in this order below.

7.2 STEPHANOECA DIPLOCOSTATA

7.2.1 *Stephanoeca diplocostata*: lorica construction

A full description of the lorica of *Stephanoeca diplocostata* can be found in Section 5.4. In summary, the lorica is barrel-shaped and consists of two chambers separated by a waist which is coincident with the position at which the lorica is attached to the cell (Figs 5.1–5.3). The anterior chamber is approximately twice the length (8–12 μm) of the posterior chamber (5–7 μm) and about 1.5–2.0 times the width. The lorica comprises two layers of costae with the outer layer containing between 15 and 24 (mode 19) longitudinal costae, each consisting of five costal strips; three in the anterior chamber and two in the posterior chamber (Table 5.1). There are four transverse costae in the anterior chamber that are enumerated from the posterior end forwards as follows: (5) posterior transverse (waist) costa; (6) posterior intermediate transverse costa; (7) anterior intermediate transverse costa; (8) anterior ring (Figs 5.1–5.3). All transverse costae except the anterior ring are on the inner surface of the longitudinal costae. The anterior ring is on the outside (see Section 4.6.1). Table 5.1 lists the numbers of costal strips comprising the various costae. The inner layer of the posterior chamber contains between 4 and 13 helical costae (Fig. 5.2), although this category is the most difficult to count. The average overall number of costal strips within a lorica is 172 ± SD 18.1 (Table 5.6). Although there is no clear-cut distinction in morphology or length between the different categories of costal strips contributing to the lorica, those comprising the posterior transverse costa are usually slightly thicker than the remainder. Strips contributing to the helical costae of the posterior chamber and the rings of the anterior chamber are usually more obviously crescentic than strips comprising the longitudinal costae (Leadbeater, 1994a).

7.2.2 *Stephanoeca diplocostata*: cell division and lorica assembly

Costal strip production is intimately linked to the cell cycle and extends throughout the greater part of interphase (see Section 5.5). This conclusion is drawn from the fact that at any time during asynchronous exponential growth 60–70% of cells have observable accumulations of costal strips on their collars (see Table 5.2) (Leadbeater, 1985). When a full complement of strips has been accumulated, the enlarged cell withdraws its flagellum and the collar microvilli shorten, drawing the strips towards the cell body (Figs

7.2, 7.3). Mitosis proceeds and at the same time the anterior end of the cell enlarges and individual sub-groupings of costal strips are rotated to stand vertically (Fig. 7.4). Mitosis is followed by cytokinesis, commencing when the anterior end of the parent cell enlarges upwards and a cleavage furrow appears between the telophasic daughter nuclei in Fig. 7.4 and has completed separation of the daughter cells in Fig. 7.5. The upper daughter cell is asymmetrically inverted through 180° with the result that the flagellar poles of the two daughters come to face each other (Figs 7.5, 7.6). The sub-groupings of costal strips are now repositioned vertically alongside the upper daughter cell (juvenile) and symmetry, except for the position of the joining thread, is restored. The juvenile, together with the covering of costal strips, is pushed backwards out of the parent lorica (Figs 7.6, 7.7). This is achieved by two overlapping threads, one arising from the collar region of each cell. Separation of the daughter cells is effected by a combination of mutual lengthening of these threads and the helical sliding of one over the other. Eventually the two threads slide off one another and the juvenile floats away. This is followed by lorica assembly, which lasts a few minutes and involves a forward movement of the bundles of strips on the juvenile (Figs 7.8, 7.9).

The organisation and movement of costal strips during interphase, cell division and lorica assembly, as glimpsed on living cells (Figs 7.2–7.9), can be clearly visualised in: (1) whole mounts for transmission electron microscopy (TEM) which reveal stages in the positioning of costal strips (Figs 7.10–7.21); (2) in whole mounts of cells treated with hydrofluoric acid which reveal the position of the collar and lorica-assembling microvilli during and after division (Figs 7.23–7.31); and (3) in sectioned material which provides greater detail of mitosis (Figs 7.32–7.34) (see Section 2.9). A combination of these observations has facilitated a reconstruction of the complex logistical events that occur during cell division and lorica assembly. Comparable views of selected stages during cell division and lorica assembly obtained by the different techniques mentioned above are listed as figure numbers in Table 7.1.

Among the first strips making their appearance at the top of the collar of a parent cell during interphase are crescentic thicker strips destined for the posterior transverse costa (Fig. 7.10). Although they appear as a continuous ring at the top of the collar, when flattened in whole mounts the ring collapses into four distinct quadrants (Figs 5.3, 7.11). The discontinuities between adjacent bundles are so distinct as to suggest that this pattern is real and not

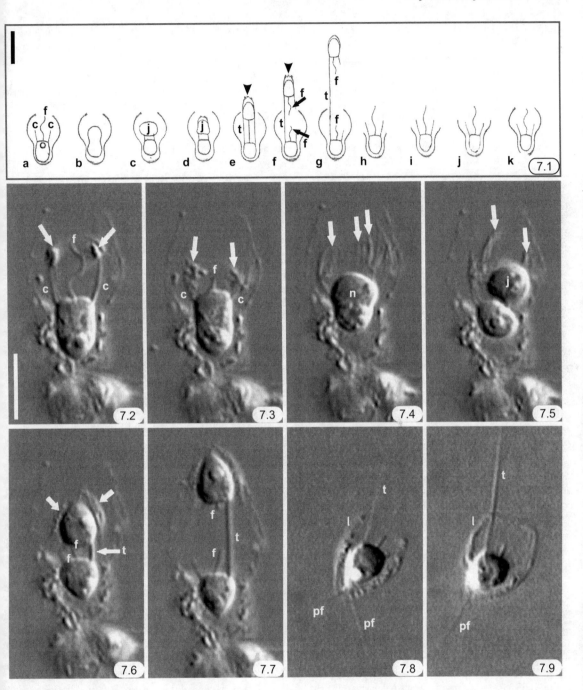

Plate 1 (Figures 7.1–7.9)

Fig. 7.1a–k *Stephanoeca ampulla*. Ellis's (1929) original drawings of 'inverted' cell division characteristic of the tectiform condition. Fig 7.1a Interphase cell with flagellum (f) and collar (c). Fig. 7.1b 'Figure-of-eight' stage without flagellum. Fig. 7.1c–d Cell division complete with inversion of upper daughter cell (juvenile) (j). Fig. 7.1e–f Juvenile, with regenerating flagellum (f) and coating of costal strips, being pushed backwards by elongating thread (t) out of parent lorica. Shimmering movement of costal strips at rear of juvenile (arrowhead).

Table 7.1 Stephanoeca diplocostata: *light and electron microscopical images illustrated in the text of cells at different stages in the cell cycle.*

	Stage in cell cycle	Living cell	Whole mount	HF acid-treated whole mount	Sectioned material
Interphase	Small accumulation of strips	–	Fig. 7.10	–	–
	Large accumulation of strips	Fig. 7.2	Figs 5.3, 7.11, 7.12,	Fig. 5.4	–
Mitosis	Prophase	Fig. 7.3	–	Fig. 7.23	Figs 2.76, 7.32
	Anaphase/telophase (figure-of-eight stage)	Fig. 7.4	Fig. 7.13	Figs 7.24, 7.25	Fig. 7.33
Cell division	Cytokinesis	Fig. 7.5	–	Figs 7.25, 7.26	–
	Daughter cells separating	Figs 7.6, 7.7	Figs 7.14, 7.15	Figs 7.27, 7.28	Fig. 7.34
	Lorica assembly	Figs 7.8–7.9	Figs 7.16–7.21	Figs 7.30, 7.31	–

artefactual. A continuum of overlapping strips, as encountered in developing transverse costae, does not subdivide in this way (Fig. 7.18). As interphase progresses, the number of costal strips within the ring increases. However, their organisation is not homogeneous but comprises numerous sub-groupings which occasionally become displaced and are immediately apparent (Fig. 7.12). These sub-groupings relate to future costae that will be assembled subsequently within the lorica of the juvenile. During cell division, at the 'figure-of-eight' asymmetric stage, the individual sub-groupings of costal strips are rotated to stand vertically by the collar microvilli (Fig. 7.13), which then shorten and move the strips towards the surface of the juvenile. Symmetry is restored when the daughter cell becomes distinct and the strips closely surround the juvenile (Fig. 7.14). As

the juvenile is pushed backwards out of the parent lorica, sub-groupings of strips that anticipate future transverse costae are rotated into the horizontal plane (Fig. 7.16).

Lorica assembly on the juvenile is achieved by the forward extension of the lorica-assembling microvilli, which interact on a one-to-one basis with the longitudinal costae (Figs 7.17–7.21). Strips destined for the transverse costae start as horizontal bundles at the anterior end of a newly released juvenile (Fig. 7.16), but as the microvilli move the longitudinal costae forward the strips within the horizontal bundles extend laterally. Evidence for lateral (rotational) movement comes from four sources. First, at an early stage there is evidence of strips sliding horizontally over each other (Fig. 7.17 arrows). Second, on juveniles where the lorica falls sideways to reveal developing transverse costae,

Fig. 7.1f–g Regrowth of flagella (f) on both daughter cells. Fig. 7.1h–k Stages in lorica assembly. Bar = 10 μm. Reprinted with permission from Ellis (1929).

Figs 7.2–7.9 *Stephanoeca diplocostata.* Stages in cell division and lorica assembly. Interference light microscopy. Bar = 5 μm (shown on Fig. 7.2). **Fig. 7.2** Interphase cell with flagellum (f) and collar (c) containing large horizontal accumulation of costal strips (arrows) at top of inner surface of collar.

Fig. 7.3 Withdrawal of flagellum (f) and shortening of collar (c) with individual sub-groupings of strips visible (arrows).

Fig.7.4 Figure-of-eight stage with enlarged prophase nucleus (n) and sub-groupings of costal strips (arrows) in vertical orientation.

Fig. 7.5 Two daughter cells, upper juvenile cell (j) undergoing inversion. Costal strips in vertical plane (arrows).

Fig. 7.6 Two daughter cells with flagellar poles facing each other. Evidence of flagellar (f) regrowth on both cells. Thick (double) thread (t) joins cells on right side. Sub-groupings of costal strips closely appressed to juvenile surface (arrows).

Fig. 7.7 Juvenile being pushed out of parent lorica by double thread (t). Flagellar regrowth (f) evident on both cells.

Figs 7.8–7.9 Lorica assembly on recently released juvenile. Thick coating of costal strips in process of being moved forwards to assemble lorica (l). Remaining single thread (t) emerging from collar and several filopodia projecting from rear (pf).

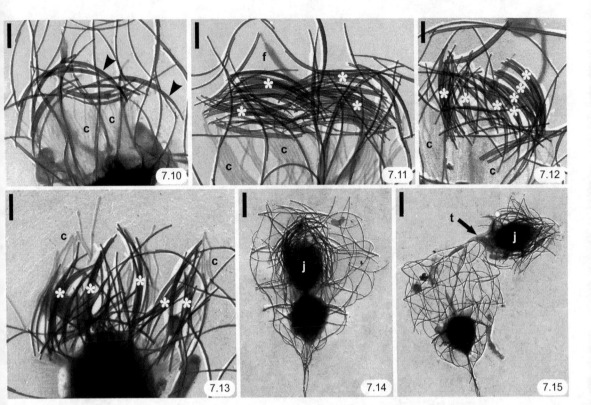

Plate 2 (Figures 7.10–7.15)

Figs 7.10–7.15 *Stephanoeca diplocostata*. Shadowcast whole mounts of cells showing stages in costal strip accumulation on collar and cell division. Figs 7.10–7.13 Bars = 1 μm. Figs 7.14–7.15 Bars = 2.5 μm. **Fig. 7.10** Early accumulation of strips on collar (c). Thick curved strips destined for future waist visible (arrowheads).

Fig. 7.11 Large accumulation of strips comprising four groupings (asterisks) on collar (c), flagellum (f).

Fig. 7.12 Disturbed accumulation of costal strips on collar (c) showing separation into sub-groupings (asterisks).

Fig. 7.13 Figure-of-eight stage in cell division showing collar microvilli (c) migrating over anterior surface of cell and sub-groupings of strips (asterisks).

Fig. 7.14 Two daughter cells with upper juvenile (j) covered in costal strips.

Fig. 7.15 Recently divided cell with juvenile (j) covered in costal strips and still attached by thread (arrow t) to daughter cell remaining in parent lorica. Figs 7.10, 7.12 and 7.13 reproduced from Leadbeater (1994a). Figs 7.14 and 7.15 reproduced from Leadbeater (1979b).

groupings of overlapping strips can be seen (Fig. 7.18). Third, in later stages of assembly the extent of overlaps between horizontal costal strips decreases until they are fully extended in the mature lorica (Figs 7.19, 7.20, 7.21). Fourth, a three-dimensional computer model of lorica assembly in a closely related species, *Stephanoeca cupula sensu* Thomsen (1988), incorporating a forward and left-handed rotation of 360°, generated a series of images closely resembling those seen in whole mounts of cells prepared for TEM (compare Fig. 7.22a–c with Figs 7.16–7.21). An important feature

illustrated in the computer-generated images is the development of helical costae in the posterior chamber. On an early-stage juvenile the sub-groupings of strips that will give rise to the helical costae are more-or-less vertical (compare Fig. 7.16 with 7.22a), but as the lorica increases in height so the slope of the helical costae within the posterior chamber becomes more diagonal (compare Fig. 7.20 with 7.22b, and 7.21 with 7.22c) (Leadbeater *et al.*, 2009). Two important points emerge from the whole mounts shown in Figs 7.16–7.21. First, the horizontal sub-groupings of strips

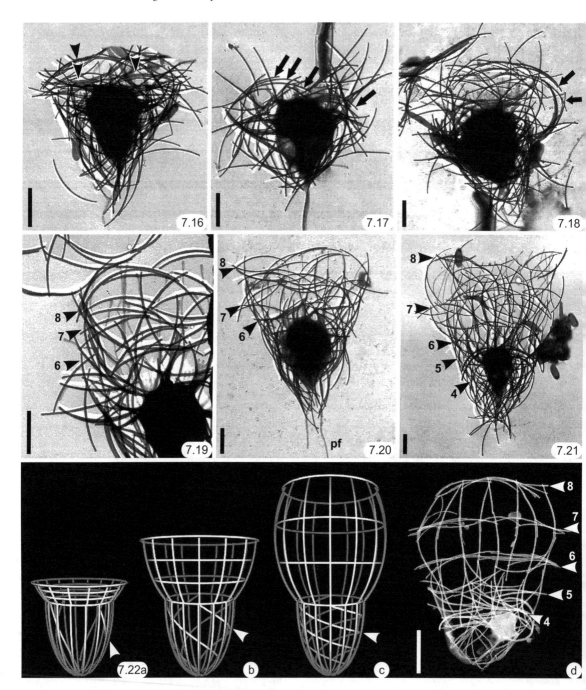

Plate 3 (Figures 7.16–7.22)

Figs 7.16–7.21 *Stephanoeca diplocostata*. Shadowcast whole mounts of cells showing lorica assembly on juvenile. Bar = 2 μm. **Fig. 7.16** Recently released juvenile with compact sub-groupings of costal strips in horizontal plane at front end (arrowheads) and in longitudinal and diagonal planes on surface of cell. Reproduced from Leadbeater (1994a).

are equally distributed around the anterior end of the juvenile. Second, at the beginning of assembly the strips for all four transverse costae are positioned closely together (Fig. 7.16). From this it is concluded that individual transverse costae do not derive from single sub-groupings of costal strips, as do the longitudinal costae, but rather they are assembled by the alignment of several (probably four) horizontal sub-groupings of strips. On relatively undisturbed juveniles, the horizontal sub-groupings of strips appear to be in quadrants (Fig. 7.16), a feature that is reminiscent of the organisation of strips at the top of the collar (Fig. 7.11).

The movement and positioning of costal strips during cell division and subsequent lorica assembly is to a large extent controlled by the collar microvilli. Their number, location, and relative lengths can be seen in whole mounts of cells treated with hydrofluoric acid to remove the silica strips (Figs 7.23–7.31) (see Table 7.2). During interphase the microvilli form a funnel-shaped collar around the flagellum (Fig. 5.4). However, after flagellar withdrawal and with the onset of mitosis, the anterior end of the cell enlarges and the bases of the collar microvilli migrate outwards (Fig. 7.23). This leads to the 'figure-of-eight' asymmetric stage (comparable to Figs 7.4, 7.13) when the microvilli appear on one side of the dividing cell (Figs 7.24, 7.25). Once the juvenile completes its inversion and the two daughter cells become distinct entities, the microvilli increase in number (Table 7.2) and migrate to their respective positions on the two daughter cells (Figs 7.26, 7.27). There are four ultimate locations for the microvilli and they are illustrated as A–D in Figs 7.26–7.31. Three locations are on the juvenile, namely: D = posterior microvilli (= filopodia); C = lorica-assembling microvilli; B = juvenile

collar microvilli. The fourth location, A, comprises the collar of the daughter cell that will remain within the parent lorica. The future lorica-assembling microvilli emerge from the juvenile at a short distance behind the collar (Figs 7.28–7.31). Separation of the two daughter cells is achieved by actin-based threads emerging from the two collars sliding over one another (Figs 7.28, 7.29). Occasionally the number of threads involved in sliding may be three or more (Fig. 7.29).

Directly upon the juvenile floating free of the parent, lorica assembly is initiated. This usually starts, and may be completed, before the separating thread has been withdrawn (Fig. 7.9). Assembly is achieved by the forward movement of the lorica-assembling microvilli interacting on a one-to-one basis with the longitudinal costae (Figs 7.30, 7.31). At the same time the flagellar shaft, which is initially a short stub, elongates (Fig. 7.31). The fully extended lorica-assembling microvilli, as seen in Figs 7.21 and 7.31, are unequivocally vertical, and it is difficult to imagine from Fig. 7.31 that there could have been a rotational movement involved in assembly. This is a challenging moment since it has been argued extensively (see Chapters 4 and 6) that a rotational movement is mandatory to achieve the helical disposition of costae in the posterior chamber and the transverse costae in the anterior chamber. Initially it was thought that assembly of a *Stephanoeca* lorica only required a forward movement (Leadbeater, 1979c, 1994a). A slight lateral separation between adjacent lorica-assembling microvilli might have been sufficient to provide the necessary lateral movement to assemble the transverse costae. However, this cannot be the case because the transformation of completely overlapping strips to a ring would require more lateral movement than a small separation can provide.

Fig. 7.17 Juvenile beginning to assemble lorica with horizontal sub-groupings of overlapping strips (arrows) at anterior end indicating an early stage in assembly of transverse costae.

Fig. 7.18 Juvenile that has fallen sideways showing circle of costal strips (arrows) in process of assembling into transverse costae.

Fig. 7.19 Juvenile about halfway through lorica assembly showing overlapping strips in three or four transverse costae (arrowheads). Extended lorica assembling microvilli also visible.

Fig. 7.20 Lorica assembly at comparable stage to Fig. 7.19. Front end of lorica is more complete, with three transverse costae visible (arrowheads), than rear end where there is still considerable overlapping of strips. Anterior lorica-assembling microvilli and posterior filopodia apparent (pf). Reproduced from Leadbeater (1979b).

Fig. 7.21 Almost completed lorica with lorica-assembling microvilli still visible.

Fig. 7.22a–d *Stephanoeca cupula sensu* Thomsen (1988). Fig. 7.22a–c Computer-generated images of a developing lorica involving a forward and 360° clockwise movement. Starting with a lorica cup comprising vertical and horizontal sub-groupings of strips (Fig. 7.22a) through an intermediate stage (Fig. 7.22b) to completed lorica (Fig. 7.22c). Note that the gradient of the helical costae to the longitudinal axis becomes shallower as assembly proceeds (arrowheads). Fig. 7.22d Actual lorica with helical and transverse costae numbered as in Fig. 4.2. Bar = 2 μm. Reproduced from Leadbeater *et al.* (2009).

Table 7.2 *Mean (± SD) numbers and lengths of microvilli in the parent collar prior to division and on daughter cells (juvenile and sister) after division but before dispersal ((n) = sample number).*

	Late interphase parent collar microvilli	Juvenile cell before lorica assembly			Sister cell (remaining in parent lorica) collar microvilli
		Posterior microvilli (= filopodia)	Lorica-assembling microvilli	Collar microvilli	
Number of microvilli ± SD	41.6 ± 2.2 (11)	3.4 ± 1.4 (16)	20.9 ± 2.7 (19)	20.6 ± 3.3 (8)	34.1 ± 5.1 (36)
Length of microvilli μm ± SD	4.7 ± 0.7 (12)	2.5 ± 0.8 (16)	1.9 ± 0.5 (16)	1.0 ± 0.2 (16)	Variable

Reproduced from Leadbeater (1994a).

Furthermore, there is a regular pattern of overlaps between adjacent strips (see Chapter 4) which can only be obtained by a unidirectional rotational movement. Thus the question is posed: if a clockwise rotation is required, how can this be achieved by lorica-assembling microvilli that are so obviously vertical? It is not plausible for individual microvilli to effect a rotation. Inevitably it appears that the complete ring of lorica-assembling microvilli must rotate in the manner of a carousel (merry-go-round) (see also Section 6.8). This, in itself, is not impossible, but how might such a movement be achieved? There is no evidence that the cell as an entity creates a turning movement – no image has ever revealed a cell with a helical twist. This leads to the conclusion that the turning movement is achieved by the intracellular cytoskeletal framework at the base of the microvilli generating the necessary torsional movement. Whether or not this involves rotation of the cytoskeletal microtubules cannot be answered with the current state of knowledge. However, whatever the mechanism might be it must be universal for all loricate species no matter whether nudiform or tectiform.

Sectioned material illustrates in detail the intracellular withdrawal of the flagellar shaft and depolymerisation of the 9 + 2 axoneme prior to mitosis (Figs 2.72–2.74). This is followed by enlargement of the anterior end of the cell (Figs 2.76, 7.32) to accommodate the expanded nucleus during division and the outward migration of the collar microvilli bringing the sub-groupings of strips towards the cell surface (Figs 7.32 and 7.33 arrowheads). Mitosis is 'open' in that the nuclear envelope disintegrates during prophase (Figs 2.76, 7.32). Fig. 7.33 shows a cell at late anaphase/early telophase (the beginning of the asymmetrical stage) when two distinct nuclear areas (n in Fig. 7.33) are visible but before re-formation of the respective nuclear envelopes. The posterior extremity of the collar microvilli has now migrated two-thirds of the distance down the side of the cell (Fig. 7.33 lower arrow c). At the anterior end of the cell many vesicles containing fibrillar material are evident (Fig. 7.33 arrow vfm). These will be passed on to the juvenile during cytokinesis and their contents discharged to the outside before and during lorica assembly. During this elongated stage, the outer organic cell investment is of note in that the lower half of the cell is held within the parent lorica by a normal lorica lining with a connection to the cell surface in the waist region. However, close to the anterior end of the cell the investment takes the form of a limited fine layer, probably comprising freshly secreted material from the vesicles containing fibrillar material (Fig. 7.33 small arrows). Following cytokinesis, newly separated daughter cells contain nuclei with restored nuclear envelopes. In the juvenile cell a few remaining vesicles containing fibrillar material remain (Fig. 7.34 arrow vfm) and the two types of organic investment can still be recognised on the respective daughter cells.

7.2.3 Discussion: *Stephanoeca diplocostata*: logistics of costal strip storage and transport throughout the cell cycle

Stephanoeca diplocostata has provided a wealth of basic information on the movement of costal strips during cell division and lorica assembly. Despite the impossibility of being able to relate individual strips during storage to their ultimate position in a future lorica, nevertheless it is apparent that they are produced and accumulated in a highly

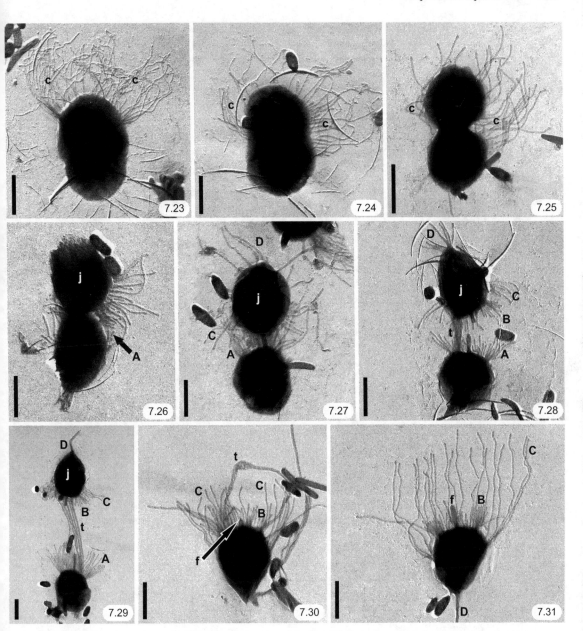

Plate 4 (Figures 7.23–7.31)

Figs 7.23–7.31 *Stephanoeca diplocostata*. Shadowcast whole mounts of cells undergoing division treated with hydrofluoric acid to remove siliceous components of loricae. Bars = 2 μm. **Fig. 7.23** Cell at prophase stage of mitosis, without flagellum and with collar microvilli (c) migrating outwards on anterior surface.

Figs 7.24–7.25 Figure-of-eight (anaphase/telophase) stages (Fig. 7.25 slightly later than Fig. 7.24) with asymmetrical arrangement of microvilli (c) migrating outwards and backwards.

Fig. 7.26 Late cytokinesis with juvenile cell (j) undergoing inversion. Microvilli are now migrating into their respective locations. A = collar microvilli of daughter that will remain within parent lorica.

ordered manner. It would appear that strips of the posterior (waist) transverse costa are formed and exocytosed early in interphase and that they are not stored in one sub-grouping but are distributed equally within the four quadrants. As the number of strips within an accumulation increases, division into sub-groupings becomes more apparent.

The combination of observations derived from light microscopy of living cells, whole mounts for TEM, hydrofluoric acid-treated specimens and sectioned material has permitted a detailed reconstruction of the complex temporal and spatial events that occur during cell division and lorica assembly (see Table 7.1 for figure numbers of

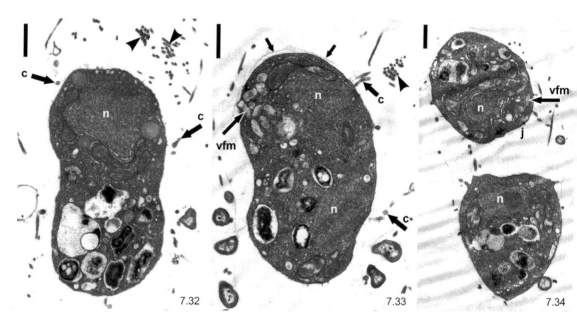

7.32   7.33   7.34

Plate 5 (Figures 7.32–7.34)

Figs 7.32–7.34 *Stephanoeca diplocostata*. TEM sections of dividing cells. Bars = 1 μm. Reproduced from Leadbeater (1994a). **Fig. 7.32** Prophase (early figure-of-eight) cell with enlarged anterior end containing nucleus (n) with partially disintegrated nuclear envelope. Collar microvilli (arrows c) have migrated over anterior surface and sub-groupings of costal strips are evident (arrowheads).
Fig. 7.33 Telophase (asymmetrical) cell with two nuclear areas (n). Collar microvilli (arrows c) with accumulated costal strips (arrowhead) have migrated onto side of cell. Group of vesicles containing fibrillar material (vfm) is present at top of cell and will be passed onto future juvenile. A thin layer of fibrillar material is evident on top of cell (small arrows).
Fig. 7.34 Two daughter cells soon after cytokinesis is complete. Upper cell is the juvenile and contains a few remaining vesicles with fibrillar material (arrow vfm). Lower cell will remain with parent lorica.

Figs 7.27–7.29 Three consecutive stages in cell division showing two cells, lower cell remaining within the parent lorica while the upper cell, juvenile (j), is pushed backwards out of the lorica. Microvilli have been shared between the two cells in four distinctive locations: A = collar of lower daughter cell; B = collar of juvenile; C = lorica-assembling microvilli on juvenile; D = posterior filopodia on juvenile. Double, triple or quadruple threads (t) separating daughter cells visible in Figs 7.28 and 7.29.
Fig. 7.30 Liberated juvenile still with separating thread (t) showing forward progression of lorica-assembling microvilli (C). Short collar microvilli (B) and flagellar stub (f) are visible.
Fig. 7.31 Juvenile at late stage of lorica formation showing fully extended lorica-assembling microvilli (C), short microvilli of the collar (B) and flagellum (f), and posterior filopodium (D). Figs 7.24 and 7.28 reproduced from Leadbeater (1994a). Fig. 7.31 reproduced from Leadbeater (1979b).

relevant illustrations). The most important feature to emerge is the relationship between the sub-groupings of strips and the behaviour of collar microvilli. The latter not only guide strips from one location to another but also provide the essential continuity of contact. While individual sub-groupings of stored strips on the collar interact with at least a quadrant of the collar microvilli, nevertheless it would appear that for each strip sub-grouping destined for a longitudinal costa, one microvillus takes a 'leading' role. This conclusion is based on the fact that the number of collar microvilli that eventually take on the lorica-assembling role is exactly the same as the number of future longitudinal costae.

While it is apparent that a single sub-grouping of strips generates a single longitudinal costa, and presumably this also applies to helical costae within the posterior chamber, the situation is more complex for transverse costae. Here it appears that each transverse costa is assembled from a portion of strips contributed by each of the four quadrants and these must be aligned precisely to create a continuous, uniformly imbricated, costal ring (see Section 4.3.5). There is no suggestion that a complete ring is assembled from one sub-grouping of strips. If this were the case there would be an asymmetrical distribution of horizontal strips at the anterior end of the juvenile and that is never observed. It is possible that quadrants of each of the four transverse costae are stored together, or in close association.

Considerable insight has been gained into the logistics and behaviour of the collar microvilli during cell division. It appears that microvilli, or at least their bases, are conserved throughout division. Thus, in common with flagellar bases (see Section 3.2), the microvilli contributing to a collar will be a mixture of generations. Of the 40 or so original microvilli in the parent collar, about half, in the form of the lorica-assembling microvilli, will pass directly to the juvenile (Table 7.2). It is not known whether the remaining 20 are shared out equally between the two daughter collars. However, the final count of microvilli on the two daughter collars is between 70 and 80 (20 collar and 20 lorica-assembling microvilli on the juvenile cell, and 35–40 in the collar of the sister cell remaining within the parent lorica). After the juvenile has assembled its lorica, the lorica-assembling and collar microvilli combine, thereby restoring the collar of the juvenile to 40 microvilli. Thus the juvenile will have microvilli of at least two generations, and this must become more complicated with each subsequent cell division.

As illustrated in Chapter 2 (Section 2.7.2), the microfilamentous bases of the collar microvilli interdigitate with the microtubular cytoskeleton. In preliminary observations of the microtubular cytoskeleton during mitosis and cell division, the daughter flagellar bases are associated with a reduced ring of cytoskeletal microtubules (Fig. 2.75) (Leadbeater, 1994b). It is possible that migration of the microvilli during division is organised and directed by these microtubules. Clearly the microvilli do not move about in an uncoordinated fashion during their complex changes in location. The possible rotation of the ring of lorica-assembling microvilli during lorica assembly on the juvenile has been discussed on several occasions elsewhere (Section 6.8). There are many precedents for rotational movements of organelles and other cell components, particularly during early embryonic development in animals and plants (Frösler and Leadbeater, 2009). For instance, in zebrafish (*Danio rerio*) embryos, the nuclei of some undifferentiated brain cells undergo a sustained 'spinning' behaviour, the mean spin being about ten rotations per hour, but the peak velocity can be two rotations per minute (Herbomel, 1999). Probably the most relevant example of a rotary movement similar to that inferred in loricate choanoflagellates is the clockwise rotation of the anterior end ('head') of certain devescovinid flagellates found in the hindgut of termites (Tamm and Tamm, 1976). A rod-like axostyle complex turns the head of the flagellate, including the plasma membrane, continuously at speeds of up to 60 revolutions per minute in a clockwise direction relative to the rest of the cell. Torque is generated from within the cell by a mechanism at present unknown (Tamm, 2008). Since head rotation includes the plasma membrane as well as the cytoplasm, this motility provides direct evidence for the fluid nature of cell membranes (Tamm, 1979).

Complex movements of sub-groupings of costal strips over the surface of microvilli must involve cell-matrix adhesion and motility mechanisms comparable to those found in many other cells, including those from mammals (Berrier and Yamada, 2007; Lock *et al.*, 2008). It is probable that microvillus–costal strip transmembrane complexes link extracellular sub-groupings of strips to the internal actin cytoskeleton of microvilli. These adhesion complexes must be highly flexible and dynamic structures, the components of which undergo rapid and regulated turnover to maintain balanced streams of mechanical and chemical information (Lock *et al.*, 2008). Such movements occur at the same time as actin in the microvilli is being polymerised and depolymerised, effecting their elongation and shortening respectively.

7.3 *DIDYMOECA COSTATA*

7.3.1 Nomenclature

Unfortunately the nomenclature of *Didymoeca costata* has a chequered history and since this species has made such an important contribution to our understanding of the tectiform condition it is essential to clarify its current taxonomic status. *Didymoeca costata* (Valkanov) Doweld is better known as *Diplotheca costata* Valkanov (1970) (see Jackson and Leadbeater, 1991). However, the name *Diplotheca* Valkanov 1970 has at least five different homonyms: while it is undesirable to have different taxa denoted by the same generic name, nevertheless according to the International Code of Zoological Nomenclature (ICZN) it is only the names of other animal taxa that are treated as homonyms for the purpose of zoological nomenclature. Since both *Diplotheca* Matthew 1885 (Mollusca) and *Diplotheca* Valkanov 1970 (Choanoflagellatea) are subject to the zoological code, the junior homonym must be rejected and a new substitute name established in its place.

Doweld (2003) established the genus *Didymoeca* as a substitute name for *Diplotheca* but unfortunately the information with an erroneous title was published in an obscure Russian journal (see Doweld, 2003) and so was completely overlooked by protistologists elsewhere. To complicate the matter further, Özdikimen (2009), unaware of the substitute name published by Doweld, also created the replacement name *Volkanus*. Regrettably the name *Volkanus costatus* (Valkanov) Özdikimen has been used recently in a publication by Leadbeater (2010). Now that Doweld's (2003) substitute name *Didymoeca* has come to light and takes precedence over *Volkanus*, the latter name must be considered invalid. This is not quite the end of the story, for Doweld (2003) included *Didymoeca* within the Craspedophyceae Christensen (1962), which was a class of algae originally erected by Chadefaud (1960) as the Craspédophycinées following Bourrelly's (1957) decision to include the choanoflagellates as an order (Craspémonadales) within the Chrysophyceae (see Section 1.8.2). Currently there are two other species of *Didymoeca*, namely *D. elongata* (Nitsche and Arndt, 2008) and *D. tricyclica* (Bergesch *et al.*, 2008).

7.3.2 *Didymoeca costata:* lorica construction

Didymoeca costata is a small tectiform species usually found in coastal water, particularly in stagnant pools on saltmarshes and rocky shores. Its remarkable claim to fame is that its lorica comprises six categories of costal strips, each category being located at a specific position within the lorica and each being easily recognisable by its distinctive morphology (Figs 4.15, 7.35, 7.36) (Leadbeater, 2010). In layman's terms, each strip category carries an identifying 'postcode'. Although there are other species in which it is possible to distinguish different strip categories, for instance *Diaphanoeca grandis* (Sections 7.4.1, 8.6.3), *Didymoeca costata* is by far the most distinctive. This feature has made it possible to determine with accuracy the order in which the various categories of strips are produced, exocytosed and stored, and to follow their subsequent movements on the surface of the cell throughout cell division and lorica assembly (Figs 7.42–7.52).

The lorica of *Didymoeca costata* contains between 60 and 75 costal strips and is immediately recognisable when viewed with electron microscopy (EM) on account of the broad 'petaloid' strips that comprise the posterior chamber and a regular array of longitudinal costae with an anterior ring that constitutes the anterior chamber (Figs 4.15, 7.35). There are three transverse costae (rings); the posterior and intermediate rings are inside the longitudinal costae and are located at the junction between the two chambers (Fig. 7.36). Each of these rings contains approximately six crescentic strips, those of the posterior ring being thicker and with a smaller radius of curvature than those comprising the intermediate ring. The anterior ring, located at the top of the lorica, also contains about six strips but differs from the two lower rings by being on the outer surface of the longitudinal costae and by containing wider, flattened strips (Figs 7.35, 7.36) (see Section 4.3.5, Table 4.3). The imbrication of strips constituting all the rings shows signs of a left-handed rotation although, particularly in the anterior ring, this pattern is often disturbed.

The petaloid strips of the posterior chamber are continuous with the longitudinal costae of the anterior chamber and their number ranges from 14 to 18 with a mode of 17 (Figs 7.35, 7.37) (Leadbeater, 2010). Each petaloid strip is broader at the anterior end and possesses a thickened midrib with lateral perforations from the mid-region backwards (Figs 4.11, 4.15, 7.37, 7.52). Each longitudinal costa of the anterior chamber comprises two strips, the lower

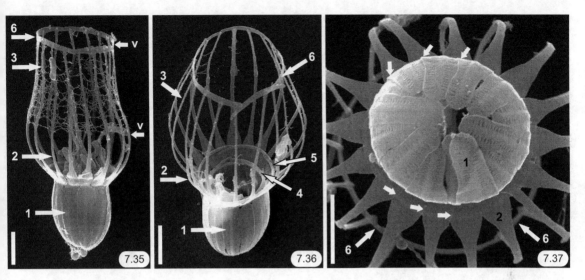

Plate 6 (Figures 7.35–7.37)

Figs 7.35–7.37 *Didymoeca costata*. SEM images of loricae showing the location and numbering of the six categories of costal strips. Numbering as follows. Longitudinal costae: (1) petaloid costal strip; (2) spatulate strip; (3) narrow strip. Transverse costae: (4) posterior transverse costa; (5) intermediate transverse costa; (6) anterior transverse costa (ring). Bars = 2 μm. **Fig. 7.35** Side view of lorica showing general disposition of costae and strips. Shorter arrows denote vertical extent of veil which has a cobweb-like appearance.
Fig. 7.36 Tilted specimen showing more clearly the positions of the two lower transverse costae.
Fig. 7.37 Lorica viewed from rear showing imbrication of petaloid and spatulate strips, respectively. Three successive anti-clockwise overlaps are apparent for both petaloid (upper arrows) and spatulate (lower arrows) strips. The anterior ring is visible (arrows 6). Reproduced from Leadbeater (2010).

bearing a spatulate flattening at the base (Figs 4.15, 7.35–7.37, 7.43–7.45); the upper strip is narrow throughout its length and terminates anteriorly with a slight flattening (Fig. 7.45). The bases of neighbouring spatulate strips overlap each other, usually in a clockwise direction, when viewed from above (Fig. 7.37 small arrows). A thin organic investment extends from about one-third of the way up the anterior chamber to the top of the lorica (Figs 4.41, 4.42, 7.35, 7.45). In SEM preparations this has a cobweb-like appearance of irregular fibrils (Figs 4.41, 7.35, 7.45) but it is, in reality, a diaphanously thin continuous investment comprising a meshwork of microfibrils (Fig. 4.42). The intervening spaces between the bottom of the investment and the spatulate bases of the longitudinal costae are open to the surrounding environment and allow an inward flow of water created by flagellar activity.

7.3.3 *Didymoeca costata*: costal strip production, exocytosis and storage

In most respects costal strip formation, exocytosis and storage in *Didymoeca costata* resemble the equivalent processes in *Stephanoeca diplocostata* (see Section 5.5). However, the distinctive morphology of the different categories of costal strips in *D. costata* has permitted a detailed analysis of the order and orientation of strips from the moment they are deposited intracellularly within silica deposition vesicles (SDVs) until they reach their final destination within the newly assembled lorica of a juvenile cell.

Costal strips are laid down individually within SDVs at regularly spaced intervals within the peripheral cytoplasm. The six categories of strips are produced one category at a time in a precise sequence. This starts with the strips of the

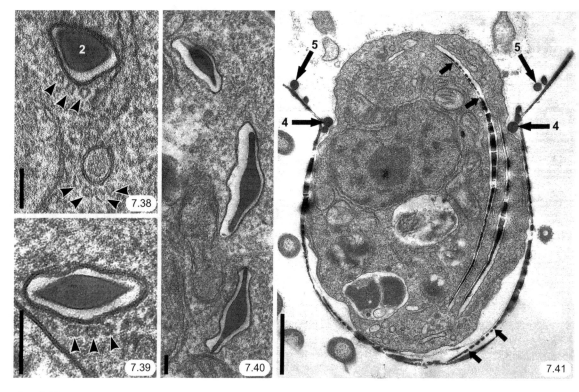

Plate 7 (Figures 7.38–7.41)

Figs 7.38–7.41 *Didymoeca costata*. TEM sections of cells showing details of costal strip formation within SDVs (see also Section 5.5).
Fig. 7.38 Transverse section of two developing strips each within an SDV with adjacent microtubules (arrowheads). The upper SDV contains a spatulate strip (2) and has the characteristic triangular shape. Bar = 0.1 μm.
Fig. 7.39 Intermediate stage in formation of a spatulate strip (lower end). The SDV is subtended by three microtubules (arrowheads).
Fig. 7.40 Three developing petaloid strips, each within an SDV. The top strip illustrates that the midrib apparently develops first. Bar = 0.1 μm.
Fig. 7.41 Longitudinal section of cell with two developing petaloid strips. The location of the posterior perforations at the upper end of the strip within the cell (upper short arrows) and at the rear end of the strip within the lorica (lower short arrows) shows that they are 'upside-down' with respect to each other. The different categories of strips within the lorica are numbered according to Fig. 7.36. Bar = 1 μm. Figs 7.38, 7.40 and 7.41 reproduced from Leadbeater (1979b).

transverse costae, proceeding sequentially from the posterior and intermediate rings to the anterior ring, and then to the strips of the longitudinal costae, again proceeding from the posterior petaloid and spatulate strips to the anterior narrow strips. The first appearance of SDVs is as elongated narrow vesicles oriented parallel to the long axis of the cell and supported on their inward-facing surface by two or more microtubules (Figs 7.38, 7.39). Initially, strips of all categories start as a cylindrical core; wing-like projections of the petaloid and spatulate strips develop later (Fig. 7.40). Once strips have acquired enough silica to become self-supporting the accompanying microtubules depolymerise. Since the orientation of the petaloid strips can be determined in section by the location of the lateral perforations at the rear end, it is apparent that these strips are formed upside-down with respect to their orientation in the lorica (Fig. 7.41). Thus in Fig. 7.41, the perforated end of the developing petaloid strip within the SDV is directed towards the front end of the cell, whereas in the lorica it is directed towards the rear (Fig. 7.41 arrows). A similar inversion is apparent for spatulate strips and must apply to all strip categories (Leadbeater, 2010). Costal strips also appear to develop in a centripetal fashion with SDVs containing mature strips being closer to the central axis of the cell and SDVs with less mature strips being towards the outside (Leadbeater, 2010).

Mature costal strips are exocytosed at the anterior end of the cell within the circumference of the collar base. The crescentic strips destined for the posterior transverse costa are the first to emerge. They move vertically on the inner surface of the collar microvilli towards the top of the collar, where they are then rotated through 90° into a horizontal position. The concave side of the strip faces inwards towards the midline of the lorica. Strips of the intermediate transverse costa and the anterior ring subsequently emerge in that order and are stored horizontally on the collar. Within an accumulation, well-defined vertical groups comprising representative strips from the three transverse costae in inverse order can be observed (Fig. 7.42). Petaloid strips are the next category to be exocytosed from the cell and these are also moved to the top of the collar and rotated through 90° where they accumulate as a ring, probably comprising four quadrants, within the ring already formed by the strips of the transverse costae (Leadbeater, 2010). Movement of strips on the surface of the collar microvilli commonly allows for one category of strips to move over another. The final two categories of strips to be exocytosed are the spatulate and narrow strips, respectively. In common with the other strips, they are exocytosed vertically from the cell, upside down, and the posterior ends form a ring within the horizontal ring of the other costal strips. They remain in a vertical position throughout the remainder of interphase and during cell division (Fig. 8.7).

7.3.4 *Didymoeca costata*: cell division and costal strip reorganisation on the juvenile cell

Once a complete set of strips has been exocytosed and stored within the parent collar, the cell undergoes mitosis and cytokinesis. The behaviour of the flagellar apparatus, nucleus and other organelles during division in *Didymoeca* is similar to that already described for *Stephanoeca* (see Sections 2.9, 7.2.2) (Leadbeater, 2010). However, while in principle the complex reorganisation of costal strips during division is similar in the two species, nevertheless there are some significant differences. One already encountered is the persisting vertical orientation of the spatulate and narrow strips of the future lorica (Fig. 8.7). In *D. costata* the juvenile resulting from cytokinesis is inverted within the confines of the parent collar and at the same time the

horizontal ring of stored costal strips is reorganised into an ensheathing cup around the juvenile as it begins its passage backwards out of the parent lorica (Fig. 7.45). As in *Stephanoeca*, the behaviour of the collar microvilli is fundamental to the strip movements that follow. At the beginning of cytokinesis the outer groups of strips, destined for the future transverse costae, move down the inner surface of the collar towards the cleavage plane between the two daughter cells (Figs 7.46a–c). At the same time the petaloid strips are rotated through 90° so that their long axes are aligned with the vertical plane of the parent lorica (Figs 7.43, 7.44, 7.46a–b). The three categories of strips of the future longitudinal costae, namely petaloid, spatulate and narrow, are now in the correct order and orientation to be passed on to an inverted juvenile as it emerges from the parent lorica. The outer groups of transverse strips continue their movement over the petaloid strips and their order becomes reversed by successive overtaking (leapfrogging) (Figs 7.43, 7.44, 7.46d–i). Thus the wider strips of the future anterior ring move downwards until they reach the leading (anterior) edge of the narrow longitudinal strips which project slightly beyond the petaloid strips (Figs 7.44, 7.45, 7.46d). They are then overtaken on the outside by the two categories of crescentic strips which move over the leading edge of the developing lorica cup and onto the inner surface formed by spatulate and narrow strips of the longitudinal costae (Figs 7.45, 7.46e–f). The crescentic strips that will form the intermediate transverse costa then come to a halt and are overtaken by the crescentic strips of the posterior costa (Fig. 7.46g–h). These come to a halt just below the strips of the previous ring (Fig. 7.46i). During this rearrangement the concave side of the crescentic strips remains facing towards the mid longitudinal line of the developing lorica cup. At the end of this rearrangement, the strips that will form the lorica of the juvenile are now correctly oriented and positioned with respect to each other, but they are telescoped to form a cup into which the inverted juvenile cell is pushed (Fig. 7.45). The collar microvilli, which until the reorganisation is complete extend the length of the outer surface of the petaloid strips, are quickly withdrawn as the juvenile cell is pushed into the cup of costal strips (Fig. 7.45). By this time a subset of the original collar microvilli, equivalent to the number of future longitudinal costae, are passed on to the juvenile, which is now fully inverted and emerges backwards from the parent lorica (Fig. 7.49). Fig. 7.46a–i summarises diagrammatically the movements of stored costal strips during cytokinesis and the inversion of the juvenile.

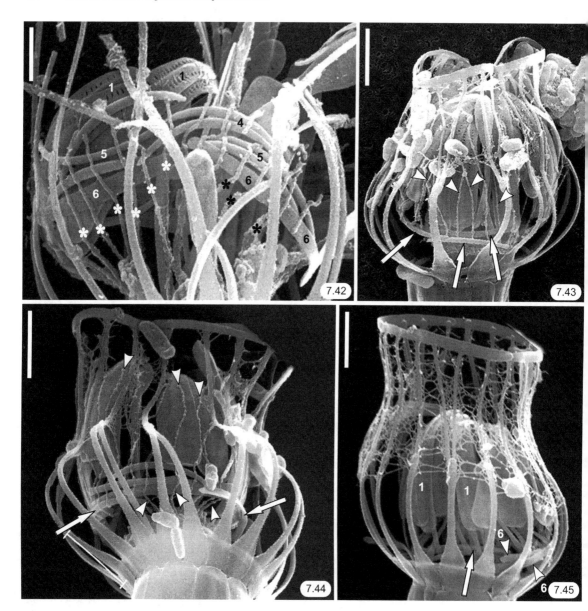

Plate 8 (Figures 7.42–7.45)

Figs 7.42–7.45 *Didymoeca costata*. SEM images of cells showing disposition of strips within an accumulation and the reorganisation of strips on the juvenile following cytokinesis while still within the parent lorica. **Fig. 7.42** Details of a horizontal accumulation of strips. Strips that will contribute to the posterior (4), intermediate (5) and anterior (6) transverse costae are outside the ring of petaloid strips (1). Asterisks denote individual collar microvilli. Bar = 1 μm.

Figs 7.43–7.44 Accumulated strips have been rearranged into a telescoped lorica cup for the developing juvenile. The collar microvilli still extend the full length of the outside of the lorica cup (arrowheads), the petaloid strips have rotated into the vertical position and the outer horizontal strips have moved over the petaloid strips to the open end of the lorica cup (arrows). Bar = 2 μm.

Fig. 7.45 The lorica cup is now fully formed; the petaloid strips (1) are on the outside, strips of the anterior ring (arrowheads 6) are on the outside of the anterior edge of the lorica cup and the strips that will form the future intermediate and posterior transverse costae are within the lorica cup (arrow). Bar = 2 μm. Reproduced from Leadbeater (2010).

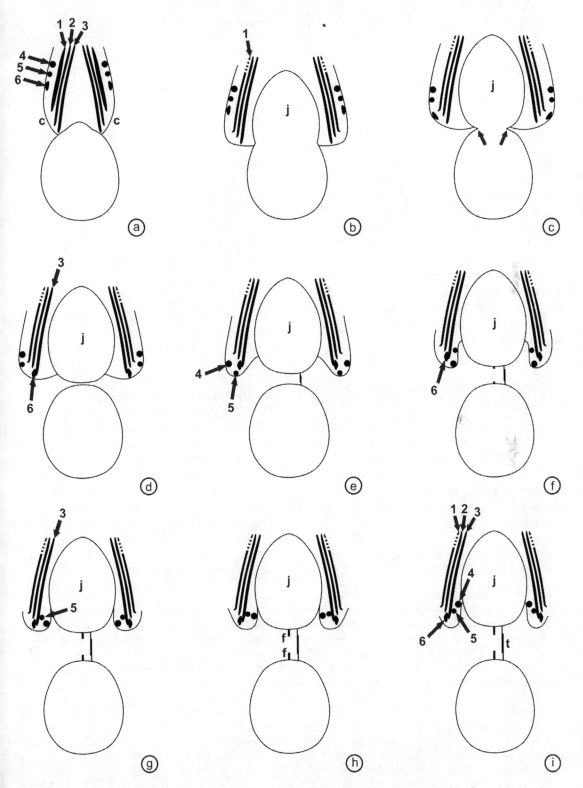

Fig. 7.46 *Didymoeca costata*. Sequence of drawings of a cell undergoing division illustrating the reorganisation of stored costal strips to form a lorica cup for the inverted juvenile cell (i). a. Parent cell about to enter division, the flagellum has been withdrawn. The collar

A transverse section in the mid-region of a juvenile, at an early stage of moving backwards within the parent lorica, shows the telescoped layering of the three categories of strips that comprise the longitudinal costae (compare Fig. 7.47 with Fig. 7.49). The narrow strips (3) of the inner layer are regularly spaced immediately outside the plasma membrane (Figs 7.47, 7.48). The spatulate strips (2) of the intermediate layer are further out and can be identified by being triangular in cross-section (Fig. 7.47). The petaloid strips (1) are on the outside and are thinner and with a thickened mid-rib (Fig. 7.47). An equivalent section in the vertical plane shows the two daughter cells facing each other (Fig. 7.49). The juvenile (upper cell) has a posterior protuberance which serves as an axis around which strips destined for the longitudinal costae are arranged. The relative positions of the strips that will give rise to the transverse costae can be seen numbered at either side of the anterior end of the juvenile cell in Fig. 7.49 (arrows). In both transverse and vertical sections the plasma membrane towards the rear of the juvenile is 'rippled' and there are regular patches of dark-staining fibrillar material (Figs 7.47–7.49).

7.3.5 *Didymoeca costata*: lorica assembly on the juvenile cell

Emergence of the juvenile with the telescoped covering of costal strips (Fig. 7.50) takes a few minutes. Once free of the parent cell, the juvenile rapidly assembles its lorica (Figs 7.51, 7.52). Assembly is dependent on the lorica-assembling microvilli that are equal in number to the longitudinal costae and involves a forward and left-handed (clockwise) rotational movement.

Evidence for the forward movement is obvious from loricae preserved at different stages during assembly. The process is analogous to pulling out a telescope from the narrow end. The inner narrow strips that form the front end of the anterior lorica chamber emerge first, followed by the spatulate strips (Fig. 7.52). This continues until all strips are fully extended. Evidence for a rotational movement is less obvious and almost certainly would be overlooked if it were not already known that it is a universal feature of lorica assembly (see Chapters 4 and 6). However, careful observation of the transverse costae shows that the component strips, which were derived from several horizontal groups when on the collar of the parent, are now imbricated and this can occasionally be seen on particularly favourable specimens of the anterior ring (Figs 7.45, 7.50, 7.52).

7.3.6 Discussion: *Didymoeca costata*: order and orientation of costal strip production

The advantage of being able to distinguish six costal strip categories in *Didymoeca costata* on the basis of morphology has permitted a unique insight into the order and orientation in which strips are formed, exocytosed and stored on a parent cell and subsequently the manner in which they are rearranged on the juvenile during cell division. Leadbeater (2010) noted four new seminal findings from the study on *D. costata*. (1) Costal strips are formed upside-down within the cell. (2) The order of costal strip production is first for the transverse (inner) costae and second for the longitudinal (outer) costae. Within both classes of costae, strips for the future posterior end are produced before those for the anterior end. (3) In contrast to longitudinal costae, where the three constituent strips are stored in a single group prior to lorica assembly, the constituent strips for a single transverse costa are stored as part of several (probably four) horizontal groups, each group containing enough strips to produce a contributing segment to all three of the transverse costae. (4) During cytokinesis, the outer horizontal

---

microvilli are in the late interphase position with a large accumulation of costal strips at the top of the inner surface of the collar. On the outside of the accumulation are the costal strips for the three transverse costae in reverse order ((4) posterior transverse costa; (5) intermediate transverse costa; (6) anterior ring). Inside are the three layers of longitudinal costal strips: (1) petaloid; (2) spatulate; (3) narrow, respectively. b–c. Cell division results in the juvenile (upper) cell (j) being inverted, at the same time the collar microvilli are drawn into the cleavage furrow (arrows in c) between the two daughter cells and the outer transverse costae move downwards. The petaloid strips (1) have rotated into the vertical position. d. The downwards movement of the outer transverse costal strips continues and the strips of the anterior ring (6) lock into position on the anterior end of the narrow strip (3). e–f. The costal strips of the intermediate (5) and posterior (4) transverse rings overtake (leapfrog) those of the anterior ring (6) and move over the anterior end of the longitudinal strips and into the lorica cup. g. Costal strips of the intermediate transverse costa (5) lock into position on the inner surface of the narrow longitudinal strips (3). h–i. Costal strips of the posterior transverse ring (4) overtake strips of the intermediate transverse ring to take up their final position within the lorica cup (compare with Fig. 7.49). By this stage the collar microvilli have shortened and the cells are separating by extension of the actin-containing threads (t). Flagella (f) on both cells are elongating.

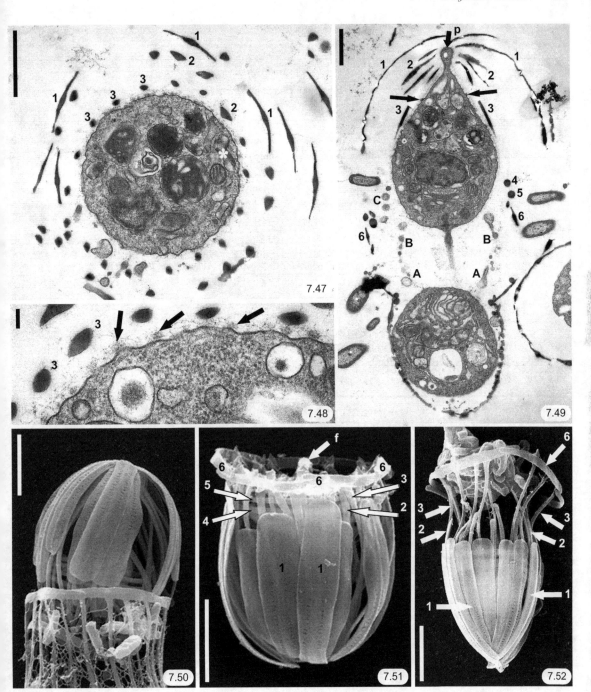

Plate 9 (Figures 7.47–7.52)

Figs 7.47–7.52 *Didymoeca costata*. TEM sections of juvenile cells during separation and SEM images of lorica assembly. Reproduced from Leadbeater (2010). **Fig. 7.47** Transverse section through juvenile that is emerging from parent lorica (compare with Fig. 7.49). The cell still contains a few vesicles of fibrillar material (asterisk). The telescoped strips around the juvenile cell are seen in concentric rings. The groups of strips at the top of the collar (destined for the future transverse costae) are moved to the anterior edge of the developing lorica cup and their order is reversed. Strips destined for the intermediate and posterior transverse costae are moved inside the cup.

Of the four points listed above, only the third was hinted at in studies on *Stephanoeca diplocostata* (Leadbeater, 1979b, c, 1994a, b), the remainder are novel and were completely unexpected. The immediate impression is of a highly derived condition and clearly the 'upside-down' orientation of costal strip production and the 'inside-out' order in which strips are exocytosed and stored must be intimately related to the inverted mode of cytokinesis. These features are diametrically opposite to those observed in nudiform species, where costal strips are deposited the correct way round and the order is from the outside inwards (Chapter 6). These matters are central to the content of Chapter 8.

Allowing for the fact that the orientation and precise order of strip production in *Stephanoeca diplocostata* could not be resolved, the perceived differences between *D. costata* and *S. diplocostata* are relatively minor. In *S. diplocostata* all the stored strips are horizontal whereas in *D. costata* the last two categories of longitudinal strips, spatulate and narrow, remain vertical after exocytosis. However, because the strips are 'upside-down' with respect to the parent cell they are correctly oriented with respect to the future juvenile. The reason for this may be a constraint of space. Since the lorica of *D. costata* is so small there is limited room for

juvenile rotation, and since these strips are the last to be exocytosed, there is no need to remove them to make way for other strips. The costal strips destined for future transverse costae remain transverse throughout division, whereas this fact is less clear in *S. diplocostata* and *Diaphanoeca grandis*, where all sub-groups of strips appear to be oriented vertically. The creation of the lorica cup in *Didymoeca costata* is earlier than it appears in *S. diplocostata* and much earlier than in *Diaphanoeca grandis*, where the transverse costal strips do not appear until the juvenile is completely free (see Section 7.4.2).

The positioning of costal strips destined for future transverse costae on the outer face of the accumulation explains how, in some genera, costal strips comprising the transverse costae, including the anterior ring, can be finally located on the outer surface of the longitudinal costae (see Table 4.3, Fig. 4.25, Sections 4.3.3, 4.3.5). In these genera the component transverse strips halt their downward progression on the outer surface of the lorica cup at the level of the respective longitudinal strips. Once in position, assembly of external transverse costae does not appear to require a direct contribution from the lorica-assembling microvilli. In genera with an inner anterior ring, the transverse strips pass over the leading edge of the lorica cup and then halt their progression (Table 4.3). However, as illustrated by A and B in Fig. 4.25, the inside or outside location of the anterior ring makes no overall difference to the pattern of costal strip overlaps at junctions between transverse and longitudinal costae.

innermost, narrow strips (3) are at regular intervals close to the surface of the plasma membrane. The triangular profiles of the spatulate strips (2) are in a disturbed intermediate ring and the petaloid strips (1) are on the outside. Bar = 1 μm.

Fig. 7.48 Higher magnification of the surface of a serial section of the cell shown in Fig. 7.49. The inner ring of narrow strips (3) is shown. The plasma membrane of the cell is rippled and there are regular patches of dense-staining fibrillar material (arrows). Bar = 0.1 μm.

Fig. 7.49 Longitudinal section through separating daughter cells showing the juvenile (top) upside-down with respect to the daughter cell that remains in the parent lorica (bottom). The juvenile is pear-shaped and has a posterior protuberance (p) around which the groups of petaloid (1), spatulate (2) and narrow (3) strips that will form the future longitudinal costae are arranged. The cell surface at the rear end is rippled and bears patches of dense-staining fibrillar material (arrows). On the right-hand side of the inverted juvenile the position of the two sets of microvilli, lorica-assembling (C) and collar (B), can be seen. The strips destined for posterior (4), intermediate (5) and anterior (6) transverse costae are visible on the left side of the inverted juvenile cell. The collar microvilli (A) of the parent cell are also present. Bar = 1 μm.

Fig. 7.50 Juvenile cell with telescoped arrangement of longitudinal strips (lorica cup), emerging backwards out of parent lorica. Bar = 2 μm.

Fig. 7.51 Juvenile cell showing relative positions of the six categories of strips comprising the lorica cup. A short flagellar stub (f) is present on the anterior end. Bar = 2 μm.

Fig. 7.52 Lateral view of partially assembled lorica showing relative positions of petaloid (1), spatulate (2) and narrow (3) strips. The anterior ring has extended to almost its mature diameter. Bar = 2 μm.

7.4 *DIAPHANOECA GRANDIS*

7.4.1 *Diaphanoeca grandis*: lorica construction and costal strip accumulation

Diaphanoeca grandis differs from both *Stephanoeca diplocostata* and *Didymoeca costata* in that the cell is not located at the base of a close-fitting lorica but is suspended by the collar microvilli from the anterior transverse costa in the middle of a voluminous lorica that extends for some distance below the cell (Fig. 4.26). *D. grandis* is a standard tectiform species; costal strips are accumulated on the collar during interphase and cytokinesis results in the inversion of the juvenile (Leadbeater and Cheng, 2010). However, the process of lorica assembly must be modified to allow for the development of a voluminous lorica at such a distance from the cell surface.

The lorica of *D. grandis* is flask-shaped with a parallel-sided neck (Fig. 4.26) and a large spherical to ovoid chamber (Figs 4.26, 4.27). It contains 12 regularly spaced outer longitudinal costae held in position by four inner transverse costae that are labelled 4 to 7, respectively, in Figs 4.26–4.28. The anterior transverse costa (7) is located at the base of the neck and comprises a double costa (Figs 4.30, 4.32, 4.33) (Leadbeater and Cheng, 2010). The three basal transverse costae, anterior (6), intermediate (5) and posterior (4), are located in the posterior third of the lorica. Each longitudinal costa contains eight strips, two above the anterior transverse costa (7), three between the anterior transverse costa and the anterior basal ring (6), two strips separating the three basal rings and one below the posterior basal ring. Superficially all strips appear rod-shaped but, unlike *S. diplocostata*, there are minor differences in length and morphology between the various strip categories. In particular, the strips of the longitudinal costae are approximately 10% longer than the strips comprising the transverse costae (Leadbeater and Cheng, 2010). With regards to morphology, four classes of strips can be distinguished: (1) strips comprising the basal transverse costae have slightly flattened tips which are usually more pronounced at the leading end (Figs 8.27–8.29) (see Section 4.3.2); (2) strips from the anterior transverse costa also have spatulate ends but one end (the leading end) is slightly larger than the other (Figs 4.30, 4.33 asterisks, 8.37, 8.39); (3) component strips of the longitudinal costae have tapering ends and are not spatulate (Fig. 4.33 arrow); (4) the anterior strip of each longitudinal costa projects forward as a spine and has a minute forked tip (Figs 4.8, 7.55). Based on this list of relatively minor characters it is possible, with experience, to categorise most strips when seen separately (Leadbeater and Cheng, 2010).

Intracellular costal strip production and storage on the collar is similar in most respects to *Stephanoeca diplocostata* (see Section 7.2.2). Occasionally nascent costal strips are visible within whole mounts of cells and, in particularly favourable specimens, the distinctive strips that contribute to the anterior costal ring can be seen. The spatulate ends of these strips are directed downwards and slightly undulated ends directed upwards (Leadbeater and Cheng, 2010). The orientation of these strips within the cell is equivalent to the direction that they will occupy on the inverted future juvenile cell after division (Fig. 8.37). This is equivalent to the coincidence in orientation of the petaloid and spatulate strips within the parent cell and on the juvenile in *Didymoeca costata* (Fig. 7.41) and thus the developing strips can be described as being 'upside-down'. Following exocytosis from within the collar base, costal strips are transported to the top of the inner surface of the collar where they are rotated through 90° and stored horizontally in a ring comprising four quadrants (Leadbeater and Cheng, 2010). The order in which strips are exocytosed and stored is first for the basal transverse costae followed by the anterior transverse ring and then the longitudinal costae from the rear to the anterior end. This sequence of production has been confirmed by growing cells in silicon-deprived conditions whereby the earlier strips are more highly silicified than those produced later (Fig. 8.8) (Leadbeater and Cheng, 2010). The opposite occurs when a culture of silicon-starved cells is re-supplied with silicon; strips produced early in the cell cycle are depleted whereas those produced after silicon addition are much thicker (Fig. 8.9) (Leadbeater and Cheng, 2010).

7.4.2 *Diaphanoeca grandis*: cell division and lorica assembly

Once a complete set of costal strips has been accumulated at the top of the collar, the cell undergoes nuclear division and cytokinesis (Fig. 7.53a–e). This is immediately preceded by withdrawal of the flagellum. The anterior end of the cell enlarges as the nucleus, which appears hyaline in Fig. 7.53a–b, divides. The bases of the collar microvilli move laterally over the front of the cell and the bundles of costal strips, which until this time were horizontal and closely packed, break up into sub-groupings and are rotated into the vertical plane (Fig. 7.53a–b arrows). This contrasts with *Didymoeca costata*, where strips destined for the future transverse costae remain in the horizontal plane (see Section

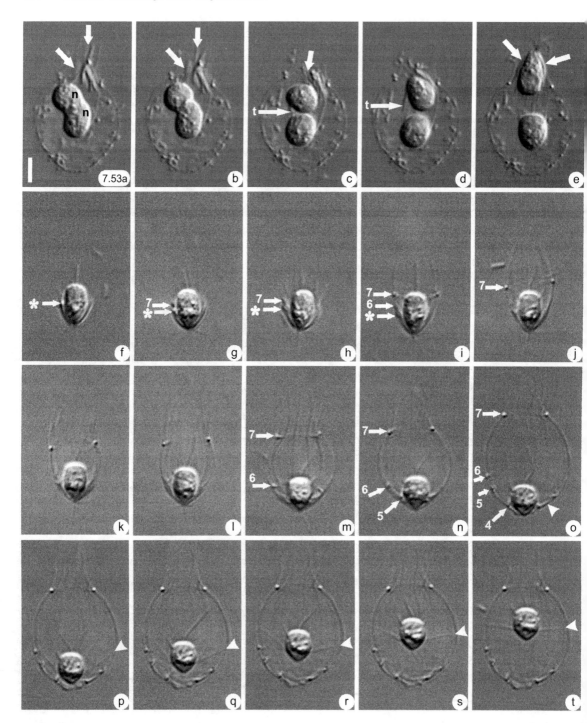

Plate 10 (Figure 7.53)

Fig. 7.53 *Diaphanoeca grandis.* Interference contrast light microscopical images of living cells. Reproduced from Leadbeater and Cheng (2010). Bar = 5 μm (shown on a). a–e. Sequence of images showing cell division starting with late anaphase/telophase (nuclei (n) in a).

7.7 for discussion). Cytokinesis proceeds diagonally and as the two daughter cells become distinct the microvilli associated with bundles of costal strips close around the upper, juvenile, cell (Fig. 7.53c–d). The microvilli retained by the lower cell, which will remain in the parent lorica, are free of costal strips (Fig. 7.53d). Once the two daughter cells have separated (Fig. 7.53c–d), the juvenile is then inverted by elongation of a thread that initially emerges from near the base of the juvenile but rapidly migrates towards the collar (Fig. 7.53c–d) (see Fig. 19 in Leadbeater and Cheng, 2010). Further movement backwards of the juvenile from the parent lorica is achieved by elongation of the two threads as they slide over one another (Fig. 7.53e). As the juvenile is pushed out of the parent lorica, the vertical bundles of costal strips become closely appressed to the surface of the juvenile (Fig. 7.53e arrows).

The newly emerged juvenile has an array of filopodia issuing from the rear (Fig. 7.54 inset, arrow pf). These may acquire temporary attachment to a surface, or in some circumstances they can be used to move the juvenile along a surface. The vertical covering of costal strips reveals two layers, an outer layer at the rear of the cell overlapping an inner one towards the front (Fig. 7.54 inset, arrowheads). On juveniles where the close packing of costal strips is disturbed, the two layers are more evident (Fig. 7.54 asterisks). There is an important distinction between the two layers in that the posterior, outer layer consists exclusively of strips that will form future longitudinal costae (Fig. 7.54 upper asterisks, 7.55); the distinctive forked tips of the anterior costal strips (spines) are clearly visible (Fig. 7.55 arrowhead). The inner layer comprises exclusively strips destined for the transverse rings. The distinctive transverse strips with a spatulate end are located at the extreme anterior end (Figs 7.54, 7.56, 7.57). As in *Stephanoeca diplocostata* (Leadbeater, 1994b) the original collar microvilli of the parent cell are shared between the collar of the

daughter cell that will remain in the parent lorica and the juvenile. On the juvenile they are divided between the new collar microvilli, which are short and emerge from around the anterior end of the cell, and the lorica-assembling microvilli which emerge from the side around the anterior end. The latter are exclusively associated with the anterior layer of costal strips that are destined for the future costal rings (Leadbeater and Cheng, 2010).

Within minutes of leaving the parent lorica the juvenile proceeds with lorica assembly (Fig. 7.53f–t). This is a single continuous movement and takes about five minutes. The anterior (inner) bundles of vertical costal strips on the juvenile, that are destined to form the four transverse costae, are rotated through 90° so that the strips become horizontal and are moved backwards beneath those of the posterior layer that will form the longitudinal costae (Fig. 7.53f–i arrow asterisk). The longitudinal strips are then pulled out anteriorly from the outer basal bundles to assemble the longitudinal costae; at the same time the diameter of the lorica cup around the juvenile enlarges (Fig. 7.53h–l). From an early stage in development the horizontal strips that will form the anterior transverse ring (Fig. 7.53g–h arrow 7) become separate from the remainder of the transverse strips that will form the basal rings (Fig. 7.53g–i arrow asterisk). The front end of the lorica develops first and soon the neck and anterior transverse ring are clearly visible (Fig. 7.53i–l). With continuing forward extension of the longitudinal costae the chamber enlarges in diameter, giving it the spherical-to-ovoid shape and the anterior basal ring becomes apparent (Fig. 7.53i–m arrow 6). The cell is still positioned at the extreme base of the developing lorica (Fig. 7.53n–o). Further elongation reveals the successive appearance of the remaining two posterior rings (Fig. 7.53n–o arrows 4 and 5). The posterior basal ring is positioned close to the surface of the juvenile cell about two-thirds of the distance towards its

---

Sub-groupings of costal strips are held vertically above the cell by the collar microvilli (a–c arrows). c–d. Diagonal separation of daughter cells and inversion of upper juvenile achieved by overlapping threads (arrow t), one thread emerging towards the base of the juvenile cell and the other from the collar of the lower daughter cell. e. Backwards emergence of juvenile cell from parent lorica. Accumulated costal strips are located vertically on the juvenile surface (arrows). f–t. Sequence of images of a juvenile assembling its lorica. f. The first stage of lorica assembly involves the inner vertical sub-groupings of strips on the juvenile being rotated through 90° to become horizontal (arrow asterisk) and being moved posteriorly between the cell surface and the outer layer of vertical strips. g–o. The anterior transverse ring (arrow 7) is the first transverse costa to separate from the horizontal ring of strips and this can be followed through to its final position at the base of the lorica neck (in o). The anterior basal ring (i, arrow 6) is the next transverse costa to appear. As the lorica chamber becomes spherical in shape, the intermediate basal ring appears (n, arrow 5) followed by the posterior basal ring (o, arrow 4). Finally the cell is lifted up towards the centre of the chamber by the lorica-assembling microvilli (o–t, arrowheads).

Plate 11 (Figures 7.54–7.57)

Figs 7.54–7.57 *Diaphanoeca grandis.* Juvenile cell (j) with covering of costal strips. Reproduced from Leadbeater and Cheng (2010).
Fig. 7.54 Vertical covering of costal strips slightly dislodged showing two layers; posterior layer of strips (upper asterisks left and right) destined for future longitudinal costae and anterior layer of strips (lower asterisks left and right) destined for transverse costae. The small spatulate tips of two strips destined for the future posterior transverse costa (arrows). Strips with larger spatulate heads destined for anterior transverse costa (arrowheads 7). Flagellar stub (arrow f). Bar = 1 μm. Fig. 7.54 inset Interference contrast light micrograph of recently liberated juvenile showing vertical layers of costal strips (arrowheads) on surface and posterior filopodia (pf). Bar = 2 μm.
Fig. 7.55 Higher magnification of tips of left-hand posterior layer of deflected strips in Fig. 7.54 showing the forked tip of a strip destined for a future longitudinal costa (arrowhead). Bar = 0.25 μm.
Fig. 7.56 Higher magnification of two strips with small spatulate tips (from Fig. 7.54 left arrows) pointing anteriorly (flagellar pole) on juvenile (arrows). Bar = 0.25 μm.
Fig. 7.57 Higher magnification of two strips with larger spatulate tips (from Fig. 7.54 arrowheads 7), destined for future anterior transverse costa, pointing anteriorly (flagellar pole) on juvenile (arrowheads 7). Bar = 0.25 μm.

base (Fig. 7.53o arrow 4). Once lorica assembly is complete, the cell is lifted upwards within the chamber by the lorica-assembling microvilli that emerge about two-thirds of the way down the cell and extend to the inner surface of the longitudinal costae (Fig. 7.53p–t arrowhead). These microvilli probably take on a suspensory role associated with the anterior transverse ring (7) (Andersen, 1988/9).

7.4.3 Discussion: *Diaphanoeca grandis*: assembly of a voluminous lorica

The similarity in the morphology of costal strip accumulation, cell division and lorica assembly in *Diaphanoeca grandis* and *Stephanoeca diplocostata* has already been noted. The order of costal strip exocytosis and storage confirms that strips attributable to the transverse costae are the first to appear during interphase, starting with the posterior rings and proceeding forwards. This is followed by strips destined for the longitudinal costae culminating with the anterior spine strips with forked tips. The repositioning of strip sub-groupings throughout cell division is similar to that in *S. diplocostata* with the exception that in *D. grandis* the strips that will form the future transverse costae are not rotated into a horizontal position on the juvenile until lorica assembly is underway. This may just be a matter of timing rather than a departure from normal practice. In *Didymoeca costata* the strips destined for the transverse costae remain horizontal throughout cell division and are never rotated into the vertical position. This is probably a reflection of the different space constraints within the loricae of the two species.

In both *Stephanoeca diplocostata* and *Didymoeca costata* lorica assembly occurs close to the cell surface within the posterior chamber and only the anterior chamber is at a distance from the cell. Since the lorica-assembling microvilli emerge from the anterior end of the cell they would appear to be ideally positioned for this forward movement. This highlights the task confronting *Diaphanoeca grandis* where the suspended cell is at some distance from the lorica in all directions, including towards the rear. However, this apparently anomalous situation is easily overcome because in *D. grandis* the cell remains at the base of the developing lorica and only when the lorica is complete do the assembling microvilli raise the cell to its rightful position at the centre of the lorica. The actin-containing microvilli can exert sufficient mechanical force to generate the flask shape of the lorica chamber. Once the strips generating the longitudinal and transverse costae are fully extended, they lock in position and the assembling microvilli can be withdrawn from their locomotory and shaping function. The large size of the *Diaphanoeca* lorica clearly demonstrates that the transverse rings are assembled in sequence rather than simultaneously.

7.4.4 General discussion

The three species detailed here, *Stephanoeca diplocostata*, *Didymoeca costata* and *Diaphanoeca grandis*, have provided valuable complementary insights into the tectiform condition. *S. diplocostata* has supplied basic information on the movement of costal strips throughout interphase, cell division and lorica assembly. In particular, the key role of collar microvilli has been demonstrated. They not only combine to provide a surface on which sub-groupings of costal strips are repositioned during cytokinesis but also, individually, provide a continuous link extending from storage on the parent collar to lorica assembly on the juvenile. *Didymoeca costata* has contributed unsurpassable insights into the order and orientation of costal strips that could only be guessed at in most other species. *Diaphanoeca grandis* has demonstrated how the process of lorica assembly can be modified to assemble a large lorica at some considerable distance from the cell. By combining these observations it is possible to draw up a list of the key features that characterise tectiform division.

7.5 FEATURES THAT CHARACTERISE THE TECTIFORM CONDITION

7.5.1 Costal strip production

(1) Costal strips are produced by a lorica-bearing parent cell throughout interphase.

(2) Individual costal strips are formed 'upside-down' with respect to their orientation within the parent lorica. The end of a costal strip that will subsequently point in the direction of extension during lorica assembly (the leading end) is directed posteriorly during development within the parent cell.

(3) Strips destined for the helical and/or transverse costae of the future lorica are produced first, followed by strips for the longitudinal costae.

(4) For all categories of costae the order of costal strip exocytosis and storage is sequentially from posterior to anterior.

(5) Costal strips are exocytosed through the anterior end of the cell, within the circumference of the collar.

(6) Costal strips are accumulated and stored horizontally within sub-groupings close to the top of the inner surface of the collar. The strip sub-groupings relate to future costae of the juvenile.

(7) In a few species, such as *Didymoeca costata*, the last categories of strips that are destined for future longitudinal costae are stored vertically, probably because of space constraints.

(8) Costal strips destined for future transverse costae are stored 'upside-down' with respect to the parent cell (see, for example, *Stephanacantha campaniformis* (Figs 45 and 47 in Thomsen *et al.*, 1991) and *Pleurasiga echinocostata* (Fig. 32 in Thomsen and Buck, 1991)).

7.5.2 Cell division

(1) Mitosis begins in the horizontal plane but during metaphase, anaphase and telophase the long axis of the dividing nucleus tilts into the diagonal/vertical plane.

(2) Cytokinesis involves inversion of the juvenile cell, which becomes covered by vertical sub-groupings of costal strips that contribute to a close-fitting 'lorica cup'.

(3) At the beginning of cytokinesis the outer horizontal strips of the accumulation on the parent collar, destined for future helical/transverse costae, are transported downwards and inside the future lorica 'cup'. In *Didymoeca* the strips remain horizontal but in *Stephanoeca diplocostata* and *Diaphanoeca* they are moved into the vertical plane.

(4) During the downward movement of the outer horizontal strips in *Didymoeca* their order is reversed by overtaking.

(5) The strips that will form the future anterior ring either terminate their downward movement before passing over the edge of the developing lorica cup, in which case the anterior ring of the future lorica will be on the outer surface of the longitudinal costae, or they halt their movement after passing over the edge, in which case the anterior ring of the future lorica will be on the inner surface of the longitudinal costae.

(6) All horizontal strips destined for future transverse costae remain with their concave surface facing the mid longitudinal line of the lorica irrespective of whether they stay on the outside of the lorica cup as an outer anterior ring or comprise inner transverse rings of the lorica cup.

(7) The collar microvilli play a key role in moving sub-groupings of costal strips throughout division and subsequently during lorica assembly.

(8) Movement of costal strips over the surface of the collar and lorica-assembling microvilli must be mediated by adhesion complexes (focal adhesions) that connect costal strips with the microvillar cytoskeleton. Shortening and lengthening of microvilli accompany the movement of costal strips, but alterations in microvillar length are not, in themselves, responsible for all the movements and changes in orientation that costal strips undergo during cell division.

(9) The transfer of sub-groupings of costal strips from the parent collar to the juvenile during cytokinesis is accompanied by the transfer of a specific sub-group of microvilli (lorica-assembling microvilli) that are numerically equivalent to the number of longitudinal costae in the future lorica of the juvenile.

(10) Sub-groupings of strips destined for longitudinal costae are positioned vertically on the juvenile at the end of cytokinesis. Sub-groupings of strips destined for the transverse costae may be oriented in the horizontal plane at the end of cytokinesis, but in some species, for instance *Diaphanoeca grandis*, they may be held in the vertical plane for some considerable time until the beginning of lorica assembly.

7.5.3 Lorica assembly

(1) Immediately prior to lorica assembly sub-groupings of costal strips destined for the longitudinal or helical costae are oriented in the vertical plane. Sub-groupings of strips destined for transverse costae are oriented in the horizontal plane.

(2) Each longitudinal and helical costa is constructed from a single sub-grouping of costal strips.

(3) Each transverse costa (ring) is constructed from several, probably four, sub-groupings of costal strips.

(4) Lorica assembly involves a forward and left-handed rotational movement that is achieved by the ring of regularly spaced lorica-assembling microvilli located a short distance behind the collar microvilli. The forward movement is achieved by extension of the lorica-assembling microvilli. The left-handed movement is

probably achieved by horizontal rotation of the ring of lorica-assembling microvilli.

(5) Costal strips within sub-groupings that give rise to longitudinal costae extend with the inside strips foremost.

(6) Once costal strips are fully extended they seal in position and the lorica-assembling microvilli are withdrawn vertically.

(7) During lorica assembly the organic investment is deposited on the inner surface of the lorica.

7.6 UNIVERSALITY OF FEATURES CHARACTERISTIC OF TECTIFORM DIVISION

There are many published electron micrographs of loricate choanoflagellates with collar-associated accumulations of costal strips, a character that unequivocally confirms the tectiform condition (Leadbeater 1972a *et seq.*; Thomsen 1973 *et seq.*; Tong 1997a). In species with morphologically distinguishable categories of strips, it is relatively easy in the light of current knowledge to recognise that strips destined for the helical and/or transverse costae of a future lorica are exocytosed and stored first and that costal strips, particularly those destined for the transverse costae, are stored upside down. Regrettably there are only a few published images showing tectiform division in species other than those described above (see Sections 7.2–7.4). An illustration of *Diaphanoeca pedicellata* at the 'figure-of-eight' stage shows the arrangement of sub-groupings of strips comprising the lorica cup (Fig. 36 in Thomsen *et al.*, 1990). Likewise, an image of *Stephanoeca supracostata* at a similar stage in division shows a lorica cup that is reminiscent of the equivalent structure in *Didymoeca costata* (Fig. 7.45) (Fig. 11 in Hara *et al.*, 1996). In both of these examples the strips destined for the transverse costae are horizontally located in quadrants at the leading edge of the lorica cup.

Members of the *Parvicorbicula–Pleurasiga* species group, including *Parvicorbicula socialis*, *Pa. quadricostata*, *Pa. circularis*, *Pa. corynocostata*, *Pleurasiga minima*, *Pl. reynoldsii* and *Pl. tricaudata*, differ from the majority of other tectiform species in having both the anterior and, more significantly, the lower transverse costae on the outer surface of the lorica. Nevertheless, the order of costal strip production is the same as in those species with internal transverse costae. For the two species of this grouping with only one external transverse costa in the form of the anterior ring, namely *Pa. corynocostata* and *Pl. echinocostata*, these strips are the first to be produced and stored on the collar (see Fig. 85 in Thomsen *et al.* (1997) and Fig. 32 in Thomsen and Buck (1991)). Presumably in species with an external lower transverse costa, the constituent strips that form this costa are attached to the outer surface of the appropriate longitudinal strips, while the collar microvilli of the parent are still in place, equivalent to the stage shown in Fig. 7.46d and between Figs 7.42 and 7.44 for *Didymoeca*. In this way the strips would remain on the outer surface of the developing lorica cup and would be able to slide into position when the lorica expands with the aid of lorica-assembling microvilli associated with the inner surface of the developing longitudinal costae (see also Sections 4.3.5 and 8.4.2).

While it appears that, apart from the five currently known nudiform species, all other loricate choanoflagellates are tectiform, nevertheless there are some genera that have loricae with relatively few costal strips and for which there is only limited information on their mode of strip storage and cell division. *Calliacantha*, *Bicosta* and *Crucispina* come into this category and in these genera an additional complicating feature is the length of their spines, which is sometimes, as in *Bicosta spinifera*, many times that of the cell (Figs 4.70–4.77). These genera are planktonic, mostly oceanic and have not to date been obtained in culture. The loricae of *Calliacantha* species can be resolved into two layers, comprising an outer layer of longitudinal and inner layer of transverse costae (Figs 4.75–4.77) (see Section 4.6.4). Images of *Calliacantha natans*, *C. simplex* and *C. multispina* are available that show accumulations of costal strips destined for future transverse costae associated within the collar (Figs 7.58–7.62) (Manton and Leadbeater, 1978; Manton and Oates, 1979a, b). In *C. simplex* the accumulated strips are of similar length to those comprising the transverse costae (Fig. 7.58) and when seen with SEM they have slightly rounded spatulate ends, another indicative feature of transverse strips (Fig. 7.59). In *C. natans*, strips comprising the two transverse costae (4 and 5) can be distinguished by their thickness (Fig. 4.77); those destined for the posterior transverse costa (4) are thinner than those destined for the anterior costa (5) (Fig. 7.62). Strips comprising the longitudinal costae, which include the spines, are formed after the transverse strips. They are grouped together within the cell and since the spines are so long they project posteriorly within a single membrane-bounded protuberance (Figs 7.60 inset, 7.61). Since the protuberance extends more-or-less parallel with

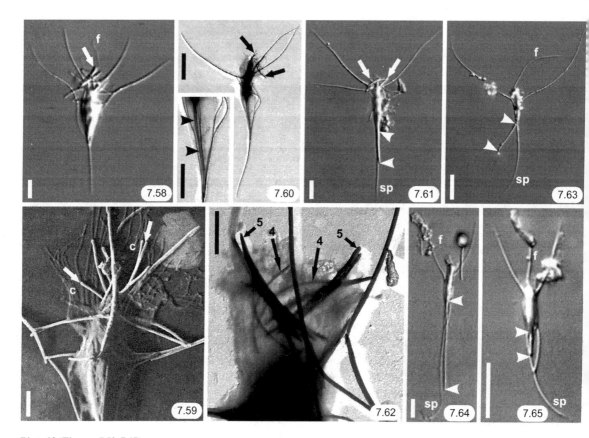

Plate 12 (Figures 7.58–7.65)

Figs 7.58–7.65 *Calliacantha* and *Bicosta* spp. Light and electron microscopy of costal strip formation and accumulation. **Fig. 7.58** *Callacantha simplex.* Loricate cell with accumulation of transverse costal strips on collar (arrow); (f) flagellum. Bar = 5 μm.
Fig. 7.59 *C. simplex.* Transverse costal strips (arrows) associated with collar microvilli (c). Bar = 2 μm. Reproduced from Manton and Oates (1979a).
Fig. 7.60 *Calliacantha natans.* Cell with transverse costal strips on collar (arrows). Bar = 5 μm. Fig. 7.60 inset Higher power of lorica chamber in Fig. 7.60 showing posterior protuberance containing developing spines (arrows). Bar = 2.5 μm.
Fig. 7.61 *C. natans.* Cell with accumulation of transverse costal strips on collar (arrows) and rear protuberance (arrowheads) containing intracellular developing spines. Posterior spine of lorica (sp). Bar = 5 μm.
Fig. 7.62 *C. natans.* Accumulation of transverse costal strips on collar. Thinner strips of posterior transverse costa (longer arrows 4). Thicker strips of anterior transverse costa (shorter arrows 5). Bar = 2 μm.
Figs 7.63–7.65 Three species of *Bicosta* each with two anterior spines and a single posterior spine (sp). Posterior protuberance containing intracellular developing spines (arrowheads). **Fig. 7.63** *B. antennigera.* **Fig. 7.64** *B. spinifera.* Bars = 10 μm. **Fig. 7.65** *B. minor.* Bar = 5 μm.

the posterior spine it is easily overlooked. Presumably the longitudinal strips remain in the vertical plane and are probably exocytosed vertically through the anterior end of the cell as it divides.

The 'minimalist' loricae of *Bicosta* species comprise two longitudinal costae that converge onto a single posterior spine (Figs 4.70–4.72, 7.63–7.65) (see Section 4.6.4). Each of the longitudinal costae contains a linear array of three strips, the anterior of which is a long attenuated spine. The two lower strips, the anterior being shorter than the posterior, are located close to the cell surface (Thomsen and Larsen, 1992). In *B. spinifera* the longitudinal costae cross over, lower-right to upper-left, in the mid-region of the cell (Fig. 4.71) (Figs 1–3 in Hara and Takahashi, 1987a; Figs 3 and 4 in Booth, 1990; Fig. 2C in McKenzie et al., 1997; Fig. 13.2C in Marchant, 2005). Division in *Bicosta* has not been recorded, but Thomsen and Larsen (1992) have deduced from individual images the intracellular arrangement and orientation of nascent costal strips in *B. spinifera*. All strips are formed and retained vertically within the cell prior to division (see also Manton et al., 1980). To accommodate the long developing spines a membrane-bounded protuberance extends from the rear of the cell. In specimens of *B. spinifera* the rear protuberance, distinguishable by its greater width, may lie parallel to the posterior spine (Fig. 7.64). In *B. antennigera*, which has anterior spines with a pronounced curve, the posterior protuberance containing nascent curved spines is deflected to one side of the posterior spine (Fig. 7.63). Thomsen and Larsen (1992, p. 55) noted that "the orientation of the strips in the 'tail' (protuberance) is reversed compared to that of the parental cell. When (after cytokinesis) the daughter cell retreats backwards from the parental lorica carrying along the set of new costal strips, these will then attain the same orientation as the original cell." In addition they recorded that the long posterior spine is "oriented upside down relative to the corresponding costal strip of the 'mother' lorica". This is a confirmation of Hara and Takahashi's (1987a) earlier observation of the same phenomenon and is in accord with the same phenomenon in *Didymoeca costata* (Fig. 7.41) (see Section 7.3.3).

While we do not currently have a complete record of cell division and lorica assembly for species of *Calliacantha* and *Bicosta*, enough is now known to confirm that these genera are tectiform. In particular, the order and orientation in which costal strips are deposited and stored meet the criteria detailed in Section 7.5. Costal strips that are destined for transverse costae are produced first, starting with the posterior costa and working forwards. They are stored on the collar as demonstrated by *Calliacantha* species. Strips destined for the longitudinal costae, including the spines, are deposited afterwards and they are upside-down with respect to the parent cell. They remain in the vertical plane and are not rotated horizontally within the collar. In this respect they are comparable to the last two categories of costal strips, spatulate and narrow, to be exocytosed in *Didymoeca costata* (see Section 7.3.3), except that in *Bicosta* and *Calliacantha* the strips are not exocytosed until cell division is underway. Other species with long spines, such as *Saroeca attenuata*, are comparable in that transverse strips, if present, are stored

on the collar and the spines are formed upside-down and extend rearwards within a membrane-bounded protuberance. Although cell division has not been observed in these long-spined species, there is no reason to think that it would not exhibit all the hallmarks of tectiform division. Thomsen and Larsen (1992) considered the absence of costal strips on the collar and the posterior extension of the cell into a 'tail' (protuberance) in *Bicosta spinifera* to be details sufficiently different from the standard tectiform condition to warrant a new term, namely 'caudiform division'. However, as explained here, the absence of costal strips on the collar of *Bicosta* species is because there are no transverse costae within the lorica and the rear protuberance of the cell is merely an elongation of the cell to accommodate spines that are longer than the cell body. Thus, in the light of current knowledge, it is possible to state that all species are tectiform, except for those that are demonstrably nudiform.

7.7 ARE TRANSVERSE COSTAE AN EXCLUSIVELY TECTIFORM FEATURE?

A transverse costa, defined in Section 4.3.5 as being an entire horizontal ring of costal strips, appears to be an exclusive feature of tectiform choanoflagellates. The nearest that a nudiform species comes to assembling a costal ring is in *Polyoeca dichotoma*, where there are horizontal bands of costal strips (Figs 6.24–6.27). The distinction between a 'horizontal' helical costa comprising a single turn and a transverse ring is difficult to defend. Within some tectiform species, such as *Stephanoeca pyxidoides* and *S. urnula* (Figs 4.49–4.51, 4.54), there are substantial overlaps between adjacent costal strips in a ring (see Section 4.6.1) which renders them almost identical in appearance to the anterior band of costal strips present on the 'nudiform' loricae of *Polyoeca dichotoma* and *Acanthoeca spectabilis* (Figs 6.24–6.28). In addition, the interaction between costal strips of a horizontal costa, whether it is part of a helix or a transverse ring, is similar. When the lorica is seen from the outer surface, the left-hand end of a horizontal strip abuts onto the inner surface of the respective longitudinal costa and points in the direction of a left-handed rotation. However, nothing precisely equivalent to the clearly defined anterior ring of *Saepicula pulchra* (Figs 4.3–4.7) has been observed in a currently known nudiform taxon. Transverse rings would appear to be an economical way of constructing a lorica in that only a relatively few strategically placed rings are required to support an arrangement of longitudinal costae in a lorica. Use of rings in this manner has permitted the construction of voluminous loricae comprising relatively few lightly silicified costae. Outstanding examples of this development are the loricae of *Crinolina*, *Diaphanoeca*, *Parvicorbicula* and *Pleurasiga* species (Figs 4.55–4.68). Pelagic choanoflagellates are almost exclusively tectiform and the majority of species are distinguished by relatively large loricae comprising a limited number of lightly silicified costal strips (see Chapter 9). However, this line of reasoning is not sufficient to justify the assertion that the tectiform condition is essential for costal ring production.

A significant feature that distinguishes transverse costae from the longitudinal and helical variety is that the latter are assembled from single sub-groupings of strips whereas transverse costae appear to be assembled by the horizontal alignment of several, probably four, sub-groupings of strips. This phenomenon was observed in *Stephanoeca diplocostata* (Figs 7.16–7.21) and *Didymoeca costata* (Figs 7.50–7.52). In *D. costata*, the horizontal orientation of strips destined for future transverse costae is established on the collar and retained throughout the subsequent reorganisation of strip sub-groupings during cytokinesis (Figs 7.42–7.45). In *Stephanoeca diplocostata* the strips destined for future transverse costae rotate into the vertical plane during cytokinesis (Fig. 7.13, 7.14) but are rapidly repositioned into the horizontal plane as the juvenile is pushed out of the parent lorica. Likewise, in *Diaphanoeca pedicellata* the horizontal orientation of these strips is established immediately after cytokinesis before the juvenile has made significant progress out of the parent lorica (Fig. 36 in Thomsen *et al.*, 1991). However, in *Diaphanoeca grandis* the horizontal alignment of strip sub-groupings destined for future transverse costae does not occur until the juvenile has been released from the parent and is about to assemble its lorica (Fig. 7.53f). Prior to this these strips

are located vertically as an under-layer on the juvenile (Fig. 7.54). What can be concluded from these observations is that costal strips, both individually and as sub-groupings, can be rotated through 90° on multiple occasions. This occurs initially as vertically exocytosed strips are accumulated in horizontal array at the top of the parent collar. They can then be returned to the vertical plane on the developing juvenile, as in *D. grandis*, and subsequently back into the horizontal plane prior to being assembled into costae. Orientation of strips in the horizontal plane would appear to be a pre-requisite to assembling a transverse costa. The alignment of neighbouring sub-groupings of strips in assembling transverse costae is remarkably precise as irregularities in the pattern of overlaps are rare.

To return to the initial question posed, namely 'Are transverse costae an exclusively tectiform feature?', the answer would appear to be affirmative for two reasons. First, the sub-groupings of strips that contribute towards transverse costae are oriented horizontally; and second, a single costa comprises strips from several, probably four, sub-groupings. No extant nudiform species generates horizontally oriented strips on the juvenile and as far as can be established the horizontal costae in *Polyoeca* and *Acanthoeca* are constructed from single sub-groupings of strips. However, there would not appear to be an intrinsic reason why nudiform species should not be capable of assembling transverse rings. Comparison of the juveniles of *Savillea* and *Helgoeca nana* (Figs 6.14 and 6.19, respectively) with the juvenile of *Diaphanoeca grandis* (Fig. 7.54) reveals equivalence in the vertical orientation of costal strips on the cell surface and yet in the two nudiform species the inner costae will assemble helical costae and in *D. grandis* they will contribute to the assembly of transverse costae.

8 • Loricate choanoflagellates: evolutionary relationship between the nudiform and tectiform conditions

One jump evolution – 'upside-down, inside-out'.

8.1 INTRODUCTION: AN EVOLUTIONARY PARADOX

It is a paradox that on the one hand loricate choanoflagellates are so consistent with respect to their lorica construction and assembly and yet on the other hand there is such a clear-cut distinction between the nudiform and tectiform conditions. As detailed in Chapters 6 and 7, the differences between these conditions are not merely matters of nuance or overlap but characters diametrically opposed to each other. At a glance, the list of features distinguishing the two conditions (see Table 6.1) appears oddly disparate, including differences in the timing of events within the cell cycle; dissimilarities in the orientation and order of costal strip production and storage on the cell surface; differences in the orientation of cytokinesis; and subsequent motility or immotility of the juvenile. The paradox centres on the ability of cells to maintain the primary logistical pathway of costal strip formation, exocytosis, storage and lorica assembly while at the same time accommodating apparently chance alterations to a suite of 'secondary' features such as timing within the cell cycle and the orientation of structures. Whether this was achieved by a series of minor changes spanning a period of evolutionary time or could have resulted from a single or limited number of abrupt 'regulatory' changes will be explored in this chapter.

8.2 BACKGROUND

Currently there are approximately 150 acanthoecid species distinguished primarily by the size and shape of the lorica and the arrangement and morphology of costae and costal strips (see Chapter 4). Distributional studies are currently available from many parts of the world and, for the time

being at least, the overall number of species recorded appears to have reached a plateau (see Chapter 9). This is not to be over-confident, for there are substantial limitations regarding the data available; many sites have not been studied with scientific rigour and knowledge of sub-tropical and tropical regions is particularly limited. At present only five species, belonging to four genera, are known to be nudiform and attributable to the Acanthoecidae (Table 6.2), while approximately 150 species are tectiform and attributable to the Stephanoecidae. Within each family there is general uniformity in the defining characters observed. All current nudiform genera are known to produce naked juvenile cells that settle, become immotile if they were flagellated, and subsequently accumulate a covering of costal strips (see Chapter 6). As discussed in Chapter 7 (see Sections 7.5 and 7.6) there is good evidence that all major 'non-nudiform' genera display characters of the tectiform condition and for those species that have not been observed with collar-associated accumulations of strips, there is no reason to doubt that this is due to limitations in sampling and observation. However, while it is possible to use these superficial characters as surrogate indicators of the two conditions, it is, nevertheless, necessary to probe more deeply into the other differentiating features (see Table 6.1). The terms 'upside-down' and 'inside-out' have been used in connection with the orientation and order of costal strip production and storage in tectiform species (see Section 7.3.6), but what precisely do these terms mean, how good is the evidence for these statements and what is their significance in an evolutionary context?

8.3 THE NUDIFORM CONDITION

Of the two conditions, nudiform is intuitively the more straightforward. The scenario in which a mature cell with a lorica undergoes division to produce a juvenile that is motile and 'naked' is also characteristic of thecate choanoflagellates (see Chapter 3), as well as many other sedentary protistan flagellates, such as chrysophytes and bicosoecids, with covering structures that are restrictive and cannot be shared between the offspring during division (Schnepf *et al.*, 1975). In nudiform species the formation of costal strips within the juvenile immediately follows cytokinesis and occupies the first period of interphase. During this time the cell is immotile, the flagellum having been withdrawn, and the collar, if visible, is minimal in length, so there is little opportunity for active feeding. The apparent inability of nudiform juveniles to feed while accumulating strips must place the juveniles at a competitive disadvantage in comparison with tectiform cells capable of feeding continuously throughout interphase (see Chapter 6).

Since costal strips produced within a nudiform juvenile are exocytosed and stored in sub-groupings laterally on the cell surface, and since they are immediately assembled into a lorica once a full complement has been produced, it seems obvious that strip orientation must be consistent from the moment of formation until the completion of lorica assembly. This can be verified by observing juveniles with strips that bear identifiable features distinguishing their polarity. In *Acanthoeca spectabilis* each anterior spine

possesses a minute forked tip at the front end (Fig. 6.30 inset). On a juvenile of *A. spectabilis* at the earliest stage of lorica assembly, when the inner anterior strips are visible, forked tips can be seen at the anterior end (Fig. 8.1 insets A and B). With respect to the order in which strips are produced and stored, the outer strips are stored first followed by the inner strips. Thus in the *micropora* form of *Savillea parva* the thinner, straighter strips destined for future longitudinal costae are stored before the inner, thicker curved strips of the helical costae (Figs 8.2, 8.3). Lorica assembly involves the forward and diagonal sliding of strips from sub-groupings to generate longitudinal and helical costae, respectively (see Chapter 6). Extension of costae involves the sliding of strips from the inside so that the most anterior strip of a costa will have been on the inside of the respective sub-grouping and therefore produced later in the silica deposition cycle than the posterior strips. Referring to the system of numbering of costal strips within a lorica (see Section 4.2.2), the order of production of the outer longitudinal costae is 1, 2 and 3, followed by the inner layer of helical costae in the order 4, 5 and 6.

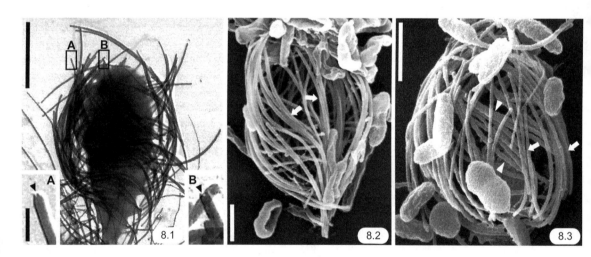

Plate 1 (Figures 8.1–8.3)

Figs 8.1–8.3 Orientation and order of costal strips on juveniles of nudiform species. **Fig. 8.1** *Acanthoeca spectabilis*. Juvenile at an early stage of lorica assembly showing locations of fork-tipped costal strips (A and B) enlarged in respective insets. Bar = 2 μm. Fig. 8.1 insets (A and B). Forked tips (arrowheads) of costal strips at higher magnification from Fig. 8.1. Bar = 0.1 μm.
Figs 8.2–8.3 *Savillea parva – micropora* form. Bars = 1 μm. **Fig. 8.2** Juvenile mid-way through accumulating costal strips showing only thinner strips (arrows) destined for future longitudinal costae.
Fig. 8.3 Juvenile with complete set of strips showing outer layer of thinner strips (arrows) destined for longitudinal costae and inner layer of thicker strips (arrowheads) destined for helical costae. Reproduced from Leadbeater (2008a).

8.4 THE TECTIFORM CONDITION

The tectiform condition is in most respects more complicated than the nudiform condition. Costal strips are
produced 'upside-down' and the order of production is
'inside-out' with the components of the inner costal layer
(outer layer in the *Parvicorbicula–Pleurasiga* species group
(see Sections 4.3.5 and 8.4.2)) being exocytosed and stored
before those of the outer layer. Costal strips are stored
horizontally at the top of the inner surface of the collar
except for a few species where certain strip categories are
maintained in the vertical plane, as, for instance, the spatulate and narrow strips of *Didymoeca costata* (Fig. 8.7), and the
longitudinal strips and spines of *Calliacantha* and *Bicosta*
(Figs 7.60, 7.62–7.64). This appears to be a secondary adaptation to accommodate strips in a confined space or to overcome the necessity of storing unwieldy strips on the collar.

8.4.1 'Upside-down' formation of costal strips

The term 'upside-down' with respect to costal strip
development within silica deposition vesicles (SDVs) refers
to a vertical rotation through 180°; there is no horizontal
rotation – the inward and outward facing surfaces of strips
continue to face in the same directions. In computer terminology this is a rotation and not a flip (mirror image).
Evidence for the upside-down orientation of developing
strips has come from several sources including *Didymoeca
costata*, where a petaloid strip within an SDV is upside-
down with respect to the equivalent strips in the parent
lorica (Fig. 7.41) (Leadbeater, 2010). Spatulate and narrow
strips are also upside-down as illustrated by their orientation when stored vertically within the collar at the end of
interphase (Fig. 8.7). An inverted orientation has been
observed with respect to the distinctive strips of the anterior transverse costa of *Diaphanoeca grandis* (Figs 8.37–8.39)
(Leadbeater and Cheng, 2010). Hara and Takahashi (1987a)
and Thomsen and Larsen (1992) also concluded that all
costal strips deposited within *Bicosta spinifera* were upside-
down with respect to the parent (mother) cell. While use of
the term 'upside-down' in the context of strips that comprise longitudinal costae is obvious, a further word of
explanation is required for strips destined for the transverse costae such as those of *D. grandis*. Here 'upside-down'
refers to the orientation of developing strips within a
parent cell as being similar to that of the equivalent strips
on an inverted juvenile (compare Figs 8.35–8.37 with
Figs 7.54–7.57). Fortunately, in *D. grandis*, strips that will

generate the future transverse costae are positioned vertically on the juvenile and so it is relatively easy to make this
comparison (Fig. 7.54) (Leadbeater and Cheng, 2010).

Inversion of strip formation is almost certainly accompanied by inversion of other related cell components,
including the SDV and probably the associated cytoskeletal
system. Inversion of the SDV-associated cytoskeleton could
explain why exocytosis of costal strips in tectiform species
occurs through the anterior surface of the cell within the
confines of the collar, as opposed to a moderately downward lateral extrusion of strips in nudiform species. In
Didymoeca costata, space constraints make the anterior
region within the collar the only realistic location through
which relatively large strips can be extruded in order to
reach the top of the inner surface of the collar (see
Fig. 7.41) (Leadbeater, 2010). Furthermore, vertical
stacking of exocytosed spatulate and narrow strips immediately above the anterior surface of the cell further supports
this conclusion (Fig. 8.7). Inversion also has some bearing
on the orientation of strips within the horizontal accumulations on the inner surface of the collar. In species, such as
Stephanacantha campaniformis and *Pleurasiga echinocostata*,
with a distinctive upturned end to each strip comprising
the anterior ring, these are stored upside down within the
accumulation of strips at the top of the collar (see Figs 45,
47 in Thomsen *et al.* (1991) and Fig. 32 in Thomsen and
Buck (1991)). In spite of there being no distinctive morphological feature to demonstrate the orientation of accumulated strips destined for the transverse costae in *Didymoeca
costata*, nevertheless they must be stored upside-down since
there is no evidence of rotation occurring during formation
of the lorica cup accompanying cytokinesis and once the
lorica cup is constructed all the component strips are in the
correct orientation with respect to the inverted juvenile
(Figs 7.42–7.45, see Section 7.3.4). This conclusion is further supported by the fact that the concave side of transverse strips always faces inwards towards the long axis of the
developing lorica cup and thus, during the reorganisation of
strips, when they 'leapfrog' each other and move over the
anterior end of the lorica cup the orientation of the strips is
not altered (see Section 7.3.4).

8.4.2 'Inside-out' order of strip formation and storage

The term 'inside-out' with respect to the order of costal
strip production and storage in *Didymoeca* and the majority
of tectiform species refers to strips comprising the inner

layer of a future lorica being formed and exocytosed before those destined for the outer layer. This contrasts with the order of strip production in nudiform species where those of the outer layer are produced first, followed by those for the inner layer (see Section 8.3). However, within each layer the order of production in both tectiform and nudiform species is posterior to anterior. Thus in *Didymoeca costata* the first category of strips to be produced is for the posterior ring, followed sequentially by the intermediate ring and anterior rings. Within the longitudinal costae, the order of strip production is petaloid, spatulate and narrow. With reference to the system of costal strip numbering (see Section 4.2.2), the order of strip production in tectiform species is inner layer of transverse costae 4, 5 and 6 followed by the outer layer of costal strips 1, 2 and 3 (Fig. 4.2a).

There are several ways of observing the order of costal strip production. The distinctive morphology of the different categories of costal strip in *Didymoeca costata* makes it possible to infer the order of production by observation of accumulations at different stages of development in whole mounts of cells for scanning electron microscopy (SEM) (Fig. 7.42). The sequence of strip accumulation can also be displayed on cells that have been starved of silicon until they fail to produce loricae, followed by an abrupt silicon spike which results in cells suddenly producing fully silicified strips. The sequence of costal strip accumulation can then be observed without the hindrance of overlying parental loricae (Figs 8.4–8.7). Using this protocol with *Didymoeca costata*, the difference in thickness and radius of curvature is apparent between the first strips destined for a posterior ring (4 in Figs 8.4, 8.5) and those of the intermediate ring (5 in Fig. 8.5). In Fig. 8.5 most of the transverse strips are present on the collar and the first petaloid strip (1 in Fig. 8.5) has made an appearance. In Fig. 8.6 a ring of petaloid strips have made their appearance and their arrangement in four sectors (quadrants) is apparent (asterisks in Fig. 8.6). However, because of flattening of the specimen it is not possible to distinguish whether all petaloid strips are oriented in the same direction or whether they face alternately to the left and right. Finally, Fig. 8.7 illustrates a cell with a full complement of stored strips. Most petaloid strips (1) are located horizontally (or diagonally) with the transverse strips (4, 5 and 6) on the outside, and in the centre of the collar is a vertical array of spatulate and narrow strips (2 and 3, respectively).

Another effective method of illustrating the order in which costal strips are produced is by manipulating the silicon content of the medium in which cells are grown in such a way that a gradient of costal strip depletion (thickness) is achieved within a single cell cycle. This has been achieved in *Diaphanoeca grandis*, where the first produced strips are the thickest and thereafter strips become sequentially thinner (Fig. 8.8). The order of strip production is for the transverse rings starting at the base (4, 5, 6 and 7 in Fig. 8.8) followed by the longitudinal costae also starting at the base (1, 2 and 3 in Fig. 8.8). An alternative effect can be obtained by re-supplying silicon to silicon-depleted cells that have not completed strip production. In this situation, strips produced before silicon addition will be much thinner than those produced afterwards (Fig. 8.9). The results of this treatment give a precise insight into the order of strip production since many combinations are obtained according to how far into the strip production cycle a particular cell has reached when silicon is replaced. In the lorica illustrated in Fig. 8.9 the four transverse rings (4, 5, 6 and 7, respectively) were produced during the period of silicon deprivation and are thin and fragmentary. Fortuitously, silicon replenishment coincided precisely with the production of strips for the longitudinal costae (1, 2 and 3) and these are fully silicified (Fig. 8.9).

Some tectiform species, such as those of *Stephanoeca*, have an inner layer that includes helical costae in the posterior chamber and a series of transverse rings in the anterior chamber (Figs 4.45–4.53, 7.22). This raises the question as to whether the helical costae, as part of the inner layer, are also produced early in the strip-deposition cycle. Loricae of silicon-depleted *S. diplocostata* cells showing a gradation of costal strip thickness reveal that the helical costae (arrows 4 in Fig. 8.10), in common with the posterior transverse (waist) costa (5 in Fig. 8.10), are more completely silicified than the other three transverse costae (6, 7 and 8 in Fig. 8.10) and that strips comprising the longitudinal costae are sequentially thinner from the posterior forwards. Thus the order of strip production is consistent irrespective of whether the inner layer consists of rings alone or rings and helical costae.

The *Parvicorbicula–Pleurasiga* species group (see Sections 4.3.5 and 7.6) contrasts with the majority of tectiform genera in that all the transverse costae, not just the anterior ring, are outside the longitudinal costae. However, this does not affect the order of costal strip production, which is those of the transverse costae first, posterior to anterior, followed by those of the longitudinal costae, posterior to anterior. It would appear that during cell division the accumulated strips contributing to the two transverse

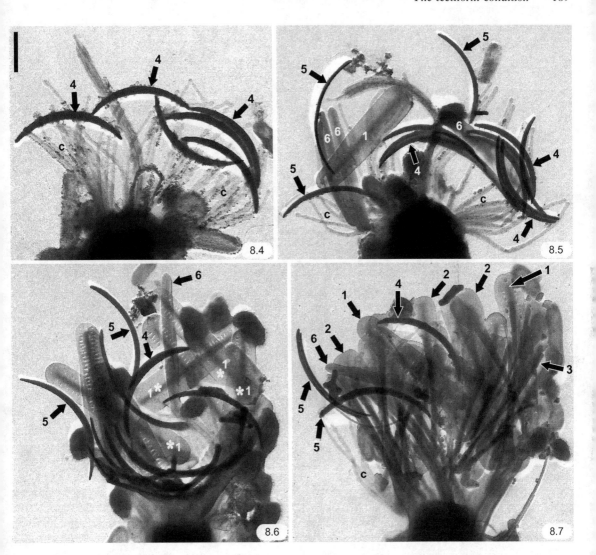

Plate 2 (Figures 8.4–8.7)

Figs 8.4–8.7 *Didymoeca costata*. Stages in costal strip accumulation at the top of the collar (c) in silicon-starved cells without loricae that have been re-supplied with silicon. Bar = 1 μm (shown on Fig. 8.4). **Fig. 8.4** Accumulation of costal strips for future posterior transverse costa (4).

Fig. 8.5 Accumulation containing strips for the posterior (4), intermediate (5) and anterior (6) transverse costae, respectively, and one petaloid strip (1).

Fig. 8.6 Accumulation containing a similar grouping of strips to Fig. 8.5 but with more petaloid strips comprising a ring of four sectors (quadrants) (asterisks 1).

Fig. 8.7 Complete accumulation of strips numbered as follows: transverse strips – posterior (4), intermediate (5), anterior (6); longitudinal strips – petaloid (1), spatulate (2) and narrow (3).

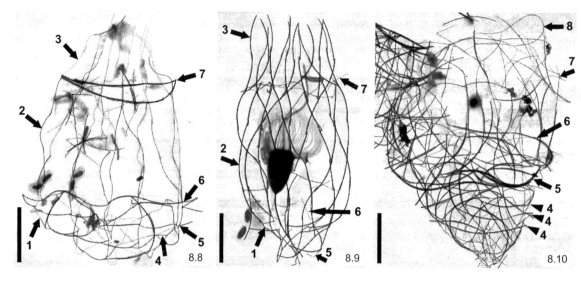

Plate 3 (Figures 8.8–8.10)

Figs 8.8–8.10 *Diaphanoeca grandis* and *Stephanoeca diplocostata*. Order of costal strip production determined from loricae displaying a gradient of silica deprivation or sudden increase in silicon content of medium. Bars = 2 μm. **Fig. 8.8** *D. grandis*. Lorica produced during gradient of silicon deprivation. Transverse costae (4–7) are thicker than longitudinal costae (1–3), which get progressively thinner towards anterior end.
Fig. 8.9 *D. grandis*. Lorica with fragmentary transverse rings (5–7) and thick, highly silicified longitudinal costae (1–3) produced after silicon addition.
Fig. 8.10 *S. diplocostata*. Lorica produced during gradient of silicon deprivation. Posterior helical (arrows 4) and transverse (waist) (5) costae thicker than intermediate (6 and 7) and anterior (8) transverse costae. Longitudinal costae are successively thinner. Figs 8.8 and 8.9 reproduced from Leadbeater and Cheng (2010).

costae (anterior ring and lower transverse costa) remain on the outer surface of the developing lorica cup and are subsequently assembled without the direct assistance of the lorica-assembling microvilli. Exactly how this is achieved is not known since it has not been observed. However, it may indicate that once the accumulated strips have been positioned by the collar microvilli on the outer surface of the lorica cup, no further direct manipulation by the lorica-assembling microvilli is required for their assembly into transverse costae.

8.4.3 Inversion of the juvenile cell during cytokinesis in tectiform species

The 'upside-down' orientation and 'inside-out' order in which costal strips are produced, exocytosed and stored in tectiform species, while executed with considerable finesse,

would be useless if during division the costal strips on the collar microvilli were not rearranged and the juvenile was not inverted to receive the newly formed lorica cup. Inversion of the juvenile together with the re-ordering of costal strips reverses the apparent complications introduced by the parent cell and returns the overall strip organisation within the lorica cup to one equivalent to that of a nudiform juvenile. It would seem highly unlikely that these two coincident instances of inversion, involving (1) costal strip production and (2) cytokinesis, could have come about independently for they are both essential for the working of the tectiform condition.

Mitosis (nuclear division) in *Stephanoeca diplocostata*, the only tectiform species to have been investigated using sectioned material, resembles that in the craspedid *Monosiga ovata* (see Section 2.9). The interphase nucleus is located just beneath the flagellar basal body at the anterior end of the cell. Immediately prior to nuclear division the emergent flagellum is withdrawn and the axoneme is depolymerised within the cell Figs (2.72–2.74). The two basal bodies serve as foci (poles) for the developing mitotic spindle (Fig. 2.77). The nuclear envelope disintegrates and a metaphase band of dense-staining chromatin appears. The chromatin segregates to the two poles accompanied by elongation of the spindle microtubules. The poles are now positioned along a diagonal plane with respect to the long axis of the cell – this represents the asymmetric stage of division (Fig. 7.33) (see Table 7.1). Once two separate daughter nuclei are present (telophase), the elongated cell undergoes cytokinesis. As far as can be determined this begins with a longitudinal separation between the two daughter nuclei; however, as cytokinesis proceeds the anterior end of the cell enlarges and an expansion of the cytoplasm to one side of the upper nucleus moves upwards and outwards (Fig. 7.4). Cytokinesis now proceeds by the development of a 'pushing' force that is exerted between projections extending from the two daughter cells. One projection appears to come from the collar region of the daughter cell that will remain with the parent lorica and the other from a region about two-thirds back from the collar of the developing juvenile (Fig. 7.5). The effect of the force generated by these projections is to push the rear of the developing juvenile upwards in an arching movement causing it to invert, with the result that the flagellar poles of the two daughter cells come to face each other (Figs 7.6, 7.14, 7.28, 7.34). The remaining connection between the two daughters, which was responsible for the pushing movement, now resolves itself into the two or more overlapping threads that will mutually serve to separate the two cells (Figs 7.7, 7.28, 7.29). A similar sequence of events is observed in *Diaphanoeca grandis* (Fig. 7.53a–e).

The semi-amoeboid behaviour of the cell during division and the use of a separating 'thread' as seen in tectiform species (Figs 7.2–7.6, 7.53a–e) are by no means exclusive to loricate species. Craspedid choanoflagellates with restrictive thecae also become amoeboid during cell division and separation of the two daughter cells is achieved by a connecting thread (Figs 3.14c, 3.27 arrow) (Leadbeater, 1977). However, what is unique to tectiform species is the inversion of the juvenile. Since cytokinesis in many eukaryotes is an actin-mediated process (Glotzer, 2001), it is possible that the coincidental inversion of SDVs and cytokinesis is causally linked by a corresponding inversion of the accompanying actin cytoskeleton (Leadbeater, 2010).

8.5 CODIFYING THE SALIENT FEATURES OF THE NUDIFORM AND TECTIFORM CONDITIONS

One way of clarifying and also visualising the similarities and differences between the nudiform and tectiform conditions is to codify them diagrammatically. Fig. 8.11 represents an attempt to achieve this. On a horizontal time axis with the mid-point denoting mitosis and cell division, which separates interphase of the parental generation (left) from interphase of the juvenile generation (right), the order in which the different categories of costal strips are produced is indicated by their respective numbers (see Fig. 4.2a–b (Section 4.2.2) and legend to Fig. 8.11 for numbering). Codifying the differences in this manner

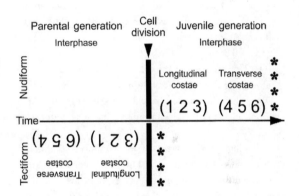

Fig. 8.11 Codified illustration of the generational (parent or juvenile) timing, and order and orientation of costal strip production characteristic of the nudiform and tectiform conditions, respectively. On the horizontal time axis, the mid-point denoting mitosis and cell division separates interphase of the parental generation (left) from interphase of the juvenile generation (right). The order in which the different categories of costal strips are produced is indicated by their respective numbers as follows: (1) posterior longitudinal costal strip; (2) intermediate longitudinal costal strip; (3) anterior longitudinal costal strip; (4) posterior transverse ring; (5) intermediate transverse ring; (6) anterior transverse ring. Brackets around a sequence of numbers indicate that the order of production within that grouping is invariable and the same in both conditions. The orientation of the numbers denotes whether the costal strips are the 'normal' way round or upside-down. The vertical columns of asterisks denote the position of lorica assembly during interphase. The sequence of numbers positioned above the axis refers to the nudiform condition and the sequence of numbers positioned below the axis refers to the tectiform condition. A full description of the numbering system is given in Section 4.2.2.

focuses attention on the changes that would be required to switch from one condition to the other.

8.6 ONE JUMP EVOLUTION OR GRADUAL CHANGE?

On morphological grounds, it would seem reasonable to argue that the more straightforward sequence of events found in nudiform species probably relates more closely to the ancestral condition of lorica production in acanthoecids. However, while molecular phylogenetic studies distinguish nudiform acanthoecids as being a distinctive clade, its relationship with tectiform acanthoecids is uncertain, depending on the data used to generate a phylogeny and, in particular, whether the third codon positions are included or excluded (for further discussion see Sections 6.9 and 10.4.1). If they are included, the nudiform and tectiform groups appear as sister clades represented by the families Acanthoecidae and Stephanoecidae, respectively, in Fig. 10.3 (see also Carr *et al.*, 2008a; Nitsche *et al.*, 2011). However, if the third codon positions are omitted the root of the nudiform clade appears within the tectiform clade, which, if correct, would indicate that the extant nudiform grouping evolved from a tectiform ancestor (Carr *et al.*, 2008a).

No single basic feature of the tectiform condition is entirely novel (see Fig. 8.11). The silica deposition apparatus is standard for all loricate species; there are precedents for the existence of the complex molecular machinery required for moving costal strips over the surface of the microvilli – for example, prey (and latex) particles are captured and moved along the microvillar surface of the collar during feeding (see Section 2.4). One of the more unusual features is the use of an actin-based thread to effect cell separation, but even this occurs in thecate choanoflagellates (see Section 3.4.4) (Leadbeater, 1977). When the tectiform condition is viewed in its entirety it seems unlikely that the differences between this and the nudiform condition could be explained by a series of stepwise morphological changes. If this had been the case then each change would have had to benefit the cell independently and one of the features that is so striking about the tectiform condition is that all of the modifications are necessary for the process to function successfully. What seems more likely is that an abrupt regulatory change occurred that fortuitously did not interfere with the primary pathway of costal strip production and lorica assembly, but did introduce changes in a suite of secondary characters. At first sight this

scenario may also seem improbable because the required changes appear somewhat disparate. However, on more careful scrutiny, rather than being disparate there is a remarkable symmetry in that all the changes in strip orientation and order brought about by the parent tectiform cell are reversed during inverted cell division, thereby producing a juvenile with a standard 'nudiform' coating of costal strips. Since this line of reasoning envisages the relationship between the nudiform and tectiform conditions being an abrupt regulatory switch, this raises the question as to whether there are precedents for such a switch within extant species.

8.6.1 The effects of silicon replenishment on silicon-starved *Stephanoeca diplocostata* cells

Loricate choanoflagellates have a facultative requirement for measurable amounts of silicon (see Chapter 5). As noted in Section 5.7.3, silicon is one of the most difficult elements to eliminate from culture media because it occurs in minute traces even in ultrapure chemicals and deionised water. Furthermore, biogenic silica is soluble in seawater medium and even in highly silicon-starved cultures fragments of loricae may be sufficient to facilitate cryptic recycling of the element. Nevertheless, cultures of *Stephanoeca diplocostata* containing completely 'naked' cells with no trace of external strips have been obtained and successfully sub-cultured (Figs 5.36, 8.13) (Leadbeater, 1989). Cells with silicon-depleted loricae and accumulations of lightly silicified strips continue to divide in the standard inverted tectiform manner. However, when silicon deprivation is extreme, cells cease to undergo inverted division but instead divide diagonally (Figs 5.42–5.44). While this mode of division resembles that observed in nudiform species (Figs 6.12, 6.18, 6.37–6.41), nevertheless there are certain differences, for example the parental flagellum appears not to be withdrawn (Fig. 5.43) and division resembles a budding process rather than an equal sharing of cell contents as in craspedid species with non-restrictive coverings (Figs 2.62–2.64, 3.2–3.4) (Karpov and Leadbeater, 1997). However, the result is to generate cultures of fully functional 'naked' cells, a small percentage of which undergo a partial division producing 'V-shaped' conjoined cells usually with two nuclei, collars and flagella (Figs 5.45, 8.23). Cleavage starts at the anterior end with continuity between the partially divided cells being restricted to the posterior end. Occasionally cells with three collars and flagella are seen.

Replenishment of silicon to mid-exponential phase cultures of silicon-starved cells results in a cessation of growth lasting 10–40 hours, depending on the particular conditions of culture (Figs 5.39, 5.40) (see Section 5.7.4). During this period silicon uptake proceeds, and costal strips are produced, exocytosed and stored outside the cell (Figs 8.14–8.17). In the first of two experiments (Fig. 5.40a) in which the fate of silicon re-supply to silicon-starved cells was studied in detail, after 48 hours, during which time there had been no increase in cell numbers, 86% of cells possessed loricae (Table 5.8). This was much higher than expected, since under normal tectiform conditions only a juvenile can assemble a lorica after cell division and during the period studied the number of cells remained more-or-less constant. In the second of these experiments (Fig. 5.40b) a detailed analysis was carried out to determine where on the cell the newly exocytosed strips were being stored. As illustrated in Table 5.9, the majority of cells, with a peak of 63% at eight hours after silicon addition, stored costal strips on the sides of the cell, while only a minority, with a peak of 14%, stored strips at the top of the collar. These results were originally reported in 1989 and, while it was appreciated that the cessation in growth triggered by silicon re-supply permitted naked cells to produce two sets of strips within a single interphase, one set for the existing cell and one set for subsequent inverted division, no explanation could be offered as to why this combination of events might have occurred (Leadbeater, 1989). Only now is it apparent that silicon re-supply to cells that have been severely silicon deprived elicits a 'nudiform' response in strip production and storage. Only after a nudiform cycle of strip production has been completed do cells return to the normal tectiform mode of strip storage and proceed with inverted cell division. This sequence of events can be summarised in a codified fashion in Fig. 8.12 using a similar format to that used in Fig. 8.11. In order to explore whether this is an acceptable explanation of these observations, it is necessary to be certain that *Stephanoeca* cells with surface clusters of strips meet the criteria for being classed as nudiform.

8.6.2 Evidence for the nudiform 'behaviour' of *Stephanoeca diplocostata*

On three separate occasions over the course of five years a series of batch culture experiments was carried out in which *S. diplocostata* cells were deprived of silicon to the point where naked cells predominated (Leadbeater, 1989).

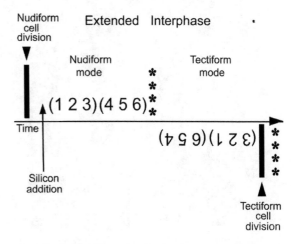

Fig. 8.12 Codified illustration showing the sequence of events with respect to costal strip production and timing of lorica assembly that occurs immediately after re-supply of silicon to severely silicon-deprived *Stephanoeca diplocostata* cells. Cell division is halted while cells take up silicon, first to deposit and store strips in the nudiform mode and then assemble a lorica. This is followed immediately (without cell division) by the production and storage of strips in the tectiform mode, after which the cell undergoes inverted tectiform division and the juvenile immediately assembles a lorica. The coding of numbers, brackets, asterisks and orientation is the same as used and recorded in the caption for Fig. 8.11 (see also Section 4.2.2).

Concentrates of naked cells were then inoculated into fresh silicon-enriched medium and on all occasions cells bearing surface clusters of strips were produced. Cultures were sampled at regular intervals so that the morphology of strip accumulation and lorica assembly could be recorded in the form of whole mounts for transmission electron microscopy (TEM) (Figs 8.13–8.22).

Figs 5.36 and 8.13 illustrate completely naked, silicon-starved cells with relatively short flagella and well-preserved collar microvilli. One of the disadvantages of repeatedly sub-culturing naked cells in silicon-free medium is that they become surrounded by bacteria (Fig. 8.13). Within an hour following the re-supply of silicon, costal strip deposition has been initiated and within two hours strips appear on the cell surface (Fig. 8.14). From an early stage extruded strips are arranged in sub-groupings parallel to the long axis of the cell (Figs 8.14–8.17 asterisks). Once a full complement has been accumulated, lorica assembly proceeds in the standard fashion involving lorica-assembling microvilli that emerge from the collar region of the cell (Figs 8.18, 8.20). The organisation of costal

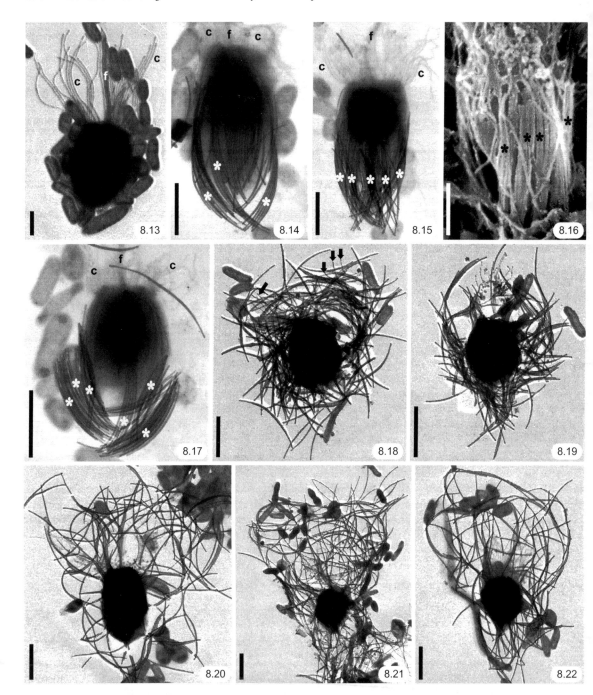

Plate 4 (Figures 8.13–8.22)

Figs 8.13–8.22 *Stephanoeca diplocostata*. 'Nudiform' costal strip storage and lorica assembly. Bars = 2 μm. **Fig. 8.13** Completely naked cell with flagellum (f) and collar microvilli (c).
Figs 8.14–8.17 Stages in the surface accumulation of costal strips. Asterisks denote sub-groupings destined for future longitudinal costae. Flagellum (f), collar (c).

strips during assembly often lacks the orderliness seen in the tectiform condition. Fig. 8.18 shows a cell at an early stage in lorica assembly, with 'swirls' of costal strips at the anterior end; the lorica-assembling microvilli can be seen just projecting beyond the anterior extremity of the strips (Fig. 8.18 arrows). The cell illustrated in Fig. 8.19 is covered by sub-groupings of strips on the surface of the cell but none are positioned horizontally as would normally be expected on a tectiform juvenile (Fig. 7.16). Fully assembled loricae have the overall shape characteristic of *S. diplocostata*, but the costal organisation is somewhat dishevelled. The longitudinal costae are more-or-less correctly positioned but the horizontal costae occur either as bands of strips in an approximately helical pattern (Fig. 8.21) or as alignments of individual strips (Fig. 8.22). This is not surprising since they have been generated from vertical sub-groupings and their mode of assembly more closely resembles that of helical costae (see Chapter 6) and not transverse costae (see Chapter 7).

All silicon-deprived cultures contain a small number of 'V-shaped' conjoined cells, which apparently result from incomplete longitudinal division of a cell that has undergone mitosis (Figs 5.45, 8.23). At the anterior end cleavage results in each of the two daughter cells possessing a flagellum surrounded by a collar of microvilli and a separate nucleus. However, partial division results in the posterior end of the two cells remaining confluent. When cultures are re-supplied with silicon, conjoined cells deposit costal strips and assemble a compound lorica (Figs 8.24–8.26). The rear end of the lorica comprises a single posterior chamber but the front end of each of the two cells assembles its own anterior chamber. The distinctness of the two anterior chambers is related to the extent of separateness of the two cells at the front end. In specimens where there is relatively little separation, the anterior chambers may be fused for part of their height (Figs 8.25, 8.26), but in specimens where there is substantial separation, the two loricae can appear to be almost separate (Fig. 8.24). In all conjoined cells, the loricae, although possessing the general *Stephanoeca* shape, contain a dishevelled arrangement of costae, particularly those aligned in the horizontal plane.

The assembly of loricae by V-shaped conjoined cells is highly informative. Since conjoined cells are only produced under conditions of severe silicon deprivation, the surrounding loricae when they appear must be a direct product of the incumbent cell. The capacity of conjoined cells to assemble a lorica that has a single posterior chamber and two anterior chambers illustrates the flexibility of the lorica-assembling process. Presumably each of the two cells develops its own lorica-assembling microvilli and is capable of moving them independently without becoming damagingly entangled with those of the other cell. Bearing in mind the unusual constraints on conjoined cells undergoing lorica assembly, there would clearly be no possibility for the cell to rotate as an entity as apparently happens during 'lorica' assembly in the chrysophyte *Dinobryon* (Herth, 1979). This provides further support for the suggestion that it is the cytoskeleton within a cell that rotates the ring of microvilli.

The ability of silicon-deprived *S. diplocostata* cells to accumulate costal strips on the cell surface and assemble their own loricae is unequivocal. This phenomenon has been observed many times and is copiously recorded by means of EM. Superficially, at least, the storage of strips in this manner followed by the immediate assembly of a lorica is typical of the nudiform condition. The morphological similarity between surface bundles of costal strips on *S. diplocostata* cells and on the nudiform *Acanthoeca spectabilis* juvenile is striking (compare Figs 8.16 and 6.44). However, two further criteria have to be satisfied before *S. diplocostata* cells with surface strips can be considered as being unequivocally nudiform. First, the orientation of costal strips should be with the anterior end pointing forwards. Second, the order of strip deposition should be for the longitudinal costae first, followed by strips for the helical and/or transverse costae. These criteria are, of course, the opposite of the tectiform condition. Unfortunately, *S. diplocostata* is not an ideal species for determining these details since it is not possible with certainty to distinguish either the orientation of individual strips or the differences between those destined for the inner and outer layers. However, the number of longitudinal costae within loricae

---

Figs 8.18–8.19 Cells with surface accumulations of strips about to assemble loricae. In Fig. 8.18 arrows indicate position of lorica-assembling microvilli.

Fig. 8.20 Lorica, appearing dishevelled, almost completely assembled; lorica-assembling microvilli just visible.

Figs 8.21–8.22 Completed loricae displaying characteristic *Stephanoeca* shape but with somewhat dishevelled 'horizontal' costae in anterior chamber. Figs 8.16 and 8.17 reproduced from Leadbeater (1989).

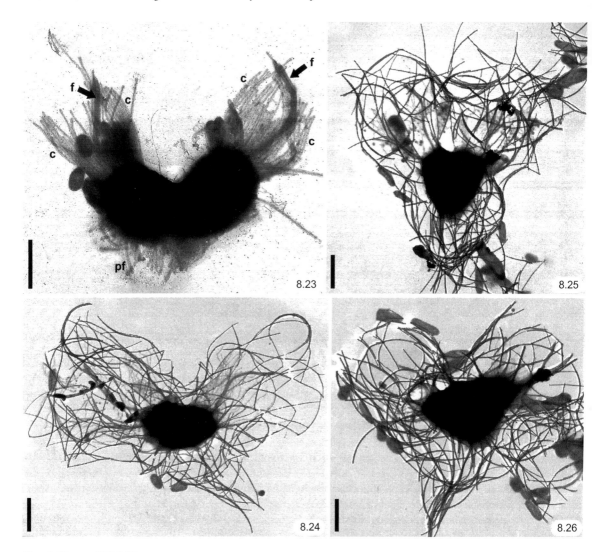

Plate 5 (Figures 8.23–8.26)

Figs 8.23–8.26 *Stephanoeca diplocostata*. Silicon-deprived, V-shaped conjoined cells re-supplied with silicon. Bars = 2 μm. **Fig. 8.23** Naked, silicon-starved conjoined cells showing substantial separation at the anterior end. Flagellum (f), collar (c), posterior filopodia (pf). **Figs 8.24–8.26** Conjoined cells that have assembled loricae after re-supply of silicon. Note that loricae are of *Stephanoeca* shape; the posterior chamber is single but the anterior chambers, which are all dishevelled, are separate. Separation of the anterior chambers is partial in Figs 8.25 and 8.26 but complete in Fig. 8.24.

assembled from surface accumulations of strips ranges from 12 to 24 with a mode of 16 ($n = 20$), which compares favourably with the equivalent values (range 15–24, mode 19 ($n = 28$) for loricae assembled on tectiform juveniles (see Table 5.1).

8.6.3 Costal strip storage on silicon-replenished *Diaphanoeca grandis* cells

One of the advantages of *Diaphanoeca grandis* is that, within broad limits, it is possible to recognise component strips belonging to the different categories of costae within the lorica (see Section 7.4.1) (Leadbeater and Cheng, 2010). Strips contributing to the longitudinal costae are approximately 10% longer than those comprising the transverse costae (Leadbeater and Cheng, 2010). Longitudinal strips are also thinner and have slightly tapering ends (Figs 4.32, 4.33). Strips attributable to the transverse costae usually have a slight sub-terminal narrowing followed by a flattened spatulate leading end (Figs 8.27–8.29). This is the end of a strip that abuts onto a longitudinal costa and points in the direction of the clockwise rotation that generated the lorica. The spatulate ends of strips comprising the anterior transverse costa are more prominent than those on strips destined for the other transverse costae (Figs 4.30, 4.33 asterisks). Nevertheless, the appearance of a spatulate end with a slight sub-terminal narrowing is a good indicator of a leading end and it is more convincing if it is also possible to show that the other end is bluntly tapered (Fig. 8.30).

When severely silicon-depleted *D. grandis* cells are re-supplied with reactive silicate, a minority accumulate costal strips on the cell surface (Fig. 8.32). The stored strips form a compact layer that extends from below the collar to the rear of the cell and is refractive when viewed with interference light microscopy (Figs 8.33, 8.34). The compaction of the costal strips, together with an organic covering that covers the rear of the cell, makes it difficult to resolve individual strips clearly with TEM. Nevertheless, on the cell illustrated in Fig. 8.32 the strips on the left-hand side have slightly flattened anterior ends (Figs 8.31, 8.32, 8.35), while the equivalent posterior ends are more tapered (Figs 8.31, 8.36). This is of significance when compared with the orientation of strips on *Diaphanoeca grandis* in the tectiform mode (Figs 7.54–7.57). In the tectiform condition, intracellular strips destined for the anterior transverse ring can be seen within a flattened, translucent parent cell and their spatulate ends are directed towards the rear of the cell (Figs 8.37–8.39). On an inverted juvenile of *D. grandis* (following inverted cell division) the equivalent strips are directed to the anterior (flagellar) end of the cell (Figs 7.54–7.57). Thus in the tectiform mode the orientation of developing strips within the parent cell is inverted with respect to their location on the nudiform mode. This would appear to provide supporting evidence that costal strips are produced and stored the 'normal' way round when produced in the nudiform mode but upside-down when produced in tectiform mode.

8.6.4 Summary of relationship between the nudiform and tectiform conditions in *Stephanoeca diplocostata*

From the evidence provided here (Section 8.6) it is concluded that *Stephanoeca diplocosta* is capable of displaying both the tectiform and nudiform conditions. With respect to the criteria relating to the nudiform condition (see Table 6.1), severely silicon-depleted naked *S. diplocostata* cells undergo a form of diagonal division and can display this type of division throughout many culture cycles. Retention of the flagellum during division by *S. diplocostata* is unusual for nudiform cells although it is not unknown in that the non-flagellate nudiform *Savillea parva* develops a flagellum immediately preceding cell division (Fig. 6.11). This feature is also observed on the non-flagellate craspedid, *Choanoeca perplexa* (Fig. 3.24) (see Section 3.4.4). Flagellar length on *S. diplocostata* cells is approximately twice the cell radius and only on rare occasions does a cell move as a result of flagellar activity.

In the nudiform mode *S. diplocostata* extrudes and stores costal strips on the surface of the cell. However, it has not been possible with certainty to establish if the strips are in the correct orientation and order to satisfy the criteria relating to the nudiform condition (see Table 6.1). Circumstantial evidence of reasonable quality supports both these criteria in the following contexts. Approximately an equivalent number of longitudinal costae are produced in both the nudiform and tectiform conditions of *S. diplocostata*, which would seem to suggest that they involve equivalent organisation of strips. The results from experiments involving silicon re-supply to silicon-depleted *Diaphanoeca grandis* cells offer support for the fact that the 'normal' nudiform orientation of the costal strips also occurs within surface accumulations (Section 8.6.3).

The precise physiological and biochemical conditions required to transform the normally tectiform *S. diplocostata* into the nudiform condition are not known. For instance, the mere absence of costal strips on the outside of the cell may not be a sufficiently good indicator as to whether the silicon deposition apparatus has completely shut down or whether there remain nascent SDVs within the cell. If these are present, and they may be segregated during division to the sibling cells, this could be sufficient for the tectiform condition to persist. It is also not clear whether the two daughter cells resulting from a division will behave in the same fashion. For instance, the daughter cell that retains external fragmentary strips may follow

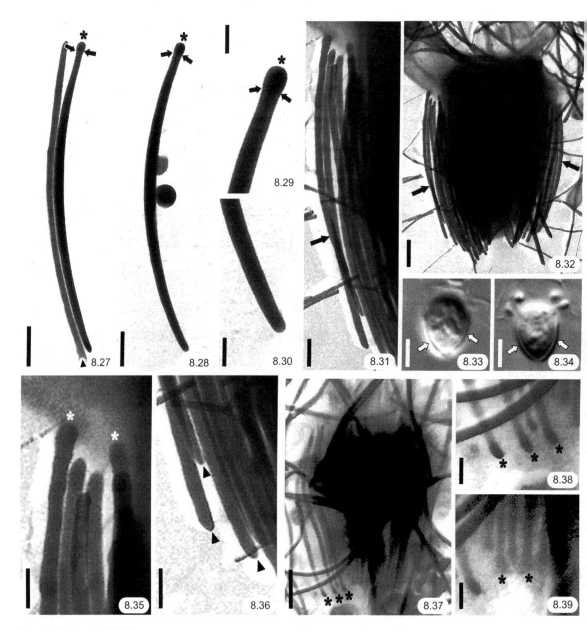

Plate 6 (Figures 8.27–8.39)

Figs 8.27–8.39 *Diaphanoeca grandis*. Nudiform and tectiform orientation, respectively, of costal strips. **Fig. 8.27** Two costal strips. Left-hand strip bears a forked tip at the end facing downwards (arrowhead) and is the anterior strip of a spine (see Fig. 4.8). Right-hand strip is attributable to one of the three basal transverse costae. Note anterior sub-terminal narrowing of strip (arrows) and slight spatulate end (asterisk). Posterior end is blunt without a sub-terminal narrowing. Bar = 0.5 μm.

Fig. 8.28 Strip from a basal transverse costa showing sub-terminal narrowing (arrows) and spatulate end (asterisk). Bar = 0.5 μm.

Figs 8.29–8.30 Higher magnifications of the two ends of the strip illustrated in Fig. 8.28 with similar labelling. Bars = 0.25 μm.

Figs 8.31–8.32 *D. grandis* cell with compact 'nudiform' covering of strips (arrows) on surface of cell. Bars = 0.5 and 1 μm, respectively.

Fig. 8.31 Higher magnification of strips visible on left side of cell in Fig. 8.32.

a different mode of division from the juvenile which is completely silicon-free.

However, what is apparent is that the definitive nudiform and tectiform conditions displayed by *S. diplocostata* are alternative scenarios that involve contrasting sets of characters and conditions for their achievement. In the nudiform condition a surface accumulation of strips needs to be directly assembled into a lorica without an intervening cell division. In the tectiform condition strips at the top of the collar can only be assembled into a lorica as a result of the parent cell undergoing inverted division. The two conditions are not a continuum; the nudiform *S. diplocostata* cell does not appear to be able to transfer a surface covering of strips to the top of the collar for a tectiform division and vice versa. These results support the suggestion that the two conditions are the result of a compound regulatory switch which can only operate exclusively in one direction or another.

8.7 EVOLUTIONARY IMPLICATIONS OF THE RELATIONSHIP BETWEEN THE NUDIFORM AND TECTIFORM CONDITIONS

8.7.1 Regulatory evolution

Since the basic construction of the lorica, comprising costae made up of costal strips arranged in linear fashion, is identical in nudiform and tectiform species, there can be no doubt that the two conditions share a common ancestry. The development of such a specialised and complex structure as the lorica must have, in itself, a lengthy evolutionary history although nothing is currently known regarding its origin. It has been argued here that the evolutionary relationship between the two conditions is based on an abrupt regulatory switch (see Section 8.6). If this is correct, then the switch would, of necessity, be achieved within a single cell generation and since it is definitive there would be no intermediate stages. Assuming that the tectiform

condition evolved from something approaching the nudiform condition (see Section 8.6), it would require only one cell to achieve the switch and subsequently survive in order for it to be able to proliferate and thereby establish a population of cells on an ever-increasing scale. However, this possibility presupposes the coincidence of many fortuitous occurrences, not the least of which is the appropriate combination of regulatory changes within the cell, the stability of the changes and the viability, survival and competitive advantage of the offspring. It is probable that over the course of evolutionary time the fortuitous switch occurred on more than one occasion but that only on one occasion did the offspring survive and proliferate. The universal consistency of the extant tectiform condition suggests that the entire tectiform lineage relates back to one successful switch. Had it occurred on more than one occasion and had the offspring survived and proliferated, it is highly likely that there would be recognisable morphological differences between the different lineages resulting from the different events.

Regulatory evolution, whereby differences between species result from changes in gene regulation rather than in the emergence of new protein-encoding genes, is now widely recognised as being a major creative force underlying species diversity across the evolutionary spectrum (Carroll *et al.*, 2005; Hunter, 2008). Little is currently known of the molecular or physiological mechanisms controlling the cellular functions involved in the nudiform/tectiform relationship of choanoflagellates. However, by reference to the codified summary of the differences between the two conditions (Fig. 8.11) and taking a parsimonious view, it is possible to identify two major categories of switch that must be involved: (1) silica deposition-related changes – timing of the costal strip deposition (silicon) cycle and the order of costal strip production; (2) cytoskeletal-related changes – inversion of costal strip formation, exocytosis and cytokinesis. Whether or not the regulatory mechanisms controlling these switches are independent of each other or are linked is unknown. However,

---

Figs 8.33–8.34 Interference contrast light micrographs of two cells with compact surface layers of costal strips (arrows). Bars = 2 μm.

Figs 8.35–8.36 Higher magnification of the two ends of the compacted strips illustrated in Fig. 8.31. Asterisks denote spatulate ends and arrowheads the tapering ends. Bars = 0.25 μm.

Fig. 8.37 Normal tectiform specimen with flattened translucent cell showing intracellular developing costal strips. Asterisks denote posteriorly facing spatulate (leading) ends. Collar (c). Bar = 1 μm.

Figs 8.38–8.39 Higher magnification of spatulate ends (asterisks) of strips from Fig. 8.37. Bar = 0.25 μm.

the apparent capacity of *Stephanoeca diplocostata* to switch from one condition to the other shows that at a higher level of regulation the combined changes can be enacted one way or another in a single action.

8.7.2 Logistical flexibility

While the overall sequence of logistical events involved in costal strip production and lorica assembly follows a common primary pathway no matter whether a species is nudiform or tectiform, nevertheless certain aspects of these processes show a high degree of flexibility which has, without doubt, been an essential contributor to the success of lorica evolution. In particular, the combination of precision and flexibility with which cells are capable of transporting costal strips on the cell surface and assembling them into loricae is quite remarkable. Cells are capable of exocytosing, transporting and assembling costal strips into loricae no matter how fragmentary, misshapen or numerically incomplete they may be (Figs 5.14, 5.34, 5.35). The molecular machinery that is responsible for these processes operates faultlessly and the result is always an external construction of some recognisable morphology (Fig. 5.15, 8.21, 8.22). The mechanism by which strips adhere to each other also operates effectively irrespective of costal strip morphology, which suggests that the surface properties of strips are unaffected by irregularities in their silicon content and morphology (Fig. 5.15).

Evidence for the flexibility of these processes has been demonstrated in connection with cells treated with poisons that affect the cytoskeleton. For instance, microtubule poisons depolymerise SDV-associated microtubules with the result that subsequent strip morphology is severely misshapen and yet cells are still capable of assembling them into loricae no matter how disorganised they might be (Fig. 5.15). Likewise silicon limitation results in the deposition and exocytosis of silica-depleted strips which may lack rigidity and length and yet they still adhere in an end-to-end fashion characteristic of costae (Fig. 5.38). Flexibility in the ability of cells to move and assemble a lorica from unusual combinations of strips must have been an essential contributor in ensuring the first formed tectiform cells were able to construct functional loricae. As illustrated by *Stephanoeca diplocostata* in the nudiform mode, the inner layer of costal strips, which on a tectiform juvenile would be assembled into transverse costae (Figs 7.16–7.21), is instead presented to the nudiform cell in a vertical or diagonal inclination (Fig. 8.18). The result is that nudiform cells of *S. diplocostata* assemble an inner layer of 'dishevelled' strips based on a pattern of helical costae (Figs 8.20–8.22).

8.7.3 Possible competitive advantage of the tectiform condition

If, as has been suggested here, evolution of the tectiform condition resulted from an abrupt switch in the timing, orientation and order of costal strip production and the inversion of the juvenile during cytokinesis from something approaching the nudiform condition, then the resulting cell would have to be at least viable, but more importantly possess some competitive advantage which allowed it to become established and proliferate within the microbial population in which it lived. The ecological role of choanoflagellates within aquatic foodwebs is as suspension feeders and within the size range 5–20 μm there are few immediate competitors (see Chapter 9). The majority of craspedid choanoflagellates are sedentary, the requirement for a substratum being essential to counteract the locomotory forces of flagellar activity (see Sections 2.3.3 and 9.3.2). The only excursion that craspedids make into the pelagic environment has been in the form of the 'temporary' motile colonies typical of *Desmarella* and *Proterospongia* or as epibionts on pelagic organisms and detritus (see Sections 3.5.2, 9.7.1). The evolution of a siliceous lorica, and the consequent diversity of loricate species, has permitted choanoflagellates to colonise a multitude of ecological niches and particularly the pelagic environment within the oceans. One direct ecological advantage of the tectiform condition, in comparison with the extant nudiform condition, is that throughout the cell cycle individuals are able to capture and ingest prey. Extant nudiform species pass through a quiescent sedentary phase during which they produce and accumulate costal strips, whereas in tectiform species it is the parent generation that produces and accumulates strips and this is achieved without interruption to the feeding process.

Whether all tectiform species are able to switch between the nudiform and tectiform conditions in the way that *Stephanoeca diplocostata* can is not known. However, *S. diplocostata* is a good exemplar because it is likely that when the evolutionary switch occurred the species involved was one with a lorica comprising rod-shaped costal strips. This would account for the universality of this type of strip in extant nudiform and tectiform species. However, once tectiform species became established, diversification of costal and lorica features occurred. This included the evolution, probably on more than one occasion, of broad, flattened costal strips typical of genera such as *Didymoeca* and

Stephanacantha. More importantly, the evolution of transverse costae by means of the horizontal alignment of sub-groupings of costal strips (see Chapter 7) permitted an increase in the size of loricae and the economical and mechanically advantageous use of silicified costal strips.

9 · Choanoflagellate ecology

And multiply each through endless years (Thomas Moore 1779–1852).

9.1 INTRODUCTION

Choanoflagellates are ubiquitous in aquatic habitats. Members of the Craspedida are distributed in freshwater, brackish and marine environments whereas, with a few exceptions (see Section 9.4.8), members of the Acanthoecida (loricate choanoflagellates) are usually restricted to marine, brackish and other saline habitats. Cell (protoplast) size (2–15 μm) and functional morphology place the choanoflagellates within a reasonably well circumscribed and universal ecological grouping known as the 'free-living heterotrophic nanoflagellates' (HNF). This is not a phylogenetically coherent grouping but rather an eclectic assemblage of heterotrophic flagellates (HF) that are of similar size and have overlapping nutritional and physiological requirements (Fenchel, 1982a; Patterson and Larsen, 1991; Eccleston-Parry and Leadbeater, 1994a; Boenigk and Arndt, 2000, 2002; Jürgens and Massana, 2008). Heterotrophy in the context used here refers principally to non-photosynthetic cells that feed phagotrophically by ingesting particles, although some autotrophic (mixotrophic) species are also capable of phagotrophic behaviour.

Choanoflagellates are classed as suspension feeders because they rely on capturing suspended prey particles for their nutrition (Fenchel, 1986a, b, 1987). Undulation of the single flagellum creates a flow field around the collar microvilli and suspended particles adhere to the outer surface of the collar where they are subsequently ingested by pseudopodia (see Section 2.4). The principal prey particles of choanoflagellates are bacteria and a range of picoplanktonic cells, including single-celled cyanobacteria, prochlorophytes, pico–eukaryotes and detritus.

The basic ecological requirement of a choanoflagellate cell is immersion in a volume of water containing suspended prey particles. For efficient operation of the feeding apparatus the locomotory effects of flagellar undulation must be counteracted and this is usually achieved by vegetative cells being secured to a surface (see Section 2.3.3) (Fenchel, 1982a; Sleigh, 1991; Christensen-Dalsgaard and Fenchel, 2003). The majority of craspedids are sedentary and attach to a surface by means of a stalk or peduncle; cells are only motile during a brief dispersal stage that immediately follows cell division or, in some species, as an intermittent colonial phase (see Chapter 3). Nevertheless, this does not necessarily confine them to the benthic environment as cells can settle onto suspended particles of flocculated debris or attach as epibionts on the surfaces of animals and plants. Some acanthoecids are also sedentary for most of their life cycle, such as *Stephanoeca cauliculata* and *Polyoeca dichotoma*, which have robust stalks that secure them to suitable substrata (Figs 4.35, 6.24). However, evolution of the basket-like lorica in acanthoecids provides a self-contained covering that in many species provides effective opposition to the locomotory effects of flagellar activity (Andersen, 1988/9). The majority of loricate species are freely suspended in the water column where they may be incorporated into the heterotrophic nanoplankton or into benthic microbial foodwebs.

Free-living HNF are an important functional group within aquatic microbial foodwebs (Sherr and Sherr, 2002). Their importance is related to their ability to make efficient use of particles the size of natural bacteria (usually <1 μm in size) and to feed more effectively than most other organisms at relatively dilute prey concentrations (10^5–10^6 bacteria ml^{-1}). Under optimal conditions they are able to consume large quantities of prey on a daily basis, often several times their own body weight, and they process it in a highly efficient manner. HNF also contribute to the regeneration of soluble inorganic nutrients, particularly those containing nitrogen and phosphorus (Fenchel, 1987;

Eccleston-Parry and Leadbeater, 1995). This dual role – consumption of bacteria and nutrient regeneration – gives these flagellates a key functional position within the 'microbial loop' (Azam et al., 1983; Laybourn-Parry and Parry, 2000). This comprises a cycle in which illuminated unicellular algae generate soluble organic extracellular products (dissolved organic matter – DOM) that provide a nutritional substrate for an actively growing population of bacteria. These are grazed by HNF, which in turn regenerate inorganic nutrients thereby enhancing the growth of unicellular algae. In addition, pico-planktonic autotrophs may be directly ingested by HNF. There has been much debate as to whether the microbial loop interacts with or is isolated from higher trophic levels within conventional aquatic foodwebs. This is summarised in the 'link or sink' debate (Fenchel, 2008). Microbial foodwebs occur to a greater or lesser extent in all natural aquatic habitats and the choanoflagellates, as suspension feeders, occupy a unique and essential niche within these foodwebs, which explains their universal distribution. The filter-feeding cell body plan has remained remarkably constant for the entire group; ecological diversification has depended on morphological variation of covering structures. In this context, evolution of the basket-like lorica was of key importance for ecological diversification of choanoflagellates.

The last 30 years have seen a substantial increase in the number of studies on various aspects of microbial ecology. Information on choanoflagellates remains patchy and unbalanced. This is partly because in many ecological studies HNF have been treated as a functional group (black box) and have not been separated into their individual systematic groupings. Some headway has been made with acanthoecids, where definitive patterns of costae within the lorica permit unequivocal species identification (Vørs, 1992, 1993a, b; Tong, 1997a, b; Tong et al., 1998; Leakey et al., 2002). The geographical locations that have been sampled are also unbalanced, with a disproportionate number of studies having been carried out in polar regions, particularly the Antarctic (Fig. 9.12), and relatively few in subtropical and tropical regions.

This chapter is divided into three major sections. Since choanoflagellates are but one of about 12 groups that comprise the 'free-living heterotrophic flagellates', the first section (Section 9.2) provides information on the systematic and functional diversity of this grouping as a whole. The second section (Section 9.3) is concerned with choanoflagellate ecophysiology. This section concentrates on experimental work, mostly of a laboratory nature, carried out on choanoflagellates in clonal culture. The third section (spanning Sections 9.4 and 9.7) reviews ecological information on marine, brackish water and freshwater choanoflagellates. Other topics considered include the variability in lorica dimensions and the possible functional role of the choanoflagellate lorica (Sections 9.5 and 9.6, respectively). For consistency and ease of comparison, throughout this chapter bacterial concentrations have been expressed as bacteria ml^{-1} and flagellate concentrations as HF or HNF l^{-1}.

9.2 SYSTEMATIC AND FUNCTIONAL DIVERSITY OF HETEROTROPHIC FLAGELLATES

Since so many physiological and ecological studies on choanoflagellates have been carried out in the broader context of 'free-living' HF, it is useful to reflect briefly on the systematic and functional diversity of this grouping as a whole. Heterotrophic flagellates (2–50 μm), comprising HNF (2–15 μm) and heterotrophic microflagellates (>15 μm), can be assigned to one of 12 broad systematic categories (Table 9.1) (Arndt et al., 2000; Jeuck and Arndt, 2013). A relatively small group of easily recognised species appears to be universal; this includes: Amastigomonas mutabilis (apusomonad), Massisteria marina (cercomonad), Bodo designis and Rhynchomonas nasuta (kinetoplastids), Jakoba libera (Jakobidae), Cafeteria roenbergensis (bicosoecid), Actinomonas mirabilis (pedinellid) and Paraphysomonas spp. (chrysomonad). Lee and Patterson (1998) undertook an analysis of the diversity and geographical distribution of free-living HF. They listed 350 species, defined by morphological criteria (morphospecies), in 31 communities from around the world, inclusive of marine and freshwater habitats. While their analysis failed to produce an overall unambiguous pattern of distribution, nevertheless communities from similar types of habitat, such as those attributable to the water column, benthic, freshwater and marine habitats, respectively, clustered together. Their conclusion was to favour a model in which there are relatively few species of free-living HF and that most of these species have a cosmopolitan distribution (see also Finlay and Fenchel, 2004). This analysis could not make allowance for the possible existence of 'cryptic speciation' (Jürgens and Massana, 2008).

For HF that feed on suspended particles, Fenchel (1986a, b, 1987) distinguished three possible feeding mechanisms: (1) filter feeding; (2) direct interception; and (3) diffusion feeding. Filter feeding requires the undulation of a

Table 9.1 *Genera and selected species commonly represented in heterotrophic flagellate communities in marine and freshwater habitats.*

Supergroup	Systematic group	Representative genera
Stramenopiles Alveolata	Stramenopiles	
Rhizaria (SAR)	Bicosoecida	*Bicosoeca Cafeteria Pseudobodo*
	Ochromonadales	*Ochromonas Paraphysomonas Spumella*
	Pedinellales	*Actinomonas Ciliophrys Pteridomonas*
	Thaumatomonadida	*Thaumatomastix*
	Cercozoa	*Cercomonas Massisteria marina Metromonas*
Opisthokonta	Choanoflagellatea	*Monosiga Salpingoeca* loricate taxa
	Filasterea	*Ministeria*
Excavata	Jakobida	*Jakoba libera*
	Euglenozoa	
	Euglenida	*Anisonema Petalomonas*
	Kinetoplastea	*Bodo Rhynchomonas nasuta*
Incertae sedis	Apusomonadida	*Amastigomonas mutabilis*
	Ancyromonadida	*Ancyromonas sigmoides*
	Cryptophyceae	*Goniomonas*

Classification after Adl *et al.* (2012).

flagellum to create a current of water that is passed through (or over) a filter. This method is employed by choanoflagellates and *Actinomonas* spp. (pedinellid), which have a ring of filopodia. The second mechanism is direct interception, sometimes called 'raptorial feeding', whereby a cell creates a current of water from which it directly intercepts particles within flow lines. This form of feeding is practised by bicosoecids, *Paraphysomonas*, *Spumella*, *Ochromonas* and *Bodo saltans* (Boenigk and Arndt, 2002). The third mechanism is diffusion feeding, which relies on mobility of prey particles by Brownian movement, swimming and/or minor water currents to cause them to collide with the cell surface. As an example, the pedinellid *Ciliophrys* possesses a slow-moving or immobile flagellum and long, narrow filopodia which radiate out from the cell and entrap colliding particles.

9.3 ECOPHYSIOLOGY OF CHOANOFLAGELLATES

9.3.1 Utilisation of dissolved organic matter by choanoflagellates

In natural environments most choanoflagellates feed by ingestion of prey particles. There is one published report of loricate choanoflagellates, *Diaphanoeca grandis* and *Acanthoecopsis* sp. (= *Acanthocorbis unguiculata*), being cultured in the absence of particulate prey (axenically) on a medium enriched with relatively high concentrations of organic supplements (dextrose, proteose peptone, liver extract) (Gold *et al.*, 1970). On a more modest scale Sherr (1988) found that field-sampled freshwater HNF, including a species of *Codosiga*, were able to take up and concentrate fluorescence within their food vacuoles when supplied with FITC-conjugated soluble dextran with a molecular mass of 2000 kDa. Furthermore, these nanoflagellates were able to utilise this dextran directly for cell growth. Marchant and Scott (1993) carried out similar experiments on field samples of loricate choanoflagellates in the Antarctic using FITC-conjugated soluble dextrans with molecular mass ranging from 3.9 to 2000 kDa at a substrate concentration of 2 mg l^{-1}. After 1 h incubation at 0 °C, six species, *Acanthocorbis unguiculata*, *Bicosta spinifera*, *Calliacantha natans*, *Crinolina aperta*, *Diaphanoeca grandis* and *Parvicorbicula socialis*, had accumulated fluorescence within their food vacuoles when supplied with dextran with a molecular mass ≥17 kDa. No indication was given as to how the fluorescent material entered the cell and made its way to the food vacuoles. Tranvik *et al.* (1993) found that a variety of FITC-labelled macromolecules including bovine milk albumin and casein (67 kDa), horse spleen ferritin (650 kDa) and succinyl-concanavalin A (55 kDa) were actively 'ingested' by marine HF, including choanoflagellates, at substrate concentrations of 1–10 mg l^{-1}.

Although dissolved organic matter does occur in nature (Sherr, 1988), its impact on the growth of actively feeding HNF is probably minimal since eukaryotes are unlikely to be able to compete with bacteria for this resource (Fenchel, 1987; Marchant and Scott, 1993). There is one possible exception to this generalisation and that is *Savillea parva*, where the *micropora* form lacks a flagellum during interphase (Figs 6.2 inset, 6.4) and has never been observed ingesting a prey particle. The lorica is so small (*c*.5 μm) (see Table 6.2), with a minute anterior pore, that it would be difficult to visualise movement of particles into the lorica.

9.3.2 Particle uptake by choanoflagellates

A brief hydrodynamic analysis was presented in Chapter 2 that related the undulation of a single flagellum on a spherical cell body to the behaviour of an 'idealised' choanoflagellate cell either freely suspended in a fluid or attached by a stalk to a substratum (see Section 2.3). The salient point to emerge from this analysis was that, for a cell with a diameter of 3–20 μm, locomotion is not compatible with the generation of feeding currents (Fenchel, 1982a; Sleigh, 1991; Christensen-Dalsgaard and Fenchel, 2003). With few exceptions, the locomotory stages in choanoflagellate life cycles are ephemeral, and motile cells either have a short collar or no collar at all (Figs 3.8, 6.17, 6.43) (Fenchel, 1982a; Dayel et al., 2011; Mah et al., 2014). Motile cells that spend longer in suspension have larger collars and are usually incorporated into colonies (Figs 3.43–3.45). One of the effects of a cell being part of a colony is to reduce swimming speed, thereby enhancing the generation of water currents for feeding purposes (Roper et al., 2013). There are a few examples of single cells that swim and feed, for example *Monosiga ovata*, but even in this instance the cell may be relatively motionless at times and it possesses a collar comprising long microvilli (Boenigk and Arndt, 2000). Perversely, the sedentary stage of *Choanoeca perplexa* has dispensed with a flagellum and relies on a large wide-angled collar for capturing prey (Leadbeater, 1977). This clearly demonstrates the effectiveness of diffusion feeding and it is probably safe to assume that this mode of feeding supplements filter feeding in other choanoflagellates (Pettitt, 2001).

Based on video recordings, Boenigk and Arndt (2000) made a study of the feeding habits of *Monosiga ovata* (mean cell volume = 24.5 μm^3). The flagellum undulated with a constant frequency of 14.4 ± 1.9 Hz and produced a weak filter current. The maximum speed of food particles within the feeding current was 9.3 ± 5.7 μm s^{-1} and the volume of water cleared was 6.4 nl cell^{-1} h^{-1}. Once particles impacted on the collar surface they were transported both up and down the collar at a speed of 58 ± 93 nm s^{-1} (see also Section 2.4). Particles that were not ingested were concentrated at the top of the collar and eventually released back to the medium. Two feeding mechanisms occurred; small particles, <0.3 μm in diameter, were transported to the base of the collar where they were ingested by pseudopodia. Larger particles were ingested by pseudopodia in the mid-region of the collar at some distance from the base. The observed ingestion rate was 24 ± 5.1 bacteria cell^{-1} h^{-1}. These results are not dissimilar to Fenchel's (1982b) observations on an unnamed species of *Monosiga*, where the frequency of flagellar beat was 30–35 Hz and the volume of water cleared about 2 nl cell^{-1} h^{-1} (Tables 9.2, 9.3).

The characteristics of choanoflagellate feeding, as highlighted by Boenigk and Arndt's (2000) study on *Monosiga ovata*, include relatively low-velocity water flow (9 μm s^{-1}) compared with interception-feeding (raptorial) nanoflagellates such as *Bodo saltans* (91 μm s^{-1}), *Cafeteria roenbergensis* (105 μm s^{-1}), *Spumella* sp. (50 μm s^{-1}) and *Ochromonas* sp. (112 μm s^{-1}). However, the specific clearance of *M. ovata* is similar to these other species because of the large filtration surface of the choanoflagellate collar. The relatively long handling times of particles (>300 s) by *Monosiga* in comparison with interception-feeding species (average 3.8–95 s) is compensated for by the simultaneous handling of many food particles and by the accumulation of smaller particles (<0.3 μm) at the cell surface outside the collar before ingestion (Boenigk and Arndt, 2000).

9.3.3 Kinetics of particle capture and ingestion

In order to make meaningful comparisons of filtration rates between species it is necessary to define certain parameters which describe particle uptake as a function of particle concentration in the surrounding medium (Fenchel, 1986a, b). Particle uptake, $U(x)$, refers to the phagocytosis of particles per cell per unit time as a function of the concentration of particles (x) in the medium. Clearance, $F = U(x) / x$, is the volume of liquid cleared of particles per unit time. U_{max} is the maximum rate of particle uptake per cell per unit time at high particle densities, which is limited by two factors: (1) the rate of particle retention on the filter, which is related to particle concentration in the medium, and (2) the rate of phagocytosis, which is related to the time it takes for the metabolic processes associated with particle ingestion and digestion to occur. At saturating particle concentrations, it is the rate of phagocytosis that will normally limit the rate of particle uptake; thus $U_{max} = 1 / \tau$, where τ is the finite time required to phagocytose one food particle. The relationship between $U(x)$ and particle concentration is a hyperbolic function – at low particle concentrations U may be more-or-less proportional to x but at higher particle concentrations F decreases with increasing x (Fenchel, 1982b). U_{max} is approached asymptotically when $x \rightarrow \infty$. F_{max} is the maximum clearance rate at low particle densities, which is realised as $x \rightarrow 0$. U_{max} / F_{max} is a constant (K_s – half saturation food density with dimensions of prey nl^{-1}) which is formally identical to the half-saturation constant of the Michaelis–Menten equation (Fenchel, 1986b, 1987). This can be interpreted as the ratio between the capacity of a predator cell to ingest and phagocytose particles and the efficiency with which particles are concentrated (filtered) from the medium. At very low particle densities the mechanism for concentrating particles (F_{max}) is likely to be the limiting factor and at high particle densities the rate of phagocytosis will be the limiting factor (U_{max}).

Maximum clearance, F_{max}, is an important parameter in that it is a direct measure of the competitive ability to secure scarce resources (bacteria at low concentrations). For heterotrophic flagellates F_{max} is expressed as $nl\,cell^{-1}\,h^{-1}$ (Tables 9.2–9.4). Volume-specific clearance, which is clearance divided by cell volume ($F_{max} /$ cell volume), is a useful value since it allows comparison between organisms of different size. Volume-specific clearance rates are expressed in units of grazer body volume (BV) cleared per grazer per unit time. Values of maximum volume-specific clearance for HF are usually around 10^4–10^5 BV $flagellate^{-1}\,h^{-1}$ (Table 9.3). This means a flagellate can clear food particles from a volume of water which is 10^4–10^5 times its own volume per hour.

Clearance can be measured in various ways (Table 9.2). A direct method of measuring particle uptake involves the use of fluorescently labelled bacteria (FLB) or fluorescent polystyrene microspheres. Using this protocol, direct counts of particles in the medium and within food vacuoles can be made. This method has been used to considerable effect by Marchant working with field samples of *Diaphanoeca grandis* from Lake Saroma, Japan and with *Acanthocorbis unguiculata* at Davis, Antarctica (Table 9.2) (Marchant, 1990; Marchant and Scott, 1993).

An alternative method of estimating clearance is to calculate values using the parameters of growth derived from batch or continuous culture experiments (Fenchel, 1986a, b). Since the yield, Y (growth efficiency), of protists is invariant over a large range of growth rates, then for balanced growth:

$$\mu(x) = U(x)\,Y$$

where $\mu(x)$ is the specific growth rate as a function of food particle concentration x, and Y is the yield constant. The specific growth rate, $\mu(x)$, can be obtained from the Monod equation (see Section 9.3.4) and Y is calculated as the increase in predator concentration divided by the decrease in bacterial concentration to give the fraction of flagellate cell produced upon the ingestion of one bacterium (see Table 9.3). If batch cultures have been used an added adjustment has to be made since once the food supply has been exhausted cells are still capable of a further cell division, thereby reducing the cell volume by approximately 50%. Thus the yield must be divided by two in order to be applicable to the exponential rate of growth (Geider and Leadbeater, 1988; Eccleston-Parry and Leadbeater, 1994a). Maximum uptake rate (U_{max}, bacteria $predator^{-1}\,h^{-1}$) is calculated as $U_{max} = \mu_{max} / 0.5Y$ and maximum clearance values (F_{max}, nl $predator^{-1}\,h^{-1}$) are calculated from the equation $F_{max} = U_{max} / K_s$ where K_s = the half-saturation (food density) constant.

With the limited kinetic data currently available on prey clearance and uptake by choanoflagellates it is not possible to do much more than tabulate the values for the various parameters and make some general comments and comparisons. Even so, there are caveats because experimental conditions, including temperature and the nature and condition of the prey particles used, vary considerably (see Table 9.2). With respect to rates of clearance, excepting the result of Cynar and Sieburth (1986) for *Codosiga gracilis*, the values fall within one order of magnitude. The values for U_{max} are particularly consistent (see Table 9.3).

Comparison of the values for maximum growth rate (μ_{max}), maximum bacterial uptake rate (U_{max}) and maximum clearance rate (F_{max}) of choanoflagellates (Table 9.3) with the equivalent parameters for a selection of HNF that usually co-exist with choanoflagellates is instructive (Table 9.4). The two kinetoplastids, *Bodo designis* and *Rhynchomonas nasuta*, and the chrysophyte, *Paraphysomonas imperforata*, exceed all the choanoflagellates investigated with respect to both U_{max} and F_{max}. *Bodo designis* and

Table 9.2 Clearance rates of a selection of marine and freshwater choanoflagellates fed with live bacteria, fluorescently labelled bacteria (FLB) or fluorescent microspheres and determined either by using laboratory cultures or freshly collected field samples.

Species	Freshwater or marine	Temp (°C)	Clearance rate (F) (nl cell^{-1} h^{-1})	Prey particle	Source of material	Reference
Monosiga sp.	M	20	2	Pseudomonas sp.	Culture (USA)	Fenchel (1982b)
Monosiga sp.	M	10–12	1.4–1.9	Planktonic bacteria	Field sample Limfjord (Denmark)	Andersen and Sorensen (1986)
Monosiga sp.	FW	15–23	2.5	Microspheres 0.25 μm	Field sample (Finland)	Salonen and Jokinen (1988)
M. ovata	FW	20	6.4	Pseudomonas putida	Culture (Russia)	Boenigk and Arndt (2000)
Codosiga gracilis	M	20	0.4	Microspheres 0.57 μm	Culture (UK)	Cynar and Sieburth (1986)
C. gracilis	M	20	4	Mesorhizobium sp.	Culture (UK)	Eccleston-Parry and Leadbeater (1994a)
Planktonic craspedids	FW	13–30	0.9–5.5	Microspheres 0.57 μm	Field samples Lake Oglethorpe (Georgia, USA)	Sanders et al. (1989)
Sedentary choanos Craspedida	FW	4–13	1.1–13	Microspheres 0.5 μm	Field sample Lake Pavin (France)	Carrias et al. (1996)
Planktonic and attached craspedids	FW	18–20	0.5–4.9	FLB	Field sample Lake Konstanz (Germany)	Cleven and Weisse (2001)
Acanthocorbis unguiculata	M	21	0.91	Vibrio sp.	Culture (USA)	Davis and Sieburth (1984)
A. unguiculata	M	0	3	Microspheres 0.25, 0.5 μm	Field sample Davis (Antarctica)	Marchant and Scott (1993)
Stephanoeca diplocostata	M	18	3	Pseudomonad	Culture (France)	Geider and Leadbeater (1988)
S. diplocostata	M	20	16	Mesorhizobium sp.	Culture (France)	Eccleston-Parry and Leadbeater (1994a)
Diaphanoeca grandis	M	5	16	Pseudomonas sp.	Culture (Denmark)	Andersen (1988/9)
D. grandis	M	0	0.7	Microspheres 0.25, 0.5 μm	Field sample Lake Saroma (Japan)	Marchant (1990)
Large and small choanoflagellates	M	−0.5	0.72–9.32	FLB	Field sample Prydz Bay (Antarctica)	Leakey et al. (1996)

Table 9.3 *Kinetic parameters with respect to prey uptake and growth for five choanoflagellate species grown in batch culture. Volume-specific clearance rates are expressed in units of grazer body volume (BV) cleared per grazer per unit time. Mesorhizobium sp. (see Thurman et al., 2010) was formerly known as Bacterium B1 (Eccleston-Parry and Leadbeater 1994a, b, 1995).*

Species	Temp (°C)	Prey	Approx. cell volume (μm^3)	Max. clearance rate		Max. bacterial uptake during exponential growth U_{max} (bacteria $cell^{-1}$ h^{-1})	Max. specific growth rate μ_{max} (h^{-1})	Half saturation constant for growth K_s (10^6 bacteria ml^{-1})	Yield stationary phase Y ($\times$ 10^{-3} cells $bact^{-1}$)	Reference
				F_{max} (nl $cell^{-1}$ h^{-1})	Volume-specific clearance (10^4 BV $cell^{-1}$ h^{-1})					
Monosiga ovata	20	Bacterium *Pseudomonas*	20	2	9.8	27	0.17	13.5	6.3	Fenchel (1982b)
Codosiga gracilis	20	*Mesorhizobium* sp.	83	4	4.5	36	0.052	9.7	2.9	Eccleston-Parry and Leadbeater (1994a)
Stephanoeca diplocostata	18	Bacterium *Pseudomonad*	20	3	16	21	0.079	6.8	6.4–8.2	Geider and Leadbeater (1988)
S. diplocostata	20	*Mesorhizobium* sp.	35	16	46	37	0.035	2.3	1.9	Eccleston-Parry and Leadbeater (1994a)
Diaphanoeca grandis	15	Bacterium *Pseudomonas*	40	16	40	40	0.12	2.4	–	Andersen (1988/9)

Paraphysomonas imperforata also have higher maximum growth rates with doubling times of 4.4 and 3.3 hours, respectively. This accords with the respective growth strategies of these three heterotrophic species and choanoflagellates in general. If field samples of water are brought into the laboratory and allowed to stand at room temperature or moderately enriched with an organic supplement, the first HNF to appear in large numbers are bodonids and species of *Paraphysomonas* – these are the classic *r*-strategists. Choanoflagellates appear in appreciable numbers only after a few days when there has been a decline in the concentrations of bacteria and opportunistic HNF.

9.3.4 Growth kinetics of choanoflagellates

While quantitative information relating to the ability of individual HF species to acquire prey under differing environmental conditions provides insights into the competitive strategies of these species, another important quantitative aspect is the ability of flagellates to utilise prey particles for growth. In order to evaluate this relationship,

once again kinetic data are required. Many of the quantitative models relating species performance to particle uptake are based on Michaelis–Menten enzyme kinetics. Here the particle is treated as being the non-replenished substrate and is therefore limiting to growth. Particle uptake is viewed as combining the mass-transfer of a substrate to biomass with saturation phenomena that are similar to non-linear Michaelis–Menten kinetics for biotransformation. A popular expression for simulating microbial growth kinetics is the Monod (1950) equation:

$$\mu = \mu_{max} \, S \, / \, (K_s + S)$$

where μ is the specific growth rate of biomass as a function of substrate concentration; μ_{max} is the maximum specific growth rate; S is the concentration of growth-limiting substrate; K_s is the half-saturation constant for growth, which is the concentration of substrate that allows the species to grow at half the maximum growth rate.

The Monod model of growth kinetics differs from the Michaelis–Menten model of enzyme kinetics in being empirical rather than mechanistic. However, Monod

Table 9.4 *Kinetic parameters relating to prey uptake, including maximum clearance (*F_{max}*) and maximum uptake (*U_{max}*), and relating to growth, including maximum specific growth rate (*μ_{max}*) and half saturation constant (*K_s*), for a selection of common heterotrophic nanoflagellates grown in laboratory culture.*

Heterotrophic flagellate	Prey bacterium	Approx cell volume (μm^3)	F_{max} (nl cell^{-1} h^{-1})	U_{max} (bacteria cell^{-1} h^{-1})	μ_{max} (h^{-1})	K_s (10^6 bacteria ml^{-1})	Reference
Bodo designis	*Mesorhizobium* sp.	54	47	160	0.16	3.4	Eccleston-Parry and Leadbeater (1994a)
Rhynchomonas nasuta	*Vibrio* sp.	33	17–37	30–210	–	–	Davis and Sieburth (1984)
Pleuromonas jaculans[*]	*Pseudomonas* sp.	50	1.4	54	0.16	38	Fenchel (1982b)
Jakoba libera	*Mesorhizobium* sp.	75	1	5	0.036	5.4	Eccleston-Parry and Leadbeater (1994a)
Ochromonas sp.[*]	*Pseudomonas* sp.	200	10	190	0.19	19	Fenchel (1982b)
Pseudobodo sp.	*Pseudomonas* sp.	90	10	84	0.15	8.4	Fenchel (1982b)
Paraphysomonas vestita	*Pseudomonas* sp.	190	17	254	0.17	14.9	Fenchel (1982b)
Paraphysomonas imperforata	*Mesorhizobium* sp.	212	58	63	0.21	1.1	Eccleston-Parry and Leadbeater (1994a)
Ciliophrys infusionum	*Mesorhizobium* sp.	220	6	259	0.045	45	Eccleston-Parry and Leadbeater (1994a)

[*] (*Pleuromonas jaculans*, *Ochromonas* sp.) denotes freshwater species, the remainder are marine.

kinetics can be thought of as describing a chain of enzymatically mediated reactions with limiting steps, each described by Michaelis–Menten kinetics.

With respect to HF, μ_{max} is a measure of predator growth in the presence of prey (bacteria) at saturating concentrations. This value represents the maximum intrinsic growth rate achievable in the prevailing environmental conditions. There are relatively few occasions in nature when ideal growth will occur; usually prey availability, nutrients or other physicochemical factors will be limiting. Values of μ_{max} for choanoflagellates range from $0.17\,h^{-1}$ for *Monosiga* sp., to $0.035\ h^{-1}$ for *Stephanoeca diplocostata*, equivalent to doubling times of 4 and 22 hours, respectively (Table 9.3). However, the value of μ_{max} for *S. diplocostata* obtained in laboratory experiments varies considerably, for example $0.079\ h^{-1}$ (Geider and Leadbeater, 1988), 0.09–$0.06\ h^{-1}$ (Section 5.7.2) (Leadbeater and Davies, 1984) and $0.035\ h^{-1}$ (Eccleston-Parry and Leadbeater, 1994a), probably as a result of the different prey bacteria and experimental conditions involved. The doubling time of four hours for *Monosiga* sp. places this species in the company of the fast-growing ubiquitous *r*-strategists such as *Actinomonas mirabilis* (2.8 h), *Paraphysomonas vestita* (3 h), *P. imperforata* (3.3 h), *Ochromonas* sp. (3.6 h), *Bodo designis* (4.3 h) and *Pleuromonas jaculans* (4.3 h) (see Table 9.4) (Fenchel, 1982b). In comparison, most choanoflagellates have longer doubling times, although *Diaphanoeca grandis* (5.8 h) and *Stephanoeca diplocostata* (8.8 h) (see Table 9.3), which are ubiquitous in inshore waters including saltmarshes and rockpools, might be considered to be *r*-strategists when grown in media with high concentrations of bacteria.

The relationship between growth rate, μ, and initial bacterial concentration in batch culture has been used to compare the competitiveness of different flagellate species (Fenchel, 1982b). This relationship can be illustrated graphically for pairs of species, such as *Codosiga gracilis* and *Stephanoeca diplocostata*, by plotting specific growth rate (μ) as a function of initial bacterial concentration (Fig. 9.1). If the resulting hyperbolic saturation curves cross each other, as they do in Fig. 9.1 at an initial bacterial concentration of 1.3×10^7 bacteria ml^{-1}, then below this crossover point *S. diplocostata* has a higher value of μ than *C. gracilis* and is therefore likely to have a competitive advantage over *C. gracilis*, whereas above this value the relationship is reversed. The half-saturation constant for growth (K_s) is also a parameter that can be used to compare the competitiveness of different flagellate species for the

same limited resource. The lower the value of K_s the greater the affinity the predator has for its prey. K_s is a measure of the efficiency with which a predator can utilise prey particles for growth under prey-limiting conditions. While the values for K_s listed in Table 9.3 are of the same order of magnitude, no obvious pattern emerges. This is also the case when wider comparisons are made with other HNF (Table 9.4) (Eccleston-Parry and Leadbeater, 1994a).

A major criticism of using laboratory experiments for obtaining kinetic data on prey uptake and growth is that the conditions of culture are often far removed from those that most species experience in nature. This is particularly true with respect to the quantity and quality of prey particles provided. All the kinetic parameters recorded in Table 9.3 were obtained from batch culture experiments. While these experiments have the benefit of being relatively quick and simple to organise and they produce results that can be used to generate kinetic data, nevertheless they have important drawbacks. In particular, conditions within a batch culture are continuously changing and it is especially difficult to obtain reliable results for low prey concentrations. Batch culture experiments are more akin to the 'feast

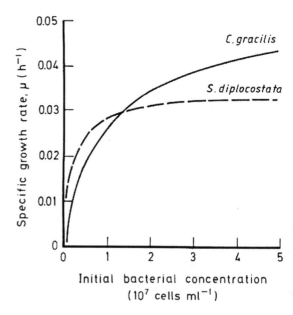

Fig. 9.1 Hyperbolic relationships obtained by plotting specific growth rate (μ) as a function of initial concentration of bacteria for two choanoflagellates, *Codosiga gracilis* and *Stephanoeca diplocostata*, grown separately in batch culture. Curves cross over each other at 1.3×10^7 bacteria ml^{-1}. Modified after Eccleston-Parry and Leadbeater (1994a).

and famine' conditions that might be expected within eutrophic environments (Eccleston-Parry and Leadbeater, 1994b). Opportunist species (*r*-strategists) are likely to thrive in batch culture conditions but the performance of

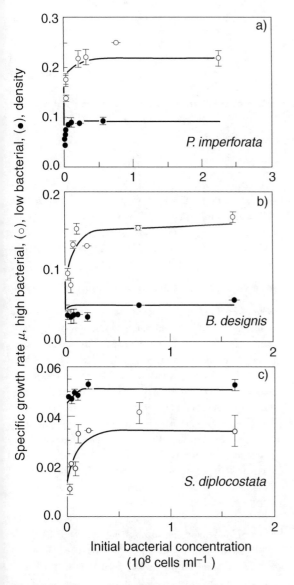

Fig. 9.2 Specific growth rates of three heterotrophic nanoflagellates as a function of initial bacterial concentration obtained from batch culture grazing experiments after the flagellate had been familiarised to high (○) and low (●) prey density, respectively. a. *Paraphysomonas imperforata*; b. *Bodo designis*; c. *Stephanoeca diplocostata*. Reproduced from Eccleston-Parry and Leadbeater (1994b).

K-strategists, which require stable low-prey conditions, are likely to be underestimated.

Eccleston-Parry and Leadbeater (1994b) examined the performance of *Stephanoeca diplocostata* and two other HNF, *Paraphysomonas imperforata* (Ochromonadales) and *Bodo designis* (kinetoplastid), under long-term conditions of high and low prey concentrations. For high prey concentration (≈ 6–8×10^7 bacteria ml^{-1}) predator flagellates were added to suspensions of prey bacteria in batch culture. For low-prey conditions flagellates were grown in continuous culture in which the flow of prey particles (bacteria) and medium was controlled to give a low prey density without altering the dilution rate. After four weeks of culturing in their respective pre-conditions, batch culture grazing experiments were carried out with a starting concentration of bacteria ranging from 8×10^5 ml^{-1} to 1.6×10^8 ml^{-1}. For each combination of flagellate and prey concentration the specific growth rate (μ) was calculated and for each flagellate the respective values of μ were plotted as a function of initial bacterial concentration. The results are presented graphically in Fig. 9.2 and the values for μ_{max} and K_s are listed in Table 9.5.

The three flagellates responded in different ways to their pre-adaptation to low-prey conditions (Fig. 9.2; Table 9.5). *Paraphysomonas impeforata* and *Bodo designis* both experienced a reduction in μ_{max}. However, while the K_s value for *Paraphysomonas imperforata* was not significantly lower than that of high-prey adapted cells, the K_s value for *Bodo designis* was significantly lower. *Stephanoeca diplocostata* experienced an increase in μ_{max} and a decrease in the value of K_s. These reduced K_s values experienced by *S. diplocostata* and *B. designis* are not unexpected because low prey density will ultimately increase the 'searching time' that a predator flagellate requires to find prey, and therefore in order for a predator species to survive in oligotrophic environments it must have a higher affinity for its prey and an efficient capture mechanism or be capable of some degree of physiological adaptation. The higher value of μ_{max} exhibited by *S. diplocostata* in low-prey conditions illustrated that this species was well adapted to survive in relatively oligotrophic waters (Eccleston-Parry and Leadbeater, 1994b).

9.3.5 Growth energetics of choanoflagellates

The bioenergetics of choanoflagellate growth has received only limited attention. Geider and Leadbeater (1988) estimated the growth efficiency of *Stephanoeca diplocostata* in

Table 9.5 *Comparison of the maximum specific growth rates (μ_{max}) and half-saturation constants for growth (K_s) for three species of heterotrophic nanoflagellates (HNF), each pre-conditioned to high and low prey density, respectively.*

		Flagellate species		
Growth parameter	Pre-conditioning prey concentration	*Stephanoeca diplocostata*	*Paraphysomonas imperforata*	*Bodo designis*
Maximum specific growth rate (μ_{max})	High	0.035 ± 0.01	0.213 ± 0.01	0.160 ± 0.01
(h^{-1})	Low	0.050 ± 0.01	0.094 ± 0.01	0.055 ± 0.01
Half saturation constant for growth	High	2.330 ± 1.92	1.124 ± 2.15	2.330 ± 1.33
(K_s) (10^6 bacteria ml^{-1})	Low	0.240 ± 0.99	0.794 ± 0.23	0.240 ± 0.18

Values ± standard error of mean.
Reproduced from Eccleston-Parry and Leadbeater (1994b).

terms of organic carbon using ^{14}C as a tracer. The prey bacterium was grown on ^{14}C-glucose until stationary phase, concentrated and added to flasks of fresh medium without organic enrichment. An aliquot of exponentially growing *S. diplocostata* cells was added to each flask containing the bacterial suspension and grazing took place. Concentrations of dissolved and particulate organic ^{14}C activity, and bacterial and *S. diplocostata* cell concentrations were determined at 12-hour intervals throughout a 110-hour period. Based on the data obtained from the control and 'grazing' flasks a carbon conversion efficiency of 40% was calculated – that is 40% of ingested 'bacterial' carbon was converted into predator flagellate biomass. This conversion value was consistent for the range of initial bacterial concentrations used and the growth rates of *S. diplocostata* recorded. However, during the same experiments the volumetric conversion efficiency (the yield by volume of predator from prey) was only 27%. Similar discrepancies between carbon and volumetric efficiencies have also been experienced with other microflagellates. Caron *et al.* (1986) considered that this discrepancy resulted from the egestion of detrital organic matter. Geider and Leadbeater (1988) concluded that the volume conversion efficiency more accurately reflected the transfer of organic carbon from bacteria to *S. diplocostata* and on this basis tentatively advanced a carbon budget for growth according to the following equation:

$$I = G + R + E$$

where I is ingested organic carbon; G is organic carbon used for growth; R is loss of organic carbon to respiration; and E is egested organic carbon. Gross growth efficiency is defined as G / I and the net growth efficiency as $G / (I - E)$. Of the bacterial carbon ingested by *S. diplocostata*, 60% was lost to respiration, 13% was egested and 27% was used for growth. Thus, a gross growth efficiency of 27% and a net growth efficiency of 31% were calculated.

9.3.6 Role of choanoflagellates in nutrient cycling

Heterotrophic flagellates are also responsible for the re-mineralisation of organic compounds, particularly those containing nitrogen and phosphorus. Unlike photosynthetic organisms and bacteria, which are capable of assimilating mineral nutrients for growth, HF are not considered able to assimilate inorganic nitrogen and phosphorus from the environment (Fenchel, 1987). Thus, whereas photosynthetic organisms and bacteria are usually net assimilators of inorganic nutrients, HF are net regenerators of these compounds. The cycling of nitrogen and phosphorus, alternately between inorganic and organic forms, is an important feature of the 'microbial loop' (Azam *et al.*, 1983; Fenchel, 2008).

Eccleston-Parry and Leadbeater (1995) investigated the regeneration of inorganic nitrogen and phosphorus by four HNF, *Stephanoeca diplocostata*, *Paraphysomonas imperforata*, *Bodo designis* and *Jakoba libera*, fed on three nutritional states of a single bacterial strain. The prey, bacterium *Mesorhizobium* sp. (formerly known as Bacterium B1 (Eccleston-Parry and Leadbeater, 1994a, b, 1995)), was grown in continuous culture in three separate nutrient regimes – balanced (C:N:P atomic ratio 50:10:1), N-rich (C:N:P atomic ratio 50:20:1) and P-rich (C:N:P atomic ratio 50:10:5). The result was to generate bacterial prey with comparable amounts of particulate carbon but with respective particulate C:N:P ratios of – balanced (C:N:P ratio 52:12:2); N-rich (C:N:P ratio 53:25:1) and P-rich (C:N:P ratio 54:12:6). A series of batch culture grazing experiments was carried out in which flagellate growth and bacterial consumption were measured. The specific growth rate values for each flagellate appeared to be independent of the C:N:P content of the bacterial prey. With respect to re-mineralisation, ammonium and total dissolved phosphate were assayed throughout the experiment. The percentage regeneration of nitrogen, when the four flagellates were fed with balanced and N-rich bacteria, respectively; and phosphorus, when the four species were fed with balanced and P-rich bacteria, respectively, are shown in Fig. 9.3a–b.

With respect to nitrogen regeneration, the performance of the four flagellates was similar in that they regenerated between 46 and 69% of bacterial-bound nitrogen as ammonium, with marginally more nitrogen from N-rich bacteria than balanced ones. However, with respect to phosphorus regeneration, *Paraphysomonas imperforata* and *Bodo designis* released significantly more bacterial-bound phosphorus with both balanced and P-rich bacteria than either *Jakoba libera* or *Stephanoeca diplocostata*. Eccleston-Parry and Leadbeater (1995) concluded that this difference probably related to the ecological survival strategies of the respective species. As already mentioned (Section 9.3.3), *Paraphysomonas imperforata* and *Bodo designis* are classic *r*-strategists with rapid growth rates and minimal facility for storage of nutrients in excess of requirement. On the other hand, *Stephanoeca diplocostata* and *Jakoba libera* are intermediate in character between *r*- and *K*-strategists in that they have slower growth rates and are more conservative with respect to the re-mineralisation of phosphorus-containing organic compounds.

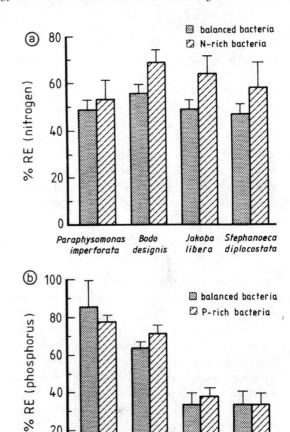

Fig. 9.3 Comparison of the percentage regeneration efficiency (% RE) for four heterotrophic nanoflagellate species with respect to (a) nitrogen and (b) phosphorus after feeding on balanced, nitrogen-rich and phosphorus-rich bacteria. Reproduced from Eccleston-Parry and Leadbeater (1995).

9.4 ECOLOGY OF MARINE AND BRACKISH WATER CHOANOFLAGELLATES

9.4.1 Introduction

Since choanoflagellates are ubiquitous in aquatic environments the scope for ecological studies is enormous. The development of choanoflagellate ecology is substantially biased towards marine locations and has been closely linked to improvements in microscopy and the techniques associated with handling living and fixed samples of flagellates and bacteria. In general terms, early work was qualitative and descriptive, whereas more recent studies have been quantitative and experimental. It is possible to distinguish four phases, and the beginnings of a fifth, in relation to the acquisition of ecological information on choanoflagellates. They are more-or-less chronologically ordered, although there are overlaps. These phases include: (1) an initial pioneer phase (nineteenth to mid twentieth century) based mainly on the light microscopy of local field collections (Section 9.4.2); (2) early records of planktonic loricate choanoflagellates obtained during sea-going cruises (1910–1970)

(Section 9.4.3); (3) improved fixation protocols and introduction of phase and interference contrast light microscopy and electron microscopy (EM) to samples collected from the field (from 1970 onwards) (Section 9.4.4); (4) quantitative studies involving the use of membrane filters and fluorescent stains for enumeration of bacteria and autotrophic and heterotrophic flagellates, and development of quantitative experiments carried out under field conditions (from 1980 onwards) (Section 9.4.5); (5) application of molecular techniques for use in field conditions (from 2000). The following account is based on the five phases of research outlined above and is treated in an approximately chronological manner.

9.4.2 Pioneer phase of choanoflagellate ecology

The 'pioneer phase' of choanoflagellate ecology is closely associated with the beginnings of choanoflagellate taxonomy and systematics. The starting point must be James-Clark's (1867b) seminal publication on the *Infusoria Flagellata* (see Chapter 1). This included the description of the first four unequivocal choanoflagellate species that he attributed to two new genera, *Codosiga* and *Salpingoeca*. If James-Clark was the initiator of the pioneer phase, then William Saville Kent (Chapter 1) was its consolidator. Throughout the 1870s Kent travelled around the UK collecting freshwater and marine samples containing filamentous algae, macrophytes and colonial hydozoa. The culmination of ten years of research resulted in the publication of the *A Manual of the Infusoria* (Kent, 1880–2). In total Kent described 46 new choanoflagellate species and six new genera: *Monosiga*, *Desmarella*, *Astrosiga*, *Lagenoeca*, *Polyoeca* and *Proterospongia*. The total number of species described, including those previously named by James-Clark (1867b), Tatem (1868) and Stein (1878), was 53. An inherent shortcoming of Kent's work, based as it is on field and aquaria samples containing mixtures of species, is the lack of reliable life cycle data on individual species.

At the end of the nineteenth and beginning of the twentieth centuries there was a considerable recycling of species descriptions and classificatory systems in the literature, with the occasional addition of taxa (Bütschli, 1882; Francé, 1897; Senn, 1900; Lemmermann, 1910). Francé (1897) provided the most cogent summary of choanoflagellate ecology at the end of the nineteenth century. In summary, he recounted that the majority of described species were sedentary, with intermittent free-swimming forms. They inhabited both fresh and saltwater, although

they appeared to reach their most intensive development in freshwater. Some species were limited to freshwater, such as the majority of *Codosiga* and *Proterospongia* species, whereas others (*Salpingoeca*) were found evenly distributed in the sea, as well as in freshwater. Choanoflagellates predominantly inhabit standing water bodies, although they have also been found in the River Danube.

Major advances of note in the twentieth century include complementary contributions by Ellis (1929, 1935) and de Saedeleer (1927, 1929). Ellis (1929) was preoccupied with marine choanoflagellates on a saltmarsh in Appledore, North Devon, UK. His resulting publication was the first to recognise the costal substructure of acanthoecid loricae. In total he described 20 new species and seven new genera, three of which are attributable to the Acanthoecida, namely *Acanthoeca*, *Stephanoeca* and *Diaphanoeca*. De Saedeleer (1927, 1929) studied freshwater choanoflagellates and, while not using cultured material, he observed aspects of cellular function and life cycles.

The final contributions of note in the pioneer phase of work are those of Norris (1965) and Boucaud-Camou (1966). Norris (1965) collected seawater samples from tide pools in the upper littoral region of the shore at Pacific Grove (Monterey), California and on the San Juan Islands, Washington State. The significance of this study is that it includes the first recognition of the choanoflagellate lorica as containing an array of costae made up of rod-shaped costal strips. Boucaud-Camou's (1966) work was based on seawater samples, collected from the Normandy coast (English Channel), to which macerations of algae were added. In these conditions many craspedids and acanthoecids were observed.

9.4.3 Early records of marine planktonic choanoflagellates

The first unequivocal published record of a planktonic loricate choanoflagellate appears to be that of Meunier (1910), who described and illustrated colonies of *Corbicula* (*Parvicorbicula*) *socialis* in plankton samples collected from the Kara Sea north of Siberia in 1907 (Fig. 9.4a–c). Meunier (1910) had previously encountered this species in plankton collections from the Greenland Sea, where he described it as being 'exceptionally abundant', but on that occasion he had confused it with colonies of *Phaeocystis pouchetii* (Haptophyceae) (see also Rat'kova and Wassmann (2002) for an account of choanoflagellates in the Barents Sea). *Parvicorbicula socialis*, as it was subsequently named

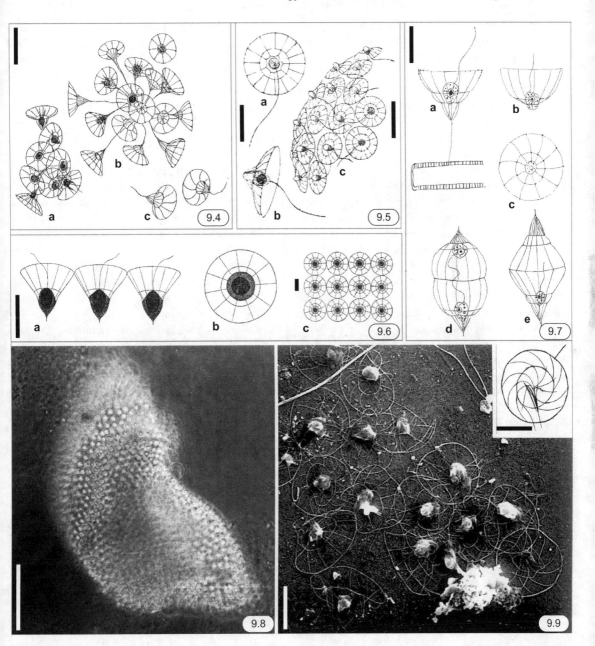

Plate 1 (Figures 9.4–9.9)

Figs 9.4–9.9 *Parvicorbicula (Corbicula) socialis*. Drawings and microscopical images of single cells and colonies. **Fig. 9.4a–c** Original illustrations of *Corbicula socialis* collected from the Kara Sea by Meunier in 1907. Bar = 20 µm. Reproduced from Meunier (1910).
Fig. 9.5a–c Illustrations of *P. socialis* collected in 1910 from the Golfe du Lion, western Mediterranean. Bars = 10 µm. Reproduced from Pavillard (1917).
Fig. 9.6a–c Illustrations of *P. socialis* collected from water around east Greenland and Franz Josef Land during the Øst expedition to the Denmark Strait in 1929. Note that neighbouring loricae adhere to each other by the extreme edges of their anterior rings. a–b. Bar = 10 µm. c. Bar = 20 µm. Reproduced from Braarud (1935).

by Deflandre (1960), was present in fixed samples collected from the North Atlantic (near Ireland) and the Mediterranean during the 'Michael Sars' North Atlantic Deep-sea Expedition in 1910 (Gaarder, 1954). *P. socialis* was also recorded in Danish waters (April–May 1899–1901) by Ostenfeld (1916) and in surface plankton (April 1910) from the Golfe du Lion, part of the western Mediterranean, by Pavillard (1917) (Fig. 9.5a–c). Braarud (1935), while on the Øst Expedition (1929), encountered *P. socialis* in the Denmark Strait between Iceland and Greenland and subsequently in net samples collected from the sea east of Greenland and between Svalbard (Spitzbergen) and Franz Josef Land in 1930 (Fig. 9.6a–c). In 1932, Gran and Braarud (1935) encountered this species at a concentration of 2×10^3 cells l^{-1} south of Nova Scotia, Canada. Bursa (1961) recorded and illustrated *P. socialis* from offshore water at Igloolik Island in the Canadian Arctic in 1955–6 (Fig. 9.7a–e). However, he was confused by the solitary nature of the specimens he collected and for this reason gave them the name *Monosiga* sp. Two of Bursa's illustrations (shown here as Fig. 9.7d–e) probably record the end product of inverted (tectiform) division. Deflandre (1960) found *Parvicorbicula socialis* in plankton samples collected in coastal water near Terre Adélie, East Antarctica (Fig. 9.12) in 1951, thereby demonstrating the global distribution of this marine colonial species. Deflandre (1960) was also responsible for the change in name to *Parvicorbicula* since the original name *Corbicula* was preoccupied. *P. socialis* has subsequently been collected many times and is distributed globally (Manton *et al.*, 1976; Hara and Tanoue, 1984; Thomsen *et al.*, 1990; McKenzie *et al.*, 1997). McKenzie *et al.* (1997) published a remarkable light micrograph of a large colony of *P. socialis* showing it to contain many hundreds of cells (reproduced here as Fig. 9.8), each of which is attached to its neighbours by the edges of the anterior transverse ring of its lorica (Figs 9.5c, 9.6c, 9.9). The combination of lorica shape, costal curvature (Figs 9.7c, 9.9 inset) and colonial habit is characteristic of *P. socialis* and for this reason confirms Deflandre's (1960) and Bursa's (1961) identification of this species.

The colonial habit of *Parvicorbicula socialis* facilitates the trapping of this species in phytoplankton nets, and explains why it was so frequently observed as an adjunct to early phytoplankton studies. Regrettably these early records are mostly qualitative with only occasional comments about abundance (Gran and Braarud, 1935). If cell concentrations were given, they were usually incorporated into global figures under the general heading 'other flagellates'. The first planktonic choanoflagellate for which extensive quantitative data are available is '*Monosiga marina*' Grøntved from the southern North Sea in 1947 (Grøntved, 1952). The correct identity of this species as described by Grøntved (1952) remains a mystery. His illustration includes seven ellipsoidal cells that vary according to the presence or absence of a 'stalk' and the length of the stalk if present (Fig. 9.10a–g). Superficially they resemble *Bicosta* cells and, in hindsight, it seems most probable that they are a mixture of species attributable to the *Bicosta/Calliacantha* consortium (see Section 4.6.4). '*Monosiga marina*' was regularly present at concentrations between 500 cells l^{-1} and 14.2×10^4 cells l^{-1} in the surface water of the southern North Sea during May 1947 (Grøntved, 1952). Paasche (1960) recorded concentrations of *Monosiga marina* as high as 1.3×10^6 cells l^{-1} for the outer edge of the Norwegian Sea (temperature 6–9 °C) at the beginning of June 1954.

In 1956 Grøntved described another planktonic choanoflagellate as *Salpingoeca natans*, which bears a close resemblance to the previously described '*Monosiga marina*' (Grøntved, 1952) (compare Fig. 9.10a–g with Fig. 9.11a–f). However, there can be little doubt that the accompanying illustrations show these specimens to be a mixture of *Bicosta* and *Calliacantha* species (Fig. 9.11a–f, see also Figs 4.70–4.72, 7.58–7.65, 9.19–9.21). An additional confusion is that Grøntved (1956) did not comment on this obvious similarity between the two sets of drawings and thus for a time there were two names in use that appeared to include a similar species consortium. Paasche and Rom (1962) provide individual counts for both *Monosiga marina* (3.5×10^3 cells l^{-1}) and *Salpingoeca natans* (1×10^3 cells l^{-1}) in surface

Fig. 9.7a–e Illustrations of *P. socialis* described as *Monosiga* sp. in offshore water at Igloolik Island in the Canadian Arctic. Bar = 10 µm. Reproduced from Bursa (1961).

Fig. 9.8 Phase-contrast micrograph of a large *P. socialis* colony collected in a 40 m sediment trap during June 1988 at Conception Bay, Newfoundland. Bar = 100 µm.

Fig. 9.9 SEM of a colony of *P. socialis* collected during a spring diatom bloom. Bar = 10 µm. Fig. 9.9 inset Image of *P. socialis* lorica showing curvature of longitudinal costae. Bar = 10 µm. Figs 9.8 and 9.9 reproduced from McKenzie *et al.* (1997).

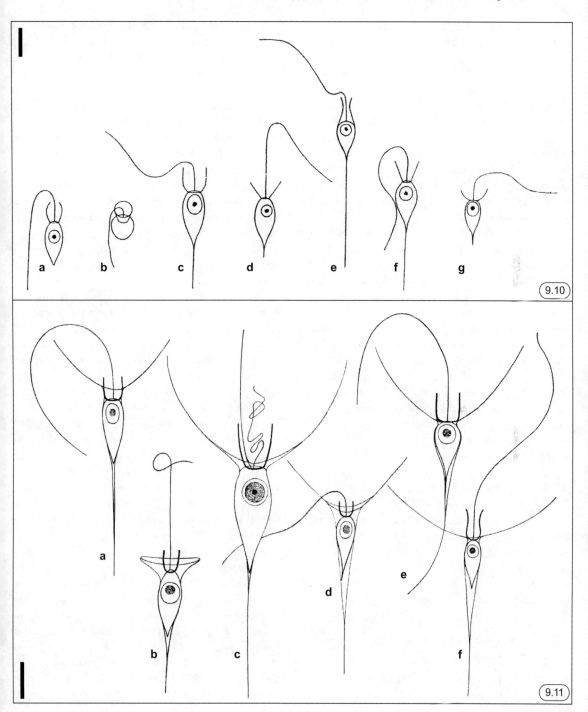

Plate 2 (Figures 9.10 and 9.11)

Figs 9.10–9.11 Grøntved's (1952, 1956) illustrations of marine planktonic choanoflagellates from the southern North Sea and Danish coastal waters. Bars = 10 μm. **Fig. 9.10a–g** Illustrations of '*Monosiga marina*' from the North Sea, Skagerrak, Norwegian Sea, Faroe-Rockall area and the Davis Strait (west Greenland).

Fig. 9.11a–f *Salpingoeca* (*Calliacantha*) *natans* from Danish coastal localities (Grøntved, 1956). In the absence of a holotype illustration for *S. natans*, Leadbeater (1978) designated Fig. 9.11c as the lectotype of *S. natans* and at the same time created a new genus *Calliacantha* for this loricate species. Figs 9.10 and 9.11 reproduced from Grøntved (1952 and 1956, respectively).

samples in May 1958 during a cruise in the Norwegian Sea. Bursa (1961) observed small populations of *Salpingoeca natans* (200 cells l⁻¹) and *Monosiga* sp. (= *Parvicorbicula socialis*) (500 cells l⁻¹) in seawater collected near Igloolik Island in the Canadian Arctic during September 1956.

The final contribution to this phase of work, and a linking theme with the following phase involving EM, was the work of Throndsen (1969, 1970a, b). Throndsen (1969) carried out an extensive study of marine flagellates at selected coastal localities between the extreme north of Norway (Varanger) and the Oslofjord. In total, seven choanoflagellate species were recorded, including four acanthoecids, *Diaphanoeca grandis*, *Stephanoeca diplocostata*, *S. ampulla* and *S. kentii*. These are well-established coastal species previously recorded by Ellis (1929) from a saltmarsh in the UK and by Norris (1965) from tidepools on the west coast of the United States. Throndsen (1970a, b) subsequently undertook a sea-going cruise to the Barents Sea between north Norway and Bear Island (Bjønøya). This time six species were identified, including *Diaphanoeca grandis* and *Parvicorbicula socialis* and four new species: *Parvicorbicula quadricostata*, *Pleurasiga reynoldsii*, *P. minima* and *Salpingoeca* (= *Bicosta*) *spinifera* (Throndsen, 1970a, b). The significance of this collection is that all the species contain full, unambiguous descriptions and that *Salpingoeca spinifera* provided a name for some of the Grøntved (1956) specimens that had two anterior spines (see Fig. 9.11a). However, the final authentication of *Salpingoeca natans* Grøntved had to await a later publication by Leadbeater (1978). Since Grøntved (1956) did not designate a holotype, this meant that all his illustrations were of equal status (syntypes). Leadbeater (1978) was therefore able to designate the specimen with three spines (Fig. 9.11c) as the lectotype, thereby fixing its status and allowing it to be used as the type of a newly created genus, *Calliacantha* Leadbeater. What these early results reveal is the ubiquity of planktonic choanoflagellates in the oceans, at least at high latitudes north and south.

9.4.4 Application of electron microscopy to nanoplankton studies

Apart from a few early incidental TEM images of choanoflagellates, the starting date of the EM phase of work was 1970, when the author visited Herdla Island, 20 miles northwest of Bergen, Norway. Nanoplankton samples were concentrated, fixed with osmium tetroxide and deposited as whole mounts on Formvar-coated grids for subsequent observation with transmission electron microscopy (TEM).

The revelation of these collections was the number and diversity of loricate choanoflagellates (Leadbeater, 1972a, b). This collecting trip was quickly followed by another with Irene Manton in June 1971 to Frederikshavn, Denmark where, with the assistance of Helge Thomsen, samples of inshore nanoplankton were collected (Leadbeater, 1972c; Manton and Leadbeater, 1974). From these modest beginnings the author, Manton and Thomsen continued to work independently with field and laboratory material for the better part of the next 20 years (see Table 9.6 for details of field locations and publications). Over a 30-year period (1970–2000), more than 50 collections were undertaken by various colleagues and species lists of acanthoecid choanoflagellates were published (Table 9.6).

The almost exclusive use of lorica morphology for acanthoecid species identification, and the relative ease with which details of costal arrangement can be determined by phase-contrast microscopy or TEM has facilitated the publication of extensive species lists for many geographical localities (see Table 9.6). Since craspedids appear to be a relatively minor component of marine nanoplankton, the listing of acanthoecids represents the major planktonic choanoflagellate fauna. The relative ease with which acanthoecid species can be recognised circumvents most of the problems that beset the identification of other HF that lack unequivocal distinguishing features (see Lee and Patterson, 1998). However, the construction of representative species lists for making biogeographical comparisons is still subject to certain extrinsic limitations. The most important of these are under-sampling and under-reporting (Lee and Patterson, 1998). These limitations apply when the number of species recorded from a site is significantly less than the number of species present. This can be caused by a failure to sample sub-habitats completely, by temporal patchiness or by a failure to include sparsely distributed species. Even if diversity is well sampled, only a small proportion of the species present may be reported. The exclusion of rare species may lead to an erroneous generalisation of cosmopolitanism, but under-reporting, which excludes species on a more random basis, is likely to lead to an erroneous sense of endemism (Lee and Patterson, 1998). Many of the species lists noted in Table 9.6 undoubtedly suffer from one or both of these shortcomings. Many of the studies lack rigorous ecological sampling protocols and habitats are often only casually described. Plankton samples from surface water in mid-ocean are likely to contain the most precisely circumscribed collections, since the environment is relatively uniform and unaffected by coastal influences. This contrasts with

Table 9.6 *Locations at which loricate choanoflagellates have been collected and subsequently identified using light and electron microscopy.*

Location	Reference	Location	Reference
Arctic Region		**North Pacific**	
Greenland and Arctic	Thomsen (1982),[*] Thomsen et al. (1995)[*]	Subarctic Pacific	Booth (1990)[*]
	Ikävalko and Gradinger (1997),[*] Throndsen 1970b	North Pacific Ocean	Hoepffner and Haas (1990)
		Californian Waters	Norris (1965), Thomsen et al. (1991)[*]
Arctic Canada	Manton et al. (1975, 1976, 1980), Vørs (1993b)	Equatorial Pacific	Vørs et al. (1995)[*]
		South Pacific	
North Atlantic		Galapagos Islands	Manton et al. (1980)
St Lawrence estuary, Canada	Bérard-Therriault et al. (1999)	Chile	Soto-Liebe et al. (2007)[*]
Newfoundland, Canada	McKenzie et al. (1997)[*]		
Pettaquamscutt River, RI, USA	Menezes (2005), Menezes and Hargraves (2005)	**SE Asia and Far East**	
Belize and Canary Islands	Vørs (1993a)[*]	Thailand	Thomsen and Boonruang (1983a, b, 1984),[*] Thomsen and Moestrup (1983)
		Taiwan and Japan	Hara et al. (1996, 1997)
Europe		Japan	Takahashi (1981b), Throndsen (1983),[*]
Denmark	Leadbeater (1972c), Thomsen (1973, 1976, 1977a)		Hara and Takahashi (1984, 1987a, b)
Norway	Throndsen (1969, 1974), Leadbeater (1972a)		
	Espeland and Throndsen (1986)[*]	**Australasia**	
Finland	Thomsen (1979),[*] Vørs (1992),[*] Ikävalko (1998),[*]	New Zealand	Moestrup (1979)[*]
	Ikävalko and Thomsen (1997)[*]	Australia	Tong (1997b),[*] Tong et al. (1998)[*]
England, Southampton Water	Tong (1997a)[*]		
		Antarctica	
		Southern Ocean	Thomsen and Larsen (1992), Leakey et al. (2002)[*]
Mediterranean and Red Sea		S. Ocean Indian Sector (75°E)	Hara et al. (1986)[*]
Adriatic Sea	Leadbeater (1973)[*]	King George Island	Chen (1994)[*]
Bay of Algiers	Leadbeater (1974)[*]	Weddell Sea	Buck (1981),[*] Buck et al. (1990), Buck and Garrison (1983), Thomsen et al. (1990,[*] 1997)[*]
Gulf of Elat	Thomsen (1978)[*]		
		Lützow-Holm Bay	Takahashi (1981a)[*]

Table 9.6 (*cont.*)

Location	Reference	Location	Reference
South Atlantic		Prydz Bay, Ellis Fjord	Marchant (1985),[*] Marchant *et al.* (1987),
Brazil	Bergesch *et al.* (2008)[*]		Marchant and Perrin (1990), Tong *et al.* (1997)
		E. Antarctica	Hara and Tanoue (1984)[*]

[*] indicates that a species list, often with comparison lists for other areas, is included.

coastal locations which often contain heterogeneous mixtures of benthic, estuarine, saltmarsh, rocky shore and open-sea planktonic species.

Table 9.7 includes a selection of species lists compiled at a variety of locations extending from the Arctic to the Antarctic and with collections made at inshore and open-sea localities, respectively. Acanthoecid species have been ordered into groupings according to lorica morphology in line with the four themes described in Chapter 4. Nudiform species head the list. Tectiform species are arranged in the following order: (1) *Stephanoeca* theme; (2) *Diaphanoeca/Crinolina* theme; (3) *Cosmoeca/Parvicorbicula* theme; (4) *Stephanacantha/Platypleura* theme (loricae with broad, flattened costal strips); (5) *Bicosta/Calliacantha* theme. Ordering species in this way reflects a morphological change from, at the top of the list, species with relatively small loricae comprising many costae (nudiform species and *Stephanoeca* theme), through intermediate species with a smaller number of costae that are lightly silicified (*Diaphanoeca/Crinolina* and *Cosmoeca/Parvicorbicula* themes) to loricae with few costal strips and a tendency towards exceptionally long spines (*Bicosta/Calliacantha* theme).

The species listed in the seven right-hand columns of Table 9.7 refer to the open ocean (North Pacific, Equatorial Pacific and Southern Ocean) or offshore waters (Icelandic offshore water, Californian waters, Weddell Sea and Lützow-Holm Bay) where there is limited influence from mixed coastal water. The Newfoundland collection (Table 9.7), gathered from north- and south-facing bays on the east side of the island, also contains a predominantly oceanic fauna, probably because of the proximity of the cold, south-flowing Labrador Current. The oceanic and offshore locations noted in Table 9.7 are widely distributed in the Northern and Southern Hemispheres. The predominant species inhabiting the surface water at all these locations come from the *Cosmoeca/Parvicorbicula*

and *Bicosta/Calliacantha* themes. *Bicosta spinifera* is common to all seven locations, and four species, namely *Cosmoeca norvegica*, *Pleurasiga minima*, *Parvicorbicula socialis* and *P. circularis*, were found in six of the seven locations. Three conclusions can be drawn from these lists of oceanic species. First, the predominant planktonic species are cosmopolitan, at least at high latitudes. Second, the relative abundance of the major species is approximately similar at all locations sampled (see Table 9.9 for references). Third, the pelagic environment favours species with loricae comprising relatively few lightly silicified costae and/or spiny projections.

The species listed in Table 9.7 for Southampton Water (UK), Port Jackson, Sydney (Australia), Shark Bay (Australia) and West Greenland are for samples collected at inshore locations. For Southampton Water, Port Jackson, Sydney and Shark Bay some of the samples were collected from the shore, and species lists also include material from enrichment cultures (Tong, 1997a, b). The species lists from these locations are more extensive than those for the open sea and include representatives from all acanthoecid themes including nudiform taxa. Species of the *Stephanoeca* theme attributable to *Acanthocorbis*, *Saepicula* and *Didymoeca* are much in evidence. *Diaphanoeca grandis* is also a common inshore species. There is a substantial representation, at least qualitatively, of species from the *Cosmoeca/Parvicorbicula* and *Bicosta/Calliacantha* themes. *Bicosta spinifera*, which is omnipresent in oceanic collections, is less frequent in inshore localities.

While Table 9.7 presents only a glimpse of the many species lists now available, nevertheless the results are universally consistent (see also Bérard-Therriault *et al.*, 1999; Menezes, 2005). Regrettably, there are few records available for tropical and subtropical waters. Thomsen and Boonruang (1983a, b, 1984) sampled the Andaman Sea off the coast of Thailand (Phuket 7°N 98°E) and it was from

Table 9.7 *Comprehensive species lists of loricate choanoflagellates collected from selected inshore or open sea locations.*

	Southampton Water, UK	Sydney, Australia	Shark Bay, Australia	West Greenland	Newfoundland, Canada	Iceland offshore	North Pacific	Californian waters	Equatorial Pacific	Southern Ocean	Weddell Sea	Lützow–Holm Bay
Acanthoeca spectabilis	+	+										
Savillea parva	+											
Polyoeca dichotoma	+		+									
Stephanoeca diplocostata	+	+	+	+	+							
S. cupula		+		+								
S. norrisii	+											
Acanthocorbis apoda		+		+								
A. unguiculata				+								
A. campanula		+		+								
Saepicula pulchra	+		+									
Didymoeca costata	+		+									
Diaphanoeca grandis	+	+		+					+			
D. multiannulata				+		+				+	+	+
D. pedicellata	+	+	+	+		+	+			+	+	
D. undulata	+	+		+		+						
Crinolina isefjordensis	+			+								
C. aperta						+				+	+	+
Cosmoeca ceratophora	+	+						+	+	+	+	
C. norvegica	+	+				+	+	+	+	+	+	
C. phuketensis		+	+					+	+			
C. takahashi												
C. ventricosa	+	+		+		+	+	+	+	+	+	
Pleurasiga minima	+	+		+		+	+	+	+	+	+	
P. echinocostata		+					+	+				
P. reynoldsii	+	+		+		+	+	+	+	+	+	
Parvicorbicula socialis			+	+	+	+	+	+	+	+	+	+
P. circularis	+	+	+	+		+	+					
P. quadricostata	+	+	+	+	+	+	+	+				+
Polyfibula spp.							+					
Nannoeca minuta		+	+			+	+	+				
Stephanacantha campaniformis							+					

Table 9.7 (cont.)

	Southampton Water, UK	Sydney, Australia	Shark Bay, Australia	West Greenland	Newfoundland, Canada	Iceland offshore	North Pacific	Californian waters	Equatorial Pacific	Southern Ocean	Weddell Sea	Lützow–Holm Bay
S. dichotoma									+			
Calotheca alata		+										
Platypleura spp.			+						+			
Bicosta spinifera	+	+			+	+	+	+	+	+	+	+
B. minor	+	+	+	+	+	+	+	+				
B. antennigera				+	+	+						+
Calliacantha natans	+	+		+	+	+		+		+	+	+
C. simplex, C. multispina	+	+	+	+	+	+	+	+	+	+	+	
C. longicaudata				+	+	+					+	
Crucispina cruciformis	+	+	+	+					+			
Apheloecion sp.									+		+	

Southampton Water (Tong, 1997a); Port Jackson, Sydney, Australia (Tong et al., 1998); Shark Bay, Australia (Tong, 1997b); West Greenland (Thomsen, 1982); Newfoundland (McKenzie et al., 1997); Iceland (Leadbeater, unpublished); North Pacific Ocean (Booth, 1990); Californian waters (combined stations 15–21 in Thomsen et al., 1991); Equatorial Pacific (Vørs et al., 1995); Southern Ocean (Leakey et al., 2002); Weddell Sea, Antarctica (offshore and nearshore in Thomsen et al., 1997); Lützow–Holm Bay, Antarctica (Takahashi, 1981a).

here that several new species with broad, flattened costal strips attributable to *Stephanacantha* and *Platypleura* were described. Bergesch *et al.* (2008) sampled South Atlantic water from the coast of Brazil (32°S 52°W) but apart from one new species with petaloid strips, *Didymoeca* (*Diplotheca*) *tricyclica*, the list of species is classic for a coastal collection with emphasis on *Stephanoeca* species. Much the same can be said with respect to samples collected from the coast of Chile (23°–45°S 72°W) (Soto-Liebe *et al.*, 2007).

A distinction can be made between open-ocean, pelagic communities, which are dominated by species from the *Cosmoeca/Parvicorbicula* and *Bicosta/Calliacantha* themes (Tables 9.7 and 9.8) to the exclusion of nudiform species and representatives of the *Stephanoeca* theme, and coastal communities that contain a broader range of species including those typical of inshore localities attributable to *Stephanoeca*, *Acanthocorbis*, *Saepicula*, *Didymoeca* and *Diaphanoeca grandis* (Table 9.8). As inshore habitats and localities become more circumscribed, for instance saltmarshes, estuaries, low- or high-salinity water bodies, ice meltpools, inland seas and endorheic lakes, there may be restrictions on the species that can survive in these conditions but, nevertheless, generally the list reflects an impoverished coastal community. Some species occupy an intermediate zone, in that they are neither major contributors to the inshore or open ocean communities but may appear in greater concentrations between these two communities (Table 9.8).

A recent study in which a partial sequence of SSU DNA (890 nt) was analysed in five loricate choanoflagellates including *Acanthocorbis unguiculata*, *Diaphanoeca grandis* and *Helgoeca nana* revealed a low intraspecific diversity (<0.5% p-distance) (Nitsche and Arndt, 2014).

9.4.5 Quantitative aspects of marine choanoflagellate ecology

ANTARCTICA AND THE SOUTHERN OCEAN
Table 9.9 summarises details of publications containing quantitative ecological information on marine choanoflagellates at a variety of locations around the globe. There is a

Table 9.8 *Approximate categorisation of acanthoecid choanoflagellates with respect to inshore, intermediate and open-ocean locations.*

Inshore/coastal	Intermediate coastal	Open ocean
Savillea parva	*Stephanacantha campaniformis*	*Diaphanoeca multiannulata*
Helgoeca nana	*S. dichotoma*	*D. pedicellata*
Acanthoeca spectabilis	*Polyfibula* spp.	*D. undulata*
Polyoeca dichotoma	*Platypleura acuta*	*Crinolina isefiordensis*
Stephanoeca diplocostata	*P. cercophora*	*C. aperta*
S. urnula	*Nannoeca minuta*	*Cosmoeca norvegica*
S. cupula	*Spinoeca buckii*	*C. ceratophora*
S. norrisii	*Crucispina cruciformis*	*C. takahashii*
S. elegans		*C. ventricosa*
S. apheles		*Pleurasiga minima*
S. cauliculata		*P. reynoldsii*
Didymoeca costata		*Parvicorbicula socialis*
D. elongata		*P. circularis*
Acanthocorbis apoda		*P. quadricostata*
A. unguiculata		*Bicosta spinifera*
A. campanula		*B. antennigera*
Saepicula pulchra		*B. minor*
S. leadbeateri		*Calliacantha natans*
Diaphanoeca grandis		*C. simplex*
		C. multispina
		C. longicaudata

preponderance of studies relating to the Antarctic and most records are for acanthoecids (see Fig. 9.12 for locations mentioned in the text). Buck and Garrison (1988), as part of the AMERIEZ 86 (Antarctic Marine Ecosystem Research: Ice Edge Zone) investigation to the Weddell Sea in March 1986 (austral autumn), have provided one of the most comprehensive quantitative records with respect to loricate species. On this cruise they collected water samples from the upper 110 m of the water column along a 400 km south-west to north-east transect comprising three southerly stations in the pack ice and four northerly stations in open water (transect 4 in Fig. 9.13). Table 9.10 provides the mean and range values for the 11 most numerous species based on pooled data obtained from the seven stations sampled. The cell concentrations, particularly the ranges, are more-or-less equivalent for all species. It is also of note that the combined concentrations of species from the three morphological themes, *Diaphanoeca/Crinolina*, *Cosmoeca/Parvicorbicula* and *Bicosta/Calliacantha*, are also equally balanced (Table 9.10). The two species that dominated open-water stations and achieved the highest concentrations, *Parvicorbicula socialis* and *Diaphanoeca pedicellata*, were commonly found as colonies as, on occasion, was *Crinolina aperta*. Buck and Garrison (1988) also used the AMERIEZ 86 data to look for possible correlations between choanoflagellate abundance, on the one hand, and primary production (as measured by chlorophyll *a* and ^{14}C uptake by photosynthesis) and secondary production (as measured by bacterial biomass and [^{3}H] thymidine incorporation) on the other. For total choanoflagellate abundance there was a significant correlation with both primary and secondary production and this also applied individually to *Parvicorbicula socialis*, *Diaphanoeca pedicellata* and *Crinolina aperta*. Buck and Garrison (1988) estimated that choanoflagellates could be consuming up to 14% of the daily bacterial cell production in open-water stations. Bacterial densities in the water column were generally low, 0.5–73.0×10^5 bacteria ml^{-1}, and yet there was significant bacterial production and a population of actively feeding choanoflagellates which suggested a closely coupled active microbial foodweb in these waters.

There have been numerous other investigations into the quantitative relationships between HF, including choanoflagellates, and other components of the microbial community. Since the pelagic microbial loop envisages a coupling between phytoplankton growth, DOM production, the growth of bacteria and their grazing by HF, many observers have looked for correlations between these

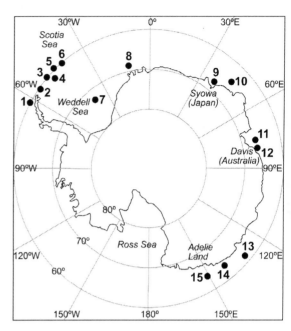

Fig. 9.12 Outline map of Antarctica with sampling locations mentioned in the text (see also Tables 9.6, 9.7 and 9.9).
(1) WINCRUISE – Garrison and Buck (1989b); (2) King George Island – Chen (1994); (3) EPOS (leg 1) – Kivi and Kuosa (1994); (leg 2) – Becquevort *et al.* (1992), Thomsen and Larsen (1992); (4) AMERIEZ 86 – Buck and Garrison (1988), Garrison and Buck (1989a, b); (5) AMERIEZ 88 Garrison *et al.* (1991, 1993); (6) AMERIEZ 83 – Garrison and Buck (1989a, b); (7) Weddell Sea Ice edge – Buck (1981), Buck and Garrison (1983); (8) Weddell Sea ANT X/3 cruise – Thomsen *et al.* (1997); (9) Lützow-Holm Bay – Takahashi (1981a); (10) Indian and Australian Sectors – Ishiyama *et al.* (1993); (11) Prydz Bay – Marchant (1985); (12) Davis (Australian Antarctic Division Station) – Marchant and Perrin (1990), Leakey *et al.* (1996); (13) Eastern Australian Antarctic Territory – Waters *et al.* (2000); (14) Terre Adélie – Deflandre (1960); (15) Western Australian Antarctic Territory – Hara and Tanoue (1984).

various components (see Sanders *et al.*, 1992 for review). Kivi and Kuosa (1994) reported a significant positive correlation between choanoflagellates on the one hand and autotrophic nanoplankton (as measured by chlorophyll *a*) on the other in the upper 40 m of the water column in the Weddell Sea during the late austral winter of 1988 (transect 3 in Fig. 9.13). Becquevort *et al.* (1992) recorded a significant positive correlation between choanoflagellates and bacterioplankton abundance in the marginal ice-zone of the north-west Weddell Sea during the austral summer of

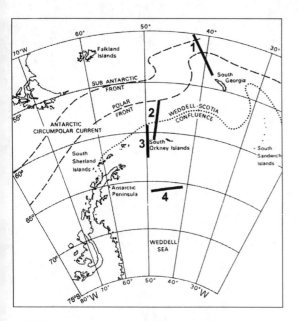

Fig 9.13 Locations of Antarctic and Southern Ocean transects mentioned in text: (1) Southern Ocean: NW/SE 725 km transect across Polar Front (Leakey *et al.*, 2002); (2) Scotia and Weddell Seas: N/S 450 km transect across Weddell-Scotia Confluence (EPOS/leg 2 cruise) (Thomsen and Larsen, 1992); (3) Western Weddell Sea: N/S 550 km transect across Weddell-Scotia Confluence (EPOS/leg 1 cruise) (Kivi and Kuosa, 1994); (4) Ice Edge Weddell Sea: SW/NE 400 km transect across ice edge (Buck and Garrison, 1988).

1988–9. Also in this context Chen (1994) investigated the distribution and abundance of choanoflagellates in Great-Wall Bay, King George Island, Antarctica during the austral summer of 1991. Four species, *Bicosta spinfera*, *Crinolina aperta*, *Diaphanoeca multiannulata* and *Parvicorbicula circularis*, accounted for 90% of the total acanthoecid population and were distributed widely at all stations sampled and at all depths from 0 to 30 m. Total choanoflagellate abundance varied between 5.1×10^3 and 2.6×10^4 cells l^{-1} with an average of 1.3×10^4 cells l^{-1}. In vertical profile the maximum abundance of total choanoflagellates was 5.3×10^4 cells l^{-1}. There was also a significant positive correlation between choanoflagellate abundance and chlorophyll *a* concentration. A similar correlation was observed by Marchant and Perrin (1990) for shallow inshore water near Davis, East Antarctica during the austral summer of 1982–3, where the concentration of choanoflagellates ranged between 10^4 and 10^6 cells l^{-1}. In winter the concentration was approximately 10^2 cells l^{-1}. A seasonal species succession was also observed with *Bicosta spinfera*, *Crinolina aperta* and *Parvicorbicula socialis* dominating the choanoflagellate population during the summer months but being below the level of detection in the winter. *Cosmoeca ventricosa* and *Saepicula leadbeateri* were detected only in winter. *Acanthocorbis unguiculata*, *Calliacantha simplex* and *Diaphanoeca grandis* were detected throughout the year. Ishiyama *et al.* (1993) recorded that loricate choanoflagellates, particularly colonial species, comprised approximately 23.5% abundance and 30% biomass of the heterotrophic nanoplankton at 8 m depth along a 500 km transect parallel to the coast of East Antarctica, starting at the Japanese station at Syowa.

Leakey *et al.* (1996) measured bacterial production using the [^{3}H]thymidine incorporation protocol, and HNF (including choanoflagellates) grazing of bacteria by measuring the uptake of FLB at weekly intervals between December 1993 and February 1994 in coastal water at Davis, East Antarctica. During this period bacterial abundance ranged from 2.2×10^5 bacteria ml^{-1} in December to 8.3×10^5 bacteria ml^{-1} in mid-January; after this the abundance declined to 4×10^5 bacteria ml^{-1} in February. Choanoflagellates dominated the HNF community at the beginning of the study period, comprising 56% of the abundance (1.5×10^6 cells l^{-1}) and 65% of the biomass (8 μg C l^{-1}) in December. Table 9.11 summarises the pooled data for choanoflagellates obtained from the seven collections during the study period. Choanoflagellates are characterised as being small (mean cell volume 16.9 μm^3) or large (mean cell volume 34.9 μm^3), although there is no indication as to which species these categories include. The peak total abundance of choanoflagellates recorded at Davis is an order of magnitude higher than the results obtained by Buck and Garrison (1988) for the Weddell Sea (see Table 9.10), although the lower values in both regions are equivalent. The clearance rates and volume-specific clearance (Table 9.11) are comfortably within the range of values obtained for laboratory experiments with cultures (see Table 9.2), although the ingestion rates obtained for the Davis material are considerably lower than the values of U_{max} recorded for cultured cells. However, the temperature of the Davis experiments was below 0 °C while the laboratory experiments ranged from 15 to 20 °C. During the initial phase of the study when choanoflagellates were particularly abundant they were capable of grazing 7% of the total bacterial biomass and almost 17% of bacterial production per day, which represented up to 68% of the

Table 9.9 *Quantitative studies of marine nanoplankton including choanoflagellates, listing: geographical location; reference; year of collection; open or ice-covered seawater samples; categorisation of quantitative data; correlations with other components of the microbial foodweb; seasonality.*

Locality	Reference	Year of collection	Map (Fig. 9.12)	Ice-covered water	Open water	Choanoflagellate Abundance	Choanoflagellate Biomass	Individual choanoflagellate species abundance and/or biomass	Correlation with bacteria and/or phytoplankton abundance and/or production	Seasonal data
Barents Sea	Rat'kova and Wassmann (2002)	1998, 1999	–		+		+	+		+
Baltic Sea	Kuuppo (1994)	1998, 1999	–		+	+			+	+
Limfjord, Denmark	Andersen and Sorensen (1986)	1983	–		+			+	+	+
Southampton Water, UK	Brandt and Sleigh (2000)	2000	–		+	+				
Newfoundland, Canada	McKenzie et al. (1997)	1988–1992	–		+			+	+	+
Greenland	Thomsen et al. (1995)	1988	–					+		
Sub-Arctic Pacific	Booth (1990)	1978–1988	–		+			+		
Californian Waters	Buck et al. (1991)	1989, 1990	–		+	+	+			
Equatorial Pacific	Vors et al. (1995)	1992	–		+	+	+			
Southern Ocean Antarctica	Leakey et al. (2002)	1996	–		+	+	+	+	+	
King George Island	Chen (1994)	1991	2		+			+	+	
Weddell Sea	Buck and Garrison (1983)	1980	7	+	+	+	+	+		
	Buck and Garrison (1988)	1986	4	+	+	+		+		
	Garrison and Buck (1989a)	1983, 1986	4, 6	+	+	+	+		+	
	Garrison and Buck (1989b)	1983, 1986	1, 4, 6	+	+	+	+		+	
	Garrison et al. (1991)	1986	5	+	+					
	Garrison et al. (1993)	1988	5	+	+			+		
	Kivi and Kuosa (1994)	1988	3	+	+		+		+	
	Becquevort et al. (1992)	1988/89	3	+	+		+		+	
Davis/Prydz Bay	Marchant and Perrin (1990)	1982/83	11	+	+			+	+	+
	Leakey et al. (1996)	1993/94	12		+	+	+		+	
Eastern Antarctica	Ishiyama et al. (1993)	1991	10		+	+	+		+	

Antarctic locations are noted on the map illustrated in Fig. 9.12.

Table 9.10 *Mean and range abundance of the most numerous choanoflagellate species identified from the AMERIEZ 86 cruise (March 1986, austral autumn).*

Species	Occurrence (number of samples *ex* 41)	Cells l^{-1}	
		Mean	Range
Diaphanoeca pedicellata	37	1.0×10^4	7×10^1–6.4×10^4
Diaphanoeca multiannulata	40	2.8×10^3	6×10^1–2.1×10^4
Crinolina aperta	24	6.8×10^3	6×10^1–7.1×10^4
Cosmoeca ventricosa	27	4.7×10^2	6×10^1–1.9×10^3
Cosmoeca takahashii[*]	20	9.2×10^2	7×10^1–4.2×10^3
Parvicorbicula socialis	37	1.8×10^4	6×10^1–1.9×10^5
Parvicorbicula quadricostata	34	1.8×10^3	6×10^1–1.1×10^4
Bicosta spinifera	25	6.7×10^2	6×10^1–2.9×10^3
B. antennigera	10	5.6×10^2	5×10^1–2.0×10^3
Calliacantha natans	36	1.0×10^3	6×10^1–1.2×10^4
Calliacantha simplex	29	5.0×10^2	7×10^1–2.6×10^3

Occurrence refers to the number of samples (total = 41) in which a particular species was recognised. *Cosmoeca takahashii*[*] (Thomsen *et al.*, 1990) was originally recorded as *Pleurasiga* sp. (*sensu* Takahashi, 1981).
Data abridged from Buck and Garrison (1988).

grazing effort of the entire HNF community (Leakey *et al.*, 1996).

The speciation and productivity of autotrophic and heterotrophic nanoflagellates across fronts between neighbouring water bodies with different physicochemical characteristics has attracted considerable attention. Mention is made elsewhere of the author's sampling of the water currents around Iceland (see Section 9.5). In the Southern Hemisphere, the Antarctic Convergence (Polar Front) is the boundary between cold, northward-flowing Antarctic water and warm southward-flowing sub-Antarctic water. A 735 km north-west to south-east transect across the Antarctic Convergence in the vicinity of South Georgia was sampled in 1996 by Leakey *et al.* (2002) (see transect 1 in Fig. 9.13). Table 9.7 (Southern Ocean) includes the list of acanthoecid species encountered along this transect, which is entirely characteristic of the open ocean. The choanoflagellate community was dominated by *Bicosta spinifera*, *Calliacantha* spp., *Diaphanoeca pedicellata* and *Parvicorbicula socialis* with respective abundance maxima of 1.6, 3.2, 0.8 and 2.7 $\times$ 10^5 cells l^{-1}. *Calliacantha* spp. exhibited a clear north to south decline in abundance with *Bicosta spinifera* being more evenly distributed. *Diaphanoeca pedicellata* and *Parvicorbicula socialis* abundance was

variable between stations, which in the case of *P. socialis* reflected the patchy distribution of relatively large colonies. The mean choanoflagellate abundance and biomass for each station, averaged over the surface 70 m, was more than twofold higher in the warmer water north of the Convergence (1.78 $\times$ 10^5 cells l^{-1} and 0.78 µg C l^{-1}, respectively) than in the colder water to the south (0.74 $\times$ 10^5 cells l^{-1} and 0.23 µg C l^{-1}, respectively) (Leakey *et al.*, 2002).

NORTHERN HEMISPHERE: OCEANIC POPULATIONS
The Northern Hemisphere has also been well served with quantitative ecological studies on marine choanoflagellates. Booth *et al.* (1982) recorded choanoflagellate abundances in surface water from the Gulf of Alaska for May and June 1978 and obtained maxima of 5.4 $\times$ 10^4 cells l^{-1} for *Parvicorbicula socialis*, 5.4 $\times$ 10^4 cells l^{-1} for *Pleurasiga reynoldsii*, 19.1 $\times$ 10^4 cells l^{-1} for *Calliacantha simplex* and 14.3 $\times$ 10^4 cells l^{-1} for *Bicosta spinifera*. These values are slightly higher than the mean values recorded by Buck and Garrison (1988) for the Weddell Sea (Table 9.10). Sukhanova (2001) recorded a maximum of 3.5 $\times$ 10^6 cells l^{-1} for *Calliacantha natans* on the south-eastern Bering Sea shelf, south of St. George Island. This intensive development of *C. natans* was related to the termination of a planktonic

Table 9.11 *Range of abundance, biomass, grazing rates and grazing impact on bacterial biomass and production, of a natural population of planktonic choanoflagellates at 5 m depth, one mile offshore from the Australian Antarctic station at Davis between 30 December 1993 and 11 February 1994.*

	Mean abundance ($\times 10^5$ cells l^{-1})	Mean biomass (μg C l^{-1})	Clearance rate (nl cell^{-1} h^{-1})	Volume-specific clearance (10^4 BV cell^{-1} h^{-1})	Ingestion rate (FLB + bacteria cell^{-1} h^{-1})	Bacterial biomass grazed (μg C cell^{-1} d^{-1})	Bacterial biomass grazed (%)	Bacterial production grazed (%)	Grazing effort (%)
Small choanoflagellates (mean cell volume 16.9 μm^3)	0.3–11.8	0.1–5.0	0.72–6.85	3.72–40.49	0.21–5.53	0.15–1.52	0.5–3.8	1.3–10.6	4–44
Large choanoflagellates (mean cell volume 34.9 μm^3)	0.1–4.1	0.1–3.3	3.05–9.32	7.87–30.54	1.03–8.28	0.07–0.88	0.2–4.9	0.7–8.7	2–40

Salinity 32.6–33.7‰; temperature –1.4 to –0.4 °C.

Abridged from Leakey *et al.* (1996).

bloom of *Phaeocystis pouchetii*, a colonial haptophycean alga (Sukhanova and Flint, 2001). Degradation of the gelatinous matrix of *P. pouchetii* resulted in the formation of micro-cosms with high concentrations of dissolved and suspended organic matter favourable for the development of bacteria and heterotrophic nanoplankton. Within these microcosms bacterial foodwebs developed that proved to be an ideal trophic niche for *Calliacantha natans*.

In a study of heterotrophic protists in the Central Arctic Ocean, choanoflagellates comprised about 4% of numerical abundance with heterotrophic dinoflagellates accounting for 16% and other flagellates 80% (Sherr *et al.*, 1997). Choanoflagellates were slightly more abundant in the Chukchi Sea, where they reached 9% of the HF popula-tion. At one station in the Nansen Basin they reached 55% of the flagellate assemblage and it was here that Sherr *et al.* (1997) were able to carry out growth experiments. Over the first two days of the experiment a doubling time of 1.7 days was recorded and over the next two days a doubling time of 1.3 days. In the Barents Sea, Rat'kova and Wassmann (2002) observed a choanoflagellate population dominated by *Parvicorbicula socialis* and *Desmarella moniliformis* during March and May 1998 and June/July 1999. A maximum choanoflagellate biomass of 7 μg C l^{-1} was observed in northern Arctic water in March 1998. This is equivalent to the values obtained by Leakey *et al.* (1996) for choano-flagellates in surface water at Davis, Antarctica in the austral summer 1993–4 (see Table 9.11).

McKenzie *et al.* (1997) sampled water from six bays around the coast of Newfoundland throughout a five-year period (1988–1992) to determine the distribution and abundance of loricate choanoflagellates. The three most abundant species recorded were *Parvicorbicula socialis*, *Bicosta spinfera* and *Calliacantha natans*. *Parvicorbicula socialis* was the most abundant of all species, reaching the highest count of 3.7 × 10^5 cells l^{-1} in Conception Bay in April 1989 (Fig. 9.8). The highest average abundance of *P. socialis* for all the bays sampled was 8 × 10^4 cells l^{-1} and this coincided with the spring diatom bloom. In the upper 100 m of the water column, *P. socialis* was capable of grazing 0.3% of the standing stock of bacteria each day in April and May, which was equivalent to 7.4% of the daily bacterial production in the water column. McKenzie *et al.* (1997) also investigated the vertical and diel distribution of *P. socialis* within the water column during a 24-hour period in mid April 1989 during the spring diatom bloom in Conception Bay. They found that during the day *P. socialis* colonies concentrated (1.6 × 10^5 cells l^{-1}) at 25 m depth,

equivalent to the chlorophyll *a* maximum for the water column. Nearer the surface, at 5 m depth, the *P. socialis* concentration was 7 × 10^4 cells l^{-1}. Samples collected at night showed a drastic decrease in the population of *P. socialis*, particularly at 25 m. The authors put this down to predation by *Calanus hyperboreus* and *Metridia lucens* (Crustacea: Copepoda), which were common zooplankton grazers in the water column at night during the same observation period. *Bicosta spinifera* and *Calliacantha natans* were also abundant in five of the bays sampled. *Bicosta spinifera* occurred from April to November, with concentrations ranging from 1.6 × 10^3 cells l^{-1} in May to 21 × 10^3 cells l^{-1} in August. *Calliacantha natans* was present from March to September with concentrations reaching over 12 × 10^3 cells l^{-1} in April and August. The less common acanthoecids, including *Bicosta minor*, *Callia-cantha longicaudata*, *C. simplex*, *Parvicorbicula quadricostata* and *Stephanoeca pedicellata*, were usually found in the summer months. In addition to the peak abundance dom-inated by *P. socialis* at the time of the spring diatom bloom, a second peak occurred in July and August. During this peak, *P. socialis* cells were absent from the water column and instead the choanoflagellate community consisted of single cells, the most abundant of which were *Bicosta spinifera* and *Calliacantha natans*.

NORTHERN HEMISPHERE: MIXED INSHORE POPULATIONS

The choanoflagellate species list for Southampton Water (south coast of England), comprising a mixture of inshore and offshore taxa, has already been mentioned (Tong, 1997a) (see Section 9.4.4, Table 9.7). At this location acanthoecid and craspedid choanoflagellates were present throughout the year with abundance maxima occurring in spring and summer (Brandt and Sleigh, 2000). Acanthoecid species abundance ranged from 2.3 to 32.5 × 10^4 cells l^{-1} and on average accounted numerically for 7% of the HF population. Craspedids were usually present in lower concentrations, ranging from 1.2 to 14.4 × 10^4 cells l^{-1} and on average accounted for 3% of the HF population. Tong (1997a) categorised acanthoecid species into four groups on the basis of their occurrence. Group 1 included those species that were present throughout the year, such as *Acanthocorbis apoda*, *Bicosta minor*, *Calliacantha natans*, *C. multispina*, *C. simplex*, *Cosmoeca norvegica*, *Crucispina cruciformis*, *Polyfibula sphyrelata*, *Parvicorbicula superpositus*. Group 2 included species that were rarely seen, such as *Calliacantha longicaudata*, *Diaphanoeca pedicellata*,

Parvicorbicula socialis and *P. quadricostata*. Group 3 included species that were sporadically present, for example *Bicosta spinifera* (usually in large numbers), *Pleurasiga reynoldsii* and *Diaphanoeca grandis*. Group 4 included species with a seasonal occurrence, such as *Parvicorbicula manubriata* and *Diaphanoeca undulata*. However, Southampton Water is a tidal estuary with a complex mixture of freshwater discharges from three rivers and seawater ingress in the form of 'double tides' from the Solent, which is confluent with the English Channel. This estuarine complexity, together with a variety of coastal habitats including extensive saltmarshes and mudflats, probably explains the heterogeneous mixture of species encountered.

A heterogeneous mixture of choanoflagellates was also sampled by Andersen and Sørensen (1986) in Limfjord, a shallow eutrophic sound in northern Jutland, Denmark. This sampling was part of a study to investigate the population dynamics and trophic coupling of pelagic microorganisms during the period March–November 1983. Eight choanoflagellate species were recognised, including six that were planktonic, namely *Monosiga* (two unnamed species), *Diaphanoeca sphaerica*, *Pleurasiga* sp., *Calliacantha* sp., *Bicosta* sp., and two that were 'benthic', namely *Diaphanoeca grandis* and *Stephanoeca* sp. During the study period the bacterial concentration ranged from 0.5 to 15.2 × 10^6 bacteria ml^{-1} (mean 6.3 × 10^6 bacteria ml^{-1}) and maximum concentrations of individual acanthoecid species ranged from 8 to 15 × 10^5 cells l^{-1}. During winter, early spring and autumn the populations of bacteria and HNF, dominated by large loricate choanoflagellates, remained stable. From May and throughout summer the populations showed successive fluctuations in size. A total of eight bacterial and seven HNF and ciliate concentration peaks were observed. Bacterial peaks were followed by HNF peaks within 3–8 days and the peaks of nanoflagellates were followed by ciliate peaks within 4–6 days. The successive peaks were of similar duration for the three respective trophic groups (range 7–23 days). The populations of planktonic HNF during these oscillations were dominated by *Monosiga* sp. (maximum concentration 8.2 × 10^6 cells l^{-1}), *Paraphysomonas* sp. (Ochromonadales) and *Pseudobodo tremulans* (bicosoecid). Andersen and Sørensen (1986) estimated that during these peaks *Monosiga* sp. must have cleared between 1.4 and 1.9 nl h^{-1} to obtain growth rates of 0.19 and 0.26 day^{-1}, respectively, equivalent to doubling times of 2.6 and 3.6 days. These estimations of clearance rates are in accordance with values obtained in the laboratory using clonal cultures (see Table 9.2).

9.4.6 Colonisation of particulate organic matter by marine choanoflagellates

The term particulate organic matter (POM) in the context of aquatic environments refers to aggregates of organic matter that may be of biotic or abiotic origin. An important feature of these particles is that they become densely colonised by microorganisms, which have key functions in their formation and decomposition (Simon *et al.*, 2002). Aggregates are subdivided according to size into microaggregates (<5 to 500 μm) and macroaggregates (>500 μm); the latter are usually known as marine or lake snow. Each particle is a unique 'hotspot' of microbial activity (see Simon *et al.*, 2002 for references).

In marine environments choanoflagellates may settle onto the surface of aggregates or they may become enmeshed within micropatches of detritus. In inshore waters, including saltmarshes, mudflats and estuaries, *Acanthoeca spectabilis*, *Savillea parva*, *Diaphanoeca grandis* and species of *Acanthocorbis* and *Stephanoeca* are often associated as epibionts on the surfaces of detritus particles (Patterson *et al.*, 1993). Enmeshment of choanoflagellates, particularly long-spined *Bicosta* and *Calliacantha* cells, within aggregates often occurs in association with the entrapment of planktonic diatoms or *Phaeocystis* colonies at the termination of their respective blooms. Association of planktonic choanoflagellates with marine snow may serve to speed their removal from the water column. Aggregates may sink at velocities of 25–200 m d^{-1} depending on their size, morphology and composition (Alldredge and Gotschalk, 1988). Velocities such as these are several orders of magnitude faster than the 0.6 μm s^{-1} (51.8 mm d^{-1}) sinking velocity of individual loricate cells of *Diaphanoeca grandis* (Andersen 1988/9). Attachment of choanoflagellates to sinking particles could explain their occurrence and proliferation throughout the entire water column (Patterson *et al.*, 1993). Nitsche *et al.* (2007) currently hold the record for collecting living choanoflagellates from the greatest depths. *Lagenoeca antarctica* was retrieved from a depth of 2551 m in the Weddell Sea, Antarctica. *Salpingoeca abyssalis* and a species of *Monosiga* were retrieved from a depth of 5038 m in the Cape Basin and between 5034 and 5084 m in the Guinea Basin of the South Atlantic.

9.4.7 Choanoflagellates from hypoxic environments

Despite choanoflagellates being aerobic and mitochondriate, there are several records of marine species occurring in hypoxic and sulphidic environments. For example, the central part of Mariager Fjord in Jutland, Denmark is approximately 40 m deep and in summer the water column is anoxic beneath 16 m. Fenchel *et al.* (1995) retrieved *Diaphanoeca grandis* cells down to a depth of 24 m in this fjord and *Acanthocorbis* sp. at about 16 m. It seems unlikely that either of these species could survive indefinitely in the absence of oxygen; the most likely explanation for their existence at these depths is that they were sedimenting through the water column, possibly associated with particulate organic matter. Neither species survived culture in the laboratory under anaerobic conditions (Fenchel *et al.*, 1995). Similarly, Marchant *et al.* (1987) recovered loricate choanoflagellates, including *Acanthoeca brevipoda*, *Acanthocorbis unguiculata*, *Diaphanoeca grandis*, *Stephanoeca complexa*, *S. norrisii* and *S. diplocostata*, from the oxic/anoxic boundary in Ellis Fjord near Davis, Antarctica (see Fig. 9.12). Samples, when collected, had a strong smell of sulphide. Wylezich *et al.* (2012) recovered two new *Codosiga* species, *C. balthica* and *C. minima*, from oxygen-depleted waters in the Gotland and Landsort Deeps of the Baltic Sea. Both species lacked the flattened mitochondrial cristae associated with choanoflagellates and instead possessed tubular or saccular cristae. Several recent metagenomic studies involving analysis of small subunit ribosomal RNA (18S rRNA) sequences in collected water samples have retrieved choanoflagellate phylotypes from marine anoxic environments (for references see Wylezich and Jürgens, 2011). These include collections from the oxic/anoxic boundary of the hypersaline L'Atalante Basin in the eastern Mediterranean (Alexander *et al.*, 2009), the Gotland Deep Basin in the Baltic Sea (Stock *et al.*, 2009) and the O$_2$/H$_2$S gradient in Framvaren Fjord, southwestern Norway (Behnke *et al.*, 2010). Misiak *et al.* (2008) found that cell death in cultures of *Diaphanoeca grandis* was 'considerably increased' at concentrations of H$_2$S over 0.1% and that 65% of cells could survive 24 hours of anoxia.

9.4.8 Salinity tolerance of marine choanoflagellates

The conventional wisdom regarding choanoflagellate biogeography has been that craspedid choanoflagellates are found in freshwater, brackish and marine environments, whereas acanthoecid species are exclusive to marine and brackish water habitats. While it is correct that the majority of recorded loricate choanoflagellates are in marine or brackish water, at least two recent findings have challenged this exclusivity. Many loricate species show considerable tolerance of elevated and reduced salinities and two species have recently been found in freshwater habitats (Paul, 2011; Nitsche, 2014).

With respect to salinities elevated above those normally recorded for the open ocean (33–38‰), Thomsen (1978) observed ten acanthoecid and two craspedid species in the Red Sea with a salinity of approximately 40‰. Buck (1981) found seven common planktonic acanthoecids within interstitial brine channels in the ice sheet of the Weddell Sea, where salinity was approximately 44‰. Even more surprising, Hoff and Franzmann (1986) collected *Acanthocorbis unguiculata* from water with a salinity of 168.5‰ and a temperature of –6.8 °C in Organic Lake in the Vestfold Hills near Davis, Antarctica. At the time of sampling in October 1984, the lake, which is not confluent with the sea, had been isolated from all other environments by ice cover for a period of seven months. *Acanthocorbis unguiculata* was present at a concentration of 5×10^5 cells l^{-1} and appeared to be indigenous within the lake.

With respect to reduced salinities, many species can survive in brackish water with salinity as low as 5‰, as experienced in the Baltic Sea near the Tvärminne peninsula, Finland (Thomsen, 1979; Vørs, 1992; Kuuppo, 1994). In the Gulf of Bothnia, a northern extension of the Baltic Sea, the salinity of the water column at 6 m below the surface covering of ice in March was 3–3.5‰. Nine loricate species, including *Stephanoeca diplocostata*, *S. urnula*, *Savillea (micropora) parva*, *Monocosta fennica* and *Diaphanoeca sphaerica*, were observed (Ikävalko, 1998). Samuelsson *et al.* (2006), also sampling the Gulf of Bothnia, observed that in spring *Diaphanoeca grandis* dominated the northern sector while *Calliacantha natans* and *C. simplex* dominated in the south. In autumn *Cosmoeca* sp. and *Acanthocorbis* sp. dominated in the north and *Stephanoeca* sp. and *Acanthocorbis* spp. in the south. Brine channels within the surface covering of ice during winter had higher salinities (10–30‰) than those recorded in the water column due to the salt concentrating process that takes place in these lacunae. These channels developed their own microbial foodwebs and craspedid and acanthoecid choanoflagellates were abundant, although the species present were similar to those found in the water column (Ikävalko and Thomsen, 1997).

There are a number of published records of loricate choanoflagellates having been collected from inland brackish and saline lakes hundreds of miles from the nearest marine location. Two acanthoecids, *Savillea parva* and a species of *Stephanoeca*, were observed and photographed in water samples collected from an inland brackish lake near Sol-Iletsk in Orenburg Oblast, Russia (Figs 9.15, 9.16).

The most remarkable observation to date is the finding of a loricate choanoflagellate in a freshwater oligo-mesotrophic lake in Mongolia (Paul, 2011). The lake is Bayan Nuur, which is located in the Uvs Nuur Basin, north-west Mongolia. Water salinity was 0.35‰ and the pH was 8.9. The loricate species was found, coincidentally, on 28 August 1987 and 1998, at concentrations of 70 and 180×10^3 cells l^{-1}, respectively. Only one species was present and on the basis of lorica construction it was attributed to *Acanthocorbis* as a new species, *A. mongolica* (Fig. 9.14) (Paul, 2011). Recently Nitsche (2014) reported finding a population of *Stephanoeca* cells in a freshwater lake in Samoa. Although the lorica resembled that of *S. diplocostata*, since this species could survive freshwater conditions it was given a new specific name, *Stephanoeca arndtii*.

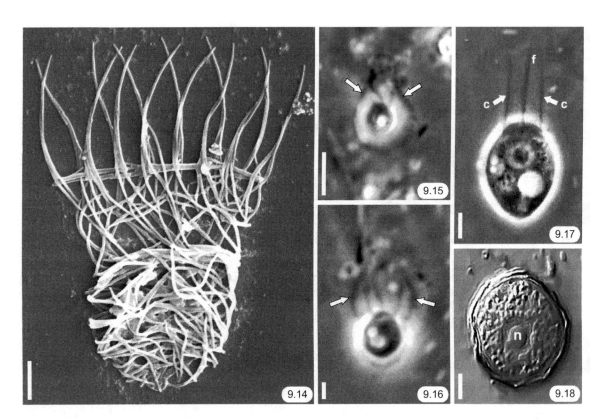

Plate 3 (Figures 9.14–9.18)

Figs 9.14–9.18 Choanoflagellates observed in unusual locations. Bars = 2 µm. **Fig. 9.14** *Acanthocorbis mongolica*. An acanthoecid species found in Bayan Nuur, a freshwater lake in Mongolia. Reproduced from Paul (2011).
Figs 9.15–9.16 Two acanthoecids observed in an inland brackish lake near Sol-Iletsk, Orenburg Oblast, Russia. **Fig. 9.15** *Savillea parva*. **Fig. 9.16** *Stephanoeca* sp. Figures kindly provided by Plotnikov A. O., Selivanova E. A. and Nemtseva N. V. (Institute of Cellular and Intracellular Symbiosis of the Urals Branch, Russian Academy of Sciences, Orenburg, Russia).
Figs 9.17 and 9.18 *Codosiga botrytis* retrieved in culture from permafrost sample. Reproduced from Stoupin *et al.* (2012); Daniel Stoupin and Áron Kiss kindly provided the original images. **Fig. 9.17** Single flagellated cell. **Fig. 9.18** Thick-walled cyst.

9.4.9 Grazing of marine choanoflagellates by microzooplankton

A matter that has exercised microbial ecologists for some time is whether the microbial loop is a closed system, and therefore serves as a 'sink', or whether it interacts with higher trophic level foodwebs. Several authors have investigated the faecal pellets of macrozooplankton for evidence of grazed acanthoecids. Since the Antarctic krill, *Euphausia superba*, is considered to be a major zooplankton grazer and an important link between primary producers and many of the top predators, including baleen whales, crabeater seals and penguins, the grazing habits of this species have been subject to considerable investigation. Tanoue and Hara (1986) and Marchant and Nash (1986) investigated the gut contents and faecal pellets of *Euphausia superba* and both reported the presence of separated costal strips, often in vast arrays, indicating the consumption of choanoflagellate prey. Urban *et al.* (1992, 1993) observed intact loricae of *P. socialis* in the faecal pellets of the pelagic tunicate *Oikopleura vanhoeffeni* in spring, and loricae of *Stephanoeca* sp. in autumn in the coastal water off Newfoundland. The extent to which loricate choanoflagellate populations are regulated by 'bottom-up' activities (availability of prey) or 'top-down' phenomena (grazing by predators) is still in need of quantification.

9.5 VARIABILITY IN LORICA SIZE OF *BICOSTA* AND *CALLIACANTHA* SPECIES

When Grøntved (1976) described *Salpingoeca natans*, his illustration showed cells with loricae that varied considerably in size (Fig. 9.11). Since the early records of *Bicosta* and *Calliacantha* were mostly based on samples collected from northerly latitudes, and Throndsen's (1970a) description of *Salpingoeca (Bicosta) spinifera* (length 50–80 μm) was based on specimens collected in the Barents Sea, there was a suggestion that cells with long-spined loricae might be characteristic of cold water currents (Manton *et al.*, 1980). Manton *et al.* (1980) listed measurements of lorica length for *Bicosta spinifera* collected from five different localities spanning a range of water temperatures from –1 to +10 °C and, at least superficially, there appeared to be a negative correlation between lorica length and temperature. However, as Manton *et al.* (1980) pointed out, disentangling possible genetic and environmental determinants is not always possible. For *Bicosta minor* and *B. antennigera* the distribution of lorica length measurements was unimodal (see Table 9.12 for

mean ± SD values). However, for *B. spinifera* the distribution is less obviously unimodal, which might suggest that this species possesses more than one biotope involving size on which environmental selection might act.

During a sea-going cruise around Iceland in 1994, the author collected samples from all the major surface-water currents in the vicinity of the island. Surface currents flow in a mainly clockwise direction. The warm saline North Atlantic Drift gives rise to the Irminger Current south of Iceland and this travels along the western and northwestern coasts until it meets the colder, less saline polar water of the East Greenland Current. This combination of water then flows along the eastern coast as the East Icelandic Current. Temperatures during May–June 1994 ranged from 7.6 °C in the North Atlantic Drift; 3–8 °C in the Irminger Current and <0–2 °C in the arctic/polar water of the East Greenland Current. Salinity was >34.9‰ in the Irminger Current and <34.9‰ in Arctic/Polar water. The principal acanthoecid species observed at the 91 stations sampled are listed in Table 9.7. *Bicosta spinifera*, *B. antennigera* and *B. minor* were observed at 60% or more of the stations sampled (Figs 9.19–9.21); *Calliacantha simplex* was observed at 20% of stations. Details of lorica length of these species are listed in Table 9.12 and for comparison equivalent details are given for these species from other (mostly cold water) locations. With the exception of *B. spinifera* from Osaka Bay, Japan, the range and mean values for individual species are similar irrespective of location. For the Icelandic material, with the exception of *B. antennigera*, there is at least a threefold difference in length between the shortest and longest loricae recorded (Table 9.12) (Figs 9.19–9.21). A similar difference also applies to values recorded for *B. minor* and *B. spinifera* by Manton *et al.* (1980) and for *Calliacantha simplex* and *C. natans* by Thomsen *et al.* (1990) (see Table 9.12). Differences of this magnitude contrast with the relatively small variations normally observed in the majority of other species. For example, the largest lorica of *Cosmoeca phuketensis* is only 1.3 times the length of the smallest and for *C. norvegica* and *C. ventricosa* the difference is a factor of 1.4 (Thomsen and Boonruang, 1984). There is a twofold difference for *Crucispina cruciformis* (Hara and Takahashi, 1987a) and a 2.4 times difference for *Diaphanoeca multiannulata* (Thomsen *et al.*, 1990). In *Bicosta minor* and *B. antennigera* the increase in overall lorica length involves a proportional increase in the length of the anterior and posterior spines as well as the chamber. In *B. spinifera* an overall increase in lorica length involves a

Table 9.12 *Lorica length (range and mean ± SD) of species of* Bicosta *and* Calliacantha *collected from various locations.*

		Bicosta minor	Bicosta antennigera	Bicosta spinifera	Calliacantha simplex	Calliacantha natans
Icelandic waters	Range (μm)	12–46	37–85	27–104	17–59	22–86
(author, unpublished	Mean ± SD (μm)	25.6 ± 5.7	61.6 ± 8.5	64.7 ± 15.1	33.1 ± 9.4	46.6 ± 11.6
data)	(*n*)	(500)	(277)	(243)	(146)	(146)
Combined data set	Range (μm)	11–42	32–80	39–118	–	–
Manton *et al.* (1980)	Mean ± SD (μm)	23.1 ± 6.1	56.4 ± 10.2	77.5 ± 16.7		
	(*n*)	(74)	(60)	(182)		
Scotia and Weddell	Range (μm)	–	–	70–100[*]	18–69	20–64
Seas, Antarctica	Mean ± SD (μm)			85.2 ± 11.1	39.8 ± 13.6	39.2 ± 13.6
Thomsen *et al.* (1990)	(*n*)			(400)	(46)	(31)
Thomsen and Larsen						
(1992)[*]						
Osaka Bay, Japan	Range (μm)	26–41	–	57–66	–	–
Hara and Takahashi	Mean (μm)	31		61		
(1987a)						

The Icelandic data were acquired by the author during a cruise around the island in 1994 and include specimens collected from all the major surrounding water masses. The combined data set from Manton *et al.* (1980) includes specimens collected from Greenland, north and south Alaska, Point Barrow and Resolute Bay, Cornwallis Island, northern Canada. With respect to specimens from the Weddell Sea, *Calliacantha simplex* and *C. natans* were collected during the AMERIEZ 86 cruise (Thomsen *et al.*, 1990) and *Bicosta spinifera* during the 1988 EPOS (leg 2) cruise to the Scotia and Weddell Seas (Thomsen and Larsen, 1992) (see also Figs 9.19–9.21).
[*] indicates that data for *Bicosta spinifera* from the Scotia and Weddell Seas are taken from Thomsen and Larsen (1992).

proportionally greater length increase of the posterior spine than the anterior spines. In *Calliacantha natans* increase in lorica length involves a proportionate increase in the length of the spines and chamber, whereas in *C. simplex* the increase in the length of the posterior and anterior spines is proportional and the chamber is only marginally longer (Thomsen *et al.*, 1990). With respect to the relationship between lorica length and water temperature, for the Icelandic material a relatively weak inverse correlation ($P < 0.01$) was obtained for *Calliacantha natans* and *C. simplex* but no significant correlation for the three *Bicosta* species. However, with relatively few exceptions there is a strong positive correlation in lorica length between individual species (see Table 9.13). This does suggest that either there is a common causal explanation for variation in lorica length or a selective pressure with a common action.

Thomsen and Larsen (1992) encountered a range of lorica lengths for *Bicosta spinifera* along a 450 km

north–south transect from the Scotia Sea to the Weddell Sea (see transect 2 in Fig. 9.13). The surface-water temperature along this transect slowly decreased with increasing latitude from 1.7 °C in the Scotia Sea to 1.0 °C at the beginning of the confluence of the two seas. Across the confluence the temperature decreased from 0.2 to –0.1 °C and in the Weddell Sea the temperature decreased from –0.9 to –1.6 °C. Based on a large number of cells ($n > 900$), Thomsen and Larsen (1992) distinguished two lorica sizes: (1) small size with lorica length 5–25 μm and (2) large size with lorica length 75–95 μm. In the Scotia Sea there was a preponderance of cells with large loricae. In the confluence between the two seas there was a preponderance of cells with small loricae and in the Weddell Sea the *B. spinifera* population consisted of a mixture of the two class sizes. The overall pattern was distinctly bimodal. As discussed in Chapter 7 (Section 7.6), the intracellular deposition of long spines by species of *Bicosta* involves the development

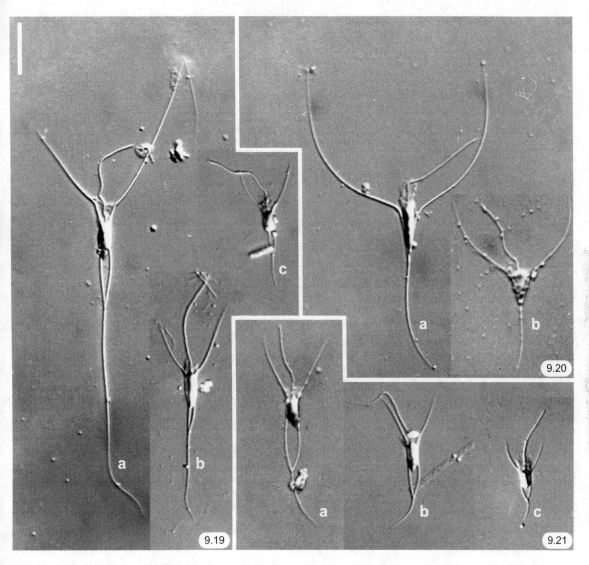

Plate 4 (Figures 9.19–9.21)

Figs 9.19–9.21 Light microscopical images to illustrate the range of lorica size (length) encountered in three species of *Bicosta* collected from seawater around the coast of Iceland. Bars = 10 μm. Fig. **9.19a–c** *Bicosta spinifera*. a. 91 μm; b. 45 μm; c. 29 μm.
Fig. **9.20a–b** *Bicosta antennigera*. a. 63 μm; b. 35 μm.
Fig. **9.21a–c** *Bicosta minor*. a. 39 μm; b. 26 μm; c. 16 μm.

of a long posterior protuberance of the cell (Figs 7.61, 7.63–7.65). The existence of a protuberance is an indicator that the cell is nearing the end of interphase and preparing for division. Thomsen and Larsen (1992) observed 'small' *B. spinifera* cells with long posterior protuberances and from this appearance they concluded that the overall variation in length could not be controlled directly by temperature. Based on this observation Thomsen and Larsen (1992) suggested that *B. spinifera* may have a 'polymorphic' life cycle and suggested that there might be a sexual stage, although they had no direct evidence to support this tentative conclusion (see also Thomsen *et al.*, 1997).

Table 9.13 *Bivariate correlation coefficients (Pearson's* r*) for lorica length between species of* Bicosta *and* Calliacantha.

	Bicosta spinifera	Bicosta antennigera	Bicosta minor	Calliacantha natans
Bicosta antennigera	0.736*** (12)			
Bicosta minor	0.593*** (28)	0.304 NS (12)		
Calliacantha natans	0.832*** (28)	0.874*** (12)	0.504** (28)	
Calliacantha simplex	0.785*** (22)	0.810*** (12)	0.501* (22)	0.838*** (22)

Probability levels of significant correlations:
* $P < 0.05$,
** $P < 0.01$,
*** $P < 0.001$;
NS – not significant. (*n*) number of samples for which average values of length are available.
Data from Leadbeater (unpublished data).

9.6 FUNCTIONAL ROLE OF CHOANOFLAGELLATE LORICAE

As the foregoing chapters testify, much is now known about the morphology, construction and assembly of the choano-flagellate lorica. However, apart from Andersen's (1988/9) study on *Diaphanoeca grandis*, relatively little is known about its functional role. Despite this serious shortcoming, it is worth reflecting on how morphological and ecological information might be combined to provide some insights into the possible functional role of the lorica. Beginning with morphology, in the majority of acanthoecid species the lorica surrounds and sometimes completely encloses the cell. The biogenic silica component of costal strips contributes sufficient rigidity to provide a skeletal support for the lorica but there is also a degree of flexibility which is probably a function of the organic matrix within individual costal strips (Gong *et al.*, 2010). As discussed and illustrated in Chapter 4, although the underlying principles (rules) governing lorica construction are universal for all species, nevertheless, within the confines of these rules, the number, thickness and morphology of costal strips and the pattern of costae vary for each morphospecies. The positioning of open spaces, either between costae or as a result of the posterior and/or anterior rings, directs the flow of water into and out of the lorica chamber. In addition, most loricae possess an organic investment which usually surrounds the cell, but in some species, such as *Diaphanoeca grandis*, *D. multiannulata*, *Crinolina aperta* and *Didymoeca costata*, covers part of the inner surface of the lorica at some distance from the cell. A function of this investment (the veil) must be to enhance the development of a sub-ambient pressure within the lorica chamber and to funnel the inflow of water through the bottom and outflow through the top of the lorica. Video recordings of *Diaphanoeca grandis* cells supplied with latex microspheres show particles accelerating as they approach the base of the lorica. They then become entrained in the water flow as it moves swiftly from the base to anterior of the lorica chamber and the particles become trapped on the outer surface of the collar microvilli. In *Didymoeca costata* the location of the spaces posterior to the organic veil directs the inflow of water and prey particles onto the bottom of the collar, where they can be entrapped and ingested; effluent water passes out through the anterior ring. The various combinations of micro- and macro-morphological features of lorica construction must serve to adapt each species to a specific ecological microniche.

An important hydrodynamic function of the lorica is to create drag which counteracts the locomotory effects of flagellar movement (Andersen, 1988/9). In some instances this may serve to reduce motility, for example in colonies of *Parvicorbicula socialis* (Fig. 9.8), but in many species it completely counteracts locomotory forces. This has the dual benefit of enhancing flagellar activity for the creation of feeding currents (see Chapter 2) and also for overcoming the necessity for a cell to be attached to a substratum. For loricate species this opens up the possibility of a cell being freely suspended within the water column, which not only benefits planktonic species but also those associated with the benthos. Prolonged periods of suspension are essential for oceanic species that are integral to pelagic microbial foodwebs. Loricate species that inhabit the most productive surface waters are principally drawn from the *Diapha-noeca/Crinolina*, *Cosmoeca/Parvicorbicula* and *Bicosta/Calliacantha* themes. What these species have in common is relatively 'lightweight' loricae containing an optimum number of thin, lightly silicifed costae. The loricae of *Crinolina aperta* and *Diaphanoeca multiannulata* are voluminous, tubular in shape and possess a veil so the undulating flagellum serves as a pump effecting an enhanced movement of water through the lorica (Andersen, 1988/9). This is highly advantageous for cells suspended in water containing low prey densities.

Loricate choanoflagellates, in common with many other non-motile planktonic species, particularly diatoms, display a number of morphological modifications that are likely to reduce their intrinsic sinking velocity. In particular, pelagic species of *Bicosta* and *Calliacantha* and *Crucispina cruciformis* possess spines that are a common means of creating 'form resistance' for the reduction of sinking velocity (Figs 9.19–9.21). The effectiveness of long spines in reducing sinking velocity was elegantly demonstrated by Walsby and Xypolyta (1977) for the marine diatom *Thalassiosira fluviatilis*. This diatom has long (60–80 μm) and slender chitin fibres that project from the margins and centre of each silica valve (see Fig. 9.22). Their removal by treatment with chitinase resulted in cells sinking more that 1.7 times faster than cells with fibres. It is likely that the spines of *Bicosta* and *Calliacantha* will have an analogous effect on the sinking velocity of these species. This is in spite of the spines containing silica, which is considerably denser than water.

9.7 ECOLOGY OF FRESHWATER CHOANOFLAGELLATES

9.7.1 Choanoflagellates and freshwater lakes

Hartmut Arndt and colleagues have been at the forefront of work on freshwater choanoflagellates (Arndt *et al.*, 2000). Arndt's work was initially based on lakes, particularly those of Germany and Austria – Müggelsee, a lake in the eastern suburbs of Berlin, 55 north German lakes (Auer and Arndt, 2001; Auer *et al.*, 2004), Neumühler See (Mathes and Arndt, 1994) and Mondsee (Salbrechter and Arndt, 1994) – but then extended to rivers, particularly the River Rhine (Weitere and Arndt, 2002a, b, 2003; Scherwass *et al.*, 2010). Table 9.14 summarises the salient data from lakes that have been studied where choanoflagellates have been recorded. Since most studies have primarily concentrated on the pelagic environment, the overall pattern is mixed and not entirely reliable. At any one time, pelagic

choanoflagellates comprise single-celled species, colonies (see Section 3.5.4) and sedentary stages that happen to be present on organic particles (lake snow) or as epibionts on suspended algae and zooplankton. Thus, in addition to the bottom-up and top-down factors that would normally be expected to regulate choanoflagellate abundance there are other extraneous factors of importance, including the presence of particulate organic matter and the species composition of planktonic algal populations (Vacqué and Pace, 1992). These extrinsic factors are to some extent related to the trophic status of the water body. Nevertheless, an overall pattern emerges of a peak of choanoflagellate cells in late spring or early summer followed by a second peak in autumn. The abundance of choanoflagellates as a percentage of heterotrophic nanoflagellates varies considerably. The average contribution is 3–50% (Table 9.14) but occasionally, as in Piburger See, Austrian Tyrol, the proportion may be as high as 75% (Pernthaler *et al.*, 1996). Peak choanoflagellate abundance in lakes may be as high as 10^5 or 10^6 cells l^{-1} (see Table 9.14), which is similar to that encountered in marine locations (Section 9.4.5). Grazing rates of between 2 and 73 bacteria cell^{-1} h^{-1} recorded for freshwater choanoflagellates are in line with U_{max} values recorded for choanoflagellate species in culture (see Table 9.3).

The epibiotic association of sedentary choanoflagellates with planktonic diatoms, particularly colonial species, is a universal phenomenon in freshwater lakes (Fig. 9.22) (Table 9.14). The most frequently reported diatoms to be colonised are *Asterionella formosa*, *Stephanodiscus hantzschii* and species of *Aulacoseira*, *Fragilaria*, *Thalassiosira* and *Tabellaria* (Table 9.14). Other planktonic algae frequently colonised include species of *Dinobryon*, the desmid *Staurastrum*, the cyanobacterium *Anabaena flos-aquae* and floating filamentous green algae such as *Cladophora* and *Conferva*. The most common epibiotic choanoflagellates are species of *Salpingoeca*, for example *S. amphoridium*, and *Codosiga*, including *C. botrytis* (Hibberd, 1975; Šimek *et al.*, 2004; Sonntag *et al.*, 2006). Other epibiotic protists that commonly accompany choanoflagellates are bicosoecids and stalked ciliates, such as *Vorticella* sp. (Carrias *et al.*, 1996, 1998; Strüder-Kypke, 1999; Sonntag *et al.*, 2006). It is not uncommon for epibiotic populations of choanoflagellates to be monospecific and occasionally they may account for 100% of epibiotic HF abundance (Šimek *et al.*, 2004; Sonntag *et al.*, 2006).

Several authors have recorded higher bacterial ingestion rates for attached cells in comparison with swimming cells,

Table 9.14 Choanoflagellates in freshwater lakes. Details of choanoflagellate seasonality; abundance as a percentage of HF; abundance $\times 10^4$ cells l^{-1}; grazing (ingestion) rate. Choanoflagellate taxa recorded and for epibiotic taxa details of host species.

Lake	Trophic status	High choano:HF ratio (seasonal occurrence)	Choano abundance (% of HF)	Abundance ($\times 10^4$ cells l^{-1})	Grazing (ingestion) rate/microspheres bacteria cell^{-1} h^{-1}	Choanoflagellate taxa	Hosts on which epibiotic choanoflagellates settle	Reference
55 lakes, north Germany	Mesotrophic Slightly eutrophic Highly eutrophic Hypereutrophic	Early summer (mesotrophic) Spring (eutrophic)	10% (mesotrophic) 5% (eutrophic)	1–690		Monosiga Codosiga Desmarella Lagenoeca Salpingoeca spp. S. amphoridium	Stephanodiscus hantzschii Aulacoseira sp.	Auer and Arndt (2001) Auer et al. (2004)
Neumühler See, north-east Germany	Mesotrophic–eutrophic	May/June (epibiotic cells)	50%* summer 8% (av)					Mathes and Arndt (1995)
Lake Konstanz, south-west Germany	Mesotrophic	Summer and autumn	14% (av) 25% max	4–25	2–12	Solitary and colonial forms	Fragilaria sp. Asterionella sp.	Cleven and Weisse (2001)
Římov Reservoir, Czech Republic	Mesotrophic Dimictic	June September	9.5–11% 3–15%	30–40 20–100	36–56 35–65	Salpingoeca amphoridium	Asterionella formosa Fragilaria crotonensis Staurastrum pingue	Šimek et al. (2004)
Mondsee, Austria	Mesotrophic		40% max 14% av surface 6% av below 6 m	0.61–23.3		Monosiga	Particles – lake snow	Salbrechter and Arndt (1994)
Piburger See, Austrian Tyrol	Oligo-mesotrophic	May	75% beginning 10% end					Pernthaler et al. (1996)
Lake Pavin, Puy-de-Dôme, France	Oligo-mesotrophic Meromictic Crater lake	Spring (free-living species) Autumn (epibiotic cells)	10% (epibiotic cells)		1.7–33.6		Aulacoseira italica Asterionella formosa Anabaena flos-aquae	Carrias et al. (1996, 1998)
Lake Oglethorpe, Georgia, USA	Eutrophic Monomictic	May/June September		3.3	8–42			Sanders et al. (1989)
Paul and Tuesday Lakes, Michigan, USA	Oligotrophic Dystrophic Meromictic	June/July (epibiotic cells)		20.4	13–30 (Paul Lake) 39–73 (Tuesday Lake)		Dinobryon sertularia D. divergens, diatoms, cyanobacteria	Vacqué and Pace (1992)
Lake Erie, USA	Oligo-mesotrophic	June	8–32%*		13–37	Monosiga spp. Codosiga spp.		Hwang and Heath (1997)

* Value based on cell volume and not cell weight

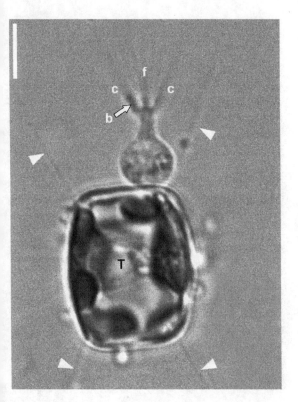

Fig. 9.22 *Freshwater* planktonic diatom, *Thalassiosira* sp. (T), with long, thin chitinous fibres (arrowheads) with attached epibiotic *Salpingoeca* cell. Collar (c), flagellum (f), bacterium on collar in process of being ingested by pseudopodium (arrow b). Bar = 5 μm. Figure kindly provided by William Gurske.

which is in accordance with the conventional wisdom that attachment is essential to achieve optimal feeding currents of water (see Section 2.3.3). Vacqué and Pace (1992) found that the ingestion rates of attached choanoflagellates in Lake Paul, Michigan, ranged from 13 to 30 bacteria cell^{-1} h^{-1} and in the nearby Tuesday Lake 39–73 bacteria cell^{-1} h^{-1}, whereas the corresponding rates for free-swimming forms were 0.24–5.21 bacteria cell^{-1} h^{-1} and 0.19–23.56 bacteria cell^{-1} h^{-1}, respectively. Similarly, in Lake Pavin, France, Carrias *et al.* (1996, 1998) also recorded higher ingestion rates for attached flagellates; 1.7–92.4 (mean 24.7 ± 20.2 SD) bacteria cell^{-1} h^{-1} compared with 1.6–20.2 (mean 8.6 ± 6.7 SD) bacteria cell^{-1} h^{-1}. Attachment of sedentary choanoflagellates to planktonic diatoms not only allows normally 'benthic' species to colonise the pelagic environment but also provides a refuge from grazing by filter-feeding rotifers, daphnids and cladoceran zooplankton (Šimek

et al., 2004). Ultimate decline of the choanoflagellate population comes from a decrease in the supply of food particles and/or sedimentation of moribund diatom colonies out of the water column.

9.7.2 Choanoflagellates in riverine environments

There is limited information on HF in general and choanoflagellates in particular with respect to riverine environments. The most intensive work to date is that of Weitere and Arndt (2002a, b, 2003) who undertook a 21-month study (May 1998 to January 2000) of planktonic HF in the lower River Rhine. They sampled the river at two locations, one at Cologne and the other at Kalkar-Grieth, 160 km downstream from Cologne. Their aim was to investigate microbial interactions within 'parcels' of water that took about 31 hours to pass between the two collecting sites. The mean bacterial abundance at both sites ranged from 0.3 to 3.5 × 10^6 bacteria ml^{-1}, with lowest abundances in winter and highest in late spring. In mid October 1998, when the water discharge in the Rhine was relatively low, the total HNF population was 20.1 × 10^4 cells l^{-1} of which choanoflagellates comprised 21% with an abundance of 4.22 × 10^4 cells l^{-1}.

A total of 20 choanoflagellate taxa were recognised, including species of *Monosiga*, *Codosiga*, *Salpingoeca*, *Lagenoeca*, *Stelexomonas*, *Desmarella* and *Proterospongia*. Choanoflagellates regularly appeared in high abundance, comprising 4–35% of the monthly mean HNF abundance in Cologne. Their highest contribution extended from May to July at both sampling sites. Species of *Monosiga* were present throughout the entire study period with an extended increase to 10–15% total HNF in summer and autumn. *Codosiga* spp. were present from March to September 1999 with a peak of approximately 10% total HNF in May. Species of *Desmarella* were recorded between June and September 1999, with a peak of approximately 10% total HNF in July.

9.7.3 Choanoflagellates in soils and permafrost

Microbial foodwebs, comprising bacteria and a variety of Protozoa, including ciliates, amoebae and HF, are essential to natural functioning of the soil environment. As in the planktonic environment, protozoa perform two important functions, namely the grazing of bacteria and the regeneration of nutrients. The equivalent of the microbial loop exists in soils except that the role of photosynthetic algae is replaced by exudates from plant root systems and decaying organic matter. Foissner (1991), in a compilation of soil HF, listed 14 choanoflagellates attributable to *Monosiga* (three species), *Salpingoeca* (eight species), *Lagenoeca globulosa*, *Codonosigopsis robini* and the ubiquitous *Codosiga botrytis*.

Many soil protozoa are capable of encystment during periods of drought or cold and this might explain why *Codosiga botrytis*, which is capable of encystment (Fisch, 1885; Stoupin *et al.*, 2012), is so common in soils (Fig. 9.18). In Arctic and sub-Antarctic regions it is not uncommon for protozoa and other soil microorganisms to become frozen into the permafrost, sometimes for long periods of time (Shatilovich *et al.*, 2009; Stoupin *et al.*, 2012). Living cells of *Codosiga botrytis* have been retrieved from permafrost cores 28 000–32 000 years old as determined by radiocarbon dating (Fig. 9.17) (Shatilovich *et al.*, 2009).

9.8 CONCLUSIONS

Choanoflagellates occupy a distinctive niche within aquatic microbial communities and foodwebs, which is reflected in their uniform functional morphology. They are suspension feeders and to function successfully they need to be anchored to a surface and held at sufficient distance from that surface to allow the free flow of fluid around the collar. They also require freely suspended individual prey particles within the medium. Their closest suspension-feeding HNF competitors are stalked chrysomonads, including interception feeding bicosoecids and species of *Paraphysomonas* and *Spumella*, and pedinellids, such as *Actinomonas*,

which like choanoflagellates possess a single flagellum surrounded by radiating filopodia (Sleigh, 1964). It is common for sedentary suspension feeding HNF to co-exist in freshwater and inshore marine habitats. Bicosoecids and choanoflagellates often share an epibiotic existence on freshwater planktonic diatoms even to the extent that they compete with each other for seasonal dominance (Carrias *et al.*, 1996, 1998).

The ecophysiological profile of choanoflagellates is also relatively consistent. From the kinetic data currently available, choanoflagellates behave in much the same way in culture as they do in the field. For example, kinetic parameters of feeding, including rates of clearance (F) and particle uptake (U), are in general accordance and vary similarly no matter whether they are obtained from culture or field experiments (Tables 9.2, 9.3, 9.11, 9.14). Cell abundance in the field with maxima of between 10^5 and 10^6 cells l^{-1} are achieved for marine species at high latitudes and freshwater species in lakes, reservoirs and rivers (Tables 9.10, 9.11, 9.14). Background concentrations of 10^3–10^4 cells l^{-1} are common. Choanoflagellate populations of 10^4–10^6 cells l^{-1}, comprising 3–56% of total HNF abundance, frequently occur in field samples containing 10^4–10^6 bacteria ml^{-1}. These figures are in accord with the general conclusion of Sanders *et al.* (1992) that there is an approximate numerical relationship between bacterioplankton and heterotrophic nanoplankton (c.1000 bacteria:1 HNF) from the euphotic zones of a variety of marine and freshwater systems. In this context it is essential to consider choanoflagellates as being only a part of the total bacterivorous nanoplankton.

10 · Choanoflagellate phylogeny: evolution of Metazoan multicellularity

Whence camest thou? and whither wilt thou go?
(Genesis 16: v8).

10.1 INTRODUCTION

It was a fortuitous coincidence that James-Clark (1867b) not only provided the first unequivocal description of a choanoflagellate but also, in the same publication, noted the striking morphological similarity between collared flagellates and sponge choanocytes. From that time onwards, in one context or another, choanoflagellates have been linked with the evolution of sponges and ultimately with the Metazoa (see Chapter 1). The nature of this possible relationship has been highly contested. James-Clark (1867b) and Kent (1880–2) argued that sponges were colonial infusoria (Protozoa). Others have viewed choanocytes as being homologous with choanoflagellate cells (Nielsen, 2008; Cavalier-Smith, 2013) and yet others have suggested that choanoflagellates may be simplified sponge-derived metazoans (Maldonado, 2004). The debate has been complicated by uncertainty surrounding the phylogeny of sponges; are they closely related to non-bilaterian animals, such as the Cnidaria, or do they comprise a separate evolutionary grouping of their own (Parazoa/Enantiozoa) independent of other multicellular animals (Sollas, 1884)? From morphological and ecological perspectives, choanoflagellates have been regarded as a distinctive group of heterotrophic nanoflagellates (see Section 1.8 and Chapter 9). Their morphological uniformity and distinctiveness – there are no obvious synapomorphies that unite choanoflagellates with any other protistan taxon – set them apart from all other protists.

For a century after the publication of Kent's (1880–2) *A Manual of the Infusoria* progress in understanding choanoflagellate systematics and phylogeny stagnated. Morphology alone, as determined by light microscopy, provided few insights beyond species identification and aspects of cell biology. In retrospect, the most significant insight was recognition of an unlikely alliance between choanoflagellates, chytrids (Chytridiomycota, Fungi) and animals (Metazoa) on account of the morphology and swimming behaviour of their motile cells (Vischer, 1945; Gams, 1947). Shared features of motile choanoflagellates, chytrid zoospores and animal spermatozoids include a radially symmetrical cell with a single 'posterior' flagellum. Movement through a fluid involves the cell body being propelled from behind by the flagellum acting as a 'pulsellum' (see Section 2.3.2). Vischer (1945) coined the term 'opistokonten' (Greek = posterior pole) to describe this type of cell morphology and flagellar behaviour and he used this character to group the choanoflagellates, chytrids and some uniflagellate algal monads within the Opistokontae (Pulselloflagellatae). This grouping was soon expanded by Gams (1947) to also include the Metazoa within an 'opistokont series'. Copeland (1956) subsequently purloined the name Opisthokonta (with changed spelling) but restricted it to the chytrids to which he gave phylum status. He dismissed Gams' (1947) opistokont series with the remark that "this interesting hypothesis must as yet be treated as far-fetched" (Copeland, 1956, p. 111). The irony of this comment only became apparent 30 years later when Cavalier-Smith (1987) resurrected the term Opisthokonta for a clade including choanoflagellates, animals and fungi. Cavalier-Smith (1998, p. 213) has intentionally retained Opisthokonta as a clade rather than a taxon for the reason that opisthokonts are "far too phenotypically diverse to be useful as a major unit of eukaryote classification". However, Patterson (1999) considered there was a case for Opisthokonta to be made a taxon but since the boundary of this group was uncertain he colloquialised the name as 'opisthokonts'. Opisthokonta is now the well-established name of a eukaryotic supergroup (Adl *et al.*, 2012).

The introduction of electron microscopy (EM) in the 1960s, while contributing valuable information on cell ultrastructure and the possible evolutionary relevance of cell coverings (Leadbeater, 2008b), provided only limited additional insight into choanoflagellate phylogeny. One of the more important features identified was that choanoflagellate mitochondria possessed flattened (non-discoidal) cristae, a character also shared with chytrids and animals cells (Cavalier-Smith, 1987). However, this character is neither consistent within these groups nor is it exclusive to them (Taylor, 1978). Recently Wylezich et al. (2012) observed tubular and saccular cristae in the mitochondria of two *Codosiga* species collected from oxygen-depleted waters in the Baltic Sea (see Section 9.4.7). Ultrastructural features associated with the flagella have also proved to be inconsistent as phylogenetic indicators. The delicate vane on some choanoflagellate flagella resembles equivalent structures on some sponge choanocyte flagella, but this feature has yet to be demonstrated universally (see Section 2.6.1). The distinctive ring of microtubules associated with the flagellar basal body, while being consistent with minor variations in choanoflagellates (Figs 2.29–2.31, 2.34), is only approximated within the zoospores of the Monoblepharidales: Chytridiomycota (Fuller and Reichle, 1968; Mollicone and Longcore, 1994). In summary, morphological and ultrastructural characters are neither sufficiently discriminating nor consistent to be reliable indicators of choanoflagellate phylogeny.

A major breakthrough came with the introduction of molecular techniques to phylogeny and systematics. The publications of Wainright et al. (1993, 1994), based on small-subunit (SSU) rDNA sequence data, demonstrated a relationship between animals, fungi and choanoflagellates, thereby vindicating earlier predictions and justifying continued use of the collective term Opisthokonta. Although the phylogenetic trees published by Wainright et al. (1993, 1994) lacked resolution and significant support for critical nodes, nevertheless the authors concluded that the Metazoa appeared as a monophyletic grouping that shared a recent common ancestry with the choanoflagellates. The fungi formed a sister grouping to the animal/choanoflagellate clade. Contemporary evidence for a common evolutionary history between animals and fungi was also provided by maximum likelihood analysis of amino acid sequences for the translation protein eukaryotic elongation factor 1A (eEF1A) (Hasegewa et al., 1993) and maximum parsimony analyses of amino acid sequences for four nuclear-encoded proteins, namely actin, α-tubulin, β-tubulin and eEF1A (Baldauf and Palmer, 1993). All analyses placed animals and fungi together within a monophyletic grouping to the exclusion of all other eukaryotes, including plants and a broad diversity of protists. More recently, phylogenetic analyses of combined multigene data sets have recovered the animal, choanoflagellate, fungal (opisthokont) clade with stronger support but it now includes an additional diverse grouping of free-living and parasitic protistan genera (Lang et al., 2002; Medina et al., 2003; Philippe et al., 2004, 2009; Steenkamp et al., 2006; Jiménez-Guri et al., 2007; Rodriguez-Ezpeleta et al., 2007; Shalchian-Tabrizi et al., 2008; Yoon et al., 2008; Parfrey et al., 2006, 2010; Baldauf et al., 2013).

Opisthokonta is now one of a small number of eukaryotic 'supergroups' universally recognised in molecular phylogenies. A consensus has emerged with respect to the circumscription of these supergroups, although their number (5–8) and names may vary (Simpson and Roger, 2004; Baldauf, 2008; Keeling et al., 2005; Hampl et al., 2009; Parfrey et al., 2010). The most commonly quoted are Opisthokonta, Amoebozoa, Archaeplastida (Plantae), SAR (Stramenopile, Alveolata, Rhizaria) and Excavata (Adl et al., 2012). Of these, Opisthokonta emerges as one of the most robustly supported by multiple data sets. In a recent evaluation of the stability and robustness of supergroups, Opisthokonta was recovered as monophyletic in 43 of the 51 published trees sampled (Parfrey et al., 2006).

Some authors combine Opisthokonta with Amoebozoa in a mega-assemblage entitled 'unikonts' (Unikonta) (Cavalier-Smith, 1998; Minge et al., 2009). However, this union and its name are controversial. Recent SSU rDNA and mitochondrial gene phylogenies have placed Apusozoa as the sister group to Opisthokonta (Cavalier-Smith and Chao, 2010; Derelle and Lang, 2012; Paps et al., 2012). Since apusozoans and some members of the Amoebozoa are biflagellate, the term 'unikont' is inappropriate and misleading. Cavalier-Smith (2013) has now accepted that a name change is necessary and introduced the term 'podiates'. At the same time Adl et al. (2012) have introduced the name Amorphea for a clade comprising Opisthokonta, Amoebozoa, Apusomonadida, *Breviata*, *Subulatomonas* and probably Ancyromonadida and *Mantamonas* (see also He et al., 2014).

10.1.1 Chapter aims

There are three aims to this chapter. The first is to examine the morphological and molecular evidence that supports the current phylogenetic delineation of Opisthokonta (Section 10.2). The second is to review the biology of individual opisthokont groups to see if an evolutionary pattern of morphological diversity can be matched onto contemporary molecular phylogenies (Sections 10.3–10.7). The third is to summarise what is currently known about the opisthokont origins of metazoan multicellularity (Section 10.8).

10.2 PHYLOGENY OF OPISTHOKONTA: MORPHOLOGICAL AND MOLECULAR PERSPECTIVES

10.2.1 Opisthokonta: shared morphological characters

The current morphological and ultrastructural justification for an opisthokont grouping is limited to two characters, as described by Adl *et al.* (2012): (1) a single 'posterior' flagellum without mastigonemes, present in at least one life cycle stage, or secondarily lost, with a pair of centrioles, sometimes modified; (2) mitochondria with flattened cristae. Unfortunately, neither of these characters is exclusive or universal to opisthokonts.

(1) Single 'posterior' flagellum. As discussed in Chapter 2, the base-to-tip planar, sinusoidal undulation of a single, 'smooth' flagellum (without mastigonemes) propels (pushes) the attached cell through a fluid with the cell body foremost (see Section 2.3.2). This form of propulsion, in which the flagellum can be described as a 'pulsellum', is an inevitable outcome of the hydrodynamics associated with an undulating 'smooth' flagellum and distinguishes the so-called 'opisthokont' condition. What is surprising is that apart from animals, chytrids and choanoflagellates, the opisthokont condition is relatively uncommon. As Vischer (1945) recorded, some green algal flagellates, such as *Pedinomonas* (*Chlorochytridion*) *tuberculata* (Chlorophyta), are also uniflagellate and pulsellate, but examples such as this are relatively unusual. However, whether a single smooth flagellum with base-to-apex undulation represents an acceptable phylogenetic indicator is open to question. It must be emphasised that the descriptive adjectives 'anterior' and 'posterior' with respect to the flagellum are entirely a matter of convention. The flagellar pole traditionally defines the anterior end of a sedentary choanoflagellate cell. However, confusingly, on a swimming cell the flagellum is often described as being 'posterior' because the cell is propelled from behind (see Sections 2.1, 2.3.2). More correctly, a swimming choanoflagellate should be described as moving with its posterior pole foremost. It is the relative orientation of the cell and not the location or functioning of the flagellum that varies.

The 'pair of centrioles' to which Adl *et al.* (2012) refer in their description of Opisthokonta comprise the flagellar basal body and an angled 'dormant' basal body. In choanoflagellates, and the ichthyosporean *Dermocystidium percae*, the two basal bodies are approximately orthogonal (at right-angles) (see Sections 2.6.2, 10.4.2) (Pekarrinen *et al.*, 2003), whereas in the majority of chytrids they lie parallel to each other (James *et al.*, 2006b). Two orthogonal basal bodies, collectively called a centrosome, are also common in animal cells (see Section 1.7), including spermatozoa (Fouquet and Kann, 1999), but there are exceptions to this general rule.

(2) Flattened mitochondrial cristae. The morphology of mitochondrial cristae is generally considered to be a conservative character and therefore of phylogenetic value, particularly in the context of higher taxa (Cavalier-Smith, 1997; Taylor, 1999). At least three categories of cristal morphology have been recognised: (1) flattened (e.g. opisthokonts, green and red algae); (2) discoidal – paddle-shaped (e.g. euglenoids and kinetoplastids); (3) tubular (e.g. diatoms, chrysophytes and Apicomplexa). However, apart from a lack of exclusivity, there are significant caveats with respect to the reliability of cristal morphology as a phylogenetic character. These include the possibility of there being more than one cristal type within an organism. For instance, tubular cristae occur in mammalian fibroblasts, hepatocytes and muscle tissue, whereas flattened cristae are present in brown adipose tissue. Preservation may also affect cristal morphology: for example, *Dictyostelium*, which is traditionally considered to possess tubular cristae, has what are interpreted as flattened cristae in material preserved by freezing methods (Fields *et al.*, 2002). Recently, Wylezich *et al.* (2012) have observed 'tubular' cristae in two craspedid species, *Codosiga minima* and *C. balthica*, collected from oxygen-depleted water in the Baltic Sea (see also Section 2.8.1).

10.2.2 Molecular phylogeny of Opisthokonta

Molecular phylogenetic studies have established unequivocally a close evolutionary relationship between animals and fungi which, together with choanoflagellates, Ichthyosporea and a diverse collection of protistan genera, comprise the Opisthokonta (Carr and Baldauf, 2011; Baldauf *et al.*, 2013). Evidence supporting this eukaryotic supergroup comes from many sources, including: (1) PCR-targeted single-gene phylogenies, particularly ribosomal DNA (rDNA); (2) concatenates of rDNA and/or combinations of PCR-targeted nuclear protein-coding genes; (3) expressed sequence tag (EST) libraries; (4) complete genome sequence data sets. The sequencing of EST libraries or whole genomes generates phylogenomic data sets comprising many thousands of amino acid positions (Table 10.1). However, each approach brings with it particular advantages and limitations (see Philippe and Telford, 2006; Philippe et al., 2011 for discussion). Limitations in taxon sampling, particularly with respect to whole-genome sequencing, have been a serious drawback to constructing detailed opisthokont phylogenies. While data sets are relatively extensive for Fungi and Metazoa, it is only lately that the sampling of choanoflagellates and ichthyosporeans has been expanded to include multiple species and genes from individual systematic groups (Cafaro, 2005; Carr et al., 2008a; Nitsche et al., 2011; Marshall and Berbee, 2011). Limited or unbalanced taxon sampling may result in species being present on long, unbroken branches that tend to group together irrespective of their true phylogenetic relationships. This artefact, known as long-branch attraction, may also result from unequal rates of change in non-adjacent lineages. Wherever possible, it is desirable to screen taxa and pick those most suitable for phylogenetic reconstruction. However, this is not always possible, particularly in the case of the opisthokont protists, where some lineages, for example Capsaspora, are only represented by a single known species.

Molecular phylogenetic analyses offer strong support for the monophyly of Opisthokonta, but at the same time describe a deep bifurcation into independent animal and fungal clades known as Holozoa and Nucletmycea, respectively (Adl et al., 2012). Both clades also include unicellular (and temporarily multicellular) protistan lineages: Holozoa contains the Metazoa, choanoflagellates, ichthyosporeans, Filasterea and Corallochytrium (Lang et al., 2002; Ruiz-Trillo et al., 2004) while Nucletmycea contains the Fungi, nucleariid amoebae and the former slime mould Fonticula alba (Steenkamp et al., 2006; Brown et al., 2009). Detailed relationships within the two clades, particularly the holozoan clade, have not been resolved completely at present. Phylogenies based on rDNA data alone give conflicting results with respect to the order of branching and key nodes often lack significant support (see Table 10.1 for references). Phylogenies based on concatenations of genes and phylogenomic data usually give stronger statistical support at the nodes but, again, topologies vary, particularly within the holozoan clade. The six most common topologies are illustrated in Fig. 10.2, and in all there is unequivocal support for a sister group relationship between the choanoflagellates and animals (Metazoa) (Shalchian-

Tabrizi et al., 2008; Ruiz-Trillo et al., 2008; Carr et al., 2008a; Nitsche et al., 2011). Analysis based on mitochondrial genome data reveals "beyond a doubt that Choanoflagellata and multicellular animals share a close sister group relationship" (Lang et al., 2002, p. 1773; see also Fig. 10.2b, Ruiz-Trillo et al., 2008; Liu et al., 2009).

10.2.3 Phylogenetic significance of rare genomic changes

An alternative to using primary sequence data for phylogenetic reconstruction is to use macromolecular characters, such as gene fusions and fission events, genomic rearrangements or large insertions and deletions (indels) within highly conserved genes, provided they are shared by two or more organisms to the exclusion of others. These relatively large-scale mutations, known as 'rare genomic changes' (Rokas and Holland, 2000), can be used as phylogenetic signatures to support particular nodes because they are infrequent, complex, largely irreversible and therefore unlikely to arise independently and be shared by unrelated taxa. However, rare genomic changes are subject to the potential problems of homoplasy, including convergence, parallelism or reversion, and must be interpreted with as much caution as is a molecular phylogeny.

INSERTION IN ELONGATION FACTOR 1A (EEF1A):
SYNAPOMORPHY FOR OPISTHOKONTA

One of the first and most robust 'rare genomic changes' to be recognised was an ~12 amino acid insertion (positions 214–225) in eEF1A (formerly eEF1α) which was shared exclusively between animals and fungi (Baldauf and Palmer, 1993) (Fig. 10.1). The animal–fungal insertion is particularly striking because the eEF1A gene is otherwise highly conserved in length and sequence (Baldauf, 1999). Subsequent work has extended the range of species sampled to include representatives of most of the protistan opisthokont lineages, including: Choanoflagellatea – Monosiga ovata, Salpingoeca amphoridium and Codosiga botrytis; Filasterea – Ministeria vibrans and Capsaspora owczarzaki; Corallochytrea – Corallochytrium limacisporum; Ichthyosporea – Amoebidium parasiticum, Ichthyophonus irregularis; and Cristidiscoidea – Nuclearia simplex and Fonticula alba (Fig. 10.1) (Baldauf, 1999; Ragan et al., 2003; Steenkamp et al., 2006; Paps et al., 2012). All were found to possess an insertion, although the number of amino acids varies, from nine in Corallochytrium to 15 in Salpingoeca amphoridium. The insertion is particularly well conserved in both length

Table 10.1 *Opisthokont molecular phylogeny: details of genes sequenced, taxa sampled and data analysis for the most complete published opisthokont phylogenetic trees (see Fig. 10.2).*

Gene sequenced	Characters	Phylogenetic inference	Holozoa						Nucletmycea		Total taxa	Reference
			Met	Choan	Caps	Minist	Ichth	Corall	Fungi	Nucl		
rDNA												
SSU	1674 nt	ML	12	9	1	1	11	1	13	3	50	Medina et al., 2003
SSU and LSU	3872 nt	ML	6	2	–	–	1	–	5	1	15	Medina et al., 2003
LSU	1513 nt	Bayes	4	2	1	1	1	–	4	1	13	Moreira et al., 2007
SSU and LSU	2574 nt	Bayes	4	2	1	1	1	–	4	1	13	Moreira et al., 2007
SSU	1645 nt	Bayes	4	1	1	1	2	1	3	2	14	Cavalier–Smith and Chao, 2010
SSU	1497 nt	ML	9	75*	–	–	12	1	–	–	97	Nitsche et al., 2011
SSU and LSU	3924 nt	Bayes	2	1	1	1	1	–	3	1	9	Yabuki et al., 2012
Concatenates of genes												
SSU, actin, tubB, EF1A	2809 nt	Bayes	19	4	–	1	1	1	2	1	29	Steenkamp et al., 2006
actin, tubB, EF1A, HSP70	1347 aa	Bayes	2	1	–	1	1	1	2	1	9	Steenkamp et al., 2006
SSU + LSU, tubA, HSP90	6415 nt	Bayes	9	17	1	–	2	1	4	1	35	Carr et al., 2008a
SSU, actin, tubB, EF1A, HSP70	1234 nt 1568 aa	Bayes	2	2	1	1	2	1	15	1 + F	26	Brown et al., 2009
rDNA, 6 protein-coding genes	6110 nt/ aa	Bayes and ML	9	16	1	1	3	1	12	1 + F	45	Paps et al., 2012
EST and genome data sets												
Nuclear protein-coding genes	31 092 aa	Bayes	47	3	1	–	2	–	7	–	60	Jiménez–Guri et al., 2007
56 nuclear encoded proteins	10 678 aa	Bayes	Many	2	1	–	2	–	11	–	many	Aleshin et al., 2007
78 protein-coding genes	17 482 aa	ML	4	3	1	1	2	–	15	–	26	Shalchian–Tabrizi et al., 2008

Table 10.1 (*cont.*)

Gene sequenced	Characters	Phylogenetic inference	Holozoa						Nucletmycea		Total taxa	Reference
			Met	Choan	Caps	Minist	Ichth	Corall	Fungi	Nucl		
110 nuclear encoded proteins	20 711 aa	Bayes	12	3	1	–	2	–	9	–	27	Ruiz–Trillo *et al.*, 2008
1487 gene matrix	270 580 aa	ML	87	2	1	–	2	–	2	–	94	Hejnol *et al.*, 2009
118 nuclear encoded genes	24 439 aa	Bayes	4	2	1	–	2	–	8	2	19	Liu *et al.*, 2009
128 genes	30 257 aa	Bayes	44	3	1	–	2	–	5	–	55	Philippe *et al.*, 2009
SSU, 16 protein-coding genes	6578 nt/aa	ML	13	1	1	–	2	–	8	–	27	Parfrey *et al.*, 2010
EST and genome data	18 106 aa	Bayes and ML	21	3	1	1	2	–	19	–	47	Toruella *et al.*, 2012
Mitochondrial data												
13 mt protein-coding genes	2619 aa	Bayes and ML	19	1	1	–	1	–	14	–	36	Ruiz–Trillo *et al.*, 2008
13 mt protein-coding genes	2710 aa	Bayes and ML	11	1	1	–	1	–	22	1	37	Liu *et al.*, 2009

Met = Metazoa; Choan = Choanoflagellates; Caps = *Capsaspora*; Minist = *Ministeria*; Ichth = Ichthyosporea; Corall = *Corallochytrium*; Nucl = *Nuclearia*; nt = nucleotides; aa = amino acids; F = *Fonticula alba*. * includes environmental sequences. ML = maximum likelihood and sequence among fungi and *Nuclearia* (Fig. 10.1). Within the Metazoa the eEF1A insertion has been identified in all the major divisions (Baldauf, 1999; Berney *et al.*, 2000). In order to confirm the exclusivity of this insertion with respect to the opisthokont clade, Steenkamp *et al.* (2006) sampled the corresponding position on the eEF1A gene in the closest relatives of opisthokonts, namely the Apusozoa and Amoebozoa. In these, as well as a wide range of other eukaryotes, the insertion was absent (Fig. 10.1). Steenkamp *et al.* (2006) also reported that a conserved seven amino acid motif in eEF1A at positions 157–163, the STEPPYS signature (serine, threonine, glutamate, proline, proline, tyrosine, serine), was shared exclusively between choanoflagellates and animals that possessed this gene (Fig. 10.1).

While the ~12 amino acid insertion in opisthokont eEF1A ranks as one of the best examples of a rare genomic change that is of phylogenetic value, nevertheless in some species eEF1A is apparently replaced by a paralogue to eEF1A, known as EFL (elongation factor like; Keeling and Iganaki, 2004). EFL protein differs from eEF1A in its primary structure but is considered to perform the same function as eEF1A in translation elongation steps. In a selection of opisthokonts, including the choanoflagellates *Monosiga brevicollis* and *Salpingoeca rosetta*, the ichthyosporeans *Sphaeroforma arctica*, *S. tapetis* and *Creolimax fragrantissima* and some fungi, as in some members of other eukaryotic groups, EFL occurs instead of eEF1A (Keeling and Iganaki, 2004; Kamikawa *et al.*, 2010; Atkinson *et al.*,

2014). The apparent displacement of eEF1A by EFL in a punctate fashion within many eukaryote groups is somewhat surprising since eEF1A has an essential role in peptide elongation cycles during protein formation. Various explanations have been suggested for this situation. The phylogenetic distribution of eEF1A is consistent with its vertical inheritance throughout eukaryotes, whereas EFL is of unknown origin and appears to have spread into distantly related eukaryotic groups/species by lateral gene transfer events. In the cells that acquired the EFL gene, the exogenous EFL could have taken over the original eEF1A function, resulting in secondary loss of the endogenous eEF1A gene (Kamikawa *et al.*, 2010). A maximum likelihood phylogenetic analysis based on the EFL gene shows that the icththyosporean *Sphaeroforma arctica* strongly groups with the choanoflagellates *Monosiga brevicollis* and *Salpingoeca rosetta* (Ruiz-Trillo *et al.*, 2006; Atkinson *et al.*, 2014). However, a sister group relationship between the fungal and choanoflagellate EFL sequences was not recovered, suggesting separate origins of the EFL orthologues in these lineages (Noble *et al.*, 2007).

10.2.4 Opisthokont mitochondrial genomics

Mitochondrial DNA (mtDNA) has been used extensively in phylogenetic studies, particularly with respect to the Metazoa and fungi, and has several advantages in comparison with nuclear DNA. mtDNA is present in a multicopy

				153	168	198	214	225	238
OPISTHOKONTA	**ANIMALIA**	Deuterostomia	*Homo*	NKMD STEPPYS QKRYE ...	GDNMLEPSANMPWFKG	WKVTRK-----	DGNASG	TTLLEALDCILPP	
			Ciona	NKMD NTEPPYS EQRFE ...	GDNMLETSENMPWFKG	WAIERK-----	EGNASG	KTLYNALDAILLP	
		Ecdysozoa	*Caenorhabditis*	NKMD STEPPFS EARFT ...	GDNMLEVSSNMPWFKG	WAVERK-----	EGNASG	KTLLEALDSIIPP	
			Drosophila	NKMD SSEPPYS EARYE ...	GDNMLEPSTNMPWFKG	WEVGRK-----	EGNADG	KTLVDALDAILPP	
		Lophotrochozoa	*Dugesia*	NKMD STEPPFS EPRFD ...	GDNMIDESSNMPWYKG	WEITRKNAKKEEIKTTG		RTLLDALDSLEPP	
			Chaetopleura	NKMD STTPPFS QPRFE ...	GDNMLEVSSNTAWFKG	WNIERK-----	EGNASG	KTLFEALDSILPP	
		Radiata	*Eugymnanthea*	NKID NTEPPYS EARFK ...	GDNMIEPSTNMSWYKG	WEIERK-----	AGKASG	KTLLEALDAVVPP	
		Parazoa	*Geodia*	NKMD STEPPYS QARYD ...	GDNMLEESPNMKWFKG	WNVERK-----	EGNASG	KTLFNPLDSILPP	
	FUNGI	Basidiomycota	*Puccinia*	NKMD TT--KWS EQRFE ...	GDNMLEESTNMGWFKG	WTKETK-----	AGVSKG	KTLLDAIDAIEPP	
		Ascomycota	*Neurospora*	NKMD TT--QWS QTRFE ...	GDNMLEPSTNCPWYKG	WEKETK-----	AGKATG	KTLLEAIDAIEPP	
		Zygomycetes	*Mucor*	NKMD TT--KWS QDRYN ...	GDNMLDESTNMPWFKG	WTKETK-----	AGSKTG	KTLLQAIDAIEPP	
		Trichomycetes	*Smittium*	NKMD SN--KYS EERFT ...	GDNMIEASTNMPWYKG	WTKETK-----	SGVSKG	VTLLDAIDAVEPP	
		Chytridiomycota	*Chytridium*	NKMD TT--NYS EDRYN ...	GDNMLEASENMPRFKG	WNKETK-----	AGSSTG	KTLLQAIDAIEPP	
		Microsporidia	*Glugea*	NKVD TIDEKNR ISRFD ...	GINIVEKGDKFEWFKG	WKPVSG-----	AG-DSI	FTLEGALNSQIPP	
	ALLIES	Choanoflagellata	*Monosiga*	NKMD STEPPYS ESRFN ...	GDNMLEESTNMGWFKG	WEITRK-----	DGNAKG	KTLLEALDAILPP	
		Ichthyosporea	*Amoebidium*	NKMD SI--KFA QDRFN ...	GDNMVEPTDNMPWYKG	WEVERK-----	EGNATG	KTLLEALDAILPP	
			Ichthyophonus	NKMD SV--KYS EDRFK ...	GDNMVAPTENMPWYKG	WTCERK-----	EGNTSG	FTLLEALDNIQAP	
		Corallochytrea	*Corallochytrium*	NKMD SI--KYS KDRFD ...	GDNMLEASTNMDWYKG	WEKD-------	GSVGG	KTLIEALDAVSPP	
		Nucleariida	*Nuclearia*	NKMD TC--KYS EERFN ...	GDNMLEPTTNMPWFQG	WEIDRK-----	NGKVMG	KTLVGALDAIEPP	
		Ministeriida	*Ministeria*	NKMD SI--KYD EARFT ...	GDNMLDASTNMPWYKG	WEVDRDK----	NGKASG	KTLIDALDAVLPP	
	APUSOZOA	Ancyromonadidae	*Ancyromonas*	NKMD DKSVNYS KARFD ...	GDNMTEPSANMPWYSG	-----------------		PTLLGALDACEVP	
		Apusomonadidae	*Apusomonas*	NKMD DKTVKYS KDRYE ...	GDNMMEPSPQMGWWKG	-----------------		GTLLEALDAFTPP	
	AMOEBOZOA	Mycetozoa	*Dictyostelium*	NKMD EKSTNYS QARYD ...	GDNMLERSDKMEWYKG	-----------------		PTLLEALDAIVEP	
		Lobosa	*Acanthamoeba*	NKMG NV--NWA ENRYN ...	GDNMVDRTDKMPWYKG	-----------------		PTLLEALDGIKPP	
			Hartmannella	NKMD SESVKYS QERYD ...	GDNMLEKSTNLPWYKG	-----------------		PTLVEALDALKEP	
		Conosa	*Entamoeba*	NKMD AI--QYK QERYE ...	GDNMIEPSTNMPWYKG	-----------------		PTLIGALDSVTPP	
		Phalansterea	*Phalansterium*	NKMD DKTVNWG EPRYQ ...	GDNMLERSANLPWYKG	-----------------		PTLLEALDNLVPP	

Fig. 10.1 The approximate 12 amino acid insertion in the EF1A gene in all lineages of Opisthokonta. The alignment illustrated is for positions 153–238 of the *Homo sapiens* eEF1A sequence. Equivalent positions are shown for members of Amoebozoa and Apusozoa, where there is no insertion. Reproduced from Steenkamp *et al.* (2006).

state (tens to hundreds of copies) in cells, whereas nuclear genes tend to be present as one or two copies. This means that there is a much greater yield of DNA per cell for PCR amplification. In metazoans mtDNA evolves faster than nuclear DNA due to higher mutation rates and possibly inefficient replication repair. Different regions of the mitochondrial genome evolve at different rates, allowing a choice of suitable regions for study. While recombination of mtDNA is common in protists and fungi where entire cells fuse; in Metazoa, with some exceptions, mtDNA is uniparentally inherited and opportunities for recombination are more limited (Rokas et al., 2003). Metazoans are usually homoplasmic for one mitochondrial haplotype and therefore the whole genome can be assumed to have the same genealogical history. It should be mentioned that all of these apparent certainties have been challenged and the extent of variation is still being assessed (Galtier et al., 2009). Nevertheless, mtDNA has been used extensively in phylogenetic studies, particularly of animals. Characteristics of mtDNA that are of particular interest include: genome size, gene content and order and nucleotide sequence.

STRUCTURAL AND COMPOSITIONAL VARIATION OF OPISTHOKONT MITOCHONDRIAL GENOMES

There is considerable variation in the size and gene content of mtDNA in opisthokonts. In general there appears to be a trend in mtDNA evolution, with both genome size and gene content being reduced during transitions from unicellular opisthokonts (choanozoa) to multicellular Metazoa and fungi (Lavrov, 2011). For example, the mtDNA of the choanoflagellate Monosiga brevicollis is 76.6 kbp in size and contains 55 genes; that of the ichthyosporean Amoebidium parasiticum is >200 kbp in size and contains 44 genes (Burger et al., 2003). The mtDNA of Trichoplax adhaerens (Placozoa, an early branching lineage of non-bilaterian animals – see Section 10.4.5) is 37–43 kbp in size and contains 39–42 genes (Signorovitch et al., 2007); that of the majority of demosponges is between 18 and 25 kbp in size and contains 40–44 genes (Lavrov, 2011). The mtDNA of bilaterian animals is reasonably consistent with a size range of 16–25 kbp and contains 36 or 37 genes. The reduction in genome size that accompanies the transition to metazoan multicellularity is due to the loss of genes, including ribosomal protein coding genes, and a decrease in intronic and intergenic spacer regions. Thus, in Monosiga brevicollis only 46.9% of mtDNA comprises the coding portion, whereas as much as 98% of metazoan mtDNA

codes for either proteins or structural RNAs (Burger et al., 2003; Signorovitch et al., 2007). The genome of Trichoplax adhaerens, which is of intermediate size, possesses some features in common with Monosiga brevicollis, and yet encodes a typical complement of metazoan mtDNA genes (Dellaporta et al., 2006). A set of well-conserved metazoan-specific insertions and deletions (indels) in mitochondrial protein-coding genes (cob and cox 1–3) distinguish the Metazoa from the choanoflagellate Monosiga brevicollis and other non-metazoan opisthokonts. In addition, an atypical R11–Y24 pair is present in mitochondrial tRNA^{Trp}/$_{UCA}$ of non-bilaterian and bilaterian animals, but absent from non-metazoan outgroups. The presence of these animal mitochondrial synapomorphies provides supporting evidence for metazoan monophyly and can serve as a guide of metazoan affinity (Lavrov, 2011).

Fungal mitochondrial genomes are variable in size, gene content and introns. They range from 19 to 94 kbp in size and contain between 30 and 40 genes. The basic fungal mitochondrial gene complement resembles that of Metazoa, although this coincidence is thought to be the result of independent gene loss in the two lineages (Burger et al., 2003). Support for the idea of independent reductions comes from analysis of mtDNA in choanoflagellates, as exemplified by Monosiga brevicollis, which comprise the deep branching sister group of Metazoa. M. brevicollis contains all the genes encoded by Metazoa in addition to several genes, including bacteria-like transfer and ribosomal RNAs, not found in the metazoan genome. The overall size and coding capacity of the M. brevicollis genome resembles the mtDNA of other protists rather than metazoan mtDNA (Gray et al., 1998). Thus it seems likely that the protistan common ancestor of the metazoan and fungal lineages had more complex, gene-rich mtDNA than extant members of either of these two opisthokont lineages (Burger et al., 2003; Bullerwell and Lang, 2005).

MITOCHONDRIAL PHYLOGENY OF OPISTHOKONTA

Complete mitochondrial genome sequences provide a wealth of protein-coding data that can be used to infer evolutionary relationships. The mtDNA tree topology (Fig. 10.2b) consistently recovers the choanoflagellates as being the sister group to the Metazoa (Ruiz-Trillo et al., 2008; Liu et al., 2009; Wang and Lavrov, 2009). Capsaspora occupies an intermediate position between ichthyosporeans and the choanoflagellate/metazoan clade. All analyses demonstrate with high confidence that the choanoflagellates are the closest sister group to the Metazoa to the exclusion of all other choanozoan groups. Increased taxon sampling is required, particularly with respect to the choanoflagellates and ichthyosporeans, each of which is represented by a single species, *Monosiga brevicollis* and *Amoebidium parasiticum*, respectively.

10.3 OPISTHOKONT CLASSIFICATION

At least two rival, but overlapping, opisthokont classificatory systems are currently available that differ according to contrasting philosophies. One system is based on 'phylogenetic systematics' or 'cladistics' and the other on 'evolutionary systematics'. Both systems, when viewed objectively, have their strengths and weaknesses, but since they have polarised opinion within the systematics community the result has been a confusion of terminology, names and hierarchies with no resolution in sight. Regrettably this makes life difficult for all but the most committed and persistent of observers.

This is neither the time nor place to enter into what is an ongoing, and at times acrimonious, debate regarding the virtues and vices of the two approaches. Both mostly accept Hennigian cladistic methodology as an adequate technique for the reconstruction of phylogenetic trees. However, they disagree profoundly over interpretation of the term monophyly with respect to the circumscription of taxa. Cladists interpret monophyly in a limited context as a grouping of *taxa* (clade) according to one or more shared 'derived characters' (synapomorphies) that are uniquely present in the most recent common ancestor and all of its descendants. Such groups are more precisely called holophyletic. Evolutionary systematists, on the other hand, interpret monophyly more widely to also include paraphyletic groups that comprise an ancestor together with some, but not all, of its descendants. A paraphyletic group is an incomplete clade, which is more appropriately called a 'grade'. It is defined by a combination of positive and negative characters. Cladists are primarily concerned with branching order and ignore morphological disparity. Evolutionary systematists, on the other hand, take into account both branching order and morphological disparity when classifying organisms. Both approaches reject polyphyletic groupings. However, evolutionary systematists accept paraphyletic groups whereas cladists, depending on their degree of purity, reject them.

The absence of consensus between the two philosophies also has major implications for classification and nomenclature. Phylogenetic systematists consider that *phylogenetic*

tree topography is fundamental to defining groups and that only *clades* (holophyletic taxa) should be formally named. In phylogenetic nomenclature, Linnaean ranks have no bearing on the spelling of taxonomic names. Ranks are not altogether forbidden, they are merely decoupled from nomenclature. They do not influence which names can be used, which taxa are associated with which names, and which names can refer to nested taxa. On the other hand, evolutionary systematists classify species into a series of higher taxa, of which only some are clades. They adopt the Linnaean rank-based system of nomenclature governed by the appropriate international codes of nomenclature. In the case of the opisthokonts this involves two codes, namely the International Code of Zoological Nomenclature (ICZN) for the Holozoa and the International Code of Nomenclature for Algae, Fungi, and Plants (ICN) for the Nucletmycea. However, these codes only refer to ranks up to the level of families. The hybrid scheme adopted by Adl *et al.* (2012), based on 'nameless ranked systematics', makes use of the oldest valid name that describes each group, irrespective of its status, and thus the scheme comprises an eclectic assortment of taxonomic nomenclature and spellings (Table 10.2).

The conflict between the two approaches to systematics can be illustrated by reference to the opisthokonts. Molecular phylogenetic studies consistently recover opisthokonts as comprising two deep branching monophyletic lineages, one based on the Metazoa and the other on the Fungi. In cladistical systematics the two clades are named without hierarchical status (Table 10.2). The term Holozoa was coined to describe the metazoan clade (Lang *et al.*, 2002) and the term Nucletmycea to describe the fungal clade (Brown *et al.*, 2009).

Cavalier-Smith (2013), an evolutionary systematist, divides the opisthokonts between three kingdoms, namely Animalia, Fungi and Protozoa. Animalia (Metazoa) and Fungi are holophyletic, whereas Protozoa is paraphyletic. However, the formal diagnosis of Protozoa is so unwieldy (Cavalier-Smith, 1993, p. 967) and has been modified on so many occasions that the boundaries of this taxon are only fully comprehensible to Cavalier-Smith (2013). Opisthokont representatives attributable to Protozoa are included within the paraphyletic phylum Choanozoa, which has expanded its content and changed its name and hierarchical status over the years. Choanozoa began its existence as the holophyletic sub-phylum Choanociliata (Cavalier-Smith, 1981) and referred exclusively to the choanoflagellates. Subsequently it was elevated to a phylum and underwent

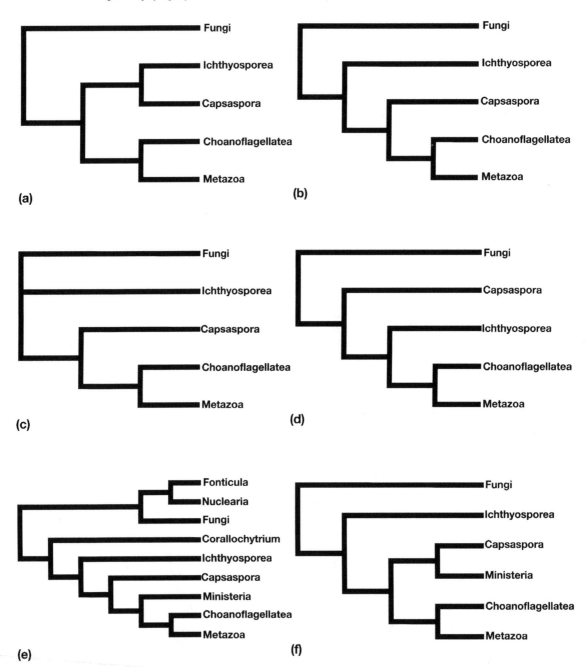

Fig. 10.2 Six postulated relationships between constituent members of Opisthokonta. a. Jiménez-Guri *et al.* (2007), Ruiz-Trillo *et al.* (2008), Parfrey *et al.* (2010), Liu *et al.* (2009), Zhao *et al.* (2012). b. Ruiz-Trillo *et al.* (2008), Liu *et al.* (2009), Lavrov (2011), Wang and Lavrov (2009). c. Philippe *et al.* (2009). d. Carr *et al.* (2008a), Hejnol *et al.* (2009). e. Brown *et al.* (2009). f. Shalchian-Tabrizi *et al.* (2008), Toruella *et al.* (2012).

a name change to Choanozoa (Cavalier-Smith, 1987). Successive expansions have meant that Choanozoa is now paraphyletic and includes the Ichthyosporea, Filasterea, Corallochytrea and Cristidiscoidea (Table 10.2) (see Cavalier-Smith, 2013 for references). There is some confusion in the general literature regarding the content of the phylum Choanozoa, which together with the heterogeneous nature of the constituent taxa – there are no obvious unifying morphological characters – has rendered the name of limited use other than as a collective term for unicellular (protistan) opisthokonts. Nevertheless, despite these confusions, most classificatory systems contain some holophyletic groups that can be agreed upon even if the nomenclature varies.

10.4 HOLOZOA

Holozoa currently comprises four protistan groups (choanozoa) – namely Choanoflagellatea, Icthyosporea, Filasterea, Corallochytrea and one multicellular group, Metazoa (animals) (Table 10.2). Multigene analyses corroborate the monophyly of Holozoa (see Table 10.1 for references) but apart from the consistent sister group relationship between the choanoflagellates and Metazoa (Fig. 10.2), uncertainty remains regarding the interrelationships between the individual groups of holozoans (Fig. 10.2) and between the four groups comprising the basal (non-bilaterian) Metazoa (Fig. 10.5).

10.4.1 Choanoflagellatea

By tradition choanoflagellates have been classified according to the morphology of their surface covering (periplast) (see Section 1.3.2 and Table 1.2) (Leadbeater, 2008b). On this basis Kent (1880–2) distinguished two families, Codonosigidae and Salpingoecidae. Species attributable to Codonosigidae possess a thin, flexible covering called a 'glycocalyx' or sheath, which permits unrestricted longitudinal cell division. Species attributable to Salpingoecidae possess a thicker, inflexible organic covering known as the theca, which results in an enclosed cell undergoing 'emergent' division (see Sections 1.3.3 and 3.4). At a later date Norris (1965) added a third family, Acanthoecidae, for species bearing loricae containing siliceous costae (Chapters 4–8).

Multi-taxon molecular phylogenies based on rDNA data and/or concatenations of nuclear protein-encoding genes recover extant choanoflagellates as being

monophyletic with strong support (100% maximum likelihood bootstrap percentage (mlBP), 1.00 Bayesian inference posterior probability (biPP)) (Carr et al., 2008a; Nitsche et al., 2011). The choanoflagellate tree divides into two well-supported clades (92–100% mlBP, 1.00 biPP) equivalent to Craspedida (Codonosigidae and Salpingoecidae) and Acanthoecida (Fig. 10.3). The craspedid grouping is

Table 10.2 *Comparison of opisthokont classificatory systems: Linnaean-based classification according to Cavalier-Smith (2013) and nameless-ranked classification according to Adl et al. (2012).*

Cavalier-Smith (2013)	Adl et al. (2012)
Opisthokonts	**Opisthokonta**
	Holozoa
Kingdom Animalia	Metazoa
Kingdom Protozoa	
Phylum Choanozoa	
Subphylum 1. Choanofila	
Class 1. Choanoflagellatea	Choanomonada
Order 1. Craspedida	Craspedida
Family Salpingoecidae	
Order 2. Acanthoecida	Acanthoecida
Family 1. Acanthoecidae	
Family 2. Stephanoecidae	
Class 2. Filasterea	Filasterea
Order Ministeriida	
Family 1. Ministeriidae	*Ministeria*
Family 2. Capsasporidae	*Capsaspora*
Class 3. Corallochytrea	*Corallochytrium*
Order Corallochytrida	
Family Corallochytriidae	
Class 4. Ichthyosporea	Ichthyosporea
Order 1. Dermocystida	
Family Rhinosporidiidae	Rhinosporidaceae
Order 2. Eccrinida	
Family Ichthyophonidae	Ichthyophonae
Subphylum 2. Paramycia	**Nucletmycea**
Class 1. Cristidiscoidea	
Order 1. Nucleariida	*Nuclearia*
Order 2. Fonticulida	*Fonticula*
Class 2. Rozellidea	
Order 1. Aphelidida	
Order 2. Rozellida	*Rozella*
Kingdom Fungi	Fungi

further divided into at least two major clades, Clade 1 and Clade 2 (Carr *et al.*, 2008a), both of which contain a mixture of representatives of the Codonosigidae and Salpingoecidae, neither of which is recovered as monophyletic.

While morphological and life cycle differences between nudiform and tectiform acanthoecids clearly distinguish the two groups, their molecular phylogenetic relationship is currently uncertain since the topology of the acanthoecids varies depending on the data used to generate a phylogeny. The disagreement centres on the use of the third position of codons in protein-coding genes. The third position of a codon is often referred to as a 'silent site'

because, in general, it does not affect the amino acid encoded by the codon. Carr *et al.* (2008a) created phylogenetic trees with data sets either including or excluding third positions and found that this recovered two different roots for the acanthoecids. The root is the hypothetical ancestor of a phylogenetic clade and allows a chronology of divergence events to be construed. The inclusion of third positions placed the root between the tectiform and nudiform clades, indicating that neither extant group had evolved from the other (Fig. 10.3). In contrast, omission of third positions resulted in a root within the tectiform clade, which, if correct, would indicate that extant nudiform taxa

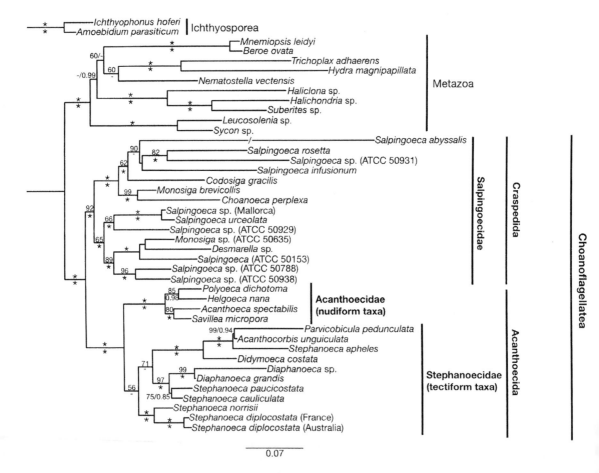

Fig. 10.3 Maximum likelihood phylogeny of choanoflagellates based on a concatenated data set of small and large subunit rDNA, *hsp90* and *tubA*. The phylogeny was created from 6415 aligned nucleotide positions. Branches are drawn proportional to the number of nucleotide substitutions per site as indicated by the scale bar at bottom centre. Branches receiving 1.00 Bayesian inference posterior probability (biPP) and 100% maximum likelihood bootstrap percentage (mlBP) support are denoted by an asterisk. mlBP and biPP values are otherwise given above and below branches, respectively. Values are omitted from weakly supported branches (i.e. mlBP <50% and biPP <0.70). Figure kindly provided by Dr Martin Carr.

evolved from a tectiform ancestor. Carr *et al.* (2008a) statistically showed that the use of third positions was appropriate for their phylogenetic analyses and furthermore showed that a nudiform:tectiform sister grouping was a more parsimonious explanation on morphological grounds than a derived position for the nudiforms (see Fig. 10.3). However, the alternative derived position of nudiform taxa cannot be discounted (Carr *et al.*, 2008a). For the time being, at least, nudiform species are attributed to the family Acanthoecidae Norris *sensu* Nitsche *et al.*, 2011 and tectiform species to the family Stephanoecidae Leadbeater (Nitsche *et al.*, 2011).

10.4.2 Ichthyosporea (Mesomycetozoea)

This is a diverse group of mostly parasitic and commensal protists known mainly from the diseases they cause to their host animals (see Table 10.3). Although the diseases caused by these protists have been known for many years, the absence of defining morphological characteristics with respect to the causative agents has meant that their systematic affinities have remained uncertain.

The first indication that a disease-causing protist was related to the choanoflagellates came from a molecular (SSU rDNA) phylogenetic study of an unclassified salmonid parasite known as the 'rosette agent' (= *Sphaerothecum destruens*) (Kerk *et al.*, 1995). Sequence data from other fish and crustacean parasitic protists soon led to the establishment of the provisional DRIP clade (an acronym for the monophyletic grouping of *Dermocystidium*, the rosette agent, *Ichthyophonus* and *Psorospermium*), a deep branching lineage within the Holozoa (Ragan *et al.*, 1996). Cavalier-Smith (1998) renamed the DRIP clade as a new class which he called Ichthyosporea because these organisms resembled unicellular walled spores and they mostly infected fish. Almost contemporaneously Herr *et al.* (1999) renamed the DRIP clade Mesomycetozoa to reflect the phylogenetic position of the group as being located between Fungi and Metazoa. Mendoza *et al.* (2002) subsequently redefined Mesomycetozoa as a paraphyletic sub-phylum comprising four classes: Choanoflagellatea, Corallochytrea, Mesomycetozoea and Cristidiscoidea. However, since the class Mesomycetozoea is synonymous with the already existing class Ichthyosporea (Cavalier-Smith, 1998), this latter name has been used throughout the current text.

Since no definitive morphological characters categorise the Ichthyosporea, current membership is based primarily on molecular phylogenetic data. As far as is known, ichthyosporeans are osmotrophs in parasitic or commensal relationships with animals (see Table 10.3). Typically, cells are spherical or ovoid with a thick cell wall. Mitochondria usually have flattened cristae, although in some *Ichthyophonus* species (*I. hoferi*) they are tubular (Ragan *et al.*, 1996; Spanggaard *et al.*, 1996). Asexual reproduction, where known, begins with a coenocytic sporangium or plasmodium, which gives rise to numerous uni- or multi-nucleate propagules which can either re-infect the host or transmit the disease to other hosts (Mendoza *et al.*, 2002; Marshall *et al.*, 2008). Sexual reproduction has not been observed to date. However, Marshall and Berbee (2010) have deduced from population genetics the existence of cryptic sexual recombination in *Pseudoperkinsus tapetis*. Based on phylogenetic data, Adl *et al.* (2012) divided the Ichthyosporea into two groupings, the Rhinosporidaceae and Ichthyophonae (Tables 10.2 and 10.3). In Rhinosporidaceae, asexual reproduction usually involves uniflagellated zoospores, while in most Ichthyophonae amoebae have been the only motile stage observed (Mendoza *et al.*, 2002). In *Dermocystidium percae*, where well-preserved zoospores have been observed with TEM, the so-called single 'posterior' flagellum is curved and emerges from the side of the cell (Pekarrinen *et al.*, 2003). A dormant basal body is located at right-angles to the flagellar basal body. The Rhinosporidaceae have yet to be cultured outside their vertebrate host cells, and are typically associated with skin, gills, mucous membranes and visceral organs. The Ichthyophonae are usually associated with the host gut, epicuticle or visceral organs and are found in both vertebrate and invertebrate animals. Several ichthyophonid species, including *Creolimax fragrantissima*, have now been obtained in culture, thereby permitting fuller investigations of their ultrastructure, cycle of asexual reproduction and phylogenetic relationships (Marshall *et al.*, 2008; Glockling *et al.*, 2013).

Phylogenetic analyses have revealed an unexpected diversity among ichthyosporeans. For instance, two morphologically diverse orders of trichomycete fungi, Amoebidiales and Eccrinales, are now placed within the Ichthyophonae (see Mendoza *et al.*, 2002; Cafaro, 2005 for references). Cavalier-Smith (2013), in a recently revised classification of the Ichthyosporea, has retained the order Dermocystida but has replaced Ichthyophonida by the order Eccrinida, which now includes further subdivisions into families to accommodate the most recent finds (see Table 10.2). There is little doubt that once further isolates are obtained in culture and with the addition of environmental sampling the icthyosporean grouping will expand from the current figure of approximately 30 genera (Marshall *et al.*, 2008).

Table 10.3 *Details of life cycles, hosts for parasitic species and mitochondrial cristae of unicellular members attributable to Opisthokonta and therefore related to the animal, fungal and choanoflagellate clades.*

	Taxa	Hosts	Life cycle traits	Mitochondrial cristae	References
Ichthyosporea					
Rhinosporidaceae	*Dermocystidium* sp.	Parasite of fish and amphibians	Cysts with endospores Uniflagellate zoospores	Flattened	Ragan *et al.* (1996)
	Rhinosporidium seeberi	Parasite of mammals and birds	Cysts with endospores No zoospores	Flattened tubulovesiculate	Herr *et al.* (1999) Fredricks *et al.* (2000) Mendoza *et al.* (2002)
	Sphaerothecum destruens (rosette agent)	Parasite of fish (salmon)	Cysts with endospores Uniflagellate zoospores	Flattened?	Arkush *et al.* (2003)
	Amphibiocystidium ranae *A. viridescens*	Parasite of frogs and newts	Cysts with endospores No zoospores	?	Pereira *et al.* (2005)
Ichthyophonae	*Amoebidium parasticum*	Parasite of freshwater crustacea and insects	Sporangium, sporangiospores Amoebic stage	Flattened	Benny and O'Donnell (2000)
	Anurofeca (Prototheca) richardsi	Parasite of anuran larvae and clams	Spherules with endospores	Flattened?	Baker *et al.* (1999)
	Creolimax fragrantissima	Marine invertebrates	Spherical cysts Amoebic stage	Flattened	Marshall *et al.* (2008)
	Ichthyophonus hopferi	Parasite of marine fish	Hyphae, plasmodium, spores Amoebic stage	Tubular	Ragan *et al.* (1996) Spanggaard *et al.* (1996)
	Pseudoperkinsus tapetis	Parasite of mussels and clams	Spherules with spores	?	Takishita *et al.* (2007)
	Psorospermium haeckeli	Parasite of crayfish	Ovoid shell-bearing spores Amoebic stage	Flattened	Vogt and Rug (1999)
	Sphaeroforma arctica	Isolated from amphipod *Gammarus setosus*	Spherules with spores	Flattened	Mendoza *et al.* (2002)
	Abeoforma whisleri	Digestive tracts of marine mussel *Mytilus* sp.	Spherules with amoebic endospores	Flattened	Marshall and Berbee (2011)

Pirum gemmata	Digestive tracts of peanut worm *Phascolosoma* sp.	Spherules with endospores No amoebic phase	Flattened	Marshall and Berbee (2011)
Filasterea				
Capsaspora / *Capsaspora owczarzaki*	Symbiont in haemolymph of pulmonate snails	Unicellular, amoeboid with long thin pseudopodia	Flattened	Hertel *et al.* (2002) Ruiz-Trillo *et al.* (2006)
Ministerida / *Ministeria vibrans*	Marine, free-living heterotrophic protist	Non-flagellate cells with narrow axopods	Flattened?	Patterson *et al.* (1993) Patterson (1999)
Corallochytrium				
Corallochytrium limacisporum	Free-living saprophyte on coral reefs	Vegetative unicells Limax-shaped amoebic spores	Flattened	Cavalier-Smith and Allsopp (1996), Sumathi *et al.* (2006)
Nucletmycea				
Nuclearida / *Nuclearia*	Marine-free living amoeboid protist	Naked filose amoebae	Flattened	Amaral Zettler *et al.* (2001) Dyková *et al.* (2003) Yoshida *et al.* (2009)

Classification based on Adl *et al.* (2012) and table based on Mendoza *et al.* (2002). See also Glockling *et al.* (2013).

10.4.3 Filasterea (*Capsaspora* and *Ministeria*)

`CAPSASPORA`

At present the genus *Capsaspora* is represented by a single species, *C. owczarzaki*, which occurs as an endosymbiont in the haemolymph of the freshwater pulmonate snail *Biomphalaria glabrata* (Owczarzak *et al.*, 1980). *B. glabrata* is also an intermediate host for the parasitic digenetic trematode *Schistosoma mansoni*, the causative agent of intestinal schistosomiasis (bilharzia) in humans. Within a host snail, amoeboid cells of *C. owczarzaki* attach by means of a peduncle to the surface of *Schistosoma* sporocysts and in a short period penetrate and lyse the sporocyst contents (Owczarzak *et al.*, 1980). Infection of a snail by *C. owczarzaki* offers the host a degree of immunity from *Schistosoma* infection (Stibbs *et al.*, 1979).

Since cultured *C. owczarzaki* cells are amoeboid with long unbranched filopodia (see Hertel *et al.*, 2002) they were initially attributed to *Nuclearia*, a genus hitherto comprising free-living naked filose amoebae (Owczarzak *et al.*, 1980). However, Amaral Zettler *et al.* (2001) carried out an ultrastructural and phylogenetic (SSU rDNA) analysis of four *Nuclearia* species in culture; three were free-living and the fourth was the symbiont previously studied by Owczarzak *et al.* (1980). Phylogenetic analysis provided strong bootstrap support for a close relationship between the three free-living *Nuclearia* species but no support for their relationship with the snail symbiont. Ultrastructural analysis also revealed differences: the symbiont was smaller (3–5 μm) than the free-living species (10–20 μm); it had flattened as opposed to discoidal mitochondrial cristae; it was not bacterivorous but fed on the cell contents of the host by means of a peduncle. On the basis of these differences, Hertel *et al.* (2002) re-classified the snail symbiont as a new taxon, *Capsaspora owczarzaki*, which they considered to be more closely related to the Ichthyosporea than to other *Nuclearia* species. The complete life cycle of *C. owczarzaki* has yet to be resolved, although it is capable of encystment in response to crowding. Flagellate cells have not been observed – this is commensurate with the absence of approximately 75% of the genes required for flagellar construction and motility (Stibbs *et al.*, 1979; Hertel *et al.*, 2002; Suga and Ruiz-Trillo, 2013).

The phylogenetic position of *C. owczarzaki* was for a long time uncertain. One major problem is that no close relative has been identified, therefore in small-scale phylogenies *C. owczarzaki* tends to be found on an unstable, long branch within Holozoa (Amaral Zettler *et al.*, 2001;

Cavalier-Smith and Chao, 2003; Hertel *et al.*, 2002; Ruiz-Trillo *et al.*, 2004; Carr *et al.*, 2008a). However, at least four recent large-scale multigene studies have recovered *C. owczarzaki* as a sister group to the sedentary marine protist *Ministeria vibrans* (see below) (Shalchian-Tabrizi *et al.*, 2008; Burger *et al.*, 2009; Paps *et al.*, 2012; Toruella *et al.*, 2012). In order to accommodate this relationship, both *Capsaspora* and *Ministeria* were placed in a newly erected group, termed Filasterea (Shalchian-Tabrizi *et al.*, 2008). This putative taxon is characterised by a naked protoplast which exhibits long, tapering tentacles, each supported by an internal skeleton of microfilaments.

`MINISTERIA`

Currently there are two described species of *Ministeria*, namely *M. marisola* and *M. vibrans* (Patterson *et al.*, 1993; Tong, 1997c). Both are marine, free-living and comprise a small spherical cell, approximately 3 μm in diameter, with a regularly arranged surface covering of 'stiff radiating arms' (filopodia) of equal length. Each arm is probably supported by a core of microfilaments (Cavalier-Smith and Chao, 2003). The most important differences between the two species are the number of radiating arms, 14 in *M. marisola* and 16–20 (up to 30) in *M. vibrans*, and the presence of a stalk in *M. vibrans* that is thicker than the arms at the proximal end. When attached to a surface, *M. vibrans* appears to vibrate, hence the specific epithet. Cells are bacterivorous, capturing and engulfing food particles between the bases of the radiating arms (Tong, 1997c). Mitochondria possess flattened cristae. Flagellated cells have been reported but not illustrated (Patterson *et al.*, 1993). Cavalier-Smith and Chao (2003) suggest that the stalk may be a modified flagellum, although their illustration is not convincing on this feature. Only *M. vibrans* has been obtained in culture.

Although phylogenetic analyses place *Ministeria* within Holozoa, its position within the group is inconsistent (Cavalier-Smith and Chao, 2003; Ruiz-Trillo *et al.*, 2006; Steenkamp *et al.*, 2006). Four recent studies strongly recovered *M. vibrans* as the sister group to *Capsaspora owczarzaki* within the Filasterea (Shalchian-Tabrizi *et al.*, 2008; Burger *et al.*, 2009; Paps *et al.*, 2012; Toruella *et al.*, 2012).

10.4.4 Corallochytrea

Corallochytrea contains a single species, *Corallochytrium limacisporum*, which is an enigmatic, marine, non-photosynthetic protist originally isolated from coral reef lagoons associated with the Lakshadweep Islands in the Arabian Sea (Raghu-Kumar *et al.*, 1987). *C. limacisporum* is a spherical, single-celled organism, 4.5–20 μm in diameter, that undergoes several rounds of binary division to produce up to 32 endospores that are subsequently released through pores in the cell wall. Released spores are elongate and amoeboid, and undergo slow, sinusoidal movements (Mendoza *et al.*, 2002). No flagellated zoospores have been observed (Raghu-Kumar *et al.*, 1987). *C. limacisporum* is a saprotroph that grows on degrading seagrass blades and bleached corals (Sumathi *et al.*, 2006). *C. limacisporum* was originally described as a thraustochytrid fungus, although cells lack most of the key thraustochytridian features such as a sagenogenetosome, an ectoplasmic net, a multilayered covering of scales and the production of flagellated zoospores (Moss, 1985). Biochemically, *C. limacisporum* displays a fungal signature in that it possesses the enzyme α-aminoadipate reductase (α-AAR) involved in the α-aminoadipate (AAA) pathway of synthesising lysine and ergosterol. However, it also possesses the sterol C-14 reductase gene involved in the sterol pathway of both animals and fungi.

Molecular phylogenetic analyses place *C. limacisporum* within Holozoa (Fig. 10.2e), but its exact relationship has yet to be confirmed. SSU rDNA sequences weakly place *C. limacisporum* as a sister group to the choanoflagellates (Cavalier-Smith and Allsopp, 1996; Cavalier-Smith and Chao, 2003) or Ichthyosporea (Hertel *et al.*, 2002). Some multigene analyses also place *C. limacisporum* with Ichthyosporea (Carr *et al.*, 2008a), although Paps *et al.* (2012) place it between Filasterea and Ichthyosporea. In all phylogenies *C. limacisporum* tends to be positioned on a long branch, which adds to the difficulty in determining its true position.

10.4.5 Metazoa (animals)

METAZOA: DESCRIPTION OF SHARED MORPHOLOGICAL CHARACTERS

Multicellularity is a basic feature of the Metazoa and is characterised by embryogenesis, whereby a single cell (the zygote) develops into numerous differentiated cells and tissue types (Section 10.8.2). Individual cells lack cell walls and are typically held together in tissue layers by intercellular junctions and an extracellular matrix comprising a basal lamina with fibrous proteins, typically collagens. Sexual reproduction involves production of an egg cell that is fertilised by a smaller, often uniflagellated sperm cell.

Metazoa are phagotrophic and/or osmotrophic. Currently there are approximately 1.3 million described species in 35–40 phyla, depending on the classification followed (Edgecombe *et al.*, 2011).

METAZOAN MONOPHYLY

Uncertainty over the systematic position of the sponges led some nineteenth-century authorities to suggest that the sponges were a separate evolutionary grouping of their own (Parazoa/Enantiozoa) independent of other multicellular animals (Sollas, 1884). However, it is now clear from both morphological and molecular analyses that all animals, including sponges, are monophyletic. Because the monophyly is robust, it can be concluded that metazoan multicellularity evolved just once within the animal lineage (Eernisse and Peterson, 2004). Cavalier-Smith (2013) acknowledges the holophyly of the Metazoa and classifies it as the Kingdom Animalia.

METAZOAN PHYLOGENY

The allocation of animals into phyla based on morphology has changed little since the mid nineteenth century. In contrast, our understanding of the relationships between animal phyla has changed dramatically within the last 20 years as a result of the reinterpretation of morphological characteristics and the contribution of molecular phylogenetics. Molecular genetic analyses have now established a widely accepted 'new animal phylogeny' (Adoutte *et al.*, 2000; Philippe and Telford, 2006; Bourlat *et al.*, 2008). A fundamental division within the Metazoa is that between diploblasts and triploblasts. Diploblasts have two distinct germ layers – endoderm/gut and ectoderm/skin – while triploblasts have an additional intermediate layer, the mesoderm. Diploblasts comprise the Porifera (sponges), Placozoa (*Trichoplax adhaerens*), Cnidaria (jellyfish and sea anemones) and Ctenophora (sea gooseberries/comb jellies). In older texts, diploblasts were known as Radiata because they were considered to display radial symmetry. However, since some representatives do not exhibit any obvious form of symmetry they are now known as 'basal metazoa' or 'non-bilaterians'. Triploblasts are synonymous with Bilateria and are defined by three tissue layers and bilateral symmetry. While the defining characteristics of the Bilateria may be discernible in some non-bilaterians (Martindale *et al.*, 2004), nevertheless bilaterians are a well-circumscribed monophyletic clade (Telford, 2006). Within the Bilateria there are three fundamental clades, namely: Lophotrochozoa – comprising rotifers, bryozoans, brachiopods, molluscs, annelids nemerteans, sipunculans and platyhelminths (Halanych *et al.*, 1995); Ecdysozoa – comprising nematodes, arthropods, onycophorans, priapulids and tardigrades (Aguinaldo *et al.*, 1997); Deuterostomia – comprising Xenoturbellida, echinoderms, hemichordates and chordates.

BASAL METAZOA (NON-BILATERIANS)

On the basis of morphological, molecular and fossil evidence it is widely accepted that non-bilaterians are the earliest-branching Metazoa. However, the four constituent groups (Porifera, Placozoa, Cnidaria and Ctenophora) display radically different adult body plans and developmental pathways from one another. The exact evolutionary relationship of these early branching taxa to each other is unresolved (Fig. 10.5). To complicate matters further there are exceptions to the two defining morphological characteristics of non-bilaterians, namely diploblasty and the absence of bilateral symmetry (Baguñà *et al.*, 2008).

PORIFERA (SPONGES)

Sponges are generally vase-shaped but may be lumpy, encrusting growths on solid substrata. They are characterised by numerous canals and chambers which open to the outside by pores (ostia). A constant stream of water flows inwards through the ostia, passing through the canals and chambers, and is expelled through a larger opening, the osculum. The body of a sponge is made up of two layers of cells separated by an intermediate layer (mesohyl) containing amoeboid cells and connective tissue reinforced with tough flexible fibres of a collagen-related protein called spongin, or with spicules containing calcium carbonate or silica. The surface layers of cells (pinacoderm) do not display bilaterian-like epithelial organisation and lack a basal membrane (see Section 10.8.2). Aquiferous chambers and canals are lined by choanocytes (choanoderm) similar in morphology to choanoflagellate cells (see Section 1.4.1). Depending on species, from 10 to 15 cell types have been recognised.

Extant sponges are distributed into four classes mainly according to the composition of their skeletons. (1) Demospongiae – marine and freshwater; possess a skeleton comprising spongin or spongin and silica spicules. (2) Hexactinellida (glass sponges) – rare, deep-sea sponges; skeleton made up of four- and/or six-pointed siliceous spicules. (3) Homoscleromorpha – originally classified within Demospongiae but on the basis of molecular phylogenetic analyses allocated to a separate class; marine; skeleton with or without silica spicules. (4) Calcarea –

exclusively marine; skeleton comprises calcium carbonate spicules.

Phylogenetic studies based solely on morphology recover extant sponges as being monophyletic, which is not surprising since they are characterised by a unique body plan and water canal system. However, molecular phylogenies are in disagreement as to whether the sponges are monophyletic or paraphyletic (for discussion, see Wörheide *et al.*, 2012). In the monophyletic model, the four sponge clades share a common ancestor and the Homoscleromorpha and Calcarea are more closely related to each other than they are to the siliceous sponges (Demospongiae and Hexactinellida) (Fig. 10.4b–c) (Philippe *et al.*, 2009; Pick *et al.*, 2010). In some paraphyletic models the siliceous sponges are the first to branch, followed by the Calcarea; the Homoscleromorpha would be closer to the Eumetazoa (placozoans, cnidarians, ctenophorans and Bilateria) than they are to the remainder of the sponges (Fig. 10.4a) (Borchiellini *et al.*, 2001; Sperling *et al.*, 2009). In this scenario the last common ancestor of all metazoans would have had a poriferan body plan (Adamska *et al.*, 2011). Wörheide *et al.* (2012) conclude that, on balance, there is stronger support for sponge monophyly rather than paraphyly.

PLACOZOA (`TRICHOPLAX ADHAERENS´)

This is an enigmatic phylum at present containing a single named species, *Trichoplax adhaerens*, although molecular analyses suggest cryptic speciation (Voigt *et al.*, 2004). Individual specimens consist of a flattened plate about 1–3 mm across with an irregular outline and a constant thickness of about 20 μm. The body comprises three layers: an upper epithelial layer containing flattened monociliated cells; an intermediate layer of star-shaped cells that form a contractile syncytial net; and a lower epithelial layer containing a mixture of narrow, monociliated cells with microvilli and gland cells. *Trichoplax* has basic cell–cell junctions between epithelial cells but is reported to lack an underlying basal lamina and an extracellular matrix (ECM). However, its genome contains a diverse set of genes that code for putative ECM proteins so it is possible that an ECM is present but not resolvable by standard microscopical means (Srivastava *et al.*, 2008; Schierwater *et al.*, 2009a). Digestion of prey occurs outside the lower epithelium and the products are pinocytosed by the monociliated cells. Movement is effected by a combination of ciliary activity and body shape changes. As far as is currently known, *Trichoplax* has a monophasic life cycle and reproduces by fission. However, *T. adhaerens* has a full

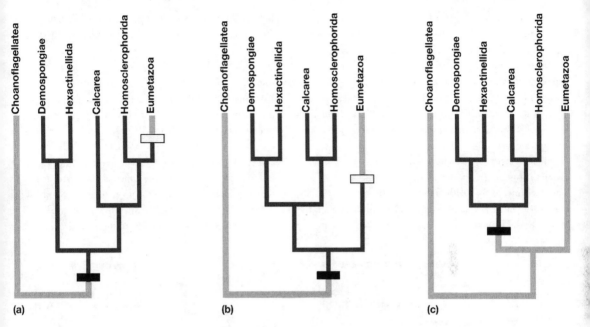

Fig. 10.4 Most parsimonious scenarios for evolution of sponge body plan characters according to: a. Sponge paraphyly; b–c. Sponge monophyly with sponge body plan acquired in stem lineage to Metazoa (b) and in stem lineage to Porifera (c). Filled horizontal bar indicates acquisition of sponge body plan; open horizontal bar indicates loss of sponge body plan. Darker branches indicate presence of sponge body plan. Based on Wörheide *et al.* (2012) and Philippe *et al.* (2009).

complement of meiosis genes which strongly suggests it can reproduce sexually (Carr *et al.*, 2010).

The phylogenetic position of Placozoa within the Metazoa is highly contentious (Fig. 10.5) (Blackstone, 2009). *Trichoplax* possesses some metazoan features – for example, the body comprises four distinct cell types and surface cells are organised into epithelia. However, it also lacks certain metazoan features, including the absence of basal lamina and ECM and there is a lack of direct evidence of sexual reproduction. Arguments have been made in favour of the 'simple' placozoan body plan being primary – that is, the evolutionary lineage leading to *Trichoplax* was always highly simplified – or secondary, which means that it is a simplified descendant of a more complex ancestor, possibly a cnidarian. The current balance of opinion taking into account morphological and molecular data is in favour of a primary scenario (see Blackstone, 2009 for references and discussion of basal relationships within non-Bilateria below).

CNIDARIA (SEA ANEMONES, CORALS AND JELLYFISH)
There are four generally recognised classes of cnidarians: Hydrozoa, Scyphozoa, Cubozoa and Anthozoa. They are represented by polyps, such as sea anemones and corals,

and by medusae such as jellyfish. A polypoid or medusoid cnidarian is a radially or biradially symmetrical, uncephalised animal with a single body opening, the 'mouth'. Tentacles surround the mouth, each containing thousands of stinging cells known as cnidocytes that serve for both offence and defence. The possession of cnidocytes, each of which possesses a cnida (e.g. a nematocyst) is the defining characteristic of the phylum, hence the name.

The body plan of Cnidaria is based on two cell layers; the inner layer is called the gastrodermis (endoderm) and the outer the epidermis (ectoderm). Between the two layers is the mesogloea, which is an extracellular matrix of connective tissue containing scattered cells. In hydrozoan medusae (jellyfish) a discrete mesoderm-like third layer called the entocodon is also present which contains striated muscle (Seipel and Schmid, 2005). The sea anemone *Nematostella vectensis*, although lacking a mesoderm, encodes most of the transcription factors crucial for mesoderm and muscle development (Martindale *et al.*, 2004). This has led some authors to suggest that Cnidaria and Bilateria are derived from a triploblast ancestor, thus confounding the traditional diploblastic classification (Seipel and Schmid, 2006).

Ctenophora are marine pelagic predators that range in size from a few millimetres to 1.5 m in size. They are morphologically diverse and are distinguished by their jelly-like appearance with biradial symmetry. Eight longitudinal rows of comb plates, each plate comprising ranks of fused cilia that beat synchronously, serve for locomotion. Two retractable tentacles and specialised adhesive cells (colloblasts) are used to trap prey. The body comprises two epithelia, each two cell layers thick, separated by a large volume of mesogloea which contains groups of muscle and mesenchyme cells. Although ctenophores superficially resemble cnidarian medusae and were traditionally classified with the Cnidaria in the Coelenterata (see Section 1.5),

both morphology and molecular phylogeny confirm that Ctenophora is a separate grouping (phylum) within the non-Bilateria.

BASAL RELATIONSHIPS WITHIN NON-BILATERIA
Five major extant lineages result from the deepest splits in metazoan phylogeny: Porifera, Placozoa, Cnidaria, Ctenophora and Bilateria (Fig. 10.5). All, with the possible exception of Porifera, are monophyletic. The topology and rooting of this five-taxon tree are still in dispute, with different studies giving either unresolved or strongly conflicting results, or they lack whole-genome data from all the groups involved (see Fig. 10.5) (Edgecombe *et al.*, 2011). This conflict is not surprising since these studies are

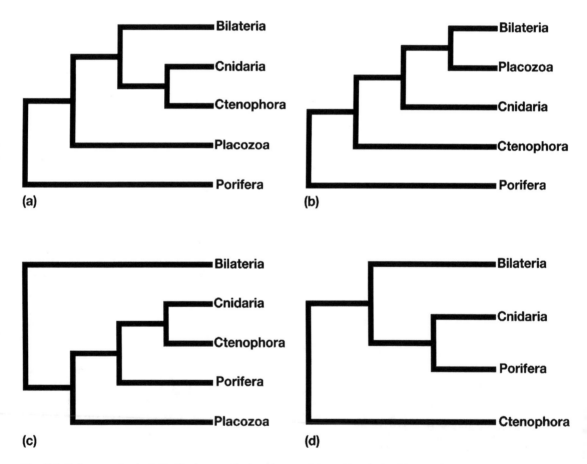

Fig. 10.5 Various postulated relationships between the five metazoan clades. a. Hypothesis with monophyletic Coelenterata (Cnidaria + Ctenophora) (Philippe *et al.*, 2009). b. Hypothesis with monophyletic Eumetazoa but with Ctenophora as sister to all eumetazoans (Pick *et al.*, 2010). c. Hypothesis with Bilateria as sister group to a clade that contains placozoans as sister to Porifera and Coelenterata (Cnidaria and Ctenophora) (Dellaporta *et al.*, 2006; Wang and Lavrov, 2007; Schierwater *et al.*, 2009b). d. Hypothesis with Ctenophora as sister group to a clade that contains Bilateria as sister to Cnidaria and Porifera (Dunn *et al.*, 2008).

attempting to reconstruct cladogenetic events that took place hundreds of millions of years ago.

The long-established morphological view that Porifera is the basal metazoan phylum and therefore the sister group to all other animals is based on the lack of tissue organisation in sponges, the absence of nervous and muscle systems and the similarity of choanocytes to choanoflagellates. Some molecular phylogenies support this relationship (Fig. 10.5a, b) (Philippe *et al.*, 2009; Pick *et al.*, 2010). However, whether or not this implies that the sponge body plan was ancestral to all metazoans depends on whether sponges are considered to be paraphyletic or monophyletic and how the latter is interpreted. Fig. 10.4a illustrates the paraphyletic scenario whereby the sponge body plan is ancestral to the Metazoa and eumetazoans are derived by loss of sponge characters. With respect to the monophyletic scenario there are two possibilities depending on whether the sponge body plan was acquired in the stem lineage of Porifera (Fig. 10.4c) or in the stem lineage of the Metazoa (Fig. 10.4b). In the latter case the eumetazoans would be derived by subsequent loss of sponge characters (Wörheide *et al.*, 2012).

Analysis of mitochondrial genomes recovered the Bilateria as sister to non-bilaterians rather than originating from within the non-Bilateria (Fig 10.5c) (Schierwater *et al.*, 2009a). One of the consequences of this hypothesis is the independent evolution of a nervous system in the Coelenterata (Cnidaria and Ctenophora) and Bilateria (Schierwater *et al.*, 2009b). This analysis also places the Placozoa as basal to other non-bilaterians, thereby resurrecting the 'placula theory' of Bütschli. There are conflicting scenarios regarding the relative order of individual non-bilaterian phyla in the various published trees. In some, the Ctenophora and Cnidaria are reunited in a 'Coelenterata' clade (Fig. 10.5a, c). Extensive discussion of the trees illustrated in Fig. 10.5 can be found in the references listed in the text above.

10.5 NUCLETMYCEA

10.5.1 Cristidiscoidea (Nucleariida and *Fonticula alba*)

NUCLEARIIDA

Nucleariids are a group of naked filose amoebae consisting of spherical or flattened cells with radiating, unbranched or branched filopodia (Amaral Zettler *et al.*, 2001; Yoshida

et al., 2009). The majority of species are between 10 and 30 μm in size and have a single nucleus, although the type species, *Nuclearia delicatula*, is 30–60 μm in size and is multinucleate (Yoshida *et al.*, 2009). Mitochondrial cristae are discoidal or flattened (Amaral Zettler *et al.*, 2001). Cells of some species are covered by an ECM and some are cyst-producing. Nucleariids are predominantly freshwater or soil amoebae that feed phagotrophically on algae and bacteria. Flagellated cells have not been observed.

On the basis of light microscopy the nucleariids were placed together with other filose amoebae, including the testate amoebae, in the Filosea, a sub-group of Amoebozoa. However, Patterson (1999) questioned the monophyly of this grouping on the grounds that filopodia were likely to have evolved on multiple occasions and because of the variety of mitochondrial cristae recorded. Amaral Zettler *et al.* (2001), in an SSU rDNA phylogeny, confirmed the polyphyletic composition of the Filosea and demonstrated that the nucleariids were placed 'near the animal–fungal divergence'. This study also revealed that one *Nuclearia* strain, later reclassified as *Capsaspora owczarzaki*, was more closely related to both Ichthyosporea and Metazoa than to the other nucleariids sampled (see Section 10.4.3). Subsequent, multiple-gene phylogenies robustly place nucleariids as the sister group to Fungi (Ruiz-Trillo *et al.*, 2004; Steenkamp *et al.* 2006; Moreira *et al.*, 2007).

`FONTICULA ALBA`

Fonticula alba was originally described as a cellular slime mould because it consists of a unicellular amoeboid protist that aggregates to form a multicellular fruiting body. Cells at the tip of the fruiting body encyst and form spores that are forcibly ejected when mature. Under suitable conditions spores germinate into amoebae that possess filopodia used for capturing bacterial prey. *F. alba* has always been an enigmatic species because its amoeboid and multicellular stages are unlike those of any other slime mould. Currently there is only one species and there are no known close relatives.

A phylogenetic analysis based on five nuclear-encoded genes recovered *F. alba* as an opisthokont that branches as a sister group to the nucleariids (Fig. 10.2e) (Brown *et al.*, 2009). *Fonticula* and *Nuclearia* are sisters to the Fungi and for this well-supported clade Brown *et al.* (2009) proposed the name Nucletmycea.

10.5.2 Fungi

Fungi are defined by their heterotrophic (absorptive) nutrition and filamentous (mycelial) growth patterns comprising multinucleate hyphae; cell walls when present contain β-glucan and usually chitin, at least in spore walls. Fungi utilise the α-aminoadipate (AAA) pathway of synthesising lysine. Motile cells, when present, possess a single, smooth flagellum.

Until recently, a combination of morphology and molecular phylogeny resolved the Fungi into four distinctive phyla divided between two early-diverging lineages, Chytridiomycota and Zygomycota, and two later-diverging lineages, Ascomycota and Basidiomycota (Liu *et al.*, 2006). Evidence from molecular phylogenies suggests that neither of the early-diverging lineages is monophyletic, while there is strong morphological and molecular support for a close relationship between Ascomycota and Basidiomycota, which are now placed in the subkingdom Dikarya (Hibbet *et al.*, 2007) (= Neomycota (Cavalier-Smith, 2013)).

In seeking the possible ancestors of the Fungi and exploring their relationship with other opisthokonts it is necessary to study the relationships among the 'basal fungal lineages' and in particular within the Chytridiomycota. Chytrids are a group of aquatic and terrestrial fungi that reproduce by means of motile cells (zoospores) typically propelled by a single posteriorly directed flagellum. Until recently all zoosporic fungi were classified within the Chytridiomycota and allocation into sub-groupings was based on their mode of reproduction and zoospore ultrastructure (see James *et al.*, 2006b for references). However, a six-gene phylogeny indicated that Chytridiomycota was polyphyletic, and consisted of four major lineages that had retained zoospore development as a means of reproduction (James *et al.*, 2006a). The earliest diverging of these lineages was *Rozella* spp., a large and diverse grouping (Lara *et al.*, 2012), some of which are parasites of chytrids and oomycetes (see also Cryptomycota (Jones *et al.*, 2011)). The zoospores of *Rozella* differ from those of other chytrids in that the flagellum emerges from the base of a relatively long cavity (Held, 1975). The second, dormant, basal body is short and is at an angle of approximately 60° to the flagellar basal body, which is unusually long. Of the other three lineages, one constitutes the 'core chytrids' (Chytridiales), one represents the Blastocladiales and one contains *Olpidium brassicae*. James *et al.* (2006b) provide a molecular phylogeny of the chytrids with superimposed illustrations of flagellar basal bodies. In most species the flagellar basal body is accompanied by a parallel dormant basal body.

Aphelidae and Microsporidia are further additions to the basal fungal lineages. In a recent molecular phylogenetic study, the aphelid *Amoeboaphelidium protococcarum* was recovered in a basal clade, together with *Rozella* and a collection of microsporidians (Karpov *et al.*, 2013). Ultrastructurally this is a mixed grouping; *Rozella* is genuinely flagellate, *Amoeboaphelidium* possesses a 'posterior' immobile pseudocilium containing microtubules and microsporidians are non-flagellate (Capella-Gutiérrez *et al.*, 2012). The coincidence of uniflagellated cells being present in most basal fungal lineages as well as in choanoflagellates, Metazoa and some Ichthyosporea has led to the general acceptance that the ancestral opisthokont was uniflagellate (the 'unikont' myth). Thus, where flagella are absent as, for example, in microsporidia and higher fungi, it would appear that the flagellar apparatus has been secondarily lost. With respect to fungal phylogeny, James *et al.* (2006a) infer that the flagellar loss could have occurred on six separate occasions.

10.6 OPISTHOKONT EVOLUTION: A CONSENSUS BETWEEN MORPHOLOGY AND MOLECULAR PHYLOGENY?

Despite early suggestions that animals, fungi and choanoflagellates are closely related (Gams, 1947; Cavalier-Smith, 1987), Opisthokonta is basically a creation of the molecular era (Wainright *et al.*, 1993 *et seq.*). In the absence of molecular phylogenetic data, it is unlikely that the current assemblage of opisthokont taxa would have ever been grouped together. This highlights a frequently encountered paradox – morphology alone is not a satisfactory indicator of higher-level phylogeny, whereas relationships inferred from molecular phylogeny are often robustly supported by high bootstrap values. Is it possible to resolve this paradox? The morphological similarity between choanoflagellate cells and sponge choanocytes remains as obvious today as it was when first recorded in the nineteenth century (James-Clark, 1867b). However, whether this similarity reflects an underlying homology or is merely a coincidence is unresolved (Mah *et al.*, 2014). Other morphological characters frequently quoted as being shared by opisthokonts, namely "possession of a single 'posterior' flagellum on motile cells and mitochondria with flattened cristae", are, unfortunately, not reliable prospective (as opposed to retrospective) phylogenetic indicators (see Section 10.2.1). Several authors have attempted the hazardous task of trying to reconcile morphological diversity with molecular phylogeny in respect of basal opisthokonts (Nielsen, 2008; Cavalier-Smith, 2013). However, the problem with such attempts is that morphological discontinuities between higher-level systematic groupings are so large that they can only be filled by speculative narratives (Cavalier-Smith, 2013; Richter and King, 2013).

In most multigene, multi-taxon phylogenies choanoflagellates are recovered as the sister group to Metazoa (Fig. 10.3). However, analyses based solely on nuclear rDNA data give more variable or inconclusive results (see Table 10.1 for references). Some molecular phylogenies recover Porifera as being the basal metazoan phylum and therefore sister to all other animals (Philippe *et al.*, 2009; Pick *et al.*, 2010). If this is correct then no matter whether the sponges are paraphyletic or monophyletic (see Fig. 10.4a, b) the sponge body plan could have been basal to all metazoa, in which case sponge choanocytes could be considered as being homologous with choanoflagellates. However, if another group is basal to the metazoa then the sponge body plan with its water canal system would be at a minimum present only in the stem group of extant sponges (Fig. 10.4c) and no immediate inferences can be made about the body plan of the metazoan ancestor (Wörheide *et al.*, 2012).

The morphological association between choanoflagellates and fungi is also based on the uniflagellate condition of their respective motile cells. Additionally, a similar radial arrangement of cytoskeletal microtubules with dark-staining circular bands surrounds the flagellar base in choanoflagellates and monoblepharid chytrids (Fuller and Reichle, 1968; Mollicone and Longcore, 1994). However, this is probably a superficial similarity in that the microtubules of monoblepharid chytrids are not cortical but are directed towards the nucleus, and the ring is not continuous but interrupted by the parallel dormant basal body. There is no evidence that the Monoblepharidales are ancestral to the remainder of the Chytridiomycota.

10.7 ESTIMATED DATING OF OPISTHOKONT DIVERSIFICATION

The 'molecular clock hypothesis' combines rates of molecular change with fossil calibrations to infer the time in geological history when two taxa diverged. However,

there are complications associated with this hypothesis in that the rates of molecular evolution are not constant but heterogeneous, phylogenetic tree topology may be uncertain and point fossil calibrations can be unreliable. Recent improvements in Bayesian molecular dating techniques allow for 'relaxation' of the molecular clock as well as incorporation of multiple and flexible fossil calibrations. Divergence times can then be estimated even when the evolutionary rate varies among lineages and fossil calibrations involve substantial uncertainties.

Two studies incorporating multi-taxon sampling and relaxed molecular frameworks provide comparative divergence dates for opisthokont taxa. Berney and Pawlowski (2006) calibrated a tree based on SSU rDNA data and obtained the following divergence dates (Mya = million years ago): opisthokont (Holozoa/Nucletmycea) divergence 960 ± 185 (95% CI) Mya; choanoflagellate/Metazoa divergence 880 Mya; chytrid/fungi divergence 810 Mya. Parfrey *et al.* (2011) calibrated a tree based on 15 protein-coding genes and obtained the following divergence dates (Mya ± 95% highest posterior density (hpd) CI): opisthokont divergence (Holozoa/Nucletmycea) 1285 ± 303 Mya; *Capsaspora* and Ichthyosporea/choanoflagellate and Metazoa divergence 1071 ± 107 Mya; choanoflagellate/Metazoa 946 ± 89 Mya; *Capsaspora*/Ichthyosporea 928 ± 125 Mya; non-Bilateria/Bilateria 785 ± 63 Mya; chytrid/fungi 1089 ± 133 Mya.

An alternative approach to using sequence-based phylogenetic analyses for inferring divergence times is to use rare genomic changes associated with conserved amino acids (Chernikova *et al.*, 2011). The assumption is that any character shared by the majority of eukaryotes and its prokaryote outgroup is the ancestral state and therefore species with a substitution in a given position possess a derived state. Amino acid changes that meet this criterion are rare and thus the frequency of parallel emergence of such characters in different lineages (homoplasy) is expected to be low. Using this approach Chernikova *et al.* (2011) estimated that the mean opisthokont (Holozoa/Nucletmycea) divergence time from all methods of analysis was 949 Mya $\pm$ 311 SD. Srivastava *et al.* (2010) estimated that 28% of amino acid substitutions between humans and their last common ancestor with the choanoflagellates occurred on the metazoan stem lineage before the divergence of sponges from other animals. This pre-metazoan period was estimated to have extended approximately 150–200 million years.

10.8 EVOLUTION OF METAZOAN MULTICELLULARITY

10.8.1 Introduction

Multicellular organisms are discrete phenotypes that consist of more than one cell (Nedelcu, 2012). Such phenotypes may be stable and represent the major part of a life cycle, as in most multicellular lineages, or they may be transient and represent only a minor, sometimes optional, portion of a life cycle, as in some craspedid choanoflagellates (Dayel *et al.*, 2011). Multicellular organisms may be relatively simple in that they lack functionally specialised cells or they may be complex with many differentiated cell types. In most lineages, multicellularity develops from a single cell (spore or zygote) whose mitotic products fail to separate. However, in a relatively few lineages, for example myxobacteria (slime bacteria) and cellular slime moulds, multicellularity may arise by aggregation of individual cells (see also Mikhailov *et al.*, 2009). Although multicellularity has originated independently on many occasions and is widely represented within the 'tree of life', the evolution of complex multicellular body plans is limited to land plants (embryophytes), red and brown algae and the two major opisthokont lineages, namely Fungi and Metazoa (animals) (Knoll, 2012; Nedelcu, 2012).

Five categories of multicellular organisation can be recognised within Opisthokonta. First, aggregative development is displayed by *Fonticula alba*, a member of Nucletmycea (Brown *et al.*, 2009). The process of fruiting in *Fonticula* involves the aggregation of individual amoeboid cells to form a multicellular body in which spores develop. Although originally described as a cellular slime mould (Amoebozoa), molecular phylogenetic analysis subsequently confirmed *F. alba* as being an opisthokont (see Section 10.5.1). Second, 'pseudomulticellular' microcolonies or spheres of cells occur within some ichthyosporeans, such as *Sphaeroforma arctica* and *Amoebidium parasiticum* (Ruiz-Trillo *et al.*, 2007). Third, a transient multicellular colony stage may occur in the life cycles of some craspedid choanoflagellates (see Section 3.5). Fourth, fungal multicellularity is characterised by mycelia that comprise interconnected, walled filamentous cells, which may exhibit varying degrees of multinuclearity. Fifth, metazoan multicellularity is an exclusive feature of animals and is characterised by embryogenesis, whereby a single cell (the zygote) develops into numerous differentiated cells and tissue types.

The development of metazoan multicellularity from a unicellular ancestor is one of the 'major transitions' of evolution, a consequence of which has been a major increase in the level of biological complexity (Grosberg and Strathmann, 2007). The 'genetic toolkit' responsible for metazoan multicellularity comprises a select set of highly conserved genes from a limited number of gene families involved in cell adhesion, intercellular communication and coordination, tissue differentiation and programmed cell death (apoptosis). Most of these gene families have their origins on the metazoan stem, prior to the evolution of extant Metazoa, and are often the result of gene novelties combining with more ancient factors present in unicellular opisthokonts and even deeper within the eukaryotic phylogenetic tree (Srivastava *et al.*, 2010; Fairclough *et al.*, 2013).

The following discussion recounts some of the more important cell adhesion and signalling protein gene families associated with metazoan multicellularity and summarises what is currently known from comparative genomics about their unicellular, and in particular choanoflagellate, origins. The genome annotation of *Monosiga brevicollis* listed 78 protein domains that are shared exclusively with metazoans but are absent from plants, fungi and slime moulds, emphasising the close relationship between choanoflagellates and metazoans (King *et al.*, 2008). Many of these domains, such as cadherin-like domains, C-type lectin domains, extracellular matrix protein domains, tyrosine kinase domains and G-protein-coupled receptor domains are central to cell adhesion and signalling processes in metazoans, suggesting a role in the origin of multicellularity. Currently, whole genome sequences are available for the following unicellular holozoan opisthokonts: Filasterea – *Capsaspora owczarzaki*; Ichthyosporea – *Sphaeroforma arctica, Amoebidium parasiticum*; Choanoflagellatea – *Monosiga brevicollis, M. ovata, Salpingoeca rosetta* (*Proterospongia* sp.).

10.8.2 Metazoan multicellularity: epithelial and mesenchymal phenotypes

Multicellular complexity in Eumetazoa is based on two tissue phenotypes, namely epithelium and mesenchyme (mesenchymal connective tissue) (Tyler, 2003). Epithelial tissue, comprising sheet-like arrangements of cells, is a primary product of early embryogenesis and provides the cell-cell cohesion necessary for maintaining the integrity of a multicellular organism. It also functions as a barrier essential for maintaining a regulated internal environment independent from that of its external surroundings. During gastrulation, the primary mesenchyme develops from embryonic epithelial tissue by a process called 'epithelial–mesenchyme transition' (EMT). The resulting mesenchymal cells provide support and structure to adjacent epithelial tissues, particularly through the production of an extracellular matrix (ECM). Mesenchyme essentially refers to the morphology of embryonic cells that are capable of developing into a diversity of tissues including connective tissues, such as bone and cartilage, the blood circulatory system and mammalian lymphatic systems. From the evolution of a primitive trilaminar body pattern (epithelium–mesenchyme–epithelium) to the complex development and organogenesis of mammals, epithelial and mesenchymal phenotypes are a basic feature of normal development and physiology.

Three criteria distinguish the epithelial tissue phenotype as displayed by bilaterians (Tyler, 2003; Fahey and Degnan, 2010). First, component cells display an aligned polarity with clearly distinguishable apical and basal surfaces aligned in parallel. Second, cells are connected by 'belt-form' junctions that form a continuous structure around the circumference of the cell and usually include tight, septate and adherens junctions (Fig. 10.6) (Tyler, 2003; Franke, 2009; Abedin and King, 2010). Third, epithelial cells are associated with extracellular matrix at their basal and apical surfaces – that is always to a basal lamina and sometimes to an apical cuticle (Tyler, 2003). These three criteria, which are also well typified functionally and genomically, are illustrated in Fig. 10.6. In general, the genes responsible for establishing cellular polarity, forming intercellular junctions and constructing and regulating adhesion to the basal lamina are metazoan-specific and highly conserved. Non-bilaterians display an apparent series of intermediate steps in the differentiation of classical bilaterian epithelia (Tyler, 2003). For example, the aquiferous chambers and canals of sponges are bounded by a monolayer of choanocytes (choanoderm) and the outer sponge surface by a layer of flattened cells (pinacoderm); both coverings are epithelial-like in that the cells display polarity. However, belt-form junctions containing permanent tight junctions or desmosomes are absent, although 'plugged' junctions are common (Mackie and Singla, 1983; Leys et al., 2007). The nearest equivalent in sponges to the bilaterian basal lamina is the mesohyl that contains amoeboid cells and connective tissue reinforced with tough, flexible fibres of spongin or spicules containing calcium carbonate or silica (see Section 10.4.5). Cnidarians are almost wholly epithelial; their epidermis and gastrodermis fulfil all the criteria of epithelia. Between the two layers is the mesogloea, which is an extracellular matrix of connective tissue containing scattered cells (Tyler, 2003).

An important product of mesenchymal tissue, and a key mediator of metazoan multicellularity, is the ECM, which comprises a meshwork of organised fibrous polymers, including collagens and glycoproteins embedded within an amorphous mixture of nonfibrous components, predominantly proteoglycans. The fibrous proteins confer tensile strength and inextensibility while the amorphous substances impart turgor (Tanzer, 2006). As well as being a substrate for cell growth, ECM proteins and structures play important roles in the polarity, adhesion, migration, proliferation and differentiation of cells. Basement membranes, which are thin layers of specialised ECM associated with eumetazoan tissues, comprise a common set of interacting proteins, namely a core network of cross-linked type IV, XV and XVIII collagens in association with laminins (a trimer of related α, β, and γ subunits); nidogens, a laminin-binding glycoprotein; perlecan, a very large and complex heparan sulphate proteoglycan; and other glycoproteins (Tanzer, 2006).

10.9 CELL–CELL AND CELL–MATRIX ADHESION

A select repertoire of supergene families of transmembrane cellular adhesion molecules (CAMs) have been identified in the Metazoa as being responsible for cell–cell and cell–matrix adhesion. These CAMs include: the cadherins, which mediate interactions between neighbouring cells and form the basis of adherens junctions (Figs 10.6, 10.7); the integrins, which regulate interactions with the ECM; the selectins, which are found on leucocytes and endothelial cells; and members of the immunoglobulin and proteoglycan superfamilies.

10.9.1 Cadherins

Cadherins are the major CAMs responsible for Ca^{2+} dependent cell–cell adhesion, primarily as adherens junctions, and can be considered as the universal adhesion machinery for the generation and maintenance of metazoan multicellularity (Figs 10.6, 10.7) (Oda and Takeichi, 2011). In structure, most cadherins are large, multi-domain, single-span transmembrane glycoproteins characterised by two or more repeat copies of an extracellular ~110 amino acid protein (EC) domain that mediates adhesion with the EC domains of other cadherins. Ca^{2+} ions are essential for adhesion and bind at sites between adjacent cadherin EC domains (Fig. 10.7). Cadherins are assigned to a large

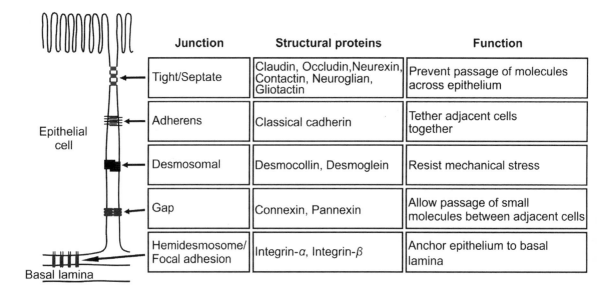

Junction	Structural proteins	Function
Tight/Septate	Claudin, Occludin, Neurexin, Contactin, Neuroglian, Gliotactin	Prevent passage of molecules across epithelium
Adherens	Classical cadherin	Tether adjacent cells together
Desmosomal	Desmocollin, Desmoglein	Resist mechanical stress
Gap	Connexin, Pannexin	Allow passage of small molecules between adjacent cells
Hemidesmosome/ Focal adhesion	Integrin-α, Integrin-β	Anchor epithelium to basal lamina

Fig. 10.6 Diagrammatic representation of vertebrate epithelial cell–cell and cell–matrix junctions. Each type of junction comprises a distinctive set of structural proteins and serves a specific function. Redrawn after Abedin and King (2010).

number of different subfamilies based on the number and arrangement of additional non-EC protein domains and sequence motifs that refine cadherin function (Nichols et al., 2012). For example, 'classical' cadherins are distinguished by the presence of a cytoplasmic region that binds p120-catenin and β-catenin at separate sites (Fig. 10.7). p120-catenin stabilises cadherins at the cell membrane, and β-catenin mediates the interactions of cadherins with the actin cytoskeleton via α-catenin; these processes play key roles in the function of adherens junctions. The classical or type-I cadherins are but one of many cadherin subfamilies, also including atypical or type-II cadherins, desmocollins, desmogleins, protocadherins and Flamingo cadherins, that are found in most metazoans. The cadherins have been shown to be involved in many biological processes other than cell adhesion; for instance, they can transfer information intracellularly by interacting with complex networks of cytoskeletal and signalling molecules and are involved in morphogenesis, polarity, cytoskeletal organisation and cell sorting/migration.

10.9.2 Pre-metazoan ancestry of cadherins

Searching currently available sequenced genomes of unicellular opisthokonts for cadherin EC domains reveals that there are 23 predicted cadherin genes in *Monosiga brevicollis*, 29 in *Salpingoeca rosetta* and only one in *Capsaspora*

owczarzaki (Nichols *et al.*, 2012). The absolute and relative abundances of cadherin genes in choanoflagellates are comparable to those of diverse metazoan genomes despite a lesser degree of morphological complexity (Abedin and King, 2008; Nichols *et al.*, 2012). In order to reconstruct the evolutionary relationships among cadherins from non-metazoans and early-branching metazoans, Nichols *et al.* (2012) grouped cadherins from choanoflagellates, *C. owczarzaki* and sponges according to shared structural features, including domain composition and arrangement. Mapping of the phylogenetic distribution of cadherin subfamilies revealed that they have origins that predate the evolution of Metazoa. In particular, three cadherin subfamilies were inferred as being present in the last common ancestor of choanoflagellates and metazoans. Of these, two subfamilies, namely lefftyrins and coherins, are shared by choanoflagellates and sponges to the exclusion of all other lineages. Members of a third premetazoan cadherin subfamily, a Hedgehog-related protein (Hedgling), are so far restricted to choanoflagellates, sponges and the cnidarian *Nematostella vectensis*, but are absent from *C. owczarzaki* and bilaterians (see Section 10.10) (Nichols *et al.*, 2012). Classical, metazoan-like cadherins with highly conserved cadherin cytoplasmic domains (see Fig. 10.7) have not been found in *M. brevicollis* or *Salpingoeca rosetta* (Abedin and King, 2008; Nichols *et al.*, 2012).

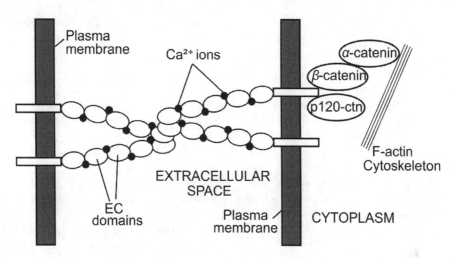

Fig. 10.7 Diagrammatic representation of the classical cadherin–catenin complex. Classical cadherins comprise transmembrane dimers and typically mediate homophilic adhesion between neighbouring cells. The extracellular portion comprises five repeat copies of an extracellular protein (EC) domain; a Ca^{2+} ion binds to each of the pockets between adjacent EC domains. The cytoplasmic region binds to p120-catenin and β-catenin at separate sites. β-catenin binds to α-catenin, which establishes a direct link between the cadherin–catenin complex and the actin cytoskeleton.

The contrast between the large number of cadherins in extant metazoan lineages and the low diversity of cadherins inferred in the metazoan stem (three subfamilies) raises the possibility that extant cadherin diversity arose from a handful of ancestral cadherin families that still exist today – although it is notable that none of the pre-metazoan cadherin families detected are present in extant Bilateria (Nichols *et al.*, 2012). Alternatively, although future studies of a broader diversity of choanoflagellates and early branching metazoans may reveal additional members of the pre-metazoan cadherin repertoire, it is also possible that cadherins present in the ancestors of metazoans and choanoflagellates were subsequently lost, or evolved beyond recognition, in both lineages.

The pre-metazoan existence of cadherin-containing (EC motif) proteins in *Monosiga brevicollis* raises the question as to their functional role in a unicellular choanoflagellate. Abedin and King (2008) attempted to answer this question by raising antibodies against an extracellular portion of a fusion protein containing cadherin EC motifs (MBCDH1). Subcellular cadherin localisation was revealed by probing cells with the antibody and staining with a rhodamine-labelled secondary antibody either before or after cell membrane permeabilisation. MBCDH1 (and MBCDH2) was detected in four regions of the choanoflagellate cell: the apical collar of actin-filled microvilli;

the basal pole of the cell which is also actin-rich; an unidentified structure at the apical end of the cell; and puncta within the cell body. Thus the association of a cadherin-containing protein with an actin cytoskeleton may pre-date diversification of choanoflagellates and Metazoa (Abedin and King, 2008). Other protein domains linked to cadherin EC repeats include: SH2 (Src homology type 2), N-hh (hedgehog N-terminal peptide), LamG (Laminin G) and EGF (epidermal growth factor), immunoglobulin and transmembrane domains. The functions of these domains suggest that some choanoflagellate cadherins may mediate intracellular signalling.

10.9.3 Integrins

Integrins are heterodimeric transmembrane receptors consisting of one α and one β subunit. They link the ECM outside a cell to the actin (microfilament) cytoskeleton within the cell (Fig. 10.8) (Hynes, 2011). The extracellular domains of integrin receptors bind divalent cations and ECM ligands, such as fibronectin, vitronectin, collagen and laminin. Adhesion between a cell and the ECM involves the formation of 'cell adhesion complexes', which comprise many integrin dimers in association with many cytoplasmic proteins, including α-actinin and talin, both of which directly bind to the intracellular domain of the integrin-β subunit, and paxillin and vinculin, both of which are scaffolding proteins that indirectly bind to integrin-β by means of talin and α-actinin (Fig. 10.8). An important element of the integrin adhesion machinery is the heterotrimer IPP complex, which is composed of ILK (integrin-linked kinase), PINCH (particularly interesting Cys-His-rich protein) and Parvin. This complex plays an important role in integrin-mediated signalling, regulating apoptosis and cell dynamics. Finally, integrin-mediated signalling occurs mainly utilising two cytoplasmic kinases, namely c-Src tyrosine kinase and FAK (focal adhesion kinase), known to be concentrated adjacent to the integrin adhesion machinery. Many other proteins are indirectly involved

with the integrin adhesion complex (Sebé-Pedrós et al., 2010).

10.9.4 Pre-metazoan ancestry of integrins

Until recently integrins were considered to be metazoan-specific. However, in a comparative genomic analysis of the integrin adhesion machinery of unicellular opisthokonts, including *Monosiga brevicollis* and *Capsaspora owczarzaki*, amoebozoans including *Dictyostelium discoideum* and the apusozoan *Amastigomonas* sp., Sebé-Pedrós et al. (2010) found that core components of the integrin adhesion complex are encoded by these taxa (Table 10.4). Of special note is the apusozoan *Amastigomonas* sp., which possesses all of the components of the canonical metazoan integrin adhesion complex, except for the signalling molecules FAK and c-Src (Table 10.4). Since Apusozoa is either the sister group to Opisthokonta or sister to the mega-assemblage Opisthokonta–Amoebozoa (see Section 10.1), this would seem to indicate that the integrin adhesion machinery evolved well before the origin of Metazoa and Fungi (Sebé-Pedrós and Ruiz-Trillo, 2010; Sebé-Pedrós et al., 2010; Hynes, 2011).

With respect to Opisthokonta, *Capsaspora owczarzaki*, a member of Filasterea and sister to the choanoflagellates, has the canonical metazoan-type integrin adhesion and signalling machinery with a full repertoire of integrin adhesion complex components including FAK and c-Src (Table 10.4). This suggests that the canonical metazoan-type integrin adhesion machinery is either specific to holozoans or that it was present before apusozoan evolution, but that the signalling molecules FAK and c-Src were subsequently lost within the apusozoan lineage.

With respect to the choanoflagellates, the situation is more complicated in that although an integrin-α orthologue was initially thought to be present in the choanoflagellate *Monosiga brevicollis* (King et al., 2008), on subsequent scrutiny of the genomes of *Monosiga brevicollis* and *Salpingoeca rosetta* a bona fide integrin-α was not detected (Sebé-Pedrós et al., 2010). Integrin-β homologues and components of the adhesion and signalling complex are also absent (Table 10.4). A possible explanation is that these domains were lost subsequent to the evolution of the opisthokont lineage. In fact, each choanoflagellate and fungal species examined possesses a distinctive repertoire of integrin adhesome components, presumably resulting from different lineage-specific losses (Sebé-Pedrós et al., 2010; Hynes, 2011).

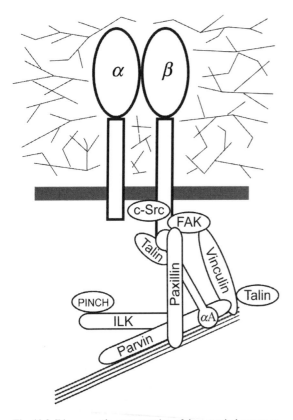

Fig. 10.8 Diagrammatic representation of the canonical metazoan integrin adhesion complex. The β-integrin tail interacts with talin and α-actinin (αA); paxillin and vinculin are scaffolding proteins. The heterotrimer IPP complex, comprising ILK, PINCH and Parvin, plays an important role in integrin-mediated signalling, regulating apoptosis and cell dynamics. c-Src and FAK are cytoplasmic kinases also involved in signalling. Redrawn after Sebé-Pedrós et al. (2010).

Table 10.4 *Distribution of different components of the integrin adhesion complex.*

	Integrin-α	Integrin-β	FAK	c-SRC	ILK	PINCH	Parvin	Paxillin	Talin	Vinculin	α-actinin
Monosiga brevicollis			±	+			±	+	+	+	+
Salpingoeca rosetta				+					+	+	+
Capsaspora owszarzaki	3–4	4	+	+	+	+	+	+	+	+	+
Dictyostelium discoideum						+		+	+	+	+
Amastigomonas sp.	1	1			+	+	+	+	+	+	+

The number of integrin α and β homologues is shown. + indicates the presence of clear homologues, whereas ± indicates the presence of putative or degenerate homologues. Absence of symbol indicates that a homologue is absent. Based on Sebé-Pedrós *et al.* (2010).

10.9.5 Other adhesion molecules

The genome of *Monosiga brevicollis* contains 12 genes that encode C-type lectins, two of which are transmembrane proteins (King *et al.*, 2008). C-type lectins mediate cell adhesion and signalling through calcium-dependent recognition and binding of specific sugars. Whereas soluble metazoan C-type lectins have functions ranging from pathogen recognition to ECM organisation, transmembrane C-type lectins mediate specific adhesive activities such as contact between leucocytes and vascular endothelial cells, cell recognition and molecular uptake by endocytosis (King *et al.*, 2008). The genome of *M. brevicollis* also contains immunoglobulin domains that have both adhesive and immune functions.

10.9.6 Extracellular matrix molecules

Monosiga brevicollis and *Salpingoeca rosetta* each encode two proteins that contain repeated collagen-like Gly–X–Y repetitive sequence motifs (where X and Y are frequently proline and hydroxyproline, respectively) and three other proteins that contain collagen C-propeptide-like (COLFI) domains (King *et al.*, 2008; Exposito *et al.*, 2008). Essential accessory proteins for collagen fibril assembly, collagen propeptidases and lysyl oxidase, are not encoded. Thus, a current model is that fibrillar collagens originated in the metazoan stem by domain shuffling (Exposito *et al.*, 2008).

Similarly, although some of the individual domains are present, for example the laminin G domain, none of the core basement membrane components are encoded in the choanoflagellates examined to date (King *et al.*, 2008). Both *M. brevicollis* and *S. rosetta* encode a fibrillin-like protein.

10.9.7 Origin of metazoan microvilli and filopodia

Recently, Sebé-Pedrós *et al.* (2013) have carried out a taxon-rich genomic survey to determine the origin and history of proteins required for the formation of metazoan filopodia. They found that homologues of key metazoan filopodial components, for example the actin cross-linking protein fascin and the motor protein myosin X, were present in other holozoans, including choanoflagellates (*Salpingoeca rosetta* and *Monosiga brevicollis*) and the filasterean *Capsaspora owczarzaki*, but absent from the Fungi, apusozoans and amoebozoans analysed.

10.10 CELL–CELL SIGNALLING TRANSDUCTION PATHWAYS

Animal signalling transduction pathways are characterised by core suites of molecules that transmit a signal from one cell (signalling cell) to another cell (receiving cell). Typically, this involves a secreted (diffusible) or cell-surface-associated ligand produced by the signalling cell being bound by a surface (transmembrane) receptor of a receiving cell (Fig. 10.9). With this binding, a cascade of molecular interactions is initiated within the receiving cell, culminating in the activation or inhibition of a target protein that is a component of a cellular process such as transcription, secretion, motility, proliferation or apoptotic cell death. In development, the most frequent target of signalling is transcription, and some pathways only affect transcription. In this manner, a signal molecule (ligand) released by a signalling cell can initiate a programme of activity or gene expression in a receiving cell (Fig. 10.9) (Gerhart, 1999; Richards and Degnan, 2009).

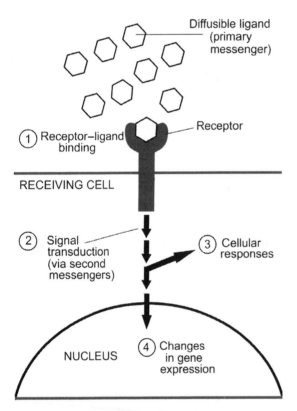

Fig. 10.9 Diagram illustrating the key stages in a signal transduction pathway. (1) An extracellular signal molecule (ligand) binds to a specific transmembrane receptor on the receiving cell. This causes a conformational change in the receptor which (2) activates transduction of the signal via a cascade of signalling proteins to a specific target protein which, either (3) results in one of a variety of cellular responses or (4) effects a change in gene expression.

Seven major intercellular signalling transduction pathways have been identified as universal to Eumetazoa (Pires-daSilva and Sommer, 2003; Nichols *et al.*, 2006). These include: wingless related (Wnt); transforming growth factor-β (TGF-β); Janus kinase/signal transducer and activator of transcription (JAK/STAT); nuclear hormone receptor (NHR); Notch; Hedgehog (Hh) and receptor tyrosine kinase (RTK) pathways. These seven pathways are a subset of a larger group of conserved signalling pathways shared by most animals (Gerhart, 1999) and are used repeatedly throughout the development of individuals and throughout eumetazoan evolution.

10.10.1 Pre-metazoan ancestry of metazoan signalling pathways

Of the seven major eumetazoan signalling pathways, no ligand or receptor orthologues were identified in the *M. brevicollis* genome from the Wnt, TGF-β and NHR pathways. The only evidence of the JAK/STAT pathway is an apparent STAT-like gene that encodes a STAT DNA-binding domain and a partial SH2 domain (King *et al.*, 2008). However, the *M. brevicollis* genome does provide an insight into the evolution of the Notch, Hedgehog and RTK signalling pathways.

10.10.2 Notch signal transduction pathway

While there is evidence that components of the Notch pathway existed prior to metazoan evolution, the canonical Notch pathway only exists in metazoans (Gazave *et al.*, 2009). Nine Notch-associated genes are specific to metazoans and none of these is found in the genome of *Monosiga brevicollis*. However, cassettes of protein domains found in metazoan Notch transmembrane receptors, including EGF (epidermal growth factor), NL (Notch/LIN-12 repeats) and ANK (ankyrin repeats), are encoded on separate *M. brevicollis* genes and in arrangements that differ from metazoan Notch proteins. King *et al.* (2008) suggest that, as a result of domain shuffling, these separate protein domains in *Monosiga* could have become linked together in a combination found in the canonical metazoan Notch receptor.

10.10.3 Hedgehog signal transduction pathway

Originally defined as a result of the genetic analysis of *Drosophila melanogaster*, the components of the Hh signalling pathway have subsequently been functionally characterised in a number of vertebrate species (mouse, zebrafish and human) and have also been identified through genome sequence analyses in species from a wide range of animal phyla. These studies have revealed a high level of conservation of the core components of the signal transduction pathway that is probably universal within Eumetazoa (Ingham *et al.*, 2011). Sponges possess a large extracellular membrane protein called Hedgling but lack the Hog domain (Bürglin, 2008). In bilaterians, the Hedgehog (Hh) intercellular signalling pathway plays a fundamental role in cell patterning, cell proliferation and participates in the development of tissues and organs during the stages of animal embryogenesis. It exerts its effect by influencing the transcription of many target genes in a concentration-dependent manner.

Hh proteins are a family of secreted signal molecules (ligands) exclusive to the Eumetazoa. They are synthesised within a signalling cell as inactive precursors composed of two distinct domains: an amino-terminal signalling 'Hedge' domain (HhN); and a carboxy-terminal, autoprocessing 'Hog' domain (HhC) (Fig. 10.11a). The Hog domain can be further separated into two regions; the first two-thirds are similar to a self-splicing intein and so this region is called the *Hint* module (Hedgehog/Intein), whereas the carboxy-terminal third binds cholesterol in Hh proteins and is called the sterol recognition region (SRR). In Hh-related proteins, this region is referred to as ARR (adduct recognition region) as the nature of this adduct is not known (Bürglin, 2008).

Processing of a full-length Hh precursor protein to generate the HhN ligand takes place within the endoplasmic reticulum (ER) of the signalling cell and involves the auto-proteolytic cleavage of the Hh precursor by the Hint module of the Hog domain (Fig. 10.10). This liberates the HhN Hedge domain (ligand), which displays all the known Hh signalling properties. The auto-cleavage reaction also provides a trigger for the transfer of the cholesterol moiety from the C-terminus of the Hog domain to the C-terminus of the HhN domain and for the addition of a palmitate molecule to the N-terminus of the HhN domain. Release of this doubly lipid-conjugated form of HhN from the signalling cell requires the activity of the multipass transmembrane protein Dispatched (DISP) (Ingham *et al.*, 2011). Transfer of the HhN ligand from a signalling cell to a receiving cell is achieved either by the transfer of plasma membrane-tethered Hh proteins from one cell to another or, alternatively, by groups of HhN ligands combining to form multimeric complexes in which the hydrophobic moieties cluster together in an inner core allowing diffusion of the ligand and long-range signalling (Fig. 10.10).

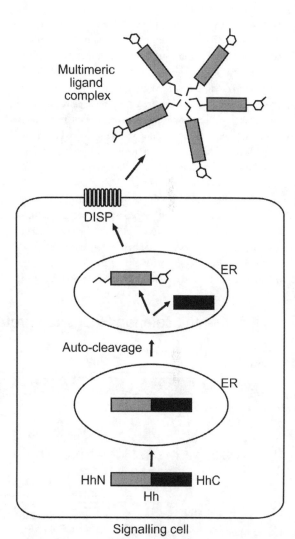

Fig. 10.10 Lipid modification and release of the Hedgehog (Hh) ligand from the signalling cell. Following its translation, full-length Hh is targeted to the endoplasmic reticulum (ER) by its signal peptide. In the ER it undergoes autoproteolysis which results in its covalent coupling to cholesterol. The N-terminal 'Hedge' domain (HhN) fragment is further modified through N-terminal palmitoylation. Release of this doubly lipid-conjugated form of Hh requires the activity of the multipass transmembrane protein Dispatched (DISP). Released particles form multimers or associate with lipoproteins in the extracellular environment. Based on Bürglin (2008).

10.10.4 Ancient origins of Hedge and Hog domains

Although Hedgehog proteins probably first arose in the common ancestor of Eumetazoa more than 650 million years ago (Ingham *et al.*, 2011), both the Hedge and Hog domains have more ancient protistan origins. Hog domain-containing proteins are probably the more ancient since they are distributed widely within the eukaryotic phylogenetic tree (Bürglin, 2008). Many protistan Hog-containing proteins have putative secreted domains upstream of the Hog domain. For example, in the choanoflagellate *Monosiga ovata* the 'Hoglet' gene encodes a protein with a C-terminal Hog domain that is coupled to an N-terminal domain with sequence similarity to cellulose-binding proteins (Fig. 10.11b) (Snell *et al.*, 2006).

Both Hedge and Hog domains are present in *Monosiga brevicollis*, but are encoded by separate genes (Fig. 10.12) (King *et al.*, 2008). In *M. brevicollis* the C-terminal Hint domain occurs in a single domain protein (King *et al.*, 2008). The Hedge domain comprises part of a transmembrane protein that also contains a von Willebrand A domain

(vWA), two cadherin repeats (CA repeats), multiple tumour necrosis factor receptor repeats (TNFR repeats), as well as single immunoglobulin (Ig), immunoglobulin I-set (I-set) and epidermal growth factor (EGF)-like repeats (Fig. 10.12a). The Hedge domain is designated HhN (King *et al.*, 2008). Proteins with a similar domain organisation, though with an expansion of the CA repeats at the expense of the TNFR repeats, are encoded by the Hedgling gene in the sponge *Amphimedon queenslandica*. The genome of the cnidarian *Nematostella vectensis* encodes the Hedgling protein as well as the true Hedgehog protein typical of bilaterians (Fig. 10.12b) (Matus *et al.*, 2008). This arrangement suggests that the Hh gene family evolved as a result of domain shuffling after divergence of the sponges and Eumetazoa (Fig. 10.12c) (King *et al.*, 2008; Nichols *et al.*, 2009; Raible and Steinmetz, 2010).

In addition to the Hh protein itself, other components of the Hh signalling pathway are present in the last common ancestor of choanoflagellates and metazoans. For example, choanoflagellates and sponges encode orthologues of the transmembrane protein Dispatched (DISP) that is

ⓐ Full-length bilaterian Hedgehog protein

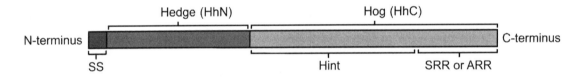

ⓑ *Monosiga ovata* Hoglet-protein

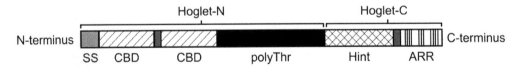

Fig. 10.11 a. The full-length Hedgehog (Hh) protein comprises two distinct domains: the N-terminal 'Hedge' domain (HhN) and the C-terminal 'Hog' domain (HhC). The Hedge domain is preceded by a terminal signal peptide sequence (SS). The Hog domain can also be separated into two regions; the first two-thirds are similar to self-splicing inteins, and this module has been named 'Hint', whereas the C-terminal one-third binds cholesterol and has been named the sterol-recognition region (SRR). In other Hog domain-containing proteins this region is referred to as the adduct recognition region (ARR), as the nature of the adduct is not known. The Hog domain promotes autocleavage to release the N-terminal Hedge domain (HhN). b. *Monosiga ovata*. Organisation of functional domains on the Hoglet gene. Hoglet can be divided into two regions: the N-terminal two-thirds and the C-terminal third. The N-terminal (Hoglet-N) can further be divided into two halves, the first half comprises two polysaccharide-binding domains (CBD) and the second half a threonine-rich protein domain (polyThr). The C-terminal (Hoglet-C) comprises two regions; the first (Hint) is homologous with the autocatalytic domain of Hedgehog proteins and is predicted to function in autocatalytic cleavage of the precursor peptide; the second is comparable to an adduct recognition region (ARR). Modified from Snell *et al.* (2006).

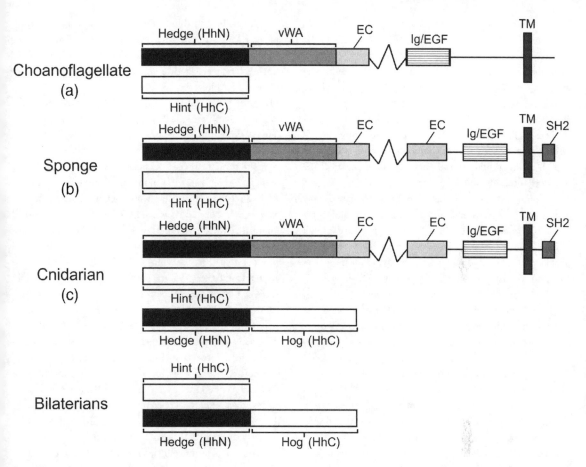

Fig. 10.12 Diagrammatic scenario of Hedgehog ligand evolution by domain shuffling. The two functional Hedgehog domains, N-terminal Hedge (signal) domain and the C-terminal Hint domain, are located on separate proteins in the ancestors of choanoflagellates and Metazoa. In choanoflagellates (a) the Hedge domain is part of a large transmembrane protein. A second gene encodes the Hint domain. The sponge *Amphimedon queenslandica* possesses both Hint/intein genes as well as a gene known as Hedgling that possesses a Hedge domain tethered to a fat-like cadherin (b). The cnidarian *Nematostella vectensis* possesses an orthologue of the Hedgling gene, three Hint/intein genes and two true Hedgehog genes (c). Modified from Nichols *et al.* (2009). von Willebrand factor A domain (vWA): extracellular cadherin domain (EC); immunoglobulin/epidermal growth factor-like domains (Ig/EGF); transmembrane domain (TM); Src homology 2 domain (SH2); Hedge domain (HhN); Hog domain (HhC).

responsible for releasing the Hh ligand from a signalling cell, and the transmembrane receptor protein Patched located on receiving cells. However, there is no evidence for the Smoothened receptor or its defining Frizzled domain (Nichols *et al.*, 2006; King *et al.*, 2008).

10.10.5 Tyrosine kinase signal transduction pathways

Tyrosine kinase signalling pathways are essential for cell–cell communication and the control of cell proliferation and differentiation in Metazoa. Tyrosine kinase signalling

comprises a three-part system of molecular components. (1) Tyrosine kinases (TKs) are enzymes that can transfer a phosphate group from ATP to a tyrosine residue on a (signal) protein in a cell – they switch on a signal. (2) Protein phosphotyrosine phosphatases are enzymes that can remove a phosphate group from a phosphorylated tyrosine-activated protein – they switch off the signal. (3) Src Homology 2 (SH2) and phosphotyrosine binding (PTB) domains bind to phosphorylated tyrosine residues on proteins and initiate signal transduction within a cell. This triad of core functions is at the heart of many cell signalling systems (Pincus *et al.*, 2008).

TKs are divided into two types: receptor TKs (RTKs) and non-receptor or cytoplasmic TKs (CTKs). RTKs mostly receive their specific ligands through their extracellular domains and initiate signal transduction cascades that are mediated by phosphorylated tyrosine residues, whereas CTKs act within cells and transmit the phosphotyrosine signals initiated by receptors. An important feature of TKs is their high degree of structural diversity (Suga *et al.*, 2012). Most TKs are made up of multiple protein domains and motifs in addition to the catalytic kinase domain. The repertoire of CTK and RTK families is highly conserved across all metazoans, including sponges, and was originally considered to be specific to animals. Plants and most unicellular organisms lack TKs, although they have a small number of dual-specificity kinases, associated tyrosine phosphatases and SH2 phosphotyrosine-binding domains generally not involved in intercellular signalling (Manning *et al.*, 2008). However, components of the tyrosine kinase signalling pathway are abundant in the *Monosiga brevicollis* genome, including approximately: 128 predicted tyrosine kinase domains comprising 88 predicted RTKs and 40 CTKs, 38 predicted phosphotyrosine phosphatases and 123 predicted phosphotyrosine-binding SH2 domains (Table 10.5) (King *et al.*, 2008; Manning *et al.*, 2008; Pincus *et al.*, 2008). In addition, two choanoflagellates, *M. brevicollis* and *M. ovata*, contain four homologues of the proto-oncogene Src and biochemical analyses reveal that these homologues conserve most of the regulatory interactions associated with metazoan Srcs (Segawa *et al.*, 2006; Li *et al.*, 2008).

M. brevicollis exhibits the largest number of tyrosine kinases and receptors, regulatory phosphatases and phosphotyrosine-binding SH2 domain proteins (signal transducers) so far discovered in a single species (see Table 10.5). However, these domains appear to be largely divergent from metazoan members of the tyrosine kinase pathway (King *et al.*, 2003, 2008; Manning *et al.*, 2008). This is evident by the lack of clear orthologues (Table 10.5 bottom row), differences in regulation of tyrosine kinase signalling and a large set of choanoflagellate tyrosine kinase domain combinations not found in metazoan proteins.

In a recent study of the two known filastereans, *Capsaspora owczarzaki* and *Ministeria vibrans*, Suga *et al.* (2012) found that their TK repertoires rivalled those of choanoflagellates and metazoans. *Capsaspora owczarzaki* possessed 103 putative TK-encoding genes in the whole genome sequence. Of these, 92 are predicted to be RTKs and 11 CTKs. A PCR-targeted survey of *Ministeria vibrans* isolated seven RTKs and eight CTKs. A detailed phylogenetic analysis of the available holozoan TK data demonstrated that the basic repertoire of metazoan CTKs was established before the divergence of Filasterea from the metazoan and choanoflagellate clades. In contrast, the complement of RTKs diversified extensively in each of the three individual clades. Suga *et al.* (2012) concluded that this difference in the divergence patterns between the two types of TKs suggests that RTKs that had been used for receiving environmental cues in unicellular choanoflagellates were subsequently recruited for cell communication purposes at the onset of metazoan multicellularity.

Table 10.5 *Number of proteins with phosphotyrosine-associated signalling domains in selected genomes.*

	Tyrosine kinase (TK) domains	Protein phosphotyrosine phosphatase domains	SH2 homology domains	Phosphotyrosine-binding (PTB) domains
Dictyostelium discoideum	0	3	13 (14)	0
Saccharomyces cerevisiae	0	7	1	0
Monosiga brevicollis	128 (136)	39 (40)	123 (143)	20 (31)
Drosophila melanogaster	33 (34)	16 (23)	28 (34)	10
Homo sapiens	90 (94)	38 (50)	110 (120)	46 (51)
H. sapiens–M. brevicollis orthologues	4	4–5	19	1

Parentheses indicate domain count due to multi-domain proteins.
Based on Manning *et al.* (2008).

Since tyrosine kinase signalling involves three distinct functional modules – TKs, phosphotyrosine phosphatases and Src homology (SH2) domains – the question is how a complex, interdependent system like this could arise. Lim and Pawson (2010) envisage phosphotyrosine phosphatases and SH2 domains evolving sequentially for separate and limited functions prior to the divergence of fungi and before the establishment of modern TK domains which occurred later. They postulate that the full tyrosine kinase signalling system was of so much greater utility than the earlier individual modules that its use expanded dramatically. This probably resulted in many more proteins in these families, as well as much more complex, multi-domain architectures than those seen in the earlier stages. This stage is represented by both the multicellular metazoan and unicellular holozoan lineages (Lim and Pawson, 2010).

10.10.6 Hippo signalling pathway

The Hippo signalling pathway controls organ size in animals (*Drosophila* and vertebrates) through the regulation of cell proliferation and apoptosis (Halder and Johnson, 2011; Sebé-Pedrós *et al.*, 2012). The pathway transduces signals from the plasma membrane of a receiving cell to the nucleus, where it then regulates gene expression. The pathway consists of a serine/threonine kinase cascade central to which is the protein kinase Hippo, which phosphorylates the protein kinase Warts. Activated Warts can then go on to phosphorylate and inactivate the transcriptional co-activator Yorkie. Yorkie is unable to bind DNA by itself. However, in its active state Yorkie binds to the transcription factor Scalloped, and the Yorkie–Scalloped complex becomes localised to the nucleus. This allows for the expression of several genes that promote organ growth, such as *cyclin E*, which promotes cell cycle progression, and *diap1* which prevents apoptosis. Thus, the inactivation of Yorkie by Warts inhibits growth through the transcriptional repression of these pro-growth regulators.

Based on a comparative genomic analysis, Sebé-Pedrós *et al.* (2012) have shown that all the core Hippo pathway components, including the kinases Hippo and Warts, the co-activator Yorkie and the transcription factor Scalloped, are present in the filasterean *Capsaspora owczarzaki*, and the choanoflagellates *Monosiga brevicollis* and *Salpingoeca rosetta*. Furthermore, in a suite of transgenic experiments, Sebé-Pedrós *et al.* (2012) were able to show that the Scalloped–Yorkie transcription factor complex of *Capsaspora* was capable of promoting tissue growth and Hippo target gene expression in *Drosophila*.

10.10.7 Calcium signalling pathways

Ca^{2+} signalling pathways play a key second messenger role in regulating many cellular processes in animal cells, including fertilisation, contraction, exocytosis, transcription, apoptosis and learning and memory (Cai, 2008). Each animal cell type selectively expresses a unique set of proteins from a comprehensive Ca^{2+} signalling 'toolkit' which allows them to transduce appropriate extracellular stimuli such as neurotransmitters, electrical signals, growth factors and hormones into spatio-temporal Ca^{2+} signals.

Analysis of the sequenced *Monosiga brevicollis* genome reveals that this species possesses homologues of various types of animal plasma membrane Ca^{2+} channels, including the store-operated channel, ligand-operated channels, voltage-operated channels, second messenger-operated channels and five out of six animal transient receptor potential channel families. Choanoflagellates also contain homologues of inositol 1,4,5-trisphosphate receptors. Furthermore, choanoflagellates possess a complete set of Ca^{2+} removal systems including plasma membrane and sarco/endoplasmic reticulum Ca^{2+} ATPases and homologues of three animal cation/Ca^{2+} exchanger families. Therefore, a complex Ca^{2+} signalling 'toolkit' probably evolved before the emergence of multicellular animals (Cai, 2008).

10.10.8 Transcription regulation

The core transcriptional apparatus of *Monosiga brevicollis* and *Capsaspora owczarzaki* contains all the major, ubiquitous classes of eukaryotic transcription factor motifs – for example, zinc-finger, homeobox and helix–loop–helix proteins (King *et al.*, 2008; Sebé-Pedrós *et al.*, 2011). The few previously metazoan-specific transcription factors found in *M. brevicollis*, for instance the p53, Myc and putative Sox/TCF families, have rather general roles in cell cycle or transcriptional control (Raible and Steinmetz, 2010). Two homeodomain proteins, both of which group with the MEIS sub-class of TALE homeodomains, are encoded by the *M. brevicollis* genome. In contrast, the *Capsaspora* genome encodes nine homeodomain-containing genes: three TALE and six non-TALE. The more extensive repertoire of transcription factors in *Capsaspora* suggests that substantial specific losses – for example, Runx, T-box, RHD domain, GRH-like and Churchill – have occurred within the *M. brevicollis* lineage (Sebé-Pedrós *et al.*, 2011). Many of the transcription factor families associated with metazoan patterning and development, for example ETS, HOX, NHR and POU, appear to be absent in both *Monosiga* and *Capsaspora* (King *et al.*, 2008; Sebé-Pedrós *et al.*, 2011). Sebé-Pedrós *et al.* (2011) conclude from a phylogenetic analysis that there appear to have been two major expansions of metazoan-specific transcription factor gene families – one prior to the divergence of *Capsaspora*, choanoflagellates and the Metazoa, for example in bZIP and bHLH, and another within the metazoan lineage. Members of the ETS, Smad, NR, bZIP, bHLH and HMG-box families and a diversity of homeobox-containing classes, including ANTP, LIM, POU, Irx, Meis, Tgif and Six TALE appear to be metazoan-specific and to have evolved prior to the divergence of the sponges and eumetazoans.

10.11 DISCUSSION

The existence of 78 protein domains in the genome of *Monosiga brevicollis* that are shared exclusively with metazoans but are absent from plants, fungi or slime moulds (although some resemble bacterial protein domains) is another indicator of the close phylogenetic relationship between choanoflagellates and metazoans (King *et al.*, 2008; Raible and Steinmetz, 2010). Furthermore, many of these domains are related to cell signalling and adhesion processes in metazoans, suggesting a role in the origin of metazoan multicellularity.

The *M. brevicollis* genome includes an extensive repertoire of metazoan cell adhesion and extracellular material (ECM) genes, including many that are predicted to encode cadherin, C-type lectin, integrin, immunoglobulin and collagen-related domains (King *et al.*, 2003, 2008). These respective functional domains therefore presumably originated before the choanoflagellate–metazoan divergence. However, many of these domains in *M. brevicollis* occur in unique arrangements, and distinctive orthologues of specific metazoan adhesion proteins are rarely found. Although these domains were present in the ancestor of choanoflagellates and metazoans, the canonical metazoan adhesion protein architectures probably evolved after divergence of the two lineages (King *et al.*, 2008).

In contrast to the significant representation of metazoan cell adhesion and ECM domains encoded by the *M. brevicollis* genome, most of the intracellular signalling cascades associated with metazoan cell–cell communication are missing or highly divergent in choanoflagellates (Raible

and Steinmetz, 2010). Of the seven major metazoan intercellular signalling transduction pathways, Wnt, TGF-β and NHR orthologues, which are present as large families in metazoans, are missing from the *M. brevicollis* genome, while other signalling pathways, for example JAK/STAT, Delta/Notch and Hedgehog, are incomplete. For some multi-domain signalling pathway components, for example Notch or Hedgehog proteins, single domains are encoded in the *M. brevicollis* genome, but as in the case of cell adhesion proteins, mostly not in metazoan-characteristic combinations (Raible and Steinmetz, 2010). With respect to tyrosine kinase signal transduction pathways, *M. brevicollis* exhibits the largest number of tyrosine kinases and receptors, regulatory phosphatases and phosphotyrosine-binding SH2 domain proteins (signal transducers) so far discovered in a single species. However, they appear to be largely divergent from metazoan members of the tyrosine kinase pathway (King *et al.*, 2003, 2008; Manning *et al.*, 2008).

The conservation of metazoan-specific protein domains associated with cell adhesion, ECM-interacting proteins and cell signalling in unique combinations within multi-domain proteins in choanoflagellates and other unicellular holozoans makes it difficult to deduce their ancestral functions in the holozoan last common ancestor. Attempts have been made to examine experimentally the functional significance of cadherin EC motifs and tyrosine kinase signalling in *M. brevicollis*. TKs may act in choanoflagellates to detect changes in the extracellular environment, as through their response to nutrient availability (King *et al.*, 2003). In addition, animal cell adhesion proteins, such as the cadherins, may mediate cell attachment or prey recognition and capture. C-type lectins might allow choanoflagellates to distinguish between and capture different bacterial species by binding specific sugar groups displayed on bacterial cell walls (King *et al.*, 2003). Snell *et al.* (2006) suggest that the two carbohydrate-binding domains (CBD) comprising part of the N-terminal domain of the Hoglet gene (Fig. 10.11b) encode a protein with cellulose-binding properties whose function could be to tether *Monosiga ovata* transiently to a cellulosic (algal) substratum.

In conclusion, this account of metazoan multicellularity summarises only the 'start of the beginning' of research that will undoubtedly expand in every way imaginable. Fortunately there are now sufficient data and discussion in the published domain to present an initial insight into what will eventually become a fascinating story. Exactly how this subject will develop is for the future. Nevertheless, we can with a reasonable degree of certainty infer that the choanoflagellates are closely related to the animals and that Opisthokonta is a stable and distinctive phylogenetic supergroup. As this present story draws to a conclusion, it is a fitting tribute to Henry James-Clark to be able to acclaim that, almost 150 years after his initial publication on collared flagellates and *Leucosolenia botryoides*, his conclusions are as relevant today as they were in 1866. If this book stands the same test of time, the author will be duly rewarded.

Glossary

Below are the meanings of some less-familiar terms and abbreviations used in this book.

acanthoecid	loricate choanoflagellate – member of order Acanthoecida
ATCC	American Type Culture Collection
basal Metazoa	equivalent to non-Bilateria; comprises Porifera, Cnidaria, Ctenophora and Placozoa
Bilateria (bilaterian animals)	animals displaying bilateral symmetry; all animals other than non-bilaterians (basal Metazoa)
biogenic silica	polymerised silicon-containing structure, often skeletal, of biological origin
C:N:P ratio	atomic ratio of carbon, nitrogen and phosphorus
CAM	cell adhesion molecules – proteins located on the cell surface involved with binding cells together or to an extracellular matrix
choanocyte	flagellated collar cell of sponge – a monolayer of choanocytes, the choanoderm, lines the aquiferous chambers and canals of sponges
Cnidaria	group of non-bilaterians that comprises sea anemones, corals and jellyfish
Coelenterata	a previously used collective name for two animal phyla – Cnidaria and Ctenophora
conjoined cells	incomplete division of a silicon-starved tectiform cell – two daughter cells, each with a full complement of organelles, remaining joined at the posterior end
costal strip accumulation	groups (bundles) of closely packed costal strips accumulated on the surface of a cell prior to lorica assembly
costal strip bundle	sub-group of closely packed costal strips that, during lorica assembly, will form a costa or part of one or more costae
craspedid	choanoflagellate with an exclusively organic covering – member of order Craspedida
CTK	cytoplasmic tyrosine kinase
cytokinesis	cell division
diploblastic	derived from two embryonic cell layers – ectoderm and endoderm
dipole / doublet	a singularity, or force, acting on a body such as a flagellum, which induces radial and transverse component velocities in the fluid flowing past the body proportional to $1 / r^3$, where r represents the distance from the application point in the body to the position of the fluid element under consideration
DOM	dissolved organic matter
dormant basal body	solitary basal body that will eventually give rise to a flagellum
ECM	extracellular material
eEF1A	eukaryotic elongation factor 1A
EM	electron microscope, electron microscopy
Eumetazoa	clade comprising all major animal groups except sponges
flagellar basal body	base of flagellum, below the transition zone, comprising a short cylinder bounded by nine regularly spaced triplet microtubules
gastrula	a metazoan embryo in an early state of germ layer formation following the blastula stage, consisting of a cup-like body of two layers of cells, the ectoderm and endoderm, enclosing a central cavity, or archenteron, that opens to the outside by the blastopore

glycocalyx	a general term referring to extracellular polymeric material, usually containing polysaccharide and/or protein, with a fine fibrillar substructure
HF acid	hydrofluoric acid – HF acid is used to dissolve silica
HF	heterotrophic flagellate
HNF	heterotrophic nanoflagellate
ICZN	International Code of Zoological Nomenclature
juvenile cell	division of a thecate or loricate cell results in two daughter cells, one of which remains with the parent covering while the other, known as the *juvenile*, is released from the parent covering. The juvenile cell subsequently produces and assembles its own theca or lorica
lorica cup	telescoped cup-shaped covering of costal strip bundles on the surface of a newly produced tectiform juvenile cell prior to lorica assembly
mastigoneme	(Fr. mastigo = flagellum; néme = thread): general term for a flagellar hair (appendage). In this text mastigoneme refers to tubular mastigonemes, which are tripartite hairs, each hair comprising a short base, long, narrow tubular shaft and one or more fine distal threads. Mastigonemes are usually present in a bilateral array on either side of the flagellar shaft. The effect of tubular mastigonemes is to reverse the direction of thrust of an undulating flagellum
Metazoa	includes all animals
mtDNA	mitochondrial DNA
Mya	million years ago
Nomenclator Zoologicus	a catalogue of zoological genera
non-Bilateria	animals that lack bilateral symmetry – includes Porifera, Placozoa, Cnidaria and Ctenorphora. Equivalent to basal Metazoa
Porifera	phylum containing the sponges
pulsellum	cell locomotion achieved by a posteriorly located flagellum – the cell is 'pushed' through a fluid by undulation of a posteriorly directed flagellum
RTK	receptor tyrosine kinase
salinity (‰)	salinity of seawater expressed as parts per mille (parts per thousand ‰), which is approximately equivalent to grams of salt per kilogram of solution.
SDV	silica deposition vesicle
SEM	scanning electron microscope, scanning electron microscopy
singularity	in general, a singularity is a point at which an equation, surface, etc. (such as a process) blows up or becomes degenerate (i.e. indeterminate); in fluid dynamics a singularity is an 'unbounded local behaviour of the velocity field or its derivatives'
SIT	silicon transporter – SIT proteins are encoded by *SIT* genes
stokeslet	a singularity, or force, peculiar to viscous motion, acting on a body such as a flagellum. It induces radial and transverse component velocities in the fluid flowing past the body proportional to $1/r$, where r represents the distance from the application point in the body to the position of the fluid element under consideration
SSU rDNA	small-subunit ribosomal DNA gene. In many texts SSU rDNA is used synonymously with 18S rDNA. However, unless measured specifically, it cannot be assumed that SSU rDNA has a sedimentation coefficient of 18S
TEM	transmission electron microscope, transmission electron microscopy
tRNA	transfer RNA

References

Abedin, M. and King, N. (2008). The premetazoan ancestry of cadherins. *Science*, **319**, 946–8.

Abedin, M. and King, N. (2010). Diverse evolutionary paths to cell adhesion. *Trends in Cell Biology*, **20**, 734–42.

Adamska, M., Degnan, B. M., Green, K. and Zwafink, C. (2011). What sponges can tell us about the evolution of developmental processes. *Zoology*, **114**, 1–10.

Adl, S. M., Simpson, A. G., Lane C. E., *et al.* (2012). The revised classification of eukaryotes. *Journal of Eukaryotic Microbiology*, **59**, 429–93.

Adoutte, A., Balavoine, G., Lartillot, N., *et al.* (2000). The new animal phylogeny: reliability and implications. *Proceedings of the National Academy of Sciences USA*, **97**, 4453–6.

Aguinaldo, A. M. A., Turbeville, J. M., Linford, L. S., *et al.* (1997). Evidence for a clade of nematodes, arthropods and other moulting animals. *Nature*, **387**, 489–93.

Alegado, R. A., Brown, L. W., Cao, S., *et al.* (2012). A bacterial sulfonolipid triggers multicellular development in the closest living relatives of animals. *elife*, **2012** (1):e00013. doi: 10.7554/eLife.00013

Aleshin, V. V., Konstantinova, A. V., Mikhailov, K. V., Nikitin, M. A. and Petrov, N. B. (2007). Do we need many genes for phylogenetic inference? *Biochemistry (Moscow)*, **72**, 1313–23.

Alexander, E. Stock, A., Breiner, H.-W., *et al.* (2009). Microbial eukaryotes in the hypersaline anoxic L'Atalante deep-sea basin. *Environmental Microbiology*, **11**, 360–81

Alldredge, A. L. and Gotschalk, C. (1988). *In situ* settling behavior of marine snow. *Limnology and Oceanography*, **33**, 339–51.

Almqvist, N., Delamo, Y., Smith, B. L., *et al.* (2001). Micromechanical and structural properties of a pennate diatom investigated by atomic force microscopy. *Journal of Microscopy*, **202**, 518–32.

Amaral Zettler, L., Nerad, T., O'Kelly, C. J. and Sogin, M. L. (2001). The nucleariid amoebae: more protists at the animal–fungal boundary. *Journal of Eukaryotic Microbiology*, **48**, 293–7.

Andersen, P. (1988/9). Functional biology of the choanoflagellate *Diaphanoeca grandis* Ellis. *Marine Microbial Food Webs*, **3**, 35–50.

Andersen, P. and Sørensen, H. M. (1986). Population dynamics and trophic coupling in pelagic microorganisms in eutrophic coastal waters. *Marine Ecology Progress Series*, **33**, 99–109.

Arkush, K. D., Mendoza, L., Adkison, M. A. and Hedrick, R. P. (2003). Observations on the life stages of *Sphaerothecum destruens* n. g., n. sp., a mesomycetozoean fish pathogen formally referred to as the rosette agent. *Journal of Eukaryotic Microbiology*, **50**, 430–8.

Arndt, H., Dietrich, D., Auer, B., *et al.* (2000). Functional diversity of heterotrophic flagellates in aquatic ecosystems. In *The Flagellates*, ed. B. S. C. Leadbeater and J. C. Green. London: Taylor and Francis, pp. 240–68.

Atkinson, G. C., Kuzmenko, A., Chicherin, I. *et al.* (2014). An evolutionary ratchet leading to loss of elongation factors in eukaryotes. *BMC Evolutionary Biology*, **14**, 35.

Auer, B. and Arndt, H. (2001). Taxonomic composition and biomass of heterotrophic flagellates in relation to lake trophy and season. *Freshwater Biology*, **46**, 959–72.

Auer, B., Elzer, U. and Arndt, H. (2004). Comparison of pelagic food webs in lakes along a trophic gradient and with seasonal aspects: influence of resource and predation. *Journal of Plankton Research*, **26**, 697–709.

Azam, F., Fenchel, T., Field, J. G. *et al.* (1983). The ecological role of water-column microbes in the sea. *Marine Ecology Progress Series*, **10**, 257–63.

Baguñà, J., Martinez, P., Paps, J. and Riutort, M. (2008). Back in time: a new systematic proposal for the Bilateria. *Philosophical Transactions of the Royal Society of London B*, **363**, 1481–91.

Baker, G. C., Beebee, T. J. and Ragan, M. A. (1999). *Prototheca richardsi*, a pathogen of anuran larvae, is related to a clade of protistan parasites near the animal–fungal divergence. *Microbiology*, 145, 1777–84.

Baldauf, S. L. (1999). A search for the origins of animals and fungi: comparing and combining molecular data. *The American Naturalist*, 154 (Suppl), S178–88.

Baldauf, S. L. (2008). An overview of the phylogeny and diversity of eukaryotes. *Journal of Systematics and Evolution*, 46, 263–73.

Baldauf, S. L. and Palmer, J. D. (1993). Animals and fungi are each other's closest relatives: congruent evidence from multiple proteins. *Proceedings of the National Academy of Sciences USA*, 90, 11558–62.

Baldauf, S. L., Romeralo, M. and Carr, M. (2013). The evolutionary origin of animals and fungi. In *Evolution from the Galapagos: Two Centuries after Darwin*, ed. G. Trueba and C. Montúfar. New York: Springer, pp. 73–106.

Barrois, M. C. (1876). L'embryologie de quelques éponges de la Manche. *Annales des Sciences Naturelles, Zoologie et Paléontologie*, Series 6, 3, 1–84.

Becquevort, S., Mathot, S. and Lancelot, C. (1992). Interactions in the microbial community of the marginal ice zone of the northwestern Weddell Sea through size distribution analysis. *Polar Biology*, 12, 211–18.

Behnke, A., Barger, K. J., Bunge, J. and Stoeck, T. (2010). Spatio-temporal variations in protistan communities along an O_2/H_2S gradient in the anoxic Framvaren Fjord (Norway). *FEMS Microbiol Ecology*, 72, 89–102.

Benny, G. L. and O'Donnell, K. (2000). *Amoebidium parasiticum* is a protozoan, not a trichomycete. *Mycologia*, 92, 1133–7.

Bérard-Therriault, L., Poulin, M. and Bossé, L. (1999). Guide d'identification du phytoplancton marin de l'estuaire et du Golfe du Saint-Laurent incluant également certains protozoaires. *Publication Spéciale Canadienne des Sciences Halieutiques et Aquatiques*, 128, 1–387.

Bergesch, M., Odebrecht, C. and Moestrup, Ø. (2008). Loricate choanoflagellates from the South Atlantic coastal zone (~32 °S) including the description of *Diplotheca tricyclica* sp. nov. *Biota Neotropica*, 8, 111–22.

Berney, C. and Pawlowski, J. (2006). A molecular time-scale for eukaryote evolution recalibrated with the continuous microfossil record. *Proceedings of the Royal Society of London B*, 273, 1867–72.

Berney, C., Pawloski, J. and Zaninetti, L. (2000). Elongation factor 1-alpha sequences do not support an early divergence of the Acoela. *Molecular Biology and Evolution*, 17, 1032–9.

Berrier, A. L. and Yamada, K. M. (2007). Cell-matrix adhesion. *Journal of Cell Physiology*, 213, 565–73.

Bhattacharyya, P. and Volcani, B. E. (1980). Sodium-dependent silicate transport in the apochlorotic marine diatom *Nitzschia alba*. *Proceedings of the National Academy of Sciences USA*, 77, 6386–90.

Bidder, G. P. (1895). The collar-cells of Heterocoela. *Quarterly Journal of Microscopical Science*, 38, 9–43.

Bidle, K. D. and Azam, F. (2001). Bacterial control of silicon regeneration from diatom detritus: significance of bacterial ectohydrolases and species identity. *Limnology and Oceanography*, 46, 1606–23.

Blackstone, N. W. (2009). A new look at some old animals. *PLoS Biology*, 7, 29–31.

Boenigk, J. and Arndt, H. (2000). Comparative studies on the feeding behaviour of two heterotrophic nanoflagellates: the filter-feeding choanoflagellate *Monosiga ovata* and the raptorial-feeding kinetoplastid *Rhynchomonas nasuta*. *Aquatic Microbial Ecology*, 22, 243–9.

Boenigk, J. and Arndt, H. (2002). Bacterivory by heterotrophic flagellates: community structure and feeding strategies. *Antonie van Leeuwenhoek*, 81, 465–80.

Bonder, E. M. and Mooseker, M. S. (1986). Cytochalasin B slows but does not prevent monomer addition at the barbed end of the actin filament. *Journal of Cell Biology*, 102, 282–8.

Booth, B. C. (1990). Choanoflagellates from the sub-arctic North Pacific Ocean with descriptions of two new species. *Canadian Journal of Zoology*, 68, 2393–402.

Booth, B. C., Lewin, J. and Norris, R. E. (1982). Nanoplankton species predominant in the subarctic Pacific in May and June 1978. *Deep Sea Research*, 29, 185–200.

Borchiellini, C., Manuel, M., Alivon, E., *et al.* (2001). Sponge paraphyly and the origin of Metazoa. *Journal of Evolutionary Biology*, 14, 171–9.

Bory de Saint-Vincent, J. B. G. M. (1822). *Dictionnaire classique d'histoire naturelle*. Vol. 1. Paris: Baudouin Frères.

Bory de Saint-Vincent, J. B. G. M. (1824). *Histoire Naturelle des Zoophytes, ou Animaux Rayonnés: Faisant Suite à l'Histoire Naturelle des Vers de Bruguière*. Encyclopédie Méthodique, par MM. Lamouroux, Bory de Saint-Vincent et Eud. Deslongchamps. Vol. 2. Paris: Mme Veuve Agasse.

Boucaud-Camou, E. (1966). Les Choanoflagellés des côtes de la Manche. I Systématique. *Bulletin Société Linnéenne Normandie*, **10**, 191–209.

Bourlat, S. J., Nielsen, C., Economou, A. D. and Telford, M. (2008). Testing the new animal phylogeny: a phylum level molecular analysis of the animal kingdom. *Molecular Phylogenetics and Evolution*, **49**, 23–31.

Bourrelly, P. (1957). Recherches sur les Chrysophycées, morphologie, phylogénie, systématique. *Revue Algologique, Mémoire Hors-Série*, **1**, 1–412.

Bourrelly, P. (1968). *Les Algues D'eau Douce. Initiation á La Systématique.* Vol. **2**: *Les Algues Jaunes et Brunes*. Paris: N. Boubée et Cie.

Braarud, T. (1935). The Øst expedition to the Denmark Strait 1929. II. The phytoplankton and its conditions of growth. *Hvalrådets Skrifter*, **10**, 1–173.

Brandt, S. M. and Sleigh, M. A. (2000). The quantitative occurrence of different taxa of heterotrophic flagellates in Southampton Water, UK. *Estuarine, Coastal and Shelf Science*, **51**, 91–102.

Brill, B. (1973). Untersuchungen zur Ultrastruktur der Choanocyte von *Ephydatia fluviatilis* L. *Zeitschrift für Zellforschung und Mikroskopische Anatomie*, **144**, 231–45.

Brown, M. W., Spiegel, F. W. and Silberman, J. D. (2009). Phylogeny of the 'forgotten' cellular slime mold, *Fonticula alba*, reveals a key evolutionary branch within Opisthokonta. *Molecular Biology and Evolution*, **26**, 2699–709.

Brown, S. S. and Spudich, J. A. (1981). Mechanism of action of cytochalasin: evidence that it binds to actin filament ends. *Journal of Cell Biology*, **88**, 487–91.

Brugerolle, G. and Bricheux, G. (1984). Actin microfilaments are involved in scale formation of the chrysomonad cell *Synura*. *Protoplasma*, **123**, 203–12.

Brunner, E., Gröger, C., Lutz, K., *et al.* (2009). Analytical studies of silica biomineralization: towards an understanding of silica processing by diatoms. *Applied Microbiology and Biotechnology*, **84**, 607–16.

Buck, K. R. (1981). A study of choanoflagellates (Acanthoecidae) from the Weddell Sea, including a description of *Diaphanoeca multiannulata* n. sp. *Journal of Protozoology*, **28**, 47–54.

Buck, K. R. and Garrison, D. L. (1983). Protists from the ice-edge region of the Weddell Sea. *Deep Sea Research*, **30**, 1261–77.

Buck, K. R. and Garrison, D. L. (1988). Distribution and abundance of choanoflagellates (Acanthoecidae) across the ice-edge zone in the Weddell Sea. *Marine Biology*, **98**, 263–9.

Buck, K. R., Marchant, H. J., Thomsen, H. A. and Garrison, D. L. (1990). *Kakoeca antarctica* gen. et sp. n., a loricate choanoflagellate (Acanthoecidae, Choanoflagellida) from Antarctic sea ice with a unique protoplast suspensory membrane. *Zoologica Scripta*, **19**, 389–94.

Buck, K. R., Chavez, F. P. and Thomsen, H. A. (1991). Choanoflagellates of the central California waters: abundance and distribution. *Ophelia*, **33**, 179–186.

Bullerwell, C. E. and Lang, B. F. (2005). Fungal evolution: the case of the vanishing mitochondrion. *Current Opinion in Microbiology*, **8**, 362–9.

Burck, C. (1909). Studien über einige Choanoflagellaten. *Archiv für Protistenkunde*, **16**, 169–86.

Burger, G., Forget, L., Zhu, Y., Gray, M. W. and Lang, B. F. (2003). Unique mitochondrial genome architecture in unicellular relatives of animals. *Proceedings of the National Academy of Sciences USA*, **100**, 892–7.

Burger, G., Yan, Y., Javadi, P. and Lang, B. F. (2009). Group I-intron trans-splicing and mRNA editing in the mitochondria of placozoan animals. *Trends in Genetics*, **25**, 381–6.

Bürglin, T. (2008). The Hedgehog protein family. *Genome Biology*, **9** (241), 1–9.

Bursa, A. S. (1961). The annual oceanographic cycle at Igloolik in the Canadian Arctic II: the phytoplankton. *Journal of the Fisheries Research Board Canada*, **18**, 563–615.

Burton, J. D., Leatherland, T. M. and Liss, P. S. (1970). The reactivity of dissolved silicon in some natural waters. *Limnology and Oceanography*, **15**, 473–6.

Bütschli, O. (1878). Beiträge zur Kenntnis der Flagellaten und einiger verwandter Organismen. *Zeitschrift für Wissenschaftliche Zoologie*, **30**, 219–81.

Bütschli, O. (1882). Protozoa. In *Dr. H. G. Bronn's Klassen und Ordungen des Thier-Reichs, wissenschaftlich dargestellt in Wort und Bild. Erster band.* Leipzig and Heidelberg: C. F. Winter'sche Verlagshandlung, pp. 321–616.

Cafaro, M. J. (2005). Eccrinales (Trichomycetes) are not fungi, but a clade of protists at the early divergence of animals and fungi. *Molecular Phylogenetics and Evolution*, **35**, 21–34.

Cai, X. (2008). Unicellular Ca^{2+} signaling 'toolkit' at the origin of Metazoa. *Molecular Biology and Evolution*, **25**, 1357–61.

Cantell, P.-E., Franzén, Å. and Sensenbaugh, T. (1982). Ultrastructure of multiciliated collar cells in the pilidium larva of *Lineus bilineatus* (Nemertini). *Zoomorphology*, **101**, 1–15.

Capella-Gutiérrez, S., Marcet-Houben, M. and Gabaldón, T. (2012). Phylogenomics supports microsporidia as the earliest diverging clade of sequenced fungi. *BMC Biology*, **10**, 47.

Caron, D. A., Davies, P. S., Madin, L. P. and Sieburth, J. M. (1986). Enrichment in microbial populations in macroaggregates (marine snow) from surface waters of the North Atlantic. *Journal of Marine Research*, **44**, 543–65.

Carr, M. and Baldauf, S. L. (2011). The protistan origins of animals and fungi. In *The Mycota: XIV Evolution of Fungi and Fungal-like Organisms*, ed. S. Pöggeler and J. Wöstermeyer. Heidelberg: Springer-Verlag, pp. 3–24.

Carr, M., Leadbeater, B. S. C., Hassan, R., Nelson, M. and Baldauf, S. L. (2008a). Molecular phylogeny of choanoflagellates, the sister group to Metazoa. *Proceedings of the National Academy of Sciences USA*, **105**, 16641–6.

Carr, M., Nelson, M., Leadbeater, B. S. C. and Baldauf, S. L. (2008b). Three families of LTR retrotransposons are present in the genome of the choanoflagellate *Monosiga brevicollis*. *Protist*, **159**, 579–90.

Carr, M., Leadbeater, B. S. C. and Baldauf, S. L. (2010). Conserved meiotic genes point to sex in the choanoflagellates. *Journal of Eukaryotic Microbiology*, **57**, 56–62.

Carrias, J. F., Amblard, C. and Bourdier, G. (1996). Protistan bacterivory in an oligomesotrophic lake: importance of attached ciliates and flagellates. *Microbial Ecology*, **31**, 249–68.

Carrias, J. F., Amblard, C., Quiblier-Lloberas, C. and Bourdier, G. (1998). Seasonal dynamics of free and attached nanoflagellates in an oligomesotrophic lake. *Freshwater Biology*, **39**, 91–101.

Carroll, S. B., Grenier, J. K. and Weatherbee, S. D. (2005). *From DNA to diversity. Molecular genetics and the evolution of animal design*. 2nd edn. Malden, MA: Blackwell Scientific.

Carter, H. J. (1857). On the ultimate structure of *Spongilla* and additional notes on freshwater Infusoria. *Annals and Magazine of Natural History*, Series 2, **20**, 21–41.

Carter, H. J. (1859). On fecundation in two Volvoces, and their specific differences: on *Eudorina*, *Spongilla*, *Astasia*, *Euglena* and *Cryptoglena*. *Annals and Magazine of Natural History*, Series 3, **3**, 1–19.

Carter, H. J. (1871). A description of two new Calciospongiae, to which is added confirmation of Prof. James-Clark's discovery of the true form of the sponge cell (animal), and an account of the polype-like pore-area of *Cliona corallinoides* contrasted with Prof. E. Häckel's view on the relationship of the sponges to the corals. *Annals and Magazine of Natural History*, Series 4, **8** (43), 1–27.

Cavalier-Smith, T. (1981). The origin and early evolution of the eukaryotic cell. In *Molecular and Cellular Aspects of Microbial Evolution*, ed. M. J. Carlile, J. F. Collins and B. E. B. Moseley. *Symposium of the Society for General Microbiology*, Vol. **33**. Cambridge: Cambridge University Press, pp. 33–84.

Cavalier-Smith, T. (1987). The origin of fungi and pseudofungi. In *Evolutionary Biology of the Fungi*, ed. A. D. M. Rayner, C. M. Brasier and D. Moore. Cambridge: Cambridge University Press, pp. 339–53.

Cavalier-Smith, T. (1993). Kingdom Protozoa and its 18 phyla. *Microbiological Reviews*, **57**, 953–94.

Cavalier-Smith, T. (1997). Amoeboflagellates and mitochondrial cristae in eukaryote evolution: megasystematics of the new protozoan subkingdoms Eozoa and Neozoa. *Archiv für Protistenkunde*, **147**, 237–58.

Cavalier-Smith, T. (1998). Neomonada and the origin of animals and fungi. In *Evolutionary Relationships among Protozoa*, ed. G. H. Coombs, K. Vickerman, M. A. Sleigh and A. Warren. Dordrecht: Kluwer Academic Publishers, pp. 375–407.

Cavalier-Smith, T. (2006). Protozoa: the most abundant predators on earth. *Microbiology Today*, November, pp. 166–9.

Cavalier-Smith, T. (2013). Early evolution of eukaryote feeding modes, cell structural diversity, and classification of the protozoan phyla Loukozoa, Sulcozoa, and Choanozoa. *European Journal of Protistology*, **49**, 115–78.

Cavalier-Smith, T. and Allsopp, M. T. E. P. (1996). *Corallochytrium*, an enigmatic non-flagellate protozoan related to choanoflagellates. *European Journal of Protistology*, **32**, 306–10.

Cavalier-Smith, T. and Chao, E. E. (2003). Phylogeny of Choanozoa, Apusozoa, and other Protozoa and early eukaryotic megaevolution. *Journal of Molecular Evolution*, **56**, 540–63.

Cavalier-Smith, T. and Chao, E. E. (2010). Phylogeny and evolution of Apusomonadida (Protozoa: Apusozoa): new genera and species. *Protist*, **161**, 549–76.

Cavalier-Smith, T., Chao, E. E. and Oates, B. (2004). Molecular phylogeny of Amoebozoa and the evolutionary significance of the unikont *Phalansterium*. *European Journal of Protistology*, **40**, 21–48.

Chadefaud, M. (1960). Les végétaux non vasculaires (Cryptogamie). In *Traité de Botanique, Systématique*, ed. M. Chadefaud and L. Emberger. Paris: Masson et Cie, pp. 1–1018.

Chen, B. (1994). Distribution and abundance of choanoflagellates in Great Wall Bay, King George Island, Antarctica in austral summer. *Proceedings of National Institute of Polar Research Symposium, Polar Biology*, **7**, 32–42.

Chernikova, D., Motamedi, S., Csürös, M., Koonin, E. V. and Rogozin, I. B. (2011). A late origin of the extant eukaryotic diversity: divergence time estimates using rare genomic changes. *Biology Direct*, **6**, 26.

Chisholm, S. A., Azam, F. and Eppley, R. W. (1978). Silicic acid incorporation in marine diatoms on light:dark cycles: use as an assay for phased cell division. *Limnology and Oceanography*, **23**, 518–29.

Christensen, T. (1962). Alger. In *Botanik, 2, Systematisk Botanik*, ed. T. W. Böcher, M. Lange and T. Sørensen. Copenhagen: Munksgaard, pp. 1–178.

Christensen-Dalsgaard, K. K. and Fenchel, T. (2003). Increased filtration efficiency of attached compared to free-swimming flagellates. *Aquatic Microbial Ecology*, **33**, 77–86.

Cienkowsky, L. (1870). Über Palmällaceen und einige Flagellaten. *Archiv für Microskopiche Anatomie*, **6**, 421–38.

Claquin, P., Martin-Jézéquel, V., Kromkamp, J. C., Veldhuis, M. J. W. and Kraay, G. W. (2002). Uncoupling of silicon compared with carbon and nitrogen metabolisms and the role of the cell cycle in continuous cultures of *Thalassiosira pseudonana* (Bacillariophyceae) under light, nitrogen and phosphorus control. *Journal of Phycology*, **38**, 922–30.

Cleven, E.-J. and Weisse, T. (2001). Seasonal succession and taxon-specific bacterial grazing rates of heterotrophic nanoflagellates in Lake Constance. *Aquatic Microbial Ecology*, **23**, 47–61.

Copeland, H. F. (1956). *The Classification of Lower Organisms*. Paolo Alto, CA: Pacific Books.

Correia, J. J. and Lobert, S. (2008). Microtubules in health and disease. In *Molecular Mechanisms of Microtubule Acting Cancer Drugs*, ed. T. Fojo. New York: Humana Press, pp. 21–46.

Curnow, P., Senior, L. and Knight, M. J. (2012) Expression, purification, and reconstitution of a diatom silicon transporter. *Biochemistry*, **51**, 3776–85.

Cynar, F. J. and Sieburth J. McN. (1986). Unambiguous detection and improved quantification of phagotrophy in apochlorotic nanoflagellates using fluorescent microspheres and concomitant phase contrast and epifluorescence microscopy. *Marine Ecology Progress Series*, **32**, 61–70.

Davis, A. K. and Hildebrand, M. (2008). Molecular processes of biosilicification in diatoms. *Metal Ions in Life Sciences*, **4**, 255–94.

Davis, P. G. and Sieburth, J. McN. (1984). Estuarine and oceanic microflagellate predation of actively growing bacteria: estimation by frequency of dividing–divided bacteria. *Marine Ecology Progress Series*, **19**, 237–46.

Dayel, M. J., Alegado, R. A., Fairclough, S. R., *et al.* (2011). Cell differentiation and morphogenesis in the colony-forming choanoflagellate *Salpingoeca rosetta*. *Developmental Biology*, **357**, 73–82.

Deflandre, G. (1960). Sur la présence de *Parvicorbicula* n. g. *socialis* (Meunier) dans le plancton de l'Antarctique (Terre Adélie). *Revue Algologique*, New Series, **5**, 183–8.

Del Amo, Y. and Brzezinski, M. A. (1999). The chemical form of dissolved Si taken up by diatoms. *Journal of Phycology*, **35**, 1162–70.

Dellaporta, S. L., Xu, A. and Sagasser, S. (2006). Mitochondrial genome of *Trichoplax adhaerens* supports placozoa as the basal lower metazoan phylum. *Proceedings of the National Academy of Sciences USA*, **103**, 8751–6.

Derelle, R. and Lang, B. F. (2012). Rooting the eukaryotic tree with mitochondrial and bacterial proteins. *Molecular Biology and Evolution*, **29**, 1277–89.

Dixit, S., Van Cappellen, P. and van Bennekom, A. J. (2001). Processes controlling solubility of biogenic silica and pore water build-up of silicic acid in marine sediments. *Marine Chemistry*, **73**, 333–52.

Dove, P. M. (1999). The dissolution kinetics of quartz in aqueous mixed cation solutions. *Geochimica and Cosmochimica Acta*, **63**, 3715–27.

Dove, P. M. and Crerar, D. A. (1990). Kinetics of quartz dissolution in electrolyte solutions using a hydrothermal mixed flow reactor. *Geochimica et Cosmochimica Acta*, **54**, 955–69.

Dove, P. M. and Elston, S. F. (1992) Dissolution kinetics of quartz in sodium chloride solutions: analysis of existing data and a rate model for 25 °C. *Geochimica et Cosmochimica Acta*, **56**, 4147–56.

Doweld, A. B. (2003). *Diplosoeca*: a new generic name in Craspedophycaceae. *Byulleten' Moskovskogo Obshchestva Ispytatelei Prirody Otdel Biologicheskii* (Bulletin of the Moscow Society of Naturalists Biological Series), **108** (2), 60.

Dujardin, F. (1841). Histoire naturelle des zoophytes. Infusoires, comprenant la physiologie et la classification de ces animaux, et la manière de les étudier à l'aide du microscope. Paris: Roret.

Dunkerly, J. S. (1910). Notes on the choanoflagellate genera *Salpingoeca* and *Polyoeca*, with description of *Polyoeca dumosa* sp. n. *Annals and Magazine of Natural History*, **5** (26), 186–91.

Dunn, C. W., Hejnol, A., Matus, D. Q., *et al.* (2008) Broad phylogenomic sampling improves resolution of the animal tree of life. *Nature*, **452**, 745–9.

Dustin, P. (1984). *Miocrotubules*. Berlin and Heidelberg: Springer-Verlag.

Dyková, I., Veverková, M., Fiala, I., Macháčková, B. and Pecková, H. (2003). *Nuclearia pattersoni* sp. n. (Filosea), a new species of amphizoic amoeba isolated from gills of roach (*Rutilus rutilus*), and its rickettsial endosymbiont. *Folia Parasitologica*, **50**, 161–70.

Eccleston-Parry, J. and Leadbeater, B. S. C. (1994a). A comparison of the growth kinetics of six marine heterotrophic nanoflagellates fed with one bacterial species. *Marine Ecology Progress Series*, **105**, 167–77.

Eccleston-Parry, J. and Leadbeater, B. S. C. (1994b). The effect of long-term low bacterial density on the growth kinetics of three marine heterotrophic nanoflagellates. *Journal of Experimental Marine Biology and Ecology*, **177**, 219–33.

Eccleston-Parry, J. and Leadbeater, B. S. C. (1995). Regeneration of phosphorus and nitrogen by four species of heterotrophic nanoflagellates feeding on three nutritional states of a single bacterial strain. *Applied Environmental Microbiology*, **61**, 1033–8.

Edgecombe, G. D., Giribet, G., Dunn, C. W., *et al.* (2011). Higher-level metazoan relationships: recent progress and remaining questions. *Organisms Diversity and Evolution*, **11**, 151–72.

Eerkes-Medrano, D. I. and Leys, S. P. (2006). Ultrastructure and embryonic development of a syconoid calcareous sponge. *Invertebrate Biology*, **125**, 177–94.

Eernisse, D. J. and Peterson, K. J. (2004). The history of animals. In *Assembling the Tree of Life*, ed. J. Cracraft and M. J. Donoghue. Oxford: Oxford University Press, pp. 197–208.

Ehlers, U. and Ehlers, B. (1977). Monociliary receptors in interstitial Proseriata and Neorhabdocoela (Turbelaria Neoophora). *Zoomorphologie*, **86**, 197–222.

Ehrenberg, C. G. (1830) (1832). Beitrage zur Kenntnis der Organisation der Infusorien und ihrer geographischen verbreitung, besonders in Sibirien. *Abhandlungen der königlichen Akademie der Wissenschaften zu Berlin gehalten in den Jahren 1830*, 1–88.

Ehrenberg, C. G. (1831). Über die Entwickelung und Lebensdauer der Infusionthiere: nebst ferneren Beiträgen zu einer Vergleichung ihrer organischen Système. *Abhandlungen der Königlichen Akademie der Wissenschaften zu Berlin gehalten in den Jahren 1831*, 1–154.

Ehrenberg, C. G. (1838). *Die Infusionsthierchen als vollkommene Organismen*. Leipzig: L. Voss.

Ehrlich, R. (1908). Ein Beitrag zur Frage von der Membran der Choanoflagellaten. *Biologisches Zentralblatt*, **28**, 117–20.

Ellis, W. N. (1929). Recent researches on the Choanoflagellata (Craspedomonadines). *Annales de la Société Royale Zoologie Belgique*, **60**, 49–88.

Ellis, W. N. (1935). Notes on the Choanoflagellata. *Journal of the Quekett Microscopical Club*, **23**, 153–72.

Entz, G. (1883). Die Flagellaten der Kochsalzteiche zu Torda und Szamosfalva. *Termeszetrajzi Fuzetek*, **7**, 139–68.

Ereskovsky, A. (2010). *The Comparative Embryology of Sponges*. Dordrecht: Springer-Verlag.

Ertl, M. (1968). Über das Vorkommen von *Protospongia haeckeli* Kent in der Donau und einige Bemerkungen zur Taxonomie dieser Art. *Archiv für Protistenkunde*, **111**, 18–23.

Ertl, M. (1981). Zur Taxonomie der Gattung *Proterospongia* Kent. Contribution to the taxonomy of the genus *Proterospongia* Kent. *Archiv für Protistenkunde*, **124**, 259–66.

Espeland, G. and Throndsen, J. (1986). Flagellates from Kilsfjorden, southern Norway, with description of two new species of Choanoflagellida. *Sarsia*, **71**, 209–66.

Exposito, J. Y., Larroux, C., Cluzel, C., *et al.* (2008). Demosponge and sea anemone fibrillar collagen diversity reveals the early emergence of A/C clades and the maintenance of the modular structure of type V/XI collagens from sponge to human. *Journal of Biological Chemistry*, **263**, 28225–35.

Fahey, B. and Degnan, B. M. (2010). Origin of animal epithelia: insights from the sponge genome. *Evolution and Development*, **12**, 601–17.

Fairclough, S. R., Dayel, M. J. and King, N. (2010). Multicellular development in a choanoflagellate. *Current Biology*, **20**, R875–6.

Fairclough, S. R., Chen, Z., Kramer, E., *et al.* (2013). Premetazoan genome evolution and the regulation of cell differentiation in the choanoflagellate *Salpingoeca rosetta*. *Genome Biology*, **14**, R15.

Feige, W. (1969). Die Feinstruktur der Epithelien von *Ephydatia fluviatilis*. *Zoologische Jahrbücher Abteilung fur Anatomie*, **86**, 177–237.

Fenchel, T. (1982a). Ecology of heterotrophic microflagellates: I. Some important forms and their functional morphology. *Marine Ecology Progress Series*, **8**, 211–23.

Fenchel, T. (1982b). Ecology of heterotrophic microflagellates: II. Bioenergetics and growth. *Marine Ecology Progress Series*, **8**, 225–31.

Fenchel, T. (1986a). Protozoan filter feeding. *Progress in Protistology*, **1**, 65–133.

Fenchel, T. (1986b). The ecology of heterotrophic microflagellates. In *Advances in Microbial Ecology*, Vol. **9**, ed. K. C. Marshall. New York: Plenum Press, pp. 57–97.

Fenchel, T. (1987). *Ecology of Protozoa*. Madison, WI: Science Tech, Inc.

Fenchel, T. (2008). The microbial loop: 25 years later. *Journal of Experimental Marine Biology and Ecology*, **366**, 99–103.

Fenchel, T., Bernard, C., Esteban, G. F., *et al.* (1995). Microbial diversity and activity in a Danish fjord with anoxic deep water. *Ophelia*, **43**, 45–100.

Fields, S. D., Arana, Q., Heuser, J. and Clarke, M. (2002). Mitochondrial membrane dynamics are altered in *cluA⁻* mutants of *Dictyostelium*. *Journal of Muscle Research and Cell Motility*, **23**, 829–32.

Finlay, B. and Fenchel, T. (2004). Cosmopolitan metapopulations of free-living microbial eukaryotes. *Protist*, **155**, 237–44.

Fisch, C. (1885). Untersuchungen über einige Flagellaten und verwandte Organismen. *Zeitschrift für Wissenschaftliche Zoologie*, **42**, 47–125.

Fischer H., Polikarpov, I. and Craievich, A. F. (2004). Average protein density is a molecular-weight dependent function. *Protein Science*, **13**, 2825–8.

Fjerdingstad, E. J. (1961a). The ultrastructure of choanocyte collars in *Spongilla lacustris* (L). *Zeitschrift für Zellforschung und Mikroskopische Anatomie*, **53**, 645–57.

Fjerdingstad, E. J. (1961b). Ultrastructure of the collar of the choanoflagellate *Codonosiga botrytis* (Ehrenb.). *Zeitschrift für Zellforschung und Mikroskopische Anatomie*, **54**, 499–510.

Foissner, W. (1991). Diversity and ecology of soil flagellates. In *The Biology of Free-living Heterotrophic Flagellates*, ed. D. J. Patterson and J. Larsen. Oxford: Clarendon Press, pp. 93–112.

Fouquet, J.-P. and Kann, M.-L. (1999). The cytoskeleton of mammalian spermatozoa. *Biologie Cellulaire*, **81**, 89–93.

Francé, R. H. (1893). Über die Organisation der Choanoflagellaten. *Zoologischen Anzeiger Canus*, **16**, 44–6.

Francé, R. H. (1897). A craspedomonadinák szervezete: der organismus der Craspedomonaden. Budapest: Kir. M. Természéttudományi Társulat.

Franke, W. W. (2009). Discovering the molecular components of intercellular junctions: a historical view. *Cold Spring Harbor Perspectives in Biology*, **1**, a003061.

Franke, W. W. and Herth, W. (1973). Cell and lorica fine structure of the chrysomonad alga *Dinobryon sertularia* Ehr. (Chrysophyceae). *Archiv für Mikrobiologie*, **91**, 323–44.

Fredricks, D. N., Jolley, J. A., Lepp, P. W., Kosek, J. C. and Relman, D. A. (2000). *Rhinosporidium seeberi*: a human pathogen from a novel group of aquatic protistan parasites. *Emerging Infectious Diseases*, **6**, 273–82.

Frenzel, J. (1891). Über einige merkwürdige Protozoen Argentiniens. *Zeitschrift für Wissenschaftliche Zoologie*, **53**, 332–60.

Fresenius, G. von (1858). Beiträge zur Kenntnis mikroskopischer Organismen. *Abhandlungen der Senckenbergischen naturforschenden Gesellschaft*, **2**, 211–42.

Fromentel, E. de (1874). *Études sur les microzoaires ou proprement dits*. Paris: G. Masson.

Frösler, J. and Leadbeater, B. S. C. (2009). Role of the cytoskeleton in choanoflagellate lorica assembly. *Journal of Eukaryotic Microbiology*, **56**, 167–73.

Fuller, M. S. and Reichle, R. E. (1968). The fine structure of *Monoblepharella* sp. zoospores. *Canadian Journal of Botany*, **46**, 279–83.

Gaarder, K. (1954). Coccolithineae, Silicoflagellatae, Pterospermataceae and other forms from the 'Michael Sars' North Atlantic Deep Sea Expedition 1910. *Report on the Scientific Research 'Michael Sars' North Atlantic Deep Sea Expedition 1910*, **2** (4), 1–20.

Galtier, N., Nabholz, B., Glémin, S. and Hurst, G. D. D. (2009). Mitochondrial DNA as a marker of molecular diversity: a reappraisal. *Molecular Ecology*, **18**, 4541–50.

Gams, H. (1947). Die Protochlorinae als autotrophe Vorfahren von Pilzen und Tieren? *Mikroskopie*, **2**, 383–7.

Garrison, D. L. and Buck, K. R. (1989a). Protozooplankton in the Weddell Sea, Antarctica: abundance and distribution in the ice-edge zone. *Polar Biology*, **9**, 341–51.

Garrison, D. L. and Buck, K. R. (1989b). The biota of Antarctic pack ice in the Weddell Sea and Antarctic Peninsula regions. *Polar Biology*, **10**, 211–19.

Garrison, D. L., Buck, K. R. and Gowing, M. M. (1991). Plankton assemblages in the ice edge zone of the Weddell Sea during the austral winter. *Journal of Marine Systems*, **2**, 123–30.

Garrison, D. L., Buck, K. R. and Gowing, M. M. (1993). Winter plankton assemblages in the ice edge zone of the Weddell and Scotia Seas: composition, biomass and spatial distributions. *Deep Sea Research*, **40**, 311–38.

Gazave, E., Lapébie, P., Richards, G. S., *et al.* (2009). Origin and evolution of the Notch signalling pathway: an overview from eukaryotic genomes. *BMC Evolutionary Biology*, **9**, 249.

Geider, R. J. and Leadbeater, B. S. C. (1988). Kinetics and energetics of growth of the marine choanoflagellate *Stephanoeca diplocostata*. *Marine Ecology Progress Series*, **47**, 169–77.

Gerhart, J. (1999). 1998 Warkany lecture: signaling pathways in development. *Teratology*, **60**, 226–39.

Glockling, S. T., Marshall, W. L. and Gleason, F. M. (2013). Phylogenetic and ecological potentials of the Mesomycetozoea (Ichthyosporea). *Fungal Ecology*, **6**, 237–47.

Glotzer, M. (2001). Animal cell cytokinesis. *Annual Review of Cell and Developmental Biology*, **17**, 351–86.

Gold, K., Pfister, R. M. and Liguori, V. R. (1970). Axenic cultivation of two species of Choanoflagellida. *Journal of Protozoology*, **17**, 210–12.

Gong, N., Wiens, M., Schröder, H. C., *et al.* (2010). Biosilicification of loricate choanoflagellate: organic composition of the nanotubular siliceous costal strips of *Stephanoeca diplocostata*. *Journal of Experimental Biology*, **213**, 3575–85.

Gonobobleva, E. L. and Maldonado, M. (2009). Choanocyte ultrastructure in *Halisarca dujardini* (Demospongiae, Halisarcida). *Journal of Morphology*, **270**, 615–27.

Gran, H. H. and Braarud, T. (1935). A quantitative study of the phytoplankton in the Bay of Fundy and the Gulf of Maine (including observations on hydrography, chemistry and turbidity). *Journal of the Biological Board of Canada*, **1**, 279–467.

Gray, J. and Hancock, G. J. (1955). The propulsion of sea urchin spermatozoa. *Journal of Experimental Biology*, **32**, 802–14.

Gray, M. W., Lang, B. F., Cedergren, R., *et al.* (1998). Genome structure and gene content in protist mitochondrial DNAs. *Nucleic Acids Research*, **26**, 865–78.

Griessmann, K. (1913). Über marine Flagellaten. *Archiv für Protistenkunde*, **32**, 1–78.

Grøntved, J. (1952). Investigations on the phytoplankton in the Southern North Sea in May 1947. *Meddelelser fra Kommissionen for Danmarks Fisheri-OG Havundersøgelser series: plankton*, **5** (5), 1–53.

Grøntved, J. (1956). Planktological contributions II: taxonomical studies in some Danish coastal localities. *Meddelelser fra Kommissionen for Danmarks Fisheri-OG Havundersøgelser. NY Serie*, **1** (12), 1–13.

Grosberg, R. K. and Strathman, R. R. (2007). The evolution of multicellularity: a minor major transition? *Annual Review of Ecology, Evolution, and Systematics*, **38**, 621–54.

Hadži, J. (1963). *The Evolution of the Metazoa*. New York: Macmillan.

Haeckel, E. H. P. A. (1866). *Generelle Morphologie der Organismen*. Berlin: G. Reimer.

Haeckel, E. H. P. A. (1870). Über der Organismus der Schwämme und ihre Verwandtschaft mit den Corallen. *Jenaische Zeitschrift für Medicin und Naturwissenschaft*, **5**, 207–54.

Haeckel, E. H. P. A. (1872). *Die Kalkschwämme, eine Monographie*. 3 vols. Berlin: G. Reimer.

Haeckel, E. H. P. A. (1874a). *Anthropogenie; oder, Entwickelungsgeschichte des Menschen*. Leipzig: Eilhelm Engelmann.

Haeckel, E. H. P. A. (1874b). Die Gastrea-Theorie die phylogenetische Classification des Thierreichs und die Homologie der Keimblätter. *Jenaische Zeitschrift für Naturwissenschaft*, **8**, 1–56.

Haeckel, E. H. P. A. (1875). Die Gastrula und die Eifurchung der Thiere. *Jenaische Zeitschrift für Naturwissenschaft*, **9**, 402–508.

Halanych, K. M., Bacheller, J. D., Aguinaldo, A. M. A., *et al.* (1995). Evidence from 18S ribosomal DNA that the lophophorates are protostome animals. *Science*, **267**, 1641–3.

Halder, G. and Johnson, R. L. (2011). Hippo signaling: growth control and beyond. *Development*, **138**, 9–22.

Hamm, C. E., Merkel, R., Springer, O., *et al.* (2003). Architecture and material properties of diatom shells provide effective mechanical protection. *Nature* **421**, 841–3.

Hampl, V., Hug, L., Leigh, J. W., *et al.* (2009). Phylogenomic analyses support the monophyly of Excavata and resolve relationships among eukaryotic 'supergroups'. *Proceedings of the National Academy of Sciences USA* **106**, 3859–64.

Hancock, G. J. (1953). The self-propulsion of microscopic organisms through liquids. *Proceedings of the Royal Society of London A*, **217**, 96–121.

Hara, S. and Takahashi, E. (1984). Re-investigation of *Polyoeca dichotoma* and *Acanthoeca spectabilis* (Acanthoecidae: Choanoflagellida). *Journal of the Marine Biological Association UK*, **64**, 819–27.

Hara, S. and Takahashi, E. (1987a). An investigation with electron microscope of marine choanoflagellates (Protozoa: Choanoflagellida) from Osaka Bay, Japan: I. Re-investigation of *Bicosta spinifera*, *B. minor*, and *Crucispina cruciformis*. *Bulletin of the Planktonic Society of Japan*, **34**, 1–13.

Hara, S. and Takahashi, E. (1987b). An investigation with the electron microscope of marine choanoflagellates (Protozoa: Choanoflagellida) from Osaka Bay, Japan: II. Two new genera and a new species of Acanthoecidae. *Bulletin of the Planktonic Society of Japan*, **34**, 15–23.

Hara, S. and Tanoue, E. (1984). Choanoflagellates in the Antarctic Ocean with special reference to *Parvicorbicula socialis* (Meunier) Deflandre. *Memoirs of National Institute Polar Research, Special Issue*, **32**, 1–13.

Hara, S., Tanoue, E., Zenimito, M., Komaki, Y. and Takahashi, E. (1986). Morphology and distribution of heterotrophic protists along 75 °E in the Southern Ocean. *Memoirs of National Institute of Polar Research, Special Issue*, **40**, 69–80.

Hara, S., Chen, Y. L., Sheu, J. C. and Takahashi, E. (1996). Choanoflagellates (Sarcomastigophora, Protozoa) from the coastal waters of Taiwan and Japan: I. Three new species. *Journal of Eukaryotic Microbiology*, **43**, 136–43.

Hara, S., Sheu, J. C., Chen, Y. L. and Takahashi, E. (1997). Choanoflagellates (Sarcomastigophora, Protozoa) from the coastal waters of Taiwan and Japan: II. Species composition and biogeography. *Zoological Studies*, **36**, 98–110.

Harrison, P. J., Waters, R. E. and Taylor, F. J. R. (1980). A broad spectrum artificial seawater medium for coastal and open ocean phytoplankton. *Journal of Phycology*, **16**, 28–35.

Hasegawa, M., Hashimoto, T., Adachi, J., Iwabe, N. and Miyata, T. (1993). Early branching in the evolution of eukaryotes: ancient divergence of *Entamoeba* that lacks mitochondria revealed by protein sequence data. *Journal of Molecular Evolution*, **36**, 380–8.

He, D., Fiz-Palacios, O., Fu, C.-J., *et al.* (2014). An alternative root for the eukaryote tree of life. *Current Biology*, **24**, 465–70.

Hejnol, A., Obst, M., Stamatakis, A., *et al.* (2009). Assessing the root of bilaterian animals with scalable phylogenomic methods. *Proceedings of the Royal Society of London B*, **276**, 4261–70.

Held, A. (1975). The zoospore of *Rozella allomycis*: ultrastructure. *Canadian Journal of Botany*, **53**, 2212–32.

Herbomel, P. (1999). Spinning nuclei in the brain of the zebrafish embryo. *Current Biology*, **9**, R627–R628.

Herr, R. A., Ajello, L., Taylor, J. W., Arseculeratne, S. N. and Mendoza, L. (1999). Phylogenetic analysis of *Rhinosporidium seeberi's* 18S small-subunit ribosomal DNA groups this pathogen among members of the protoctistan Mesomycetozoa clade. *Journal of Clinical Microbiology*, **37**, 2750–4.

Hertel, L. A., Bayne, C. J. and Loker, E. S. (2002). The symbiont *Capsaspora owczarzaki*, nov. gen. nov. sp., isolated from three strains of the pulmonate snail *Biomphalaria glabrata* is related to members of the Mesomycetozoea. *International Journal for Parasitology*, **32**, 1183–91.

Herth, W. (1979). Behaviour of the chrysoflagellate alga, *Dinobryon divergens*, during lorica formation. *Protoplasma*, **100**, 345–51.

Herth, W. (1980). Calcofluor white and Congo red inhibit chitin microfibril assembly of *Poterioochromonas*: evidence for a gap between polymerization and microfibril formation. *Journal of Cell Biology*, **87**, 442–50.

Herth, W., Kuppel, A. and Schnepf, E. (1977). Chitinous fibrils in the lorica of the flagellate chrysophyte *Poterioochromonas stipitata* (syn. *Ochromonas malhamensis*). *Journal of Cell Biology*, **73**, 311–21.

Hibberd, D. J. (1975). Observations on the ultrastructure of the choanoflagellate *Codosiga botrytis* (Ehr.) Saville-Kent with special reference to the flagellar apparatus. *Journal of Cell Science*, **17**, 191–219.

Hibberd, D. J. (1983). Ultrastructure of the colonial colourless flagellate *Phalansterium digitatum* Stein (Phalansteriida ord. nov.) and *Spongomonas uvella* Stein (Spongomonadida ord. nov.). *Protistologica*, **19**, 523–35.

Hibbett, D. S., Binder, M., Bischoff, J. F., *et al.* (2007). A higher-level phylogenetic classification of the Fungi. *Mycological Research*, **111** (5), 509–47.

Higdon, J. J. L. (1979a). A hydrodynamic analysis of flagellar propulsion. *Journal of Fluid Mechanics*, **90**, 685–711.

Higdon, J. J. L. (1979b). The generation of feeding currents by flagellar motions. *Journal of Fluid Mechanics*, **94**, 305–30.

Hildebrand, M. and Wetherbee, R. (2003). Components and control of silicification in diatoms. In *Progress in Molecular and Subcellular Biology, Silicon Biomineralization*, Vol. 33, ed. W. E. G. Müller. Heidelberg: Springer-Verlag, pp. 11–57.

Hoepffner, N. and Haas, L. W. (1990). Electron microscopy of nanoplankton from the North Pacific central gyre. *Journal of Phycology*, **26**, 421–39.

Hoff, J. van den and Franzmann, P. D. (1986). A choanoflagellate in a hypersaline Antarctic lake. *Polar Biology*, **6**, 71–3.

Hollande, A. (1952). Ordre des Choanoflagellés ou Craspédomonadines. In *Traité de Zoologie*, Vol. 1, ed. Grassé P.-P. Paris: Masson et Cie, pp. 579–98.

Hunter, P. (2008). The great leap forward: major evolutionary jumps might be caused by changes in gene regulation rather than the emergence of new genes. *EMBO Reports*, **9**, 608–11.

Hurd, D. C. (1983). Physical and chemical properties of sileceous skeletons. In *Silicon Geochemistry and Biogeochemistry*, ed. S. R. Aston. London: Academic Press, pp. 187–244.

Hwang, S.-J. and Heath, R. T. (1997). The distribution of protozoa across a trophic gradient, factors controlling their abundance and importance in the plankton food web. *Journal of Plankton Research*, **19**, 491–518.

Hynes, R. O. (2011). The evolution of metazoan extracellular matrix. *Journal of Cell Biology* **196**, 671–9.

Ikävalko, J. (1998). Further observations on flagellates within sea ice in northern Bothnian Bay, the Baltic Sea. *Polar Biology*, **19**, 323–9.

Ikävalko, J. and Gradinger, R. (1997). Flagellates and heliozoans in the Greenland Sea ice studied alive using light microscopy. *Polar Biology*, **17**, 473–81.

Ikävalko, J. and Thomsen, H. A. (1997). The Baltic Sea ice biota (March 1994): a study of the protistan community. *European Journal of Protistology*, **33**, 229–43.

Ingham, P. W., Nakano, Y. and Seger, C. (2011). Mechanisms and functions of Hedgehog signalling across the metazoa. *Nature Reviews Genetics*, **12**, 393–406.

Ishiyama, M., Hiromi, J., Tanimura, A. and Kadota, S. (1993). Abundance and biomass distribution of microbial assemblages at the surface in the oceanic province of Antarctic Ocean. *Proceedings of the NIPR Symposium on Polar Biology*, **6**, 6–20.

Iyengar, M. O. P. and Ramanathan, K. R. (1940). *Cladospongia*, a new member of the Craspedomonadaceae from Madras. *Journal of the Indian Botanical Society*, **19**, 241–5.

Jackson, S. M. and Leadbeater, B. S. C. (1991). Costal strip accumulation and lorica assembly in the marine choanoflagellate *Diplotheca costata* Valkanov. *Journal of Protozoology*, **38**, 97–104.

James, T. Y., Kauff, F., Schoch, C. L., *et al.* (2006a) Reconstructing the early evolution of fungi using a six-gene phylogeny. *Nature*, **443**, 818–22.

James, T. Y., Letcher, P. M., Longcore, J. E., *et al.* (2006b). A molecular phylogeny of the flagellated fungi (Chytridiomycota) and description of a new phylum (Blastocladiomycota). *Mycologia*, **98**, 860–71.

James-Clark, H. (1866a). On the nature of sponges. *Proceedings of the Boston Natural History Society*, **11**, 16–17.

James-Clark, H. (1866b). Conclusive proofs on the animality of the ciliate sponges, and their affinities with the Infusoria Flagellata. *American Journal of Science and Arts*, Series 2, **42**, 320–5.

James-Clark, H. (1867a). Conclusive proofs on the animality of the ciliate sponges, and their affinities with the Infusoria Flagellata. *The Annals and Magazine of Natural History*, Series 3, **19**, 13–19.

James-Clark, H. (1867b). On the Spongiae Ciliatae as Infusoria Flagellata: or observations on the structure, animality and relationship of *Leucosolenia botryoides* Bowerbank. *Memoirs of the Boston Society of Natural History*, **1**, 305–40.

James-Clark, H. (1868). On the Spongiae Ciliatae as Infusoria Flagellata: or observations on the structure, animality and relationship of *Leucosolenia botryoides* Bowerbank. *Annals and Magazine of Natural History*, Series 4, **1**, 133–42; 188–215; 250–64.

James-Clark, H. (1871a). Note on the Infusoria Flagellatae and the Spongiae Ciliatae. *American Journal of Science and Arts*, Series 3, **1**, 113–14.

James-Clark, H. (1871b). The American *Spongilla*, a craspedote, flagellate, infusorian. *American Journal of Science and Arts*, Series 3, **2**, 426–36.

James-Clark, H. (1872). The American *Spongilla*, a craspedote, flagellate, infusorian. *Monthly Microscopical Journal*, **7**, 104–14.

Jeuck, A. and Arndt, H. (2013). A short guide to common heterotrophic flagellates of freshwater habitats based on the morphology of living organisms. *Protist*, **164**, 842–60.

Jeuck, A., Arndt, H. and Nitsche, F. (2014). Extended phylogeny of the Craspedida (Choanomonada). *European Journal of Protistology*, **50**, 430–43.

Jiménez-Guri, E., Philippe, H., Okamura, B. and Holland, P. W. H. (2007). *Buddenbrockia* is a cnidarian worm. *Science*, **317**, 116–18.

Johnson, R. E. and Brokaw, C. J. (1979). Flagellar hydrodynamics: a comparison between the resistive-force theory and the slender-body theory. *Biophysical Journal*, **25**, 113–27.

Johnson, R. E. and Wu, T. Y. (1979). Hydromechanics of low-Reynolds-number flow. Part 5: motion of a slender torus. *Journal of Fluid Mechanics*, **95**, 263–77.

Jones, M. D. M., Forn, I., Gadelha, C., *et al.* (2011). Discovery of novel intermediate forms redefines the fungal tree of life. *Nature*, **474**, 200–3.

Jorgensen, C. B. (1983). Fluid mechanical aspects of suspension feeding. *Marine Ecology Progress Series*, **11**, 89–103.

Jürgens, K. and Massana, R. (2008). Protistan grazing on marine bacterioplankton. In *Microbial Ecology of the Oceans*, ed. D. L. Kirchman. 2nd edn. New York: John Wiley and Sons, Inc., pp. 383–441.

Kamatani, A. (1982). Dissolution rates of silica from diatoms decomposing at various temperatures. *Marine Biology*, **68**, 91–6.

Kamatani, A. and Riley, J. P. (1979). Rate of dissolution of diatom cell walls in seawater. *Marine Biology*, **55**, 29–35.

Kamikawa, R., Sakaguchi, M., Matsumoto, T., Hashimoto, T. and Inagaki, Y. (2010). Rooting for the root of elongation factor-like protein phylogeny. *Molecular Phylogenetics and Evolution*, **56**, 1082–8.

Karpov, S. A. (1982). The ultrastructure of a fresh-water flagellate *Monosiga ovata* Kent (Choanoflagellida: Monosigidae). *Tsitologiia*, **24**, 400–4.

Karpov, S. A. and Coupe, S. J. (1998). A revision of choanoflagellate genera *Kentrosiga* Schiller, 1953 and *Desmarella* Kent, 1880. *Acta Protistologica*, **37**, 23–7.

Karpov, S. A. and Leadbeater, B. S. C. (1997). Cell and nuclear division in a freshwater choanoflagellate, *Monosiga ovata* Kent. *European Journal of Protistology*, **33**, 323–34.

Karpov, S. A. and Leadbeater, B. S. C. (1998). Cytoskeleton structure and composition in choanoflagellates. *Journal of Eukaryotic Microbiology*, **45**, 361–7.

Karpov, S. A. and Mylnikov, A. P. (1993). Preliminary observations on the ultrastructure of mitosis in choanoflagellates. *European Journal of Protistology*, **29**, 19–23.

Karpov, S. A., Mikhailov, K. V., Mirzaeva, G. S., *et al.* (2013). Obligately phagotrophic aphelids turned out to branch with earliest-diverging fungi. *Protist*, **164**, 195–205.

Keeling, P. J. and Iganaki, Y. (2004). A class of eukaryotic GTPase with a punctuate distribution suggesting multiple functional replacements of translation elongation factor 1α. *Proceedings of the National Academy of Sciences USA*, **101**, 15380–5.

Keeling, P. J., Burger, G., Durnford, D. G., *et al.* (2005). The tree of eukaryotes. *Trends in Ecology and Evolution*, **20**, 670–6.

Kent, W. S. (1870). Haeckel on the relationship of the sponges to the corals. *Annals and Magazine of Natural History*, Series 4, **5**, 204–18.

Kent, W. S. (1871a). Affinities of the sponges. *Nature*, **4**, 184–5.

Kent, W. S. (1871b). Notes on James-Clark's Flagellate Infusoria with description of new species. *Monthly Microscopical Journal*, **6**, 261–7.

Kent, W. S. (1878a). Observations upon Prof. Ernst Haeckel's 'Physemaria' and on the affinity of the sponges. *Annals and Magazine of Natural History*, Series 5, **1**, 1–17.

Kent, W. S. (1878b). Notes on the embryology of sponges. *Annals and Magazine of Natural History*, Series 5, **2**, 139–56.

Kent, W. S. (1878c). A new field for the microscopist. *Popular Science Review*, New Series, **2** (6), 113–32.

Kent, W. S. (1880–2). *A Manual of the Infusoria*. Vols **1–3**. London: D. Bogue.

Kerk, D., Gee, A., Standish, M., *et al.* (1995). The rosette agent of Chinook salmon (*Oncorhynchus tshawytscha*) is closely related to choanoflagellates, as determined by phylogenetic analyses of its small ribosomal subunit RNA. *Marine Biology*, **122**, 187–92.

King, N., Hittinger, C. T. and Carroll, S. B. (2003). Evolution of key cell signaling and adhesion protein families predates the origin of animals. *Science*, **301** (5631), 361–3.

King, N., Westbrook, M. J., Young, S. L., *et al.* (2008). The genome of the choanoflagellate *Monosiga brevicollis* and the origin of metazoan multicellularity. *Nature*, **451**, 783–8.

Kivi, S. and Kuosa, H. (1994). Late winter microbial communities in the western Weddell Sea. *Polar Biology*, **14**, 389–99.

Klaveness, D. and Guillard, R. R. L. (1975). The requirement for silicon in *Synura petersenii* (Chrysophyceae). *Journal of Phycology*, **11**, 349–55.

Knoll, A. H. (2012). The multiple origins of complex multicellularity. *Annual Review of Earth and Planetary Sciences*, **39**, 217–39.

Kroger, N. (2007). Prescribing diatom morphology: toward genetic engineering of biological nanomaterials. *Current Opinion in Chemical Biology*, **11**, 662–9.

Kümmel, G. and Brandenburg, J. (1961). Die Reusengeisselzellen (Cyrtocyten). *Zeitschrift für Naturforschung*, **16b**, 692–7.

Kuuppo, P. (1994). Annual variation in the abundance and size of heterotrophic nanoflagellates on the SW coast of Finland, the Baltic Sea. *Journal of Plankton Research*, **16**, 1525–42.

Lackey, J. B. (1940). Some new flagellates from the Woods Hole area. *American Midland Naturalist*, **23**, 463–71.

Lackey, J. B. (1959). Morphology and biology of a new species of *Protospongia*. *Transactions of the American Microscopical Society*, **78**, 202–6.

Lang, B. F., O'Kelly, C., Nerad, T., Gray, M. W. and Burger, G. (2002). The closest unicellular relatives of animals. *Current Biology*, **12**, 1773–8.

Lapage, G. (1925). Notes on the choanoflagellate, *Codosiga botrytis*, Ehrbg. *Quarterly Journal of Microscopical Science*, **69**, 471–508.

Lara, E., Moreira, D. and López-Garcia, P. (2012). The environmental clade LKM11 and *Rozella* form the deepest branching clade of fungi. *Protist*, **161**, 116–21.

Lauterborn, R. (1894). Über die Winterfauna einiger Gewässer der Obertrheinebene mit Beschreibungen neuer Protozoën. *Biologisches Centralblat*, **14**, 390–8.

Lavrov, D. (2011). Key transitions in animal evolution: a mitochondrial DNA perspective. In *Key Transitions in Animal Evolution*, ed. B. Schierwater and R. DeSalle. Enfield, NH: Science Publishers and CRC Press, pp. 35–54.

Laybourn-Parry, J. E. M. and Parry, J. D. (2000). Flagellates and the microbial loop. In *The Flagellates*, ed. B. S. C. Leadbeater and J. C. Green. London: Taylor and Francis, pp. 216–39.

Leadbeater, B. S. C. (1972a). Fine structural observations on some marine choanoflagellates from the coast of Norway. *Journal of the Marine Biological Association UK*, **52**, 67–79.

Leadbeater, B. S. C. (1972b). Identification, by means of electron microscopy, of flagellate nanoplankton from the coast of Norway. *Sarsia*, **49**, 107–24.

Leadbeater, B. S. C. (1972c). Ultrastructural observations on some marine choanoflagellates from the coast of Denmark. *British Phycological Journal*, **7**, 195–211.

Leadbeater, B. S. C. (1973). External morphology of some marine choanoflagellates from the coast of Jugoslavia. *Archiv für Protistenkunde*, **115**, 234–52.

Leadbeater, B. S. C. (1974). Ultrastructural observations on nanoplankton collected from the coast of Jugoslavia and the Bay of Algiers. *Journal of the Marine Biological Association UK*, **54**, 179–96.

Leadbeater, B. S. C. (1975). A microscopical study of the marine choanoflagellate *Savillea micropora* (Norris) comb. nov., and preliminary observations on lorica development in *S. micropora* and *Stephanoeca diplocostata* Ellis. *Protoplasma*, **83**, 119–29.

Leadbeater, B. S. C. (1977). Observations on the life-history and ultrastructure of the marine choanoflagellate *Choanoeca perplexa* Ellis. *Journal of the Marine Biological Association UK*, **57**, 285–301.

Leadbeater, B. S. C. (1978). Renaming of *Salpingoeca sensu* Grøntved. *Journal of the Marine Biological Association UK*, **8**, 511–15.

Leadbeater, B. S. C. (1979a). Developmental and ultrastructural observations on two stalked marine choanoflagellates, *Acanthoecopsis spiculifera* Norris and *Acanthoeca spectabilis* Ellis. *Proceedings of the Royal Society of London B*, **204**, 57–66.

Leadbeater, B. S. C. (1979b). Developmental studies on the loricate choanoflagellate *Stephanoeca diplocostata* Ellis: I. Ultrastructure of the non-dividing cell and costal strip production. *Protoplasma*, **98**, 241–62.

Leadbeater, B. S. C. (1979c). Developmental studies on the loricate choanoflagellate *Stephanoeca diplocostata* Ellis: II. Cell division and lorica assembly. *Protoplasma*, **98**, 311–28.

Leadbeater, B. S. C. (1980). Four new species of loricate choanoflagellates from South Brittany, France. *Cahiers de Biologie Marine*, **21**, 345–53.

Leadbeater, B. S. C. (1983a). Life history and ultrastructure of a new marine species of *Proterospongia* (Choanoflagellida). *Journal of the Marine Biological Association UK*, **63**, 135–60.

Leadbeater, B. S. C. (1983b). Distribution and chemistry of microfilaments in choanoflagellates, with special reference to the collar and other tentacle systems. *Protistologica*, **19**, 157–66.

Leadbeater, B. S. C. (1985). Developmental studies on the loricate choanoflagellate *Stephanoeca diplocostata* Ellis: IV. Effects of silica deprivation on growth and lorica production. *Protoplasma*, **127**, 171–9.

Leadbeater, B. S. C. (1987). Developmental studies on the loricate choanoflagellate *Stephanoeca diplocostata* Ellis: V. The cytoskeleton and the effects of microtubule poisons. *Protoplasma*, **136**, 1–15.

Leadbeater, B. S. C. (1989). Developmental studies on the loricate choanoflagellate *Stephanoeca diplocostata* Ellis: VI. Replenishment of silica to silica-starved cells. *Protoplasma*, **153**, 71–84.

Leadbeater, B. S. C. (1994a). Developmental studies on the loricate choanoflagellate *Stephanoeca diplocostata* Ellis: VII. Dynamics of costal strip accumulation and lorica assembly. *European Journal of Protistology*, **30**, 111–24.

Leadbeater, B. S. C. (1994b). Developmental studies on the loricate choanoflagellate *Stephanoeca diplocostata* Ellis. VIII. Nuclear division and cytokinesis. *European Journal of Protistology*, **30**, 171–83.

Leadbeater, B. S. C. (2006). The 'mystery' of the flagellar vane in choanoflagellates. *Nova Hedwigia*, **130**, 213–24.

Leadbeater, B. S. C. (2008a). Choanoflagellate lorica construction and assembly: the nudiform condition. I. *Savillea* species. *Protist*, **159**, 259–68.

Leadbeater, B. S. C. (2008b). Choanoflagellate evolution: the morphological perspective. *Protistology*, **5**, 256–67.

Leadbeater, B. S. C. (2010). Choanoflagellate lorica construction and assembly: the tectiform condition. *Volkanus costatus* (= *Diplotheca costata*). *Protist*, **161**, 160–76.

Leadbeater, B. S. C. and Barker, D. A. N. (1995). Biomineralization and scale production in the Chrysophyta. In *Chrysophyte Algae, Ecology, Phylogeny and Development*, ed. C. D. Sandgren, J. P. Smol and J. Kristiansen. Cambridge: Cambridge University Press, pp. 141–64.

Leadbeater, B. S. C. and Cheng, R. (2010). Costal strip production and lorica assembly in the large tectiform choanoflagellate *Diaphanoeca grandis* Ellis. *European Journal of Protistology*, **46**, 96–110.

Leadbeater, B. S. C. and Davies, M. E. (1984). Developmental studies on the loricate choanoflagellate *Stephanoeca diplocostata* Ellis: III. Growth and turnover of silica, preliminary observations. *Journal of Experimental Biological Ecology*, **81**, 251–68.

Leadbeater, B. S. C. and Karpov, S. A. (2000). Cyst formation in a freshwater strain of the choanoflagellate *Desmarella moniliformis* Kent. *Journal of Eukaryotic Microbiology*, **47**, 433–9.

Leadbeater, B. S. C. and Manton, I. (1974). Preliminary observations on the chemistry and biology of the lorica in a collared flagellate (*Stephanoeca diplocostata* Ellis). *Journal of the Marine Biological Association UK*, **54**, 269–76.

Leadbeater, B. S. C. and Morton, C. (1974a). A microscopical study of a marine species of *Codosiga* James-Clark (Choanoflagellata) with special reference to the ingestion of bacteria. *Biological Journal of the Linnean Society*, **6**, 337–47.

Leadbeater, B. S. C. and Morton, C. (1974b). A light and electron microsope study of the choanoflagellates *Acanthoeca spectabilis* Ellis and *A. brevipoda* Ellis. *Archiv für Microbiologie*, **95**, 279–92.

Leadbeater, B. S. C. and Riding, R. (ed.) (1986). *Biomineralization in Lower Plants and Animals*. Oxford: Oxford University Press.

Leadbeater, B. S. C., Hassan, R., Nelson, M., Carr, M. and Baldauf, S. L. (2008a). A new genus, *Helgoeca* gen. nov., for a nudiform choanoflagellate. *European Journal of Protistology*, **44**, 227–37.

Leadbeater, B. S. C., Henouil, M. and Berovic, N. (2008b). Choanoflagellate lorica construction and assembly: the nudiform condition. II. *Acanthoeca spectabilis*. *Protist*, **159**, 495–505.

Leadbeater, B. S. C., Yu, Q., Kent, J. and Stekel, D. G. (2009). Three-dimensional images of choanoflagellate loricae. *Proceedings of the Royal Society of London B*, **276**, 3–11.

Leakey, R. J. G., Archer, S. D. and Grey, J. (1996). Microbial dynamics in coastal waters of East Antarctica: bacterial production and nanoflagellate bacterivory. *Marine Ecology Progress Series*, **142**, 3–17.

Leakey, R. J. G., Leadbeater, B. S. C., Mitchell, E., McCready, S. M. M. and Murray, A. W. A. (2002). The abundance and biomass of choanoflagellates and other nanoflagellates in waters of contrasting temperature to the north-west of South Georgia in the Southern Ocean. *European Journal of Protistology*, **38**, 333–50.

Lee, W. J. and Patterson, D. J. (1998). Diversity and geographic distribution of free-living heterotrophic flagellates: analysis by PRIMER. *Protist*, **149**, 229–43.

Lemmermann, E. (1910). *Kryptogamenflora der Mark Brandenburg. Algen I (Schizophyceen, Flagellaten, Peridineen)*. Leipzig: Gebruder Borntrager.

Leuckart, R. (1854). Bericht über die Leistungen in der Naturgeschichte der niederen Thiere während der Jahre 1848–1853. *Archiv für Naturgeschichte*, **20** (2), 289–473.

Levin, T. C. and King, N. (2013). Evidence for sex and recombination in the choanoflagellate *Salpingoeca rosetta*. *Current Biology*, **23**, 1–5.

Lewin, J. C. (1961). The dissolution of silica from diatom walls. *Geochimica et Cosmochimica Acta*, **21**, 182–98.

Leys, S. P. (2004). Gastrulation in sponges. In *Gastrulation: From Cells to Embryo*, ed. C. Stern. New York: Cold Spring Harbor Press, pp. 23–31.

Leys, S. P. and Eerkes-Medrano, D. (2005). Gastrulation in calcareous sponges: in search of Haeckel's Gastraea. *Integrative and Comparative Biology*, **45**, 342–51.

Leys, S. P. and Eerkes-Medrano, D. (2006). Feeding in a calcareous sponge: particle uptake by pseudopodia. *Biological Bulletin*, **211**, 157–71.

Leys, S. P. and Ereskovsky, A. (2006). Embryogenesis and larval differentiation in sponges. *Canadian Journal of Zoology*, **84**, 262–87.

Leys, S. P., Mackie, G. O. and Reiswig, H. M. (2007). The biology of glass sponges. *Advances in Marine Biology*, **52**, 1–145.

Leys, S. P., Yahel, G., Reidenbach, M. A., *et al.* (2011). The sponge pump: the role of current induced flow in the design of the sponge body plan. *PLoS One*, **6** (12), e27787.

Li, W., Young, S. L., King, N. and Miller, W. T. (2008). Signaling properties of a non-metazoan Src kinase and the evolutionary history of Src negative regulation. *Journal of Biological Chemistry*, **283**, 15491–501.

Lighthill, J. (1976). Flagellar hydrodynamics. *SIMA Reviews*, **18**, 161–230.

Lim, W. A. and Pawson, T. (2010). Phosphotyrosine signaling: evolving a new cellular communication system. *Cell*, **142**, 661–7.

Liron, N. and Blake, J. R. (1981). Existence of viscous eddies near boundaries. *Journal of Fluid Mechanics*, **107**, 109–29.

Liu, Y. J., Hodson, M. C. and Hall, B. D. (2006). Loss of the flagellum happened only once in the fungal lineage: phylogenetic structure of Kingdom Fungi inferred from RNA polymerase II subunit genes. *BMC Evolutionary Biology*, **6**, 74.

Liu, Y., Steenkamp, E. T., Brinkmann, H., *et al.* (2009). Phylogenomic analyses predict sistergroup relationship of nucleariids and Fungi and paraphyly of zygomycetes with significant support. *BMC Evolutionary Biology*, **9**, 272.

Lock, J. G., Wehrle-Haller, B. and Strömblad, S. (2008). Cell-matrix adhesion complexes: master control machinery of cell migration. *Seminars in Cancer Biology*, **18**, 65–76.

Loucaides, S., Van Cappellen, P. and Behrends, T. (2008). Dissolution of biogenic silica from land to ocean: role of salinity and pH. *Limnology and Oceanography*, **53**, 1614–21.

Lyons, K. M. (1973). Evolutionary implications of collar cell ectoderm in coral planula. *Nature*, **245**, 50–1.

Mackie, G. O. and Singla, C. L. (1983). Studies on hexactinellid sponges: I. Histology of *Rhabdocalyptus dawsoni* (Lambe, 1873). *Philosophical Transactions of the Royal Society of London B*, **301**, 365–400.

Mah, J. L., Christensen-Dalsgaard, K. K. and Leys, S. P. (2014) Choanoflagellate and choanocyte collar-flagellar systems and the assumption of homology. *Evolution and Development*, **16**, 25–37.

Maldonado, M. (2004). Choanoflagellates, choanocytes, and animal multicellularity. *Invertebrate Biology*, **123**, 1–22.

Mann, S. (2001). *Biomineralization: Principles and Concepts in Bioinorganic Materials Chemistry*. Oxford: Oxford University Press.

Mann, S. and Williams, R. J. P. (1982). High resolution electron microscopy studies of the silica lorica in the choanoflagellate *Stephanoeca diplocostata* Ellis. *Proceedings of the Royal Society of London B*, **216**, 137–46.

Manning, G., Young, S. L., Miller, W. T. and Zhai, Y. (2008). The protist, *Monosiga brevicollis*, has a tyrosine kinase signaling network more elaborate and diverse than found in any known metazoan. *Proceedings of the National Academy of Sciences USA*, **105**, 9674–9.

Manton, I. and Leadbeater, B. S. C. (1974). Fine-structural observations on six species of *Chrysochromulina* from wild Danish marine nanoplankton, including a description of *C. campanulifera* sp. nov. and a preliminary summary of the nanoplankton as a whole. *Det Kongelige Danske Videnskabernes Selskab Biologiske Skrifter*, **20**, 1–26.

Manton, I. and Leadbeater, B. S. C. (1978). Some critical qualitative details of lorica construction in the type species of *Calliacantha* Leadbeater (Choanoflagellata). *Proceedings of the Royal Society of London B*, **203**, 49–57.

Manton, I. and Oates, K. (1979a). Further observations on *Calliacantha* Leadbeater (Choanoflagellata), with special reference to *C. simplex* sp. nov. from many parts of the world. *Proceedings of the Royal Society of London B*, **204**, 287–300.

Manton, I. and Oates, K. (1979b). Further observations on choanoflagellates in the genus *Calliacantha* Leadbeater, with special reference to *C. multispina* sp. nov. from South Africa and Britain. *Journal of the Marine Biological Association UK*, **59**, 207–13.

Manton, I., Sutherland, J. and Leadbeater, B. S. C. (1975). Four new species of choanoflagellates from Arctic Canada. *Proceedings of the Royal Society of London B*, **189**, 15–27.

Manton, I., Sutherland, J. and Leadbeater, B. S. C. (1976). Further observations on the fine structure of marine collared flagellates (Choanoflagellata) from Arctic Canada and west Greenland: species of *Parvicorbicula* and *Pleurasiga*. *Canadian Journal of Botany*, **54**, 1932–55.

Manton, I., Sutherland, J. and Oates, K. (1980). A re-investigation of collared flagellates in the genus *Bicosta* Leadbeater with special reference to correlations with climate. *Philosophical Transactions of the Royal Society of London B*, **290**, 431–47.

Manton, I., Bremer, G. and Oates, K. (1981). Problems of structure and biology in a large collared flagellate (*Diaphanoeca grandis* Ellis) from Arctic seas. *Proceedings of the Royal Society of London B*, **213**, 15–26.

Marchant, H. J. (1985). Choanoflagellates in the Antarctic marine food chain. In *Antarctic Nutrient Cycles and Food Webs*, ed. W. R. Siegfried, P. R. Condy and R. M. Laws. Berlin and Heidelberg: Springer-Verlag, pp. 271–6.

Marchant, H. J. (1990). Grazing rate and particle size selection by the choanoflagellate *Diaphanoeca grandis* from the sea-ice of Lake Saroma, Hokkaido. *Proceedings of the NIPR Symposium on Polar Biology*, **3**, 1–7.

Marchant, H. J. (2005). Choanoflagellates. In *Antarctic Marine Protists*, ed. F. J. Scott and H. J. Marchant. Canberra and Hobart: Australian Biological Resources Study, Canberra, and Australian Antarctic Division, pp. 326–46.

Marchant, H. J. and Nash, G. V. (1986). Electron microscopy of gut contents and faeces of *Euphausia superba* Dana. *Memoirs of the Natural Institute of Polar Research*, Special Issue **40**, 167–77.

Marchant, H. J. and Perrin, R. (1990). Seasonal variation in abundance and species composition of choanoflagellates (Acanthoecidae) at Antarctic coastal sites. *Polar Biology*, **10**, 499–505.

Marchant, H. J. and Scott, F. J. (1993). Uptake of sub-micrometre particles and dissolved organic material by Antarctic choanoflagellates. *Marine Ecology Progress Series*, **92**, 59–64.

Marchant, H. J., Hoff, J. van den, and Burton, H. R. (1987). Loricate choanoflagellates from Ellis Fjord, Antarctica including the description of *Acanthocorbis tintinnabulum* sp. nov. *Proceedings of the NIPR Symposium on Polar Biology*, **1**, 10–22.

Marron, A. O., Alston, M. J., Heavens, D., *et al.* (2013). A family of diatom-like silicon transporters in the siliceous loricate choanoflagellates. *Proceedings of the Royal Society of London B*, **280**, doi: 10.1098/rspb.2012.2543.

Marshall, W. L. and Berbee, M. L. (2010). Population-level analyses indirectly reveal cryptic sex and life history traits of *Pseudoperhinsus tapetis* (Ichthyosporea, Opisthokonta): a unicellular relative of the animals. *Molecular Biology and Evolution*, **27**, 2014–26.

Marshall, W. L. and Berbee, M. L. (2011). Facing unknowns: living cultures (*Pirum gemmata* gen. nov., sp. nov., and *Abeoforma whisleri*, gen. nov., sp. nov.) from invertebrate digestive tracts represent an undescribed clade within the unicellular opisthokont lineage Ichthyosporea (Mesomycetozoea). *Protist*, **162**, 33–57.

Marshall, W. L., Celio, G., McLaughlin, D. J. and Berbee, M. L. (2008). Multiple isolations of a culturable, motile Ichthyosporean (Mesomycetozoa, Opisthokonta), *Creolimax fragrantissima* n. gen., n. sp., from marine invertebrate digestive tracts. *Protist*, **159**, 415–33.

Martindale, M. Q., Pang, K. and Finnerty, J. R. (2004). Investigating the origins of triploblasty: 'mesodermal' gene expression in a diploblastic animal, the sea anemone *Nematostella vectensis* (phylum, Cnidaria; class, Anthozoa). *Development*, 310, 2463–74.

Martin-Jézéquel, V. and Lopez, P. J. (2003). Silicon: a central metabolite for diatom growth and morphogenesis. In *Progress in Molecular and Subcellular Biology, Silicon Biomineralization*, Vol. 33, ed. W. E. G. Müller. Heidelberg: Springer-Verlag, pp. 99–124.

Martin-Jézéquel, V., Hildebrand, M. and Brzezinski, M. A. (2000). Silicon metabolism in diatoms: implications for growth. *Journal of Phycology*, 36, 821–40.

Mathes, J. and Arndt, H. (1994). Biomass and composition of protozooplankton in relation to lake trophy in north German lakes. *Marine Microbial Food Webs*, 8, 357–75.

Mathes, J. and Arndt, H. (1995). Annual cycle of protozooplankton (ciliates, flagellates and sarcodines) in relation to phyto- and metazooplankton in Lake Neumühler See (Mecklenburg, Germany). *Archiv für Hydrobiologie*, 134, 337–58.

Matus, D. Q., Magie, C. R., Pang, K., Martindale, M. Q. and Thomsen, G. H. (2008). The Hedgehog gene family of the cnidarian, *Nematostella vectensis*, and implications for understanding metazoan Hedgehog pathway evolution. *Developmental Biology*, 313, 501–18.

McKenzie, C. H., Deibel, D., Thompson, R. J., MacDonald, B. A. and Penney R. W. (1997). Distribution and abundance of choanoflagellates (Acanthoecidae) in the coastal cold ocean of Newfoundland, Canada. *Marine Biology*, 129, 407–16.

Medina, M., Collins, A. G., Taylor, J. W., *et al.* (2003). Phylogeny of Opisthokonta and the evolution of multicellularity and complexity in Fungi and Metazoa. *International Journal of Astrobiology*, 2, 203–11.

Mehl, D. and Reiswig, H. M. (1991). The presence of flagellar vanes in choanomeres of Porifera and their possible phylogenetic implications. *Zeitschrift für Zoologische Systematik und Evolutionsforschung*, 29, 312–19.

Melkonian, M., Reize, I. B. and Preisig, H. R. (1987). Maturation of a flagellum/basal body requires more than one cell cycle in algal flagellates: studies on *Nephroselmis olivacea* (Prasinophyceae). In *Algal Development: Molecular and Cellular Aspects*, ed. W. Wiessner, D. G. Robinson and D. G. Starr. Berlin: Springer-Verlag, pp. 102–13.

Mendoza, L., Taylor, J. and Ajello, L. (2002). The class Mesomycetozoea: a heterogeneous group of microorganisms at the animal–fungal boundary. *Annual Revue of Microbiology*, 56, 315–44.

Menezes, S. (2005). Nanoplankton biodiversity in the Pettaquamscutt River estuary, Ph.D. thesis, University of Rhode Island.

Menezes, S. and Hargraves, P. (2005). Species diversity patterns of choanoflagellates in the Pettaquamscutt River Estuary, Rhode Island, USA. *Journal of Eukaryotic Microbiology*, 52, 7S–27S

Meunier, A. (1910). *Microplankton des Mers de Barents et de Kara. Duc d'Orleans: Campagne Arctique de 1907*. Bruxelles: Bullen.

Mignot, J.-P. and Brugerolle, G. (1982). Scale formation in chrysomonad flagellates. *Journal of Ultrastructure Research*, 81, 13–26.

Mikhailov, K. V., Konstantinova, A. V. and Nikitin, M. A. (2009). The origin of Metazoa: a transition from temporal to spatial cell differentiation. *BioEssays*, 31, 758–68.

Miklucho-Maclay, N. (1868). Beiträge zur Kenntnis der Spongien I. *Jenaische Zeitschrift für Medizin und Naturwissenschaft*, 4, 221–40.

Minge, M. A., Silberman, J. D. and Orr, R. J. S. (2009). Evolutionary position of breviate amoebae and the primary eukaryote divergence. *Proceedings of the Royal Society of London B*, 276, 597–604.

Misiak, K., Dunlap, K., Rodriguez, S. S. and Julian, D. (2008). Tolerance of a choanoflagellate to environmental stressors. *The FASEB Journal*, 22, (648.21).

Moestrup, Ø. (1979). Identification by electron microscopy of marine nanoplankton from New Zealand, including the description of four new species. *New Zealand Journal of Botany*, 17, 61–93.

Mollicone, M. R. N. and Longcore, J. E. (1994). Zoospore ultrastructure of *Monoblepharis polymorpha*. *Mycologia*, 86, 615–25.

Monod, J. (1950). La technique de la culture continue: théorie et applications. *Annales de l'Institut Pasteur*, 79, 390–410.

Moreira, D., von der Heyden, S., Bass, D., *et al.* (2007). Global eukaryote phylogeny: combined small- and large-subunit ribosomal DNA trees support monophyly of Rhizaria, Retaria and Excavata. *Molecular Phylogenetics and Evolution*, 44, 255–66.

Mortlock, R. A. and Froelich, P. N. (1989). A simple method for the rapid determination of biogenic opal in pelagic marine sediments. *Deep Sea Research*, 36, 1415–26.

Moss, S. T. (1985). An ultrastructural study of taxonomically significant characters of the Thraustochytriales and Labyrinthulales. *Botanical Journal of the Linnean Society*, 91, 329–57.

Müller, O. F. (1786). *Animalcula Infusoria Fluviatilis et Marina*. Copenhagen: Havniae et Lipsiae.

Müller, W. E. G. (2003). Silicon biomineralization: biology, biochemistry, molecular biology, biotechnology. *Progress in Molecular Subcellular Biology*, Vol. 33, Berlin: Springer-Verlag.

Nedelcu, A. M. (2012). Evolution of multicellularity. In *Encyclopedia of Life Sciences*. Chichester: John Wiley and Sons.

Nichols, S. A., Dirks, W., Pearse, J. S. and King, N. (2006). Early evolution of animal cell signaling and adhesion genes. *Proceedings of the National Academy of Sciences USA*, **103**, 12451–6.

Nichols, S. A., Dayel, M. J., and King, N. (2009). Genomic, phylogenetic, and cell biological insights into metazoan origins. In *Animal Evolution: Genes, Genomes, Fossils and Trees*, ed. M. J. Telford and D. T. J. Littlewood. Oxford: Oxford University Press, pp. 24–32.

Nichols, S. A., Roberts, B. W., Richter, D. J., Fairclough, S. R. and King, N. (2012). Origin of metazoan cadherin diversity and the antiquity of the classical cadherin/β-catenin complex. *Proceedings of the National Academy of Sciences USA*, **109**, 13046–51.

Nielsen, C. (2001). *Animal Evolution: Interrelationships of the Living Phyla*. 2nd edn. Oxford: Oxford University Press.

Nielsen, C. (2008). Six major steps in animal evolution: are we sponge-derived larvae? *Evolution and Development*, **10**, 241–57.

Nitsche, F. (2014). *Stephanoeca arndtii* spec. nov.: first cultivation success including molecular and autecological data from a freshwater acanthoecid choanoflagellate from Samoa. *European Journal of Protistology*, **50**, 412–21.

Nitsche, F. and Arndt, H. (2008). A new choanoflagellate species from Taiwan: morphological and molecular biological studies of *Diplotheca elongata* nov. sp. and *D. costata*. *European Journal of Protistology*, **44**, 220–6.

Nitsche, F. and Arndt, H. (2014). Comparison of similar Arctic and Antarctic morphotypes of heterotrophic protists regarding their genotypes and ecotypes. *Protist*. In press.

Nitsche, F., Weitere, M., Scheckenbach, F., *et al.* (2007). Deep sea records of choanoflagellates with a description of two new species. *Acta Protozoologica*, **46**, 99–106.

Nitsche, F., Carr, M., Arndt, H. and Leadbeater, B. S. C. (2011). Higher level taxonomy and molecular phylogenetics of the Choanoflagellatea. *Journal of Eukaryotic Microbiology*, **58**, 452–62.

Noble, G. P., Rogers, M. B. and Keeling, P. J. (2007). Complex distribution of EFL and EF-1α proteins in a green algal lineage. *BMC Evolutionary Biology*, **7**, 82.

Nogales, E. (2000). Structural insights into microtubule function. *Annual Review of Biochemistry*, **69**, 277–302.

Nohýnková, E., Tumova, P. and Kulda, J. (2006). Cell division of *Giardia intestinalis*: flagellar developmental cycle involves transformation and exchange of flagella between mastigonts of a diplomonad cell. *Eukaryotic Cell*, **5**, 753–61.

Nørrevang, A. and Wingstrand, K. G. (1970). On the occurrence and structure of choanocyte-like cells in some echinoderms. *Acta Zoologica*, **51**, 249–70.

Norris, R. E. (1965). Neustonic marine Craspedomonadales (Choanoflagellata) from Washington and California. *Journal of Protozoology*, **12**, 589–612.

Oda, H. and Takeichi, M. (2011). Structural and functional diversity of cadherin at the adherens junction. *Journal of Cell Biology*, **193**, 1137–46.

Orme, B. A. A., Otto, S. R. and Blake, J. R. (2001). Enhanced efficiency of feeding and mixing due to chaotic flow patterns around choanoflagellates. *IMA Journal of Mathematics Applied in Medicine and Biology*, **18**, 293–325.

Orme, B. A. A., Blake, J. R. and Otto, S. R. (2003). Modelling the motion of particles around choanoflagellates. *Journal of Fluid Mechanics*, **475**, 333–55.

Ostenfeld, C. H. (1916). De Danske farvandes plankton i aarene 1898–1901: Phytoplankton og protozoer. 2. Protozoer; Organismere med usikker stilling; Parasiter i phytoplanktonter. *Det Kongelige Danske Videnskabernes Selskab Biologiske Skrifter. 8, Raekke II*, **2**, 115–95.

Owczarzak, A., Stibbs, H. H. and Bayne, C. J. (1980). The destruction of *Schistosoma mansoni* mother sporocysts *in vitro* by amoebae isolated from *Biomphalaria glabrata*: an ultrastructural study. *Journal of Invertebrate Pathology*, **35**, 26–33.

Oxley, F. (1884). On *Protospongia pedicellata*, a new compound infusorian. *Journal of the Royal Microscopical Society*, **2**, 530–2.

Özdikimen, H. (2009). Substitute names for some unicellular animal taxa (Protozoa). *Munis Entomology and Zoology*, **4**, 233–56.

Paasche, E. (1960). Phytoplankton distribution in the Norwegian Sea in June, 1954, related to the hydrography and compared with primary production data. *Fiskeridirektoratets skrifter, Serie Havundersøkelser*, **12**, 1–77.

Paasche, E. (1973). Silicon and the ecology of marine plankton diatoms: II. Silicate-uptake kinetics in five diatom species. *Marine Ecology*, **19**, 262–9.

Paasche, E. and Rom, A. M. (1962). On the phytoplankton vegetation of the Norwegian Sea in May 1958. *Nytt Magasin for Botanikk*, **9**, 33–60.

Paps, J., Medina-Chacón, L. A., Marshall, W., Suga, H. and Ruiz-Trillo, I. (2012). Molecular phylogeny of unikonts: new insights into the position of apusomonads and ancyromonads and internal relationships of opisthokonts. *Protist*, **164**, 2–12.

Parfrey, L. W., Barbero, E., Lasser, E., *et al.* (2006). Evaluating support for the current classification of eukaryotic diversity. *PloS Genet* **2** (12): e220.

Parfrey, L. W., Grant, J., Tekle, Y. I., *et al.* (2010). Broadly sampled multigene analyses yield a well-resolved eukaryotic tree of life. *Systematic Biology*, **59**, 518–33.

Parfrey, L. W., Lahr, D. J. G., Knoll, A. H. and Katz, L. A. (2011). Estimating the timing of early eukaryotic diversification with multigene molecular clocks. *Proceedings of the National Academy of Sciences USA*, **108**, 13624–9.

Patterson, D. J. (1999). The diversity of eukaryotes. *American Naturalist*, **154** (Suppl), S86–S124.

Patterson, D. J. and Larsen, J. (eds) (1991). *The Biology of Free-Living Heterotrophic Flagellates*. Oxford: Clarendon Press.

Patterson, D. J., Nygaard, K., Steinberg, G. and Turley, C. M. (1993). Heterotrophic flagellates and other protists associated with oceanic detritus throughout the water column in the mid North Atlantic. *Journal of the Marine Biological Association UK*, **73**, 67–95.

Paul, M. (2011). *Acanthocorbis mongolica* nov. spec.: a loricate choanoflagellate (Acanthoecida) from a Mongolian freshwater lake. *European Journal of Protistology*, **48**, 1–8.

Pavillard, M. J. (1917). Protistes nouveaux ou peu connus du plankton Mediterraneen. *Comptes Rendus Hebdomadaires des Séances de l'Académie des Sciences Paris*, **164**, 925–8.

Peck, R. K. (2010). Structure of loricae and stalks of several bacterivorous chrysomonads (Chrysophyceae): taxonomical importance and possible ecological significance. *Protist*, **161**, 148–59.

Pekarrinen, M., Lom, J., Murphy, C. A., Ragan, M. A. and Dyková, I. (2003). Phylogenetic position and ultrastructure of two *Dermocystidium* species (Ichthyosporea) from the common perch (*Perca fluviatilis*). *Acta Protozoologica*, **42**, 287–307.

Pereira, C. N., Di Rosa, I., Fagotti, A., *et al.* (2005). The pathogen of frogs *Amphibiocystidium ranae* is a member of the Order Dermocystida in the Class Mesomycetozoea. *Journal of Clinical Microbiology*, **43**, 192–8.

Pernthaler, J., Šimek, K., Sattler, B., *et al.* (1996). Short-term changes of protozoan control on autotrophic picoplankton in an oligo-mesotrophic lake. *Journal of Plankton Research*, **18**, 443–62.

Petersen, J. B. (1929). Beiträge zur Kenntnis der Flagellatengeisseln. *Botanisk Tidsskrift*, **40**, 312–19.

Petersen, J. B. and Hansen, J. B. (1954). Electron microscope observations on *Codonosiga botrytis* (Ehr.) James-Clark. *Botanisk Tidsskrift*, **51**, 281–91.

Pettitt, M. E. (2001). Prey capture and ingestion in choanoflagellates, Ph.D. thesis, University of Birmingham.

Pettitt, M. E., Orme, B. A. A., Blake, J. R. and Leadbeater, B. S. C. (2002). The hydrodynamics of filter feeding in choanoflagellates. *European Journal of Protistology*, **38**, 313–32.

Philippe, H. and Telford, M. J. (2006). Large-scale sequencing and the new animal phylogeny. *Trends in Ecology and Evolution*, **21**, 614–20.

Philippe, H., Snell, E. A., Bapteste, E., *et al.* (2004) Phylogenomics of eukaryotes: impact of missing data on large alignments. *Molecular Biology and Evolution*, **21**, 1740–52.

Philippe, H., Derelle, R., Lopez, P., *et al.* (2009). Phylogenomics revives traditional views on deep animal relationships. *Current Biology*, **19**, 706–12.

Philippe, H., Brinkmann, H., Lavrov, D. V., *et al.* (2011). Resolving difficult phylogenetic questions: why more sequences are not enough. *PLoS Biol* **9** (3): e1000602.

Pick, K. S., Philippe, H., Schreiber, F., *et al.* (2010). Improved phylogenomic taxon sampling noticeably affects non-bilaterian relationships. *Molecular Biology and Evolution*, **27**, 1987–2010.

Pickett-Heaps, J., Schmid, A.-M. and Edgar, L. A. (1990). The cell biology of diatom valve formation. *Progress in Phycological Research*, **7**, 1–168.

Pincus, D., Letunic, I., Bork, P. and Lim, W. A. (2008). Evolution of the phospho-tyrosine signaling machinery in premetazoan lineages. *Proceedings of the National Academy of Sciences USA*, **105**, 9680–4.

Pires-daSilva, A. and Sommer, R. J. (2003). The evolution of signalling pathways in animal development. *Nature Reviews Genetics*, **4**, 39–49.

Pitelka, D. R. (1974). Basal bodies and root structure. In *Cilia and Flagella*, ed. M. A. Sleigh. London: Academic Press, pp. 437–69.

Pozdnyakov, I. R. and Karpov, S. A. (2013). Flagellar apparatus structure of choanocyte in *Sycon* sp. and its significance for phylogeny of Porifera. *Zoomorphology* 132, 351–7.

Preisig, H. R., Anderson, O. R. and Corliss, J. O., *et al.* (1994). Terminology and nomenclature of protist cell surface structures. *Protoplasma*, 181, 1–28.

Ragan, M. A., Goggin, C. L., Cawthorn, R. J., *et al.* (1996). A novel clade of protistan parasites near the animal–fungal divergence. *Proceedings of the National Academy of Sciences USA*, 93, 11907–12.

Ragan, M. A., Murphy, C. A. and Rand, T. G. (2003). Are Ichthyosporea animals or fungi? Bayesian phylogenetic analysis of elongation factor 1a of *Ichthyophonus irregularis*. *Molecular Phylogenetics and Evolution*, 29, 550–62.

Raghu-Kumar, S., Chandramohan, D. and Ramaiah, N. (1987). Contribution of the thraustochytrid *Corallochytrium limacisporum* Raghu-Kumar to microbial biomass in coral reef lagoons. *Indian Journal of Marine Sciences*, 16, 122–5.

Raible, F. and Steinmetz, P. R. H. (2010). Metazoan complexity. In *Introduction to Marine Genomics*, ed. J. M. Cock, K. Tessmar-Raible, C. Boyen and F. Viard. Dordrecht: Springer-Verlag, pp. 143–78.

Rat'kova, T. N. and Wassmann, P. (2002). Seasonal variation and spatial distribution of phyto- and protozooplankton in the central Barents Sea. *Journal of Marine Systems*, 38, 47–75.

Raven, J. A. (1983). The transport and function of silicon in plants. *Biology Reviews*, 58, 179–207.

Raven, J. A. and Waite, A. M. (2004). The evolution of silicification in diatoms: inescapable sinking and sinking as escape? *New Phytologist*, 162, 45–61.

Reimann, B. E. F., Lewin, J. C. and Volcani, B. E. (1966). Studies on the biochemistry and fine structure of silica shell formation in diatoms: II. The structure of the cell wall of *Navicula pelliculosa* (Breb.) Hilse. *Journal of Phycology*, 2, 74–84.

Reiswig, H. M. (1975). Bacteria as food for temperate-water marine sponges. *Canadian Journal of Zoology*, 53, 582–9.

Richards, G. S. and Degnan, B. M. (2009). The dawn of developmental signaling in the Metazoa. *Cold Spring Harbor Symposia on Quantitative Biology*, 74, 81–90.

Richter, D. J. and King, N. (2013). The genomic and cellular foundations of animal origins. *Annual Review of Genetics*, 47, 509–37.

Rieger, R. M. (1976). Monociliated epidermal cells in Gastrotricha: significance for concepts of early metazoan evolution. *Zeitschrift für Zoologische Systematische und Evolutionsforschung*, 14, 198–226.

Rodríguez-Ezpeleta, N., Brinkmann, H., Burger, G., *et al.* (2007). Toward resolving the eukaryotic tree: the phylogenetic positions of jakobids and cercozoans. *Current Biology*, 17, 1420–5.

Rokas, A. and Holland, P. W. H. (2000). Rare genomic changes as a tool for phylogenetics. *Tree*, 15, 454–9.

Rokas, A., King, N., Finnerty, J. and Carroll, S. B. (2003). Conflicting phylogenetic signals at the base of the metazoan tree. *Evolution and Development*, 5, 346–59.

Roper, M., Dayel, M. J., Pepper R. E. and Koehl, M. A. R. (2013). Cooperatively generated stresslet flows supply fresh fluid to multicellular choanoflagellate colonies. *Physical Review Letters*, 110, 228104.

Roubeix, V., Becquefort, S. and Lancelot, C. (2008). Influence of bacteria and salinity on diatom biogenic silica dissolution in estuarine systems. *Biogeochemistry*, 88, 47–62.

Round, F. E., Crawford, R. M. and Mann, D. G. (1990). *The Diatoms: Biology and Morphology of the Genera*. Cambridge: Cambridge University Press.

Ruiz-Trillo, I., Inagaki, Y., Davis, L. A., *et al.* (2004). *Capsaspora owczarzaki* is an independent opisthokont lineage. *Current Biology*, 14, R946–947.

Ruiz-Trillo, I., Lane, C. E., Archibald, J. M. and Roger, A. J. (2006). Insights into the evolutionary origin and genome architecture of the unicellular opisthokonts *Capsaspora owczarzaki* and *Sphaeroforma arctica*. *Journal of Eukaryotic Microbiology*, 53, 379–84.

Ruiz-Trillo, I., Burger, G. and Holland, P. W. H., *et al.* (2007). The origins of multicellularity: a multi-taxon genome initiative. *Trends in Genetics*, 23, 113–18.

Ruiz-Trillo, I., Roger, A. G., Burger, G., Gray, M. W. and Lang, B. F. (2008). A phylogenomic investigation into the origin of Metazoa. *Molecular Biology and Evolution*, 25, 664–72.

Saedeleer, H. de (1927). Notes de protistologie: I. Craspédomonadines; matérial systématique. *Annales de la Société Royale Zoologique de Belgique*, 58, 242–88.

Saedeleer, H. de (1929). Notes de protistologie: II. Craspédomonadines. Morphologie et physiologie. *Recueil de l'Institut Zoologique Torley-Rosseau*, 2, 241–87.

Salbrechter, M. and Arndt, H. (1994). The annual cycle of protozooplankton in the mesotrophic, alpine Lake Mondsee (Austria). *Marine Microbial Food Webs*, 8, 217–34.

Salonen, K. and Jokinen, L. (1988). Flagellate grazing on bacteria in a small dystrophic lake. *Hydrobiologia*, 161, 203–9.

Salvini-Plawen, L. von (1978). On the origin and evolution of the lower Metazoa. *Zeitschrift für Zoologische Systematik und Evolutionsforschung*, 16, 40–88.

Samuelsson, K., Berglund, J. and Andersson, A. (2006). Factors structuring the heterotrophic flagellate and ciliate community along a brackish water primary production gradient. *Journal of Plankton Research*, 28, 345–59.

Sanders, R. W., Porter, K. G., Bennett, S. J. and DeBiase, A. E. (1989). Seasonal patterns of bacterivory by flagellates, ciliates, rotifers, and cladocerans in a freshwater planktonic community. *Limology and Oceanography*, 34, 673–87.

Sanders, R. W., Caron, D. A. and Berninger, U.-G. (1992). Relationship between bacteria and heterotrophic nanoplankton in marine and fresh-waters: an inter-ecosystem comparison. *Marine Ecology Progress Series*, 86, 1–14.

Sandgren, C. D. and Barlow, S. B. (1989). Siliceous scale production in chrysophye algae: II. SEM observations regarding the effects of metabolic inhibitors on scale regeneration in laboratory populations of scale-free *Synura petersenii* cells. *Nova Hedwigia*, 95, 27–44.

Sandgren, C. D., Hall, S. A. and Barlow, S. B. (1996). Siliceous scale production in chrysophyte and synurophyte algae: I. Effects of silica-limited growth on cell silica content, morphology and construction of the scale layer of *Synura petersenii*. *Journal of Phycology*, 32, 675–92.

Scherwass, A., Bergfeld, T., Schöl, A., Weitere, M. and Arndt, H. (2010). Changes in the plankton community along the length of the River Rhine: Lagrangian sampling during a spring situation. *Journal of Plankton Research*, 32, 491–502.

Schierwater, B., de Jong, D. and DeSalle, R. (2009a). Placozoa and the evolution of Metazoa and introsomatic cell differentiation. *International Journal of Biochemistry and Cell Biology*, 41, 370–9.

Schierwater, B., Eitel, M., Jakob, W., *et al.* (2009b). Concatenated analysis sheds light on early metazoan evolution and fuels a modern 'Urmetazoon' hypothesis. *PLoS Biology*, 7 (1).

Schiller, J. (1953). Über neuen Craspedomonaden (Choanoflagellaten). *Archiv für Hydrobiologie*, 48, 248–59.

Schnepf, E., Röderer, G. and Herth, W. (1975). The formation of the fibrils in the lorica of *Poterioochromonas stipitata*: tip growth, kinetics, site, orientation. *Planta*, 125, 45–62.

Schouteden, H. (1908). Notes sur les choanoflagellates. *Annales de la Société Royale Zoologique et Malacologique de Belgique*, 43, 169–81.

Schulze, F. E. (1885). Über das Verhältniss der Spongien zu den Choanoflagellaten. *Sitzungsberichte der Königlich Preussischen Akademie der Wissenschaften zu Berlin*, pp. 179–91.

Sebé-Pedrós, A. and Ruiz-Trillo, I. (2010). Integrin-mediated adhesion complex: cooption of signaling systems at the dawn of Metazoa. *Communicative and Integrative Biology*, 3, 475–7.

Sebé-Pedrós, A., Roger, A. J., Lang, F. B., King, N. and Ruiz-Trillo, I. (2010). Ancient origin of the integrin–mediated adhesion and signaling machinery. *Proceedings of the National Academy of Sciences USA*, 107, 10142–7.

Sebé-Pedrós, A., Mendoza, A. de, Lang, B. F., Degnan, B. and Ruiz-Trillo, I. (2011). Unexpected repertoire of metazoan transcription factors in the unicellular holozoan *Capsaspora owczarzaki*. *Molecular Biology and Evolution*, 28, 1241–54.

Sebé-Pedrós, A., Zheng, Y., Ruiz-Trillo, I. and Pan, D. (2012). Premetazoan origin of the Hippo signalling pathway. *Cell Reports*, 1, 13–20.

Sebé-Pedrós, A., Burkhardt, P., Sánchez-Pons, N., *et al.* (2013). Insights into the origin of metazoan filopodia and microvilli. *Molecular Biology and Evolution*, 30, 2013–23.

Segawa, Y., Suga, H., Iwabe, N., *et al.* (2006). Functional development of Src tyrosine kinases during evolution from a unicellular ancestor to multicellular animals. *Proceedings of the National Academy of Sciences USA*, 103, 12021–6.

Seipel, K. and Schmid, V. (2005). Evolution of striated muscle: jellyfish and the origin of triploblasty. *Developmental Biology*, 282, 14–26.

Seipel, K. and Schmid, V. (2006). Mesodermal anatomies in cnidarian polyps and medusae. *International Journal of Developmental Biology*, 50, 589–99.

Senn, G. (1900). Flagellaten. In *Die Natürlichen Pflanzenfamilien* Vol. 1, Part 1a, ed. H. G. A. Engler and K. A. E. Prantl. Leipzig: Wilhelm Engelmann Verlag, pp. 93–188.

Shalchian-Tabrizi, K., Minge, M. A., Espelund, M., *et al.* (2008). Multigene phylogeny of Choanozoa and the origin of animals. *PLoS ONE*, **3** (5).

Shatilovich, A. V., Shmakova, L. A., Mylnikov, A. P. and Gilichinsky, D. A. (2009). Ancient protozoa isolated from permafrost. In *Permafrost Soils*, ed. R. Margesin. Berlin and Heidelberg: Springer-Verlag, pp. 97–118.

Sherr, E. B. (1988). Direct use of high molecular weight polysaccharide by heterotrophic flagellates. *Nature*, **335**, 348–51.

Sherr, E. B. and Sherr, B. F. (2002). Significance of predation by protists in aquatic microbial food webs. *Antonie van Leeuwenhoek*, **81**, 293–308.

Sherr, E. B., Sherr, B. F. and Fessenden, L. (1997). Heterotrophic protists in the central Arctic Ocean. *Deep-Sea Research II*, **48**, 1665–82.

Signorovitch, A. Y., Buss, L. W. and Dellaporta, S. L. (2007). Comparative genomics of large mitochondria in placozoans. *PLoS Genetics*, **3**, 44–50.

Šimek, K., Jezbera, J., Horňák, K., Vrba, J. and Sed'a, J. (2004). Role of diatom-attached choanoflagellates of the genus *Salpingoeca* as pelagic bacterivores. *Aquatic Microbial Ecology*, **36**, 257–69.

Simon, M., Grossart, H. P., Schweitzer, B. and Ploug, H. (2002). Microbial ecology of organic aggregates in aquatic ecosystems. *Aquatic Microbial Ecology*, **28**, 175–211.

Simpson, A. G. B. and Roger, A. J. (2004). The real 'kingdoms' of eukaryotes. *Current Biology*, **17**, R693–696.

Simpson, T. L. and Volcani, B. E. (eds) (1981). *Silicon and Siliceous Structures in Biology Systems*. New York: Springer-Verlag.

Skuja, H. (1932). Beitrag zur Algenflora Lettlands I. *Acta Horti Botanici Universitatis Latviensis*, **7**, 25–85.

Skuja, H. (1956). Taxonomische und biologischen Studien über das Phytoplankton Schwedischer Binnengewässer. *Nova Acta Regiae Societatis Scientiarum Upsaliensis*, Series 4, **16**, 1–404.

Sleigh, M. A. (1964). Flagellar movement of the sessile flagellates *Actinomonas*, *Codonosiga*, *Monas* and *Poteriodendron*. *Quarterly Journal of Microscopical Science*, **105**, 405–14.

Sleigh, M. A. (1991). Mechanisms of flagellar propulsion; a biologist's view of the relation between structure, motion, and fluid mechanics. *Protoplasma*, **164**, 45–53.

Smith, D. J. (2009). A boundary element regularised Stokeslet method applied to cilia and flagella-driven flow. *Proceedings of the Royal Society of London A*, **465**, 3605–26.

Snell, E. A., Brooke, N. M., Taylor, W. R., *et al.* (2006). An unusual choanoflagellate protein released by Hedgehog autocatalytic processing. *Proceedings of the Royal Society of London B*, **273**, 401–7.

Sollas, W. J. (1884). On the development of *Halisarca lobularis* (O. Schmidt). *Quarterly Journal of Microscopical Science*, Series 2, **24**, 603–21.

Sollas, W. J. (1888). Report on the *Tetractinellida* collected by HMS *Challenger* during the years 1873–1876: Report on the scientific results of the voyage of HMS *Challenger*, 1873–1876. *Zoology* **25** (63), 1–458.

Sonntag, B., Posch, T., Klammer, S., Teubner, K. and Psenner, R. (2006). Phagotrophic ciliates and flagellates in an oligotrophic, deep, alpine lake: contrasting variability with seasons and depths. *Aquatic Microbial Ecology*, **43**, 193–207.

Soto-Liebe, K., Collantes, G. and Kuznar, J. (2007). New records of marine choanoflagellates off the Chilean coast. *Investigaciones Marinas*, **35**, 113–20.

Spanggaard, B., Skouboe, P., Rossen, L. and Taylor, J. W. (1996). Phylogenetic relationships of the intercellular fish pathogen *Ichthyophonus hoferi* and fungi, choanoflagellates and the rosette agent. *Marine Ecology*, **126**, 109–15.

Sperling, E. A., Peterson, K. J. and Pisani, D. (2009). Phylogenetic-signal dissection of nuclear housekeeping genes supports the paraphyly of sponges and the monophyly of Eumetazoa. *Molecular Biology and Evolution*, **26**, 2261–74.

Srivastava, M., Begovic, E., Chapman, J., *et al.* (2008). The *Trichoplax* genome and the nature of placozoans. *Nature*, **454**, 955–60.

Srivastava, M., Simakov, O., Chapman, J., *et al.* (2010). The *Amphimedon queenslandica* genome and the evolution of animal complexity. *Nature*, **466**, 720–7.

Steenkamp, E. T., Wright, J. and Baldauf, S. L. (2006). The protistan origins of animals and fungi. *Molecular Biology and Evolution*, **23**, 93–106.

Stein, F. R. von (1849). Untersuchungen über die Entwickelung der Infusorien. *Archiv für Naturgeschichte*, **15**, 92–148.

Stein, F. R. von (1878). *Der Organismus der Infusionthiere*. Leipzig: W. Englemann.

Stern, C. D. (ed.) (2004). *Gastrulation: From Cells to Embryo*. New York: Cold Spring Harbor Press.

Stibbs, H. H., Owczarzak, A., Bayne, C. J. and DeWan, P. (1979). Schistosome sporocyst-killing amoebae isolated from *Biomphalaria glabrata*. *Journal of Invertebrate Pathology*, **33**, 159–70.

Stock, A., Jürgens, K., Bunge, J. and Stoeck, T. (2009). Protistan diversity in suboxic and anoxic waters of the Gotland Deep (Baltic Sea) as revealed by 18S rRNA clone libraries. *Aquatic Microbial Ecology*, **55**, 267–84.

Stoupin, D., Kiss, A. K., Arndt, H., *et al.* (2012). Cryptic diversity within the choanoflagellate morphospecies complex *Codosiga botrytis*: phylogeny and morphology of ancient and modern isolates. *European Journal of Protistology*, **48**, 263–73.

Strickland, J. D. H. and Parsons, T. R. (1968). A practical handbook of seawater analysis. *Bulletin of the Fisheries Research Board of Canada*, **167**, 1–311.

Strüder-Kypke, M. C. (1999). Periphyton and sphagnicolous protists of dystropic bog lakes (Brandenburg, Germany). *Limnologica*, **29**, 393–406.

Subhash, G., Yao, S., Bellinger, B. and Gretz, M. R. (2005). Investigation of mechanical properties of diatom frustules using nanoindentation. *Journal of Nanoscience and Nanotechnology*, **5**, 50–6.

Suga, H. and Ruiz-Trillo, I. (2013). Development of ichthyosporeans sheds light on the origin of metazoan multicellularity. *Developmental Biology*, **377**, 284–92.

Suga, H., Dacre, M., de Mendoza, A., *et al.* (2012). Genomic survey of premetazoans shows deep conservation of cytoplasmic tyrosine kinases and multiple radiations of receptor tyrosine kinases. *Science Signaling*, **5**, 1–9.

Sukhanova, I. N. (2001). Choanoflagellida on the southeastern Bering Sea shelf. *Oceanology*, **41**, 227–31.

Sukhanova, I. N. and Flint, M. V. (2001). *Phaeocystis pouchetii* at the eastern Bering Sea shelf. *Oceanology*, **41**, 75–85.

Sullivan, C. W. (1977). Diatom mineralization of silicic acid: II. Regulation of $Si(OH)_4$ transport rates during the cell cycle of *Navicula pelliculosa*. *Journal of Phycology*, **13**, 86–91.

Sullivan, C. W. and Volcani, B. E. (1981). Silicon in the cellular metabolism of diatoms. In *Silicon and Siliceous Structures in Biological Systems*, ed. T. L. Simpson and B. E. Volcani. New York, Springer-Verlag, pp. 15–42.

Sumathi, J. C., Raghukumar, S., Kasbekar, D. P. and Raghukumar, C. (2006). Molecular evidence of fungal signatures in the marine protist *Corallochytrium limacisporum* and its implications in the evolution of animals and fungi. *Protist*, **157**, 363–76.

Takahashi, E. (1981a). Loricate and scale-bearing protists from Lützow-Holm Bay, Antarctica: I. Species of the Acanthoecidae and the Centrohelida found at a site selected on the fast ice. *Antarctic Record, National Institute of Polar Research, Tokyo, Japan*, **73**, 1–22.

Takahashi, E. (1981b). Floristic study of ice algae in the sea ice of a lagoon, Lake Saroma, Hokkaido, Japan. *Memoirs of the National Institute of Polar Research. Series E. Biology and Medical Science*, **34**, 49–56.

Takishita, K., Fujiwara, Y., Kawato, M., *et al.* (2007). Molecular identification of the ichthyosporean protist '*Pseudoperkinsus tapetis*' from the mytilid mussel *Adipicola pacifica* associated with submerged whale carcasses in Japan. *Marine Biotechnology*, **10**, 13–18.

Tamada, K. and Fujikawa, H. (1957). The steady two-dimensional flow of viscous fluid at low Reynolds numbers passing through an infinite row of equal parallel cylinders. *Quarterly Journal of Mechanics and Applied Mathematics*, **10**, 425–32.

Tamm, S. L. (1979). Membrane movements and fluidity during rotational motility of a termite flagellate. *Journal of Cell Biology*, **80**, 141–9.

Tamm, S. L. (2008). Unsolved motility looking for an answer. *Cell Motility and the Cytoskeleton*, **65**, 435–40.

Tamm, S. L. and Tamm, S. (1976). Rotary movements and fluid membranes in termite flagellates. *Journal of Cell Science*, **20**, 619–39.

Tanoue, E. and Hara, S. (1986). Ecological implications of fecal pellets produced by the Antarctic krill *Euphausia superba* in the Antarctic Ocean. *Marine Biology*, **91**, 359–69.

Tanzer, M. L. (2006). Current concepts of extracellular matrix. *Journal of Orthopaedic Science*, **11**, 326–31.

Tatem, T. G. (1868). On a new species of microscopical animals. *Transactions of the Royal Microscopical Society*, **16**, 31–3.

Taylor, F. J. R. (1978). Problems in the development of an explicit hypothetical phylogeny of the lower eukarotes. *Biosystems*, **10**, 67–89.

Taylor, F. J. R. (1999). Ultrastructure as a control for protistan molecular phylogeny. *American Naturalist*, **154** (Suppl), S125–S136.

Taylor, G. (1952a). Analysis of the swimming of microscopic organisms. *Proceedings of the Royal Society of London A*, **209**, 407–61.

Taylor, G. (1952b). The action of waving cylindrical tails in propelling microscopic organisms. *Proceedings of the Royal Society of London A*, **211**, 225–39.

Telford, M. J. (2006). Animal phylogeny. *Current Biology*, **16**, R981–985.

Thomsen, H. A. (1973). Studies on marine choanoflagellates: I. Silicified choanoflagellates of the Isefjord (Denmark). *Ophelia*, **12**, 1–26.

Thomsen, H. A. (1976). Studies on marine choanoflagellates: II. Fine structural observations on some silicified choanoflagellates from the Isefjord (Denmark), including the description of two new species. *Norwegian Journal of Botany*, **23**, 33–51.

Thomsen, H. A. (1977a). Studies on marine choanoflagellates: III. An electron microscopical survey of the genus *Acanthoecopsis*. *Archiv für Protistenkunde*, **119**, 86–99.

Thomsen, H. A. (1977b). External morphology of the choanoflagellate *Salpingoeca gracilis* James-Clark. *Journal of the Marine Biological Association UK*, **57**, 629–34.

Thomsen, H. A. (1978). Nanoplankton from the Gulf of Elat (= Gulf of Aqaba) with particular emphasis on choanoflagellates. *Israel Journal of Zoology*, **27**, 34–44.

Thomsen, H. A. (1979). Electron microscopical observations on brackish water nannoplankton from the Tvärminne area, S.W. coast of Finland. *Acta Botanica Fennica*, **110**, 11–37.

Thomsen, H. A. (1982). Planktonic choanoflagellates from Disco Bugt, West Greenland, with a survey of the marine nanoplankton of the area. *Meddelelser om Grønland, Bioscience*, **8**, 3–35.

Thomsen, H. A. (1988). An electron microscopical study of marine loricate choanoflagellates: *Nannoeca minuta* (Leadbeater) gen. et comb. n. and *Stephanoeca cupula* (Leadbeater) comb. nov. *Zoologica Scripta*, **17**, 315–23.

Thomsen, H. A. and Boonruang, P. (1983a). A microscopical study of marine collared flagellates (Choanoflagellida) from the Andaman Sea, S.W. Thailand; species of *Stephanacantha* gen. nov. and *Platypleura* gen. nov. *Protistologica*, **19**, 193–214.

Thomsen, H. A. and Boonruang, P. (1983b). Ultrastructural observations on marine choanoflagellates (Choanoflagellida, Acanthoecidae) from the coast of Thailand; species of *Apheloecion* gen. nov. *Journal of Plankton Research*, **5**, 729–53.

Thomsen, H. A. and Boonruang, P. (1984). A light and electron microscopical investigation of loricate choanoflagellates (Choanoflagellida, Acanthoecidae) from the Andaman Sea, S.W. Thailand and Denmark; species of *Cosmoeca* gen. n. *Zoologica Scripta*, **13**, 165–81.

Thomsen, H. A. and Buck, K. (1991). Choanoflagellate diversity with particular emphasis on the Acanthoecidae. In *The Biology of Free-living Heterotrophic Flagellates*, ed. D. J. Patterson and J. Larsen. Oxford: Clarendon Press, pp. 259–83.

Thomsen, H. A. and Larsen, J. (1992). Loricate choanoflagellates of the Southern Ocean with new observations on cell division in *Bicosta spinifera* (Throndsen, 1970) from Antarctica and *Saroeca attenuata* Thomsen, 1979, from the Baltic Sea. *Polar Biology*, **12**, 53–63.

Thomsen, H. A. and Moestrup, Ø. (1983). Electron microscopical investigations on two loricate choanoflagellates (Choanoflagellida) *Calotheca alata* gen. et. sp. nov. and *Syndetophyllum pulchellum* gen. et. comb. nov. from Indo-Pacific localities. *Proceedings of the Royal Society of London B*, **219**, 41–52.

Thomsen, H. A., Buck, K. R., Coale, S. L., Garrison, D. L. and Gowing, M. M. (1990). Loricate choanoflagellates (Acanthoecidae Choanoflagellida) from the Weddell Sea, Antarctica. *Zoologica Scripta*, **19**, 367–87.

Thomsen, H. A., Buck, K. R. and Chavez, F. P. (1991). Choanoflagellates of the central California waters: taxonomy, morphology and species assemblages. *Ophelia*, **33**, 131–64.

Thomsen, H. A., Østergaard, J. B. and Hansen, L. E. (1995). Loricate choanoflagellates from West Greenland (August 1988) including the description of *Spinoeca buckii* gen. et sp. nov. *European Journal of Protistology*, **31**, 38–44.

Thomsen, H. A., Garrison, D. L. and Kosman, C. (1997). Choanoflagellates (Acanthoecidae, Choanoflagellida) from the Weddell Sea, Antarctica: taxonomy and community structure with particular emphasis on the ice biota; with preliminary remarks on choanoflagellates from Arctic sea ice (North East Water Polynya, Greenland). *Archiv für Protistenkunde*, **148**, 77–114.

Throndsen, J. (1969). Flagellates of Norwegian coastal waters. *Nytt Magasin for Botanikk*, **16**, 161–216.

Throndsen, J. (1970a). *Salpingoeca spinifera* sp. nov., a new planktonic species of the Craspedophyceae recorded in the Arctic. *British Phycological Journal*, **5**, 87–9.

Throndsen, J. (1970b). Marine planktonic acanthoecaceans (Craspedophyceae) from Arctic waters. *Nytt Magasin for Botanikk*, **17**, 104–10.

Throndsen, J. (1974), Planktonic choanoflagellates from North Atlantic waters. *Sarsia*, **56**, 95–122.

Throndsen, J. (1983). Ultra- and nanoplankton flagellates from coastal waters of Southern Honshu and Kyushu, Japan (including some results from the western part of the Kuroshio off Honshu). *Fisheries Agency, Japan*, **4**, 1–62.

Thurman, J., Drinkall, J. and Parry, J. D. (2010). Digestion of bacteria by the freshwater ciliate *Tetrahymena pyriformis*. *Aquatic Microbial Ecology*, **60**, 163–74.

Tong, S. M. (1997a). Choanoflagellates in Southampton Water including the description of three new species. *Journal of the Marine Biological Association UK*, **77**, 929–58.

Tong, S. M. (1997b). Heterotrophic flagellates from the water column in Shark Bay, Western Australia. *Marine Biology*, **128**, 517–36.

Tong, S. M. (1997c). Heterotrophic flagellates and other protists from Southampton Water, U.K. *Ophelia*, **47**, 71–131.

Tong, S. M., Vørs, N. and Patterson, D. J. (1997). Heterotrophic flagellates, centrohelid heliozoans and filose amoebae from marine and freshwater sites in the Antarctic. *Polar Biology*, **18**, 91–106.

Tong, S. M., Nygaard, K., Bernard, C., Vørs, N. and Patterson, D. J. (1998). Heterotrophic flagellates from the water column in Port Jackson, Sydney, Australia. *European Journal of Protistology*, **34**, 162–94.

Toruella, G., Derelle, R., Paps, J., *et al.* (2012). Phylogenetic relationships within the Opisthokonta based on phylogenomic analyses of conserved single-copy protein domains. *Molecular Biology and Evolution*, **29**, 531–44.

Towlson, P. C. (1992). Utilization of silica by *Stephanoeca diplocostata* Ellis (Choanoflagellida) and *Synura petersenii* Korsh: (Synurophyceae); Ph.D. thesis, University of Birmingham.

Tranvik, L. J., Sherr, E. B. and Sherr, B. F. (1993). Uptake and utilization of 'colloidal DOM' by heterotrophic flagellates in seawater. *Marine Ecology Progress Series*, **92**, 301–9.

Tréguer, P., Nelson, D. M., van Bennekom, A. J., *et al.* (1995). The silica balance in the world ocean: a reestimate. *Science*, **268**, 375–9.

Tyler, S. (2003). Epithelium: the primary building block for metazoan complexity. *Integrative and Comparative Biology*, **43**, 55–63.

Urban, J. L., McKenzie, C. H. and Diebel, D. (1992). Seasonal differences in the content of *Oikopleura vanhoeffeni* and *Calanus finmarchicus* faecal pellets: illustrations of zooplankton food web shifts in coastal Newfoundland waters. *Marine Ecology Progress Series*, **84**, 255–64.

Urban, J. L., McKenzie, C. H. and Diebel, D. (1993). Nanoplankton found in fecal pellets of macrozooplankton in coastal Newfoundland waters. *Botanica Marina*, **36**, 267–81.

Vacqué, D. and Pace, M. L. (1992). Grazing on bacteria by flagellates and cladocerans in lakes of contrasting food-web structure. *Journal of Plankton Research*, **14**, 307–21.

Vischer, W. (1945). Über einen pilzähnlichen, autotrophen Mikroorganismus, *Chlorochytridion*, einige neue Protococcales und die systematische Bedeutung der Chloroplasten. *Verhandlungen der Naturforschenden Gesellschaft in Basel*, **56**, 41–59.

Vlk, W. (1938). Über den Bau der Geissel. *Archiv für Protistenkunde*, **90**, 448–88.

Vogt, G. and Rug, M. (1999). Life stages and tentative life cycle of *Psorospermium haeckeli*, a species of the novel DRIPs clade from the animal–fungal dichotomy. *Journal of Experimental Zoology*, **42**, 31–42.

Voigt, O., Collins, A. G., Pearse, V. B., *et al.* (2004). Placozoa: no longer a phylum of one. *Current Biology*, **14**, R944–945.

Vørs, N. (1992). Heterotrophic amoebae, flagellates and heliozoa from the Tvärminne area, Gulf of Finland, in 1988–1990. *Ophelia*, **36**, 1–109.

Vørs, N. (1993a). Marine heterotrophic amoebae, flagellates and heliozoa from Belize (Central America) and Tenerife (Canary Islands), with descriptions of new species. *Luffisphaera bulbochaete* n. sp., *L. longihastis* n. sp., *L. turriformis* n. sp. and *Paulinella intermedia* n. sp. *Journal of Eukaryotic Microbiology*, **40**, 272–87.

Vørs, N. (1993b). Heterotrophic amoebae, flagellates and heliozoa from Arctic marine waters (North West Territories, Canada and West Greenland). *Polar Biology*, **13**, 113–26.

Vørs, N., Buck, K. R., Chavez, F. P., *et al.* (1995). Nanoplankton of the equatorial Pacific with emphasis on the heterotrophic protists. *Deep-Sea Research*, **42**, 585–602.

Wainright, P. O., Hinkle, G., Sogin, M. L. and Stickel, S. K. (1993). Monophyletic origins of the Metazoa: an evolutionary link with Fungi. *Science*, **260**, 340–2.

Wainright, P. O., Patterson, D. J. and Sogin, M. L. (1994). Monophyletic origin of animals: a shared ancestry with Fungi. In *Molecular Evolution of Physiological Processes*, ed. D. M. Fambrough. New York: Rockefeller University Press, pp. 39–53.

Walsby, A. E. and Xypolyta, A. (1977). The form resistance of chitan fibres attached to the cells of *Thalassiosira fluviatilis* Hustedt. *British Phycological Journal*, **12**, 215–23.

Wang, X. and Lavrov, D. V. (2007). Mitochondrial genome of the homoscleromorph *Oscarella carmela* (Porifera, Demospongiae) reveals unexpected complexity in the common ancestor of sponges and other animals. *Molecular Biology and Evolution*, **24**, 363–73.

Wang, X. and Lavrov, D. V. (2009). Seventeen new complete mtDNA sequences reveal extensive mitochondrial genome evolution within the Demospongiae. *PLoS One*, **3** (7), e2723.

Waters, R. L., van den Enden, R. and Marchant, H. J. (2000). Summer microbial ecology off East Antarctica (80–150 °E): protistan community structure and bacterial abundance. *Deep-Sea Research II*, **47**, 2401–35.

Weitere, M. and Arndt, H. (2002a). Top-down effects on pelagic heterotrophic nanoflagellates (HNF) in a large river (River Rhine): do losses to the benthos play a role? *Freshwater Biology*, **47**, 1437–50.

Weitere, M. and Arndt, H. (2002b). Water discharge-regulated bacteria/heterotrophic nanoflagellate (HNF) interactions in the water column of the River Rhine. *Microbial Ecology*, **44**, 19–29.

Weitere, M. and Arndt, H. (2003). Structure of the heterotrophic flagellate community in the water column of the River Rhine (Germany). *European Journal of Protistology*, **39**, 287–300.

Wiese, C. and Zheng, Y. (2006). Microtubule nucleation: g-tubulin and beyond. *Journal of Cell Science*, **119**, 4143–53.

Wilmer, P. (1996). *Invertebrate Relationships: Patterns in Animal Evolution*. Cambridge: Cambridge University Press.

Woollacott, R. M. and Pinto, R. L. (1995). Flagellar basal apparatus and its utility in phylogenetic analyses of the porifera. *Journal of Morphology*, **226**, 247–65.

Wörheide, G., Dohrmann, M., Erpenbeck, D., *et al.* (2012). Deep phylogeny and evolution of sponges (Phylum Porifera). *Advances in Marine Biology*, **61**, 1–77.

Wylezich, C. and Jürgens, K. (2011). Protist diversity in suboxic and sulfidic waters of the Black Sea. *Environmental Microbiology*, **13**, 2939–56.

Wylezich, C., Karpov, S. A., Mylnikov, A. P., Anderson, R. and Jürgens, K. (2012). Ecologically relevant choanoflagellates collected from hypoxic water masses of the Baltic Sea have untypical mitochondrial cristae. *BMC Microbiology*, **12**: 271.

Yabuki, A., Chao, E. E., Ishida, K.-I. and Cavalier-Smith, T. (2012). *Microhelia maris* (Microhelida ord. n.), an ultrastructurally highly distinctive new axopodial protist species and genus, and the unity of phylum Heliozoa. *Protist*, **163**, 356–88.

Yoon, H. S., Grant, J., Tekle, Y. I., *et al.* (2008). Broadly sampled multigene trees of eukaryotes. *BMC Evolutionary Biology*, **18**, 8–14.

Yoshida, M., Nakayama, T. and Inouye, I. (2009), *Nuclearia thermophila* sp. nov. (Nucleariidae), a new nucleariid species isolated from Yunoko Lake in Nikko (Japan). *European Journal of Protistology*, **45**, 147–55.

Zhao, S., Burki, F., Bråte, J., *et al.* (2012). *Collodictyon*: an ancient lineage in the tree of eukaryotes. *Molecular Biology and Evolution*, **29**, 1557–68.

Zhukov, B. F. and Karpov, S. A. (1985). *Freshwater Choanoflagellates*. Leningrad: Nauka.

Figure and table credits

Some of the illustrations and tables in this book are reproduced from sources noted in the individual captions. The author and publishers acknowledge the sources of copyright material below and are grateful for the permission granted. While every effort has been made, it has not always been possible to identify the sources of all material used, or to trace all copyright holders. If any omissions are brought to our notice, we will be happy to include the appropriate acknowledgements on reprinting.

Figures 2.1a and b, 2.3, 2.5, 2.6, 2.39, 2.53, 3.28, 3.30, 3.33, 3.35, 3.37–3.40, 3.43, 3.47 and Table 3.1 are already copyrighted to Cambridge University Press.

Figures 2.9, 2.15, 2.20, 2.22, 2.23, 2.25, 2.27–2.29, 2.31, 2.32, 2.35, 2.38, 2.40, 2.42, 2.44–2.48, 2.62–2.65, 2.67–2.70 and 3.50–3.58 are reproduced with permission from Professor Sergey Karpov.

Figures 1.2 and 1.3 are reproduced with permission of the Museum für Naturkunde, Berlin.

Figure 1.5 and 1.5 inset are reproduced from *Abhandlungen der Senckenbergischen naturforschenden Gesellschaft* with permission from © Senckenberg Gesellschaft fur Naturforschung.

Figures 1.7 inset, 1.15, 1.16, 1.20 and 6.1c are reproduced from the *Annals and Magazine of Natural History* with permission from Taylor & Francis Ltd.

Figures 1.10–1.12, 1.14, 2.16, 3.14c, 4.62, 4.64 and 4.74 are reproduced from *Archiv für Protistenkunde* with permission of Elsevier B.V.

Figures 1.13 and 2.4 are reproduced from the *Quarterly Journal of Microscopical Science* with permission from The Company of Biologists.

Figures 2.2, 4.34, 5.5, 5.6, 5.8–5.10, 5.14, 5.15, 5.17, 5.18, 5.34, 5.48, 7.14, 7.15, 7.20, 7.31, 8.16, 8.17 and Tables 5.2, 5.3, 5.8 and 5.9 are reproduced from *Protoplasma* with permission from Springer-Verlag GmbH.

Figures 2.33 and 2.34 are reproduced from the *Journal of Cell Science* with permission from The Company of Biologists and Dr David Hibberd.

Figures 2.7, 2.8, 2.10, 2.68, 2.70, 2.72–2.75, 2.77, 4.20, 4.27, 4.28, 4.30, 4.33, 6.16–6.19, 7.10, 7.12, 7.13, 7.16, 7.21, 7.24, 7.28, 7.32–7.34, 7.53–7.57, 9.14, 9.17, 9.18 and Table 7.2 are reproduced from the *European Journal of Protistology* with permission from Elsevier B.V. Marcus Paul gave permission to reproduce Figure 9.14; Daniel Stoupin and Áron Kiss provided original images of Figures 9.17 and 9.18, respectively.

Figures 2.9, 2.20, 2.28, 2.30, 2.36, 2.42, 2.48, 3.49–3.51, 3.53–3.58, 6.45–6.50, 6.53, 6.57–6.59, 8.8, 8.9 and Table 6.4 are reproduced from the *Journal of Eukaryotic Microbiology* with permission from © 2009 by the International Society of Protistologists.

Figure 2.16a and b are reproduced from *Botanisk Tidsskrift* with permission from John Wiley & Sons Ltd.

Figure 2.19 is reproduced from *Nova Hedwigia* with permission from © 2013 by Schweizerbart Science Publishers.

Figures 2.43, 2.52 and 4.17 are reproduced from *Protistologica* with permission from Elsevier B.V.

Figure 3.6 is redrawn from the *Biological Journal of the Linnean Society* with permission from John Wiley & Sons Ltd.

Figure 3.47c is redrawn from *Archiv für Hydrobiologie* with permission from Fundamental and Applied Limnology/Archiv für Hydrobiologie.

Figures 3.47h–j are redrawn from the *Journal of the Royal Microscopical Society* with permission from John Wiley & Sons Ltd.

Figure 3.47k is redrawn from the *Transactions of the American Microscopical Society* with permission from John Wiley & Sons Ltd.

Figure 3.47l is redrawn with no objections from *Biologisches Centralblatt*.

Figures 4.1d–g and 7.1 are reproduced with permission from *Annales de la Société Royale Zoologie Belgique*.

Figures 4.3, 4.5, 4.7, 4.35 and 4.50 are reproduced from *Cahiers de Biologie Marine* with permission from Cahiers de Biologie Marine.

Figures 4.18, 4.21, 6.5, 6.7–6.12, 6.14, 6.31, 6.36–6.42, 6.44, 6.51, 6.52, 6.54–6.56, 7.35–7.38, 7.40–7.45, 7.47–7.52 and 8.3 are reproduced from *Protist* with permission from Elsevier B.V.

Figures 4.44 and 4.57 are reproduced from the *Philosophical Transactions of the Royal Society B* with permission from Royal Society Publishing.

Figure 4.68 is reproduced from *Zoologica Scripta* with permission from John Wiley & Sons Ltd and Professor Helge Thomsen.

Figure 4.77 is reproduced from the *British Phycological Journal* with permission from Taylor & Francis Ltd (www.tandf.co.uk/journals).

Figures 5.24, 5.26, 5.27, 9.2 and Table 9.5 are reproduced from the *Journal of Experimental Marine Biology and Ecology* with permission from Elsevier B.V.

Figures 6.21–6.23, 6.25, 6.26, 7.22, 7.59 and 10.11b are reproduced from the *Proceedings of the Royal Society B* with permission from Royal Society Publishing.

Figure 9.1 and Table 9.11 are reproduced from *Marine Ecology Progress Series* with permission from John Wiley & Sons Ltd. Dr Ray Leakey also gave permission for a modified reproduction of Table 9.11.

Figure 9.3 is reproduced from *Applied Environmental Microbiology* with permission of Copyright © American Society for Microbiology.

Figure 9.7a–e are reproduced from the *Journal of the Fisheries Research Board Canada* with permission from the Fisheries Research Board of Canada.

Figures 9.8 and 9.9 are reproduced from *Marine Biology* with permission from Springer-Verlag GmbH and Dr Cynthia McKenzie.

Figure 10.1 is reproduced from *Molecular Biology and Evolution* with permission from Oxford University Press. Professor Emma Steenkamp also gave permission to reproduce Figure 10.1.

Table 10.3 is modified from *Annual Revue of Microbiology* with permission from Annual Revues Inc.

Table 10.5 is modified from Manning *et al.* (2008). Copyright (2008) National Academy of Sciences, USA.

Figure 10.4a is modified from *Advances in Marine Biology* with permission from Elsevier B.V.

Figures 10.4b and c are modified after *Current Biology* with permission from Elsevier B.V.

Figure 10.6 is redrawn from *Trends in Cell Biology* with permission from Elsevier B.V.

Figure 10.8 and Tables 10.4 and 10.5 are reproduced from the *Proceedings of the National Academy of Sciences of the USA* with permission from PNAS.

Table 10.5 is reproduced from the *Proceedings of the National Academy of Sciences USA*. Copyright (2008) National Academy of Sciences, USA.

Figure 10.10 is redrawn from *Genome Biology* with permission from © 2008 BioMed Central Ltd. Professor Thomas Bürglin also gave permission to reproduce Figure 10.10.

Figure 10.12 is modified from *Animal Evolution: Genes, Genomes, Fossils and Trees*, ed. M. J. Telford and D. T. J. Littlewood, with permission from Oxford University Press.

Choanoflagellate species index

Acanthocorbis Hara and Takahashi
  Acanthocorbis apoda (Leadbeater) Hara
    and Takahashi, 75, 87, 221, 223, 229
  Acanthocorbis campanula (Espeland)
    Thomsen, 223
  Acanthocorbis haurakiana Thomsen, 89,
    97
  Acanthocorbis mongolica Paul, 232
  Acanthocorbis unguiculata (Thomsen)
    Hara and Takahashi, 71, 77, 204,
    206–7, 221, 223, 225, 231
Acanthoeca Ellis, 142–53
  Acanthoeca brevipoda Ellis, 75, 133, 142,
    144, 231
  Acanthoeca spectabilis Ellis, 20, 66, 75, 84,
    87, 133, 141–53, 186, 221, 223, 230
Apheloecion Thomsen, 222
Astrosiga Kent, 13
  Astrosiga disjuncta (Fromentel) Kent, 58

Bicosta Leadbeater, 216
  Bicosta antennigera Moestrup, 183,
    222–3, 227, 234–6
  Bicosta minor (Reynolds) Leadbeater, 89,
    97, 222–3, 229, 234–6
  Bicosta spinifera (Throndsen)
    Leadbeater, 9, 73, 87, 89, 96–7, 182,
    187, 204, 220, 222–3, 227, 230, 233–7

Calliacantha Leadbeater
  Calliacantha longicaudata (Leadbeater)
    Leadbeater, 89, 97, 222–3, 229
  Calliacantha multispina Manton and
    Oates, 89, 97, 181, 222–3, 229
  Calliacantha natans (Grøntved)
    Leadbeater, 31, 89, 97, 181–2, 204,
    222–3, 227, 230–1, 234, 236
  Calliacantha simplex Manton and Oates,
    89, 95, 97, 181–2, 222–3, 227, 230–1,
    234
Calotheca Thomsen and Moestrup
  Calotheca alata Thomsen and Moestrup,
    222
Choanoeca Ellis
  Choanoeca perplexa Ellis, 4, 18, 20, 37,
    52, 54–5, 205

Cladospongia Iyengar and Ramathan
  Cladospongia elegans Iyengar and
    Ramathan, 60
Codonodesmus Stein
  Codonodesmus phalanx Stein, 58
Codonosigopsis (Robin) Senn
  Codonosigopsis robini Senn, 240
Codosiga James-Clark, 3, 9, 238
  Codosiga balthica Wylezich and Karpov,
    37, 45, 231, 243
  Codosiga botrytis (Ehrenberg) Bütschli,
    4–6, 9, 21–2, 28, 30, 34, 45, 49, 61,
    232, 237, 240, 244
  Codosiga gracilis (Kent) de Saedeleer,
    23–5, 29, 45–6, 207–8, 210
  Codosiga minima Wylezich and Karpov,
    37, 45, 231, 243
  Codosiga pulcherrima James-Clark, 3, 45
  Codosiga pulcherrimus James-Clark.
    See *Codosiga pulcherrima* James-Clark
Cosmoeca Thomsen, 231
  Cosmoeca ceratophora Thomsen, 221, 223
  Cosmoeca norvegica Thomsen, 23, 88, 92,
    95, 221, 223, 229, 233, 237
  Cosmoeca phuketensis Thomsen, 88, 94–5,
    221, 233
  Cosmoeca takahashii Thomsen, 88, 95,
    221, 223, 227
  Cosmoeca ventricosa Thomsen, 23, 88,
    94–5, 221, 223, 225, 227, 233
Crinolina Thomsen, 184
  Crinolina aperta (Leadbeater) Thomsen,
    85, 92–3, 204, 221, 223–5, 227, 236–7
  Crinolina isefiordensis Thomsen, 88,
    92–3, 221, 223
Crucispina Espeland and Throndsen
  Crucispina cruciformis (Leadbeater)
    Espeland, 97, 222–3, 229, 233, 237

Desmarella Kent, 9, 238–9
  Desmarella moniliformis Kent, 58, 61–2,
    229
  Desmarella phalanx (Stein) Kent, 58
Diaphanoeca Ellis
  Diaphanoeca grandis Ellis, 25, 29, 66, 69,
    80–1, 85–6, 88, 92, 101, 107–8, 122–3,

175–9, 188, 190, 196, 198, 204, 206–8,
  210, 218, 220–1, 223, 230–1, 236
  Diaphanoeca multiannulata Buck, 85, 92,
    221, 223, 225, 227, 233, 236–7
  Diaphanoeca pedicellata Leadbeater, 181,
    184, 221, 223, 227, 230
  Diaphanoeca sphaerica Thomsen, 231
  Diaphanoeca spiralifurca Hara, 83–4
  Diaphanoeca undulata Thomsen, 221,
    223, 230
Didymoeca Doweld
  Didymoeca costata (Valkanov) Doweld,
    69, 85, 87, 166–75, 184, 187–9, 221,
    223, 236
  Didymoeca elongata (Nitsche and Arndt)
    Doweld, 166, 223
  Didymoeca tricyclica (Bergesch et al.)
    Doweld, 166, 223
Diplosiga Frenzel
  Diplosiga socialis Frenzel, 5
Diplosigopsis Francé, 5

Helgoeca Leadbeater
  Helgoeca nana (Thomsen) Leadbeater,
    20, 77, 133, 137–40, 223

Kakoeca Buck and Marchant
  Kakoeca antarctica Buck and Marchant, 84

Lagenoeca Kent, 238–9
  Lagenoeca antarctica Nitsche et al., 230
  Lagenoeca globulosa Francé, 240

Monocosta Thomsen
  Monocosta fennica Thomsen, 68, 231
Monosiga Kent, 21, 25, 207, 230, 238–9
  Monosiga brevicollis Ruinen, 10, 30,
    247–9, 266–76
  Monosiga gracilis Kent, 6, 21
  Monosiga marina Grøntved, 216–17
  Monosiga ovata Kent, 29, 31–5, 40–1, 205,
    207–8, 244, 272, 274

Nannoeca Thomsen
  Nannoeca minuta (Leadbeater) Thomsen,
    221, 223

Parvicorbicula Deflandre, 184
  Parvicorbicula circularis Thomsen, 82,
    181, 221, 223
  Parvicorbicula corynocostata Thomsen,
    82, 181
  Parvicorbicula manubriata Tong, 82, 95, 230
  Parvicorbicula quadricostata Throndsen,
    82, 89, 95, 181, 218, 221, 223, 227,
    229–30
  Parvicorbicula socialis (Meunier)
    Deflandre, 56, 72, 82, 88, 92, 94–5,
    181, 204, 214–16, 218, 221, 223, 227,
    230, 233, 236–7
  Parvicorbicula superpositus Booth, 229
Platypleura Thomsen, 222–3
  Platypleura acuta Thomsen, 223
  Platypleura cercophora Thomsen, 223
  Platypleura infundibuliformis
    (Leadbeater) Thomsen, 74
Pleurasiga Schiller, 184
  Pleurasiga echinocostata Espeland, 72, 82,
    88, 94–5, 187, 221
  Pleurasiga minima Throndsen, 23, 82, 88,
    92, 94–5, 181, 218, 221, 223, 237
  Pleurasiga reynoldsii Throndsen, 82, 181,
    218, 221, 223, 230
  Pleurasiga tricaudata Booth, 82, 181
Polyfibula Manton, 221, 223
  Polyfibula sphyrelata Thomsen, 229
Polyoeca Kent
  Polyoeca dichotoma Kent, 54, 130–1, 133,
    141–2, 145, 183, 202, 221, 223
Proterospongia Kent, 9, 14, 16, 20
  Proterospongia choanojuncta Leadbeater,
    36–7, 52, 55–6, 60
  Proterospongia dybsoeënsis (Grøntved)
    Loeblich III, 60
  Proterospongia haeckeli Kent, 3, 12, 60
  Proterospongia haeckeli var. *clarki*
    Schiller, 60
  Proterospongia lackeyi Bourrelly, 60
  Proterospongia nana (Braarud)
    Throndsen, 60

Proterospongia pedicellata (Oxley) Ertl, 60
Proterospongia skujae (Skuja) Bourrelly,
  60
Proterospongia skujae var. *gracilis* (Skuja)
  Bourrelly, 60

Saepicula Leadbeater
  Saepicula leadbeateri Takahashi, 223, 225
  Saepicula pulchra Leadbeater, 70–1, 77,
    183, 221, 223
Salpingoeca James-Clark, 3, 239
  Salpingoeca abyssalis Nitsche *et al.*, 230
  Salpingoeca amphoridium James-Clark, 3,
    23–30, 56, 237–8, 244
  Salpingoeca ampulla Kent, 66–7, 85
  Salpingoeca angulosa de Saedeleer, 49
  Salpingoeca gracilis James-Clark, 3, 48,
    63
  Salpingoeca infusionum Kent, 6, 47–8, 54
  Salpingoeca inquillata Kent, 47, 54
  Salpingoeca marina James-Clark, 3
  Salpingoeca marinus James-Clark.
    See *Salpingoeca marina* James-Clark
  Salpingoeca natans Grøntved, 216–18
  Salpingoeca oblonga Stein, 47, 54, 61
  Salpingoeca pyxidium Kent, 8
  Salpingoeca rosetta Dayel *et al.*, 10, 45,
    48–9, 51, 57, 247, 266, 269, 275
  Salpingoeca spinifera Throndsen, 218
  Salpingoeca urceolata Kent, 52
Saroeca Thomsen
  Saroeca attenuata Thomsen, 183
Savillea Loeblich, 138
  Savillea micropora (Norris) Leadbeater.
    See *Savillea parva* (Ellis) Loeblich
    (*micropora* form)
  Savillea parva (Ellis) Loeblich, 18, 69,
    84, 186, 221, 223, 230, 232
  Savillea parva (Ellis) Loeblich (*micropora*
    form), 26, 77, 84, 135, 137–8, 186,
    205, 231
  Savillea parva (Ellis) Loeblich (*parva*
    form), 133, 136, 145

Sphaeroeca Lauterborn, 13
  Sphaeroeca salina Bourrelly, 60
Spinoeca Thomsen
  Spinoeca buckii Thomsen, 223
Stelexomonas Lackey, 239
Stephanacantha Thomsen, 223
  Stephanacantha campaniformis
    (Leadbeater) Thomsen, 74–5, 80, 187,
    221, 223
  Stephanacantha dichotoma Thomsen,
    222–3
Stephanoeca Ellis
  Stephanoeca ampulla (Kent) Ellis, 155,
    157, 218
  Stephanoeca apheles Thomsen, 223
  Stephanoeca arndtii Nitsche and Arndt,
    232
  Stephanoeca cauliculata Leadbeater,
    83–4, 202, 223
  Stephanoeca complexa (Norris)
    Throndsen, 231
  Stephanoeca cupula (Leadbeater)
    Thomsen, 69, 77–8, 84, 88, 221, 223
  Stephanoeca diplocostata Ellis, 25–6, 31,
    36, 38, 40, 42, 66, 88, 101–3, 106, 110,
    114, 127, 156–66, 184, 190, 192–7,
    200, 207–8, 210, 213, 221, 223, 231
  Stephanoeca elegans (Norris) Throndsen,
    223
  Stephanoeca kentii Ellis, 218
  Stephanoeca norrisii Thomsen, 87–8, 91,
    221, 223, 231
  Stephanoeca pedicellata Leadbeater, 229
  Stephanoeca pyxidoides Leadbeater, 83,
    87–8, 91, 183
  Stephanoeca supracostata Hara, 181
  Stephanoeca urnula Thomsen, 83, 87–8,
    91, 183, 223, 231
Stylochromonas Lackey
  Stylochromonas minuta Lackey, 16
Syndetophyllum Thomsen and Moestrup
  Syndetophyllum pulchellum (Leadbeater)
    Thomsen and Moestrup, 72, 80

Other species index

Abeoforma Marshall and Berbee (Opisthokonta, Ichthyosporea)
  Abeoforma whisleri Marshall and Berbee, 254
Actinomonas Kent (Stramenopiles, Dictyochophyceae), 204
  Actinomonas mirabilis Kent, 203, 210
Algoriphagus Bowman *et al.* emend Nedashkovskaya *et al.* (Cyclobacteriaceae, Bacteroidetes)
  Algoriphagus machipongonensis Alegado *et al.*, 57
Amastigomonas de Saedeleer (Incertae sedis, Apusomonadida)
  Amastigomonas mutabilis (Griessmann) Molina and Nerad, 203–4
Amoebidium Cienkowsky (Opisthokonta, Ichthyosporea)
  Amoebidium parasiticum Cienkowsky, 244, 248, 254, 264
Amoeboaphelidium Scherffel (Opisthokonta, Aphelidea)
  Amoeboaphelidium protococcarum Gromov and Mamkaeva, 262
Amphibiocystidium Pascolini (Opisthokonta, Ichthyosporea)
  Amphibiocystidium ranae (Guyénot and Naville) Pascolini, 254
  Amphibiocystidium viridescens Raffel *et al.*, 254
Amphimedon Duchassaing and Michelotti (Porifera, Demospongiae)
  Amphimedon queenslandica Hooper and Van Soest, 272
Anabaena Bory de St-Vincent (Cyanobacteria, Nostocales)
  Anabaena flos-aquae Brébisson ex Bornet and Flauhault, 237–8
Ancyromonas Kent (Incertae sedis, Ancyromonadida)
  Ancyromonas sigmoides Kent, 204
Anisonema Dujardin (Excavata, Euglenozoa), 204
Anthophysa Bory de St-Vincent (Stramenopiles, Ochromonadales), 1
  Anthophysa dichotoma Bory de St-Vincent, 1

Anthophysa solitaria Bory de St-Vincent, 1, 3
Anthophysa vegetans (Müller) Stein, 1
Anurofeca Baker *et al.* (Opisthokonta, Ichthyosporea)
  Anurofeca (Prototheca) richardsi (Wong and Beebee) Baker *et al.*, 254
Ascidiella Roule (Tunicata, Ascidiacea)
  Ascidiella aspersa Müller, 25
Asterionella Hassall (Stramenopiles, Bacillariophyceae)
  Asterionella formosa Hassall, 237–8
Aulacoseira Thwaites (Stramenopiles, Melosirids), 237–8
  Aulacoseira italica (Ehrenberg) Simonsen, 238

Bicosoeca James–Clark (Stramenophiles, Bicosoecida), 204
Biomphalaria Preston (Mollusca, Gastropoda)
  Biomphalaria glabrata Say, 256
Bodo Ehrenberg (Excavata, Kinetoplastea), 204
  Bodo designis Skuja, 203, 206, 209–13
  Bodo saltans Ehrenberg, 204–5

Cafeteria Fenchel and Patterson (Stramenopiles, Bicosoecida), 204
  Cafeteria roenbergensis Fenchel and Patterson, 203, 205
Calanus Leach (Crustacea, Copepoda)
  Calanus hyperboreus Kroyer, 229
Capsaspora Hertel *et al.* (Opisthokonta, Filasterea)
  Capsaspora owczarzaki Hertel *et al.*, 244, 255–6, 261, 266, 268–9, 274–5
Cercomonas Dujardin, (Rhizaria, Cercozoa), 204
Chaetoceros Ehrenberg (Stramenopiles, Mediophyceae)
  Chaetoceros simplex Ostenfeld, 118
Ciliophrys Cienkowsky (Stramenopiles, Pedinellales), 204
  Ciliophrys infusionum Cienowsky, 209
Ciona Linnaeus (Tunicata, Ascidiacea)
  Ciona intestinalis Linnaeus, 25

Cladophora Kützing (Chloroplastida, Cladophoraceae), 237
Corallochytrium Raghu-Kumar (Opisthokonta, Corallochytrium)
  Corallochytrium limacisporum Raghu-Kumar, 244, 255–6
Creolimax Marshall *et al.* (Opisthokonta, Ichthyosporea)
  Creolimax fragrantissima Marshall *et al.*, 247, 253–4
Cyrtophora Pascher (Stramenopiles, Dictyochophyceae)
  Cyrtophora pedicellata Pascher, 16

Danio Hamilton (Actinopterygii, Cyprinidae)
  Danio rerio Hamilton, 165
Dermocystidium Perez (Opisthokonta, Ichthyosporea), 254
  Dermocystidium percae Reichenbach-Klinke, 253–4
Dictyostelium Brefeld (Amoebozoa, Dictyostelia)
  Dictyostelium discoideum Raper, 269
Dinobryon Ehrenberg (Stramenopiles, Chromulinales), 63, 140, 150, 195, 237
  Dinobryon divergens Imhof, 238
  Dinobryon sertularia Ehrenberg, 238

Epistylis Ehrenberg (Ciliophora, Peritrichia), 1
  Epistylis botrytis Ehrenberg, 1, 3–4

Fonticula Worley *et al.* (Opisthokonta, Nucletmycea)
  Fonticula alba Worley *et al.*, 244, 261, 264
Fragilaria Lyngbe (Stramenopiles, Fragilariophyceae), 237–8
  Fragilaria crotonensis Kitton, 238

Goniomonas Stein (Incertae sedis, Cryptophyceae), 204
Guancha Miklucho-Maclay (Porifera, Calcarea)
  Guancha blanca Miklucho-Maclay, 13

Haliclona Grant (Porifera, Demospongiae)
Haliclona permollis Bowerbank, 25

Ichthyophonus Plehn and Mulsow
(Opisthokonta, Ichthyosporea)
Ichthyophonus hopferi Plehn and Mulsow,
254
Ichthyophonus irregularis Rand *et al.*, 244

Jakoba Patterson (Excavata, Jakobida)
Jakoba libera (Ruinen) Patterson, 203–4,
209, 212–13

Leucosolenia Ellis and Solander (Porifera,
Calcarea)
Leucosolenia botryoides Ellis and
Solander, 1, 11

Massisteria Larsen and Patterson (Rhizaria,
Cercozoa)
Massisteria marina Larsen and Patterson,
203–4
Metridia Boeck (Crustacea, Copepoda)
Metridia lucens Boeck, 229
Metromonas Larsen and Patterson
(Rhizaria, Cercozoa), 204
Ministeria Patterson *et al.* (Opisthokonta,
Filasterea), 204
Ministeria marisola Patterson *et al.*, 256
Ministeria vibrans Tong, 244, 255–6, 274

Nematostella Stephenson (Cnidaria,
Anthozoa)
Nematostella vectensis Stephenson, 259,
266, 272
Nuclearia Cienkowsky (Opisthokonta,
Nucletmycea), 255
Nuclearia delicatula Cienkowsky, 261
Nuclearia simplex Cienkowsky, 244

Ochromonas Wyssotzki (Stramenopiles,
Ochromonadales), 204–5, 209–10
Oikopleura Mertens (Urochordata,
Appendicularia)
Oikopleura vanhoeffeni Lohmann, 233
Olpidium (Braun) Schröt (Opisthokonta,
Nucletmycea, Chytridiomycota)
Olpidium brassicae (Woronin) Dang, 262

Paraphysomonas de Saedeleer
(Stramenopiles,
Ochromonadales), 128, 203–4, 230

Paraphysomonas imperforata Lucas, 206,
209–13
Paraphysomonas vestita (Stokes) de
Saedeleer, 209–10
Petalomonas Stein (Excavata, Euglenozoa),
204
Phaeocystis Lagerheim (Haptophyta,
Prymnesiophyceae)
Phaeocystis pouchetii (Hariot) Lagerheim,
214, 227
Phalansterium Cienkowsky (Amoebozoa,
Incertae sedis), 16
Phalansterium consociatum (Fresenius)
Cienkowsky, 16
Phalansterium digitatum Stein, 16
Phalansterium intestinum Cienkowsky, 16
Pirum Marshall and Berbee (Stramenopiles,
Ichthyosporea)
Pirum gemmata Marshall and Berbee, 255
Pleuromonas Perty (Excavata,
Parabodonida)
Pleuromonas jaculans Perty, 209–10
Poterioochromonas Scherffel (Stramenopiles,
Chromulinales), 63
Poterioochromonas stipitata Scherffel,
48–50
Pseudobodo Griessmann (Stramenopiles,
Bicosoecida), 204, 209
Pseudobodo tremulans Griessmann,
230
Pseudoperkinsus Figueras (Opisthokonta,
Ichthyosporea)
Pseudoperkinsus tapetis Figueras,
254
Psorospermium Hilgendorf (Opisthokonta,
Ichthyosporea)
Psorospermium haeckeli Hilgendorf,
254
Pteridomonas Pennard, (Stramenopiles,
Pedinellales), 204

Rhinosporidium Minchin and Fantham
(Opisthokonta, Ichthyosporea)
Rhinosporidium seeberi (Wernicke) Seeber,
254
Rhynchomonas Klebs (Excavata,
Neobodonida)
Rhynchomonas nasuta (Stokes) Klebs,
203–4, 206, 209
Rozella Cornu (Opisthokonta,
Nucletmycea, Chytridiomycota), 262

Schistosoma Weinland (Platyhelminthes,
Digenea)
Schistosoma mansoni Sambon, 256
Sphaeroforma Jøstensen *et al.*
(Opisthokonta, Ichthyosporea)
Sphaeroforma arctica Jøstensen *et al.*,
247, 254, 264
Sphaeroforma tapetis (Figueras) Marshall
et al., 247
Sphaerothecum Arkush *et al.* (Opisthokonta,
Ichthyosporea)
Sphaerothecum destruens Arkush *et al.*,
253–4
Spongilla Linneans (Porifera,
Demospongiae)
Spongilla arachnoidea James-Clark,
11
Spongilla lacustris Lamarck, 11
Spumella Cienkowsky (Stramenopiles,
Ochromonadales), 204–5
Staurastrum Meyen ex Ralfs (Charophyta,
Zygnematophyceae)
Staurastrum pingue Teiling, 238
Stephanodiscus Ehrenberg (Stramenopiles,
Mediophyceae)
Stephanodiscus hantzschii Grunow,
237–8
Sycon Risso (Porifera, Calcarea)
Sycon compressum Fabricius, 5
Synura Ehrenberg (Stramenopiles,
Synurales)
Synura petersenii Korshikov, 128

Tabellaria Ehrenberg (Stramenopiles,
Bacillariophyceae), 237
Thalassiosira Cleve (Stramenopiles,
Mediophyceae), *237*, 237
Thalassiosira fluviatilis Hustedt,
237
Thalassiosira weissflogii (Grunow) Fryxell
and Hasle, 118
Thaumatomastix Lauterborn (Rhizaria,
Thaumatomonadidae), 204
Trichoplax Schultze (Metazoa, Placozoa)
Trichoplax adhaerens Schultze, 248,
258

Volvox Linnaeus (Chloroplastida,
Chlorophyceae)
Volvox vegetans Müller, 1
Vorticella Linnaeus (Ciliophora,
Peritrichia), 237

General index

Acanthoeca spectabilis
  cell division, 145–6
  costal strip accumulation, 145
  cytoskeletal poisons, 150–2
  dwarf cells, 150
  helical costae, 75, 144
  lorica assembly, 146, 149
  lorica construction, 142–3
  lorica dimensions, 142
  organic investment, 144
acanthoecid phylogeny, 252
acanthoecid themes, 85–98
  Bicosta/Calliacantha theme, 95–6, 220,
    223–4
  Cosmoeca/Parvicorbicula theme, 92, 94,
    220, 223–4
  Diaphanoeca/Crinolina theme, 92–3, 224
  Stephanacantha/Platypleura theme, 220
  Stephanoeca theme, 85, 91, 220
Acanthoecida, 8, 44, 130, 251
Acanthoecidae, 8, 131, 154, 185, 251, 253
Amoebidiales, 253
Amoebozoa, 242, 261
Amorphea, 242
Antarctica collecting sites, 224
anterior ring
  compound, 87
  location, 70, 72, 78, 80, 93, 98, 102, 174
Aphelidae, 262
Apusozoa, 242, 268
aquiferous chambers, 1, 11, 258, 265, 278

basal bodies
  cell division, 40, 44
  choanocyte, 11
  Choanoeca perplexa, 52
  chytrid, 243
  classification, 32
  collar cells, 14
  dormant (non-flagellar), 32, 243, 253,
    262
  flagellar withdrawal, 40
  ichthyosporean, 243
  Monosiga ovata, 34
Bicosta spinifera
  left-handed rotation, 97

lorica size variation, 233
Bicosta/Calliacantha theme, 95, 220, 223–4
  costal strip accumulation, 182
  lorica size variation, 233–4
biogenic silica, 100–2, 107, 115, 117–18
Boucaud-Camou, Eve, 131, 140, 214
Bütschli, Otto, 3, 261

cadherins
  classical, 266
  hedgling, 266
  premetazoan subfamilies, 266
Capsaspora, 256
cell adhesion molecules (CAMs), 265–9
  integrins, 267
cell division, 8, 41–2
  codifying characters, 191
  diagonal, 9, 40, 127, 131, 136–8, 140, 146
  emergent, 5, 9, 47, 49, 52, 54
  inverted, 9, 40, 155, 157
  longitudinal, 9, 40, 44
  non-restrictive, 5, 47
  role of microvilli, 165, 169
  separating threads, 156, 191
cell–cell signalling, 275
  calcium mediated pathways, 275
  Hedgehog pathway, 270
  Hippo signalling pathway, 275
  Notch pathway, 270
  pathways, 269
  pre-metazoan ancestry, 270
  transcriptional regulation, 275
  tyrosine kinase domains, 274
  tyrosine kinase mediated pathways, 273
chirality (handedness), 70, 132
choanocytes
  choanoflagellate relatedness, 1, 10, 241,
    262
  early illustrations, 7
  flagellar vane, 30
  microvilli, 5
choanoderm, 258, 265, 278
Choanoflagellata, 4
choanoflagellate ecology, 213–40
  abyssal depths, 230
  acanthoecid species, 223

Antarctic species, 227–8
bacterial production, 225
biogeography. *see* geographical records
early quantitative records, 216
electron microscopy, 218
epibiotic species, 237
freshwater acanthoecids, 232
freshwater ecology, 237–40
hypoxic environments, 231
lorica size variation, 233–6
marine studies, 226
phases of study, 213
planktonic suspension, 236
polymorphic life cycle, 234
quantitative aspects, 223
rivers, 239
salinity tolerance, 231
sampling limitations, 218
soil and permafrost, 239
trophic coupling, 224–30
choanoflagellate ecophysiology, 204–13
choanoflagellate predation
  faecal pellets, 129, 233
choanoflagellate/choanocyte similarity
  evolutionary link, 14
  functional, 10
  morphological, 10
  ultrastructural, 11
Choanoflagellatea, 8, 16, 251
  multi-taxon, multi-gene phylogeny,
    251
choanoflagellates
  axenic culture, 204
  classed as algae, 16
  classification, 15
  distinguishing criteria, 4, 29
  early illustrations, 2
  first record, 1
Choanozoa, 249, 251
Chytridiomycota, 241, 262
  Monoblepharidales, 242, 263
classificatory systems, 249–51
  choanoflagellate, 17
Cnidaria
  phylogeny, 259
Codonosigidae, 8, 15, 44, 251

Codosiga botrytis
  basal bodies, 32
  encystment, 61, 240
  flagellar vane, 28
  nomenclature, 4
  permafrost, 232, 240
collar
  cell division, 165
  early illustrations, 6
  fluid flow, 25
  length, 45, 205
  microvilli, 5
  morphology, 24
  particle capture, 5
  particle ingestion, 5
collar cells (cyrtocytes), 14
collar microvilli
  behaviour, 36
  cell division, 165
  characteristics, 24
  pressure drop, 25
  ultrastructure, 25, 35
colonial choanoflagellates, 54–61
  Codosiga, 45
  colony form, 57, 60
  culture, 56
  definition, 54
  Diaphanoeca pedicellata, 224
  ecological niche, 61
  life cycles, 59
  linear colonies, 58
  natural conditions, 58
  nomenclature, 59
  Parvicorbicula socialis, 216, 224, 229
  Proterospongia choanojuncta, 54, 56
  Proterospongia haeckeli, 60
  Salpingoeca amphoridium, 56
  Salpingoeca rosetta, 57
  swimming behaviour, 58
conjoined cells, 192, 195
Cosmoeca/Parvicorbicula theme, 92, 94, 220, 223–4
costae
  definition, 68
  first record, 66
  helical, 68
  junctions, 78
  longitudinal, 72
  numbering, 69
  transverse, 68
  transverse ring, 68
costal strip accumulation
  Calliacantha species, 181
  inside-out, 187
  microvillar relationship, 165
  nudiform morphology, 186, 193, 197–8
  quadrants, 161, 165, 169, 175, 188

sequence, 156, 187, 189
costal strip production
  Bicosta species, 183
  codified, 191, 193
  exocytosis, 105, 169
  logistical flexibility, 200
  SDV-associated microtubules, 105, 168
  sequence, 167, 175
  silica deposition vesicles (SDVs), 104, 107, 167
  silicalemma, 104
  silicon gradient, 188, 190
  tectiform, 179
  upside-down, 168, 187
costal strips
  abutment, 68, 99
  accumulation, 104, 122, 156, 182
  adhesion, 107
  biochemical analysis, 107
  definition, 68
  dissolution, 119–20
  first description, 66
  imbrication, 74, 99
  leading end, 68
  misshapen, 110–11, 150
  morphology, 72–3
  nudiform accumulation, 136, 138, 145, 193, 197
  organic scaffold, 107
  orientation, 175
  silicon deficient, 122
  X-ray diffraction, 105
Craspedida, 44, 251
  cell cycle, 44, 46
Craspedomonadina, 4
Craspedophyceae, 16, 166
Ctenophora, 260
Cylicomastiges, 4
cytokinesis
  separating threads, 52, 161, 191
cytoplasmic bridge, 57–8
cytoskeleton
  filopodia, 34
  microfilament/microtubule interactions, 36, 153, 162, 165
  microfilaments, 34–6
  microtubule ring, 32–3
  microtubules, 32–8, 149–50
  microvilli, 34, 149
  S1 muscle myosin, 34, 36

Diaphanoeca grandis, 175–9
  anterior transverse costa, 80, 82
  cell division, 175–6
  costal strip categories, 175, 196
  lorica assembly, 177
  lorica construction, 80, 175

nudiform behaviour, 197–8
  sinking velocity, 230
  veil, 85–6, 122
Diaphanoeca/Crinolina theme, 92–3, 224
diatoms
  frustule synthesis, 101
  planktonic, 237
  silicon requirement, 100
  silicon uptake, 100
  spring bloom, 229
Didymoeca (Diplotheca) elongata, 166
Didymoeca (Diplotheca) tricyclica, 166
Didymoeca costata, 166–74
  cell division, 169, 171
  costal strip accumulation, 189
  costal strip categories, 166
  costal strip orientation, 168
  costal strip production, 167–8
  left-handed rotation, 167
  lorica assembly, 172
  lorica construction, 166–7
  lorica cup, 169–70
  nomenclature, 166
  veil, 167
diffusion feeding, 27, 203
Diplotheca costata. see Didymoeca costata
dissolved organic matter (DOM), 203, 205
  microbial loop, 203
DRIP clade, 253

Eccrinales, 253
Ellis, William Neale, 66, 131, 155, 214
elongation factor 1A (eEF1A)
  amino acid insertion, 244
  early phylogeny, 242
  STEPPYS motif, 247
elongation factor like (EFL)
  choanoflagellates, 247
encystment, 61–2
excystment, 62
extracellular coverings, 5, 8
  gel périphérique, 9
  glycocalyx (sheath), 5, 44
  lorica, 5
  microfibrils, 48
  mucous sheath, 9
  Schleimhülse, 9
  stalk development, 49
  theca, 5, 44

Filasterea, 256
  Capsaspora, 256
  Ministeria, 256
filopodia, 45, 56–7, 177, 256
  ancestry, 269
  behaviour, 36
flagellar basal body, 243

during division, 40
  maturation, 44, 52
flagellar hydrodynamics, 19–26
  feeding currents, 21
  flow fields, 21–4
  influence of collar, 24
  influence of lorica, 25, 236
  motile cells, 19
  multiple toroidal eddies, 24
  optimal swimming motions, 20
  optimal feeding motions, 22
  particle velocity, 205
  pedicel length, 23
  resistive-force model, 19
  singularities, 19
  slender-body theory, 19, 22, 25
  stationary cells, 21–2
  swimming cells, 20–1
flagellar vane, 28–31
flagellum
  absence, 18, 52, 136
  axoneme, 30–1
  central filament, 32
  development, 45
  length, 28
  single posterior, 18, 243, 253, 262
  striated fibrous root, 32
  surface, 28
  transitional region, 30, 32
  vane, 16, 28–31
  withdrawal, 40, 162
Fonticula alba, 244
Fresenius, Georg, 3
freshwater acanthoecids, 232
fungal phylogeny, 262
fungal signature, 257, 262
Fungi, 244

Gastraea theory, 13
geographical records, 219, 221
  Andaman Sea, 220
  Antarctic Convergence, 227
  Antarctica, 220, 224–5
  Arctic Ocean, 229
  Barents Sea, 229
  Californian waters, 220
  Equatorial Pacific, 220
  Gulf of Alaska, 227
  Iceland, 220
  Limfjord, Denmark, 230
  Newfoundland, 220, 229
  North Pacific, 220
  Port Jackson, Sydney, 220
  Southampton Water, 220, 229
  Southern Ocean, 220, 225
  West Greenland, 220
glycocalyx (sheath), 5, 9, 251

Golgi apparatus, 39
  cell covering secretion, 40, 52
  cisternae, 37
  secretory activity, 63
  secretory vesicles, 37, 52
Griessmann, Karl, 5, 47, 52
growth kinetics, 205–13
  batch culture, 113
  bioenergetics, 211
  clearance, 205–7, 225
  continuous culture, 211
  growth parameters, 210
  growth strategies, 210
  heterotrophic nanoflagellates (HNF), 212
  Monod equation, 209
  nutrient cycling, 212–13
  particle capture and growth, 208–9
  particle capture and uptake, 205–13, 237
  population growth, 209–11
  response to prey concentration, 211

Haeckel, Ernst, 10, 13
Helgoeca nana
  helical costae, 138
  lorica assembly, 139–40
  lorica construction, 137–8
helical costae, 75
  assembly, 147
  chirality (handedness), 70, 135
  clockwise rotation. see left-handed
    rotation
  left-handed, 75, 82
  left-handed rotation, 82, 99, 138, 149
  number of turns, 75, 83, 135, 144, 150
  occurrence, 77
  production sequence, 188
  recognition, 77
  variety, 98
helix
  chirality (handedness), 132
  definition, 132
  left-handed, 132, 144
heterotrophic flagellates (HF), 202–4
  dimensions, 203
  feeding strategies, 203
heterotrophic nanoflagellates (HNF),
    202–13
  dimensions, 203
  prey particles, 202
heterotrophy, 202
Holozoa, 244, 249, 251

Ichthyosporea, 253–6
  Ichthyophonae, 253
  life cycles, host, mitochondrial cristae,
    254
  Rhinosporidaceae, 253

Infusoria Flagellata, 1, 214
integrin-mediated signalling, 268
  components, 269
  pre-metazoan ancestry, 268
International Code of Nomenclature for
    Algae, Fungi, and Plants (ICN), 249
International Code of Zoological
    Nomenclature (ICZN), 3, 249

James-Clark, Henry, 1, 10–11, 214, 241
  publications, 3–4
juvenile
  Acanthoeca spectabilis, 145
  costal strips, 138, 145, 156, 158, 169, 177,
    179
  craspedid, 44, 52
  definition, 9
  Didymoeca costata, 169
  Helgoeca nana, 138
  lorica assembly, 110
  nudiform taxa, 104, 130, 147, 153, 185
  Polyoeca dichotoma, 140
  posterior filopodia, 56, 177
  Savillea parva (micropora form), 136
  Stephanoeca diplocostata, 156
  tectiform taxa, 104, 131, 155, 190

Kent, William Saville, 3, 11, 21, 47, 58, 66,
    214
Kingdom Animalia, 249
Kingdom Fungi, 249
Kingdom Protozoa, 249

left-handed rotation
  Bicosta spinifera, 97
  Stephanoeca cauliculata, 83–4
left-handed helix
  definition, 68
longitudinal costae, 72
  assembly, 83, 165
  spines, 75
  Stephanoeca cauliculata, 83
lorica, 5
  3D computer model, 76
  chamber, 68
  definition, 68
  functions, 25, 98, 236–7
  length, 234
  organic investment, 83
  preservation, 69
  Saepicula pulchra, 70–1
  size variation, 233–6
  suspensory membrane, 84
  terminology, 8, 67
  veil, 85–6, 92, 167
lorica assembly
  3D computer model, 138–9, 159, 161

lorica assembly (cont.)
  Acanthoeca spectabilis, 145–53
  Diaphanoeca grandis, 177
  Didymoeca costata, 172
  left-handed rotation, 84, 161, 172
  mechanism, 158
  rotational movement, 149, 165
  rules, 67, 70, 99
  Stephanoeca diplocostata, 158
lorica cup, 169
loricate choanoflagellate
  first record, 66–7

Manual of the Infusoria, 11, 21, 214, 241
Mesomycetozoea, 253
Metazoa, 249, 257–61
  ancestry, 10
  Bilateria, 257
  diploblasts, 257
  monophyly, 248, 257
  non-bilateria, 257–8, 260
  phylogeny, 257
  triploblasts, 257
microbial foodwebs, 202, 224
microbial loop, 203, 224, 233
microfibrils
  Acanthoeca spectabilis, 84, 87
  carbohydrate-containing, 48–9
  chemical composition, 48–9
  cup-shaped theca, 48
  Diaphanoeca grandis, 85, 87
  Didymoeca costata, 85, 87
  helical arrangement, 49
  meshwork, 135–6, 142, 144
  Poterioochromonas stipitata, 49
  robustness, 48
  Savillea parva, 84
  stalk (peduncle), 48
  veil, 85
microfilament poisons
  cytochalasin D, 27, 151
Microsporidia, 262
microtubule poisons
  colchicine, 109–10, 150
  effect on costal strips, 109
  mode of action, 109
  nocodazole, 27
  vinca alkaloids, 27
microtubules
  categories, 108
  chemical composition, 108
microvilli
  actin microfilaments, 36
  ancestry, 269
  categories, 165
  collar, 5, 152

lorica-assembling, 147, 149, 158, 161, 164, 172
  microfilament poisons, 152
  tectiform cell division, 162
Ministeria, 256
mitochondria
  constricted, 38
  cristal morphology, 243
  flattened cristae, 37–8, 242, 253, 256
  tubular cristae, 37, 242–3
mitochondrial genomics
  fungi, 248
  gene content, 248
  opisthokont phylogeny, 248
mitochondrial reticulum, 37–8
  ER association, 37
Monod equation, 209
multicellularity, 257
  cadherins, 265
  categories, 264
  cell–cell signalling, 275
  c-type lectins, 269
  epithelial tissue, 265
  extracellular matrix molecules, 269
  integrins, 267–9
  mesenchymal tissue, 265
  opisthokont, 264
  tissue phenotypes, 264

Nomenclator Zoologicus, 3
non-bilateria
  phylogeny, 261
non-restrictive cell coverings, 44
non-restrictive cell division, 5
Norris, Rich, 16, 214
nucleariid amoebae, 244
Nucletmycea, 249, 261–2
  Fonticula alba, 261
  Fungi, 262
  Nucleariida, 261
nucleus
  division, 40, 156
  heterochromatin, 40
  nuclear envelope, 40
  spindle, 40
nudiform condition, 185
nudiform species, 132
  characters, 132
  evolutionary significance, 153
  lorica dimensions, 133
  phylogenetic relations, 154, 192
nudiform/tectiform relationship, 191, 197, 199, 252

opisthokont classification
  cladistics (phylogenetic systematics), 249
  classificatory systems, 249

evolutionary systematics, 249
  Linnaean ranks, 249
  nomenclature, 249
opisthokont diversification, 263
Opisthokonta
  classification, 251
  early phylogeny, 242
  eEF1A insertion, 244
  eukaryotic supergroup, 241–3
  mitochondrial genomics, 247
  molecular phylogeny, 243–9
  monophyly, 244
  origin of name, 241
  phylogeny, 244–5, 262
opistokont series, 241
Opistokontae, 241
opistokonten, 241

particle capture, 26
  bidirectional transport, 21, 26–7, 205
  diffusion feeding, 205
  early illustrations, 6
  particle selection, 27
  spiral pathway, 5
  video recordings, 26
particle ingestion
  collar, 4–5
  early views, 26
  phagocytosis, 27, 37, 205
  pseudopodia, 5, 38
particulate organic matter (POM), 230
  enmeshment of choanoflagellates, 230
  sinking velocities, 230
Parvicorbicula socialis, 215–16, 224, 227, 229
Parvicorbicula/Pleurasiga species group, 82, 93, 95, 99, 130, 181, 188
Phalansteriidae, 16
Phalansterium, 16
phylogenetic reconstruction
  rare genomic changes, 244
Placozoa
  phylogeny, 259
  Trichoplax adhaerens, 258
podiates, 242
Polyoeca dichotoma
  branching morphology, 140
  early records, 131, 140
  lorica construction, 140–1
  lorica dimensions, 142
  organic investment, 142
Proterospongia
  colony morphology, 59–60
  controversy, 13
  missing link, 12, 60
  sponge ancestor, 12
pulsellum, 18, 241, 243

regulatory evolution, 199
restrictive cell coverings, 44, 47
Reynolds number (Re), 18
rosette agent, 253
rosette inducing factor (RIF), 58

Salpingoeca rosetta
  colony induction, 57
  comparative genomics, 264–9
  cytoplasmic bridges, 45
  EFL, 247
  life cycle, 57
  sexual reproduction, 10
  theca, 48
Salpingoecidae, 16, 44, 251
Savillea, 133–8
  species, 133
Savillea micropora, 133
Savillea parva, 133
  3D computer models of loricae, 135
  nudiform features, 138
Savillea parva (*micropora* form), 135
  cell division, 136–7
  costal strip accumulation, 136
  flagellum absent, 136
  helical costae, 135
  lorica assembly, 136, 139
  lorica construction, 135
  organic investment, 135–6
Savillea parva (*parva* form), 135
  lorica construction, 136
Schlingvacuoles (gullet vacuoles), 5
SDV-associated microtubules, 105, 168
  depolymerisation, 105
  Didymoeca costata, 168
  function, 105
  Stephanoeca diplocostata, 128
sexual reproduction, 9
  definition, 9
  gametes, 10
  meiotic detection toolkit, 10
  ploidy level, 10
  polymorphic life cycle, 235
  repeat retrotransposons, 10
  single nucleotide polymorphisms
    (SNPs), 10
silica deposition vesicles (SDVs), 28, 34,
    101, 107, 128, 167–8
silicic acid transporter proteins (SITs), 101
silicon
  amorphous, 105
  biogenic silica, 102

biogeochemical cycling, 101
  bound, 102
  cryptic recycling, 113, 122
  definition, 102
  depletion, 121
  dissolution, 118
  facultative requirement, 100
  mass per cell, 115
  monosilicic acid, 100
  protists, 100
  reactive silicate, 102
  starvation, 123, 127, 188
  terminology, 102
  uptake, 113
SIT-type genes
  Diaphanoeca grandis, 101
sponge ancestry
  conflicting views, 12
  from Cnidaria, 13
sponges (Porifera), 258
  amphiblastulae, 12
  ancestry, 10
  as colonial protozoa, 11
  Calcarea, 258
  choanoflagellate ancestry, 14
  Demospongiae, 258
  evolution, 15
  Hexactinellida, 258
  Homoscleromorpha, 258
  phylogeny, 258
Spongiae Ciliatae, 1
Stein, Friedrich Ritter von, 1, 47, 214
Stephanacantha / Platypleura theme, 220
Stephanoeca diplocostata, 156–66
  cell division, 156, 158
  colchicine treatment, 109
  conjoined cells, 192, 195–6
  costal strip accumulation, 156, 159
  costal strip categories, 103
  costal strip formation, 104
  experimental species, 102
  growth constants, 113
  helical costae, 188
  lorica assembly, 158, 160
  lorica construction, 102, 106, 156
  lorica dissolution, 119–20
  mitosis, 162, 164
  nudiform behaviour, 193–4
  population growth, 112, 114–15
  silicon dissolution constants, 118
  silicon per cell, 117
  silicon replenishment, 124–5

silicon starvation, 123, 127, 192
  silicon turnover, 114–16
  silicon-deficient growth, 121
  silicon-enriched growth, 112, 122
  SIT-type genes, 101
Stephanoeca theme, 85, 91
Stephanoecidae, 131, 154, 185, 253
suspension feeding, 18, 202
swarm gemmules (amphiblastulae), 12
Synzoospore hypothesis, 14

tectiform cell division
  Bicosta species, 182
  Calliacantha species, 181
  characteristics, 131, 156, 179–80
  costal strip movements, 172
  costal strip reorganisation, 169
  cytokinesis, 191
  Diaphanoeca grandis, 175–6, 178
  first description, 155
  inverted juvenile, 155–6, 169,
    177, 190
  lorica assembly, 158, 172, 177, 180
  separating threads, 156, 177, 191
  Stephanoeca diplocostata, 159
  uniformity of characters, 174, 181
tectiform condition, 187
  competitive advantage, 200
tectiform species
  characters, 132, 155
  phylogenetic relations, 192
theca, 5
  carbohydrate composition, 52
  cup-shaped, 48, 51
  definition, 44
  flask-shaped, 48, 50, 54
  morphology, 47, 50
third position of codons, 154, 252
transverse costae
  assembly, 161, 165
  distinguishing features, 77
  external, 78, 82, 181
  left-handed rotation, 83
  occurrence, 183
  Stephanoeca cupula, 77–8

Unikonta, 242
unikonts, 242
urmetazoan, 14

veil, 85–6, 92, 167
  functional role, 236

Printed in the United States
by Baker & Taylor Publisher Services